PARTICLE AND NUCLEAR PHYSICS

Proceedings of the
International Symposium on

PARTICLE AND NUCLEAR PHYSICS

Beijing
Sept 2-7 1985

Edited by

Hu Ning
Wu Chong-shi

Organized by

Institute of Theoretical Physics,
Peking University

Sponsored by

United Nations Educational,
Scientific and Cultural Organization
Chinese National Commission for UNESCO
Asian Physics Education Networks

World Scientific

Published by

World Scientific Publishing Co Pte Ltd
P. O. Box 128, Farrer Road, Singapore 9128.
242 Cherry Street, Philadelphia PA 19106-1906, USA.

Library of Congress Cataloging-in-Publication Data

International Symposium on Particle and Nuclear
Physics (1985 : Peking, China)
Proceedings of the International Symposium on
Particle and Nuclear Physics, Beijing September 2-7, 1985.

1. Particles (Nuclear physics) – Congresses.
2. Nuclear physics – Congresses. I. Hu, Ning.
II. Wu, Chong-shi. III. Pei-ching ta hsueh. Institute
of Theoretical Physics. IV. Asian Physics Education
Networks. V. Hu, Ning. VI. Wu, Chong-shi. VII. Title.
QC793.I5614 1985 539.7'21 86-23385
ISBN 9971-50-175-9

Printed in Singapore by General Printing and Publishing Services Pte Ltd.

Opening Address

Ladies and Gentlemen

It is my great pleasure to declare the opening of the International Symposium on Particle and Nuclear Physics, Beijing, September 2-7, 1985, and express our hearty welcome to all participants, especially those from abroad. This symposium is sponsored by ASPEN, UNESCO, Chinese National Commission for UNESCO, ICTP and Peking University. Many of their representatives are present at this session.

I think it is a very good idea to have this joint symposium on particle and nuclear physics. These two branches of physics of structure of matter seem to converge into a single field of study when it was discovered that proton and neutron are nuclear matter containing least number of quarks. Deep inelastic scattering experiments have been carried out to observe the structure function of atomic nuclei in the same way as the structure function of proton was observed.

As is the case in other countries, the largest number of theoretical physicists are teachers in various universities. One important task of the Institute of Theoretical Physics of Peking University is to provide facilities for scientific interchange and personal contact among physicists both inside China and abroad. This symposium is one of our first attempts in recent years. I hope our cooperation with UNESCO, ICTP and ASPEN will be more fruitful in the future.

Ladies and gentlemen, I hope all of you will have a good time in Beijing and come back to Beijing again.

Ning Hu
Chairman, Organization Committee

CONTENTS

Opening Address v
Ning Hu

PART A: PARTICLE PHYSICS

An Investigation of Multiplicity-Distributions and -Correlations in PP Collisions at $\sqrt{s} = 540$ GeV for Various Pseudorapidity Intervals 3
Cai Xu

Bifurcation and Dynamical Symmetry Breaking 4
Ngee-Pong Chang

Status of Electroweak Theory for Heavy Quark Decays and CP Violation 13
Ling-Lie Chau

Geometrical Integrability and Equations of Motion: A Unifying Description of Equations of Motion 52
Ling-Lie Chau

The Charged Particle Multiplicity in $\overline{P}\overline{P}$ Annihilation 76
Chi-jiang Chen and Jing-fa Lu

Negative Binomial Distribution for Multiplicity Distributions in e^+e^- Annihilation 77
C. K. Chew and Y. K. Lim

Spontaneous Breakdown of Supersymmetry and Goldstone Fermion at Nonzero Temperatures 80
Swee-Ping Chia

The β-Functions of the SU(2) Lattice Gauge Theories with SIN and EXP Actions 87
Dong Shao-jing and Li Wen-zhou

Relativistic Effects, E1 and M1 Transitions for Heavy Quarkonia 88
Yi-Bing Ding, Ju He, Shen-Ou Cai, Dan-Hua Qin and Kuang-Ta Chao

Basic Features of String Theories 92
Jean-Loup Gervais

Variational Study of Lattice QCD 127
Guo Shuo-hong

The Mixing Effects of Z° and Toponium 132
T. H. Ho, J. Liu, X. Zhang and Z. Y. Zhu

Gauge Group Cocycles Hirarchy 133
Hou Bo-Yu

A Unified Treatment of Ordinary Hadrons and New Particles as Bound States in Quark Model 134
Ning Hu

Multi-Particle Production and the Three Fire-Ball Model 146
Huang Chao-shang

The Hadronic Structure and Sum Rules in QCD 151
Tao Huang

Potential-Model Fitting of Quarkonium States 155
C. J. Koh and W. Hellman

Multiplicity Distributions in Limited Pseudorapidity Intervals 159
C. S. Lam and M. S. Zahir

Four Quark States and Glueballs 173
Bing-An Li

Stochastic Quantization at Positive Temperature 176
S. C. Lim

EMC Effect and the Distortion of Vacuum in Nuclei 181
Liu Lian-sou

The Fractional Fermion Number in (1 + 1) Dimensions 185
Guang-jiong Ni and Rong-tai Wang

Gluon and Massless Gluino Scattering Using N = 2 Supersymmetry 189
Stephen J. Parke

Signatures of Exotic Fermions and Other New "Low-Energy" Phenomena in Superstring E_6 196
Jonathan L. Rosner

QCD Analysis of the Neutrino Charged Current Structure Function XF_3 in Deep Inelastic Scattering 209
Mohammad Saleem and Fazal-e-Aleem

Analysis of the Compton Scattering of Protons 213
Mohammad Saleem and Fazal-e-Aleem

A Unifying Method of Obtaining the Feynman, Kac, Cameron and Albevario-Hoegh-Krohn Solution to the Heat or Schroedinger Equation 217
Shaharir

Present Status of Composite Models in Particle Physics — Minimal Composite Model of "Elementary" Particles and Fields 226
Hidezumi Terazawa

The Analytical Approach for Calculating the Effect of the Virtual Quark Loops to SU_2 Average Plaquette 236
Chi Min Wu and Pei Ying Zhao

Raphasing Invariants and CP in Neutral B Mesons 240
Dan-di Wu

The General Chern-Simons Characteristic Classes and their Physical Applications 244
Wu Yue-liang, Xie Yan-bo and Zhou Guang-zhao

General Solutions of Wess-Zumino Consistency Conditions for Axial-Vector Current Divergence Anomaly 245
Xiong Chuan-Sheng and Zhu Zhong-Yuan

On Vacuum Solutions of Conformally Symmetric Theories 246
Xu Bo-Wei

Dynamical Equations and a New Analytical Approach in Lattice Gauge Theories 250
She-sheng Xue

Implications of Spontaneously Broken Conformal Supergravity 255
Zha Chao-zheng

The Status of the Beijing Electron-Positron Collider (BEPC) and the Heavy Ion Research Facility of Lanzhou (HERFL) Under Construction 256
H. Y. Zhu

Rescaling for Kaon Structure Function 265
Zhu Wei, Shen Jian-guo and Qiu Xi-jun

PART B: NUCLEAR PHYSICS

SDG Boson Model and its Application — 271
Y. Akiyama

Shape Phase Transition in the Region of Z = 40 — 279
A. Arima and M. Sugita

Structure of the Transitional Nuclei ^{149}Pm, ^{151}Eu and ^{153}Tb — 288
S. Bhattacharya, R. K. Guchhait and S. Sen

Energies and Electric Quadrupole Transition Strengths of Ground State Bands of Even Nuclei — 292
S. Bhattacharya and S. Sen

Pion Scattering and Charge Exchange from Deformed Nuclei — 296
H. C. Chiang

Configuration Mixing in the Ground State of ^{96}Mo — 300
M. Shafi Chowdhury

F-Spin Breaking and Gd Isotopes Energy Spectra — 308
Ding Xiao-Nan

Fermion Dynamical Symmetry and the Nuclear Shell Model — 309
Joseph N. Ginocchio

The Short Range Effective Interaction and the Spectra of Oxygen Isotopes in (s-d) Space — 322
Li Xian-ying, Yao Shi-huai and Zhang Qing-ying

Physical Subalgebra Chains in Semisimple Lie Algebra A_n — 323
Yinsheng Ling and Xiaoqian Zhou

The Non-Analog Double Charge Exchange Transition and the Relationship of Energy Dependence — 327
Ma Wei-hsing, Wang Si-wen, Zhang Gao-you, Chen Chung-kuang, Yang Zhen-rong and Zhao Shu-ping

The Non-Analog Double Charge Exchange Transition $^{12}C(\pi^+,\pi^-)^{12}O_{(g.s.)}$ and $^{16}O(\pi^+,\pi^-)^{16}Ne_{(g.s.)}$ — 328
Ma Wei-hsing, Wang Si-wen, Zhang Gao-you, Chen Chung-kuang, Yang Zhen-rong and Zhao Shu-ping

Recent Developments in the Theory of Nuclear Molecules — 329
Jae Young Park and Moon Hoe Cha

Phase Transitions and Shape Changes in Finite Nuclei and the Importance of Fluctuations — 338
Peter Ring

Giant Resonances in Hot Rotating Nuclei 351
Peter Ring

The Pair-Aligned Intrinsic Wave Function in Single-j Configuration 365
Shen Hongqing, Cui Haiyuan, Yao Shihuai, Li Xianyin and Yang Bangjun

Nuclear Structure Parameters from Coulomb Excitation with 2.5 - 4.5 MeV Protons 367
K. P. Singh, D. C. Tayal, Gulzar Singh and H. S. Hans

Application of Dyson Boson Mapping to the Analysis of Mode-Mode Coupling in Ge and Se Isotopes 371
Kenjiro Takada

On the Structure Functions of Nucleons and Nuclei 382
Anthony W. Thomas

Calculation of Dynamical Effects Within the Framework of Folded Diagram Theory 394
Wang Zixing, Huang Weizhi, Song Hongqiu and Cai Yanhuang

A Unified Microscopic Description of Many Particle Systems Especially Nuclear Systems 398
Karl Wildermuth

Intrinsic Structure of the Cranking Shell Model Wave Functions and Nuclear Pairing Phase Transition 412
C. S. Wu and J. Y. Zeng

Preliminary Plan of Experimental Area and Heavy Ion Nuclear Physics Research on HIRFL 424
Wu Enjiu and Shen Wenqing

Nuclear Phase Transition Illustrated with Schematic Models 428
Xu Gong-ou, Li Fu-li and Fu De-ji

Generator Coordinate Method and Dynamic Group Representation 438
Xu Gong-ou, Wang Shun-jin and Yang Ya-tian

Self-Consistent Structure of Bosons in Nuclei 442
Li-ming Yang, Zhi-ning Zhou and Da-hai Lu

A Pair Aligned Intrinsic State Method to Determine the IBM Bosons and Hamiltonian 449
Yang Ze-sen, Qi Hui, Liu Yong and Deng Wei-zhen

The Short Range Effective Interaction and the Spectra of Calcium Isotopes in (f-p) Space 450
Zhang Qing-ying, Li Shen-wu and Wei Jian-xin

Concluding Remarks 451
Ling-Lie Chau

PART A: PARTICLE PHYSICS

AN INVESTIGATION OF MULTIPLICITY-DISTRIBUTIONS AND CORRELATIONS IN PP COLLISIONS AT s=540 GeV FOR VARIOUS PSEUDORAPIDITY INTERVALS

Cai Xu

Institute of Particle Physics, Huazhong Normal University, Wuhan

Several theoretical models like geometrical-dynamical model, QCD clusters, dual parton models, QCD jet calculus, generalized Bose-Einstein distributions, soft QCD bremsstrahlung, statistical model have been proposed to understand the KNO scaling behaviour of charged particle multiplicity distributions in high-energy non-single-diffractive hadron-hadron collisions. The new results measured by the UA5 Collaboration on multiplicities for various intervals of pseudorapidity at centre-of-mass energy $\sqrt{s}$=540 GeV[1] as well as such data at higher energies[2] are expected to be able differentiate between the competing models. It is shown that the recently observed pseudorapidity-interval-dependence in charged multiplicity-distributions and -correlations are nature consequence of the statistical model[3]. It is pointed out in particular: For sufficiently high total c.m.s. energy, there exists a pseudorapidity interval W in which the KNO-plot satisfies an uniform type $\psi(z) = 4z \exp(-2z)$.

REFERENCES

[1] G.J. Alner et al., (UA5 Collaboration), preprint, CERN-EP/85-61 (1985).

[2] J.G. Rushbrooke, (UA5 Collaboration), preprint, CERN-EP/85-124 (1985).

[3] Liu Lian-sou and Meng Ta-chung, Phys. Rev. D27(1983)2640;
Chou Kuang-chao, Liu Lian-sou and Meng Ta-chung, Phys. Rev. D28 (1983)1080;
Cai Xu, Liu Lian-sou and Meng Ta-chung, Phys. Rev. D29(1984)869;
Cai Xu and Liu Lian-sou, Lett. Nuovo Cimento 37(1983)495;
Cai Xu, Huang Chao-shang, Meng Ta-chung and Sa Ban-hao, Chinese Phys. Lett. 2(1985)101;
Cai Xu, Liu Lian-sou and Meng Ta-chung, Comm. in Theor. Phys. 4 (1985)847;
Cai Xu, Chao Wei-qin, Huang Chao-shang and Meng Ta-chung, preprint, FUB/HEP 85-4(1985).

Bifurcation
&
Dynamical Symmetry Breaking*

by

Ngee-Pong Chang

Physics Department
City College of the City University of New York
New York, N.Y. 10031

(Invited talk at the International Symposium on Particle and Nuclear Physics, held September 2-6. 1985, Beijing, China)

In physics, we find many examples of bifurcation phenomenon that we have all come to accept as commonplace and as part of standard folk lore. While the physics of phase transitions have been studied for a very long time, the connection with bifurcation theory is a fairly recent development. In field theory, my brother, L.N. Chang, and I have found[1] in the course of our studies on Dynamical Symmetry Breaking that it can also be understood as a phase transition. This connection is sufficiently new and interesting that I am very pleased with this opportunity to bring to the attention of this Symposium.

Bifurcation theory[2] is a new branch of mathematics that studies mappings containing a control parameter λ . As λ is varied, the mapping behavior changes dramatically at certain critical values of λ_c , called the bifurcation points. If $f(x, \lambda)$ is the mapping, then x is the **fixed point** of the mapping whenver x satisfies the equation

$$x = f(x, \lambda) \qquad (1)$$

For λ below some critical value, there may be only one **stable** fixed point at x=0, say. For λ above that critical value, **bifurcation** may occur and the mapping then yields two fixed points, one of which ($x \neq 0$) will be the new **stable** fixed point while the old fixed point at x = 0 becomes unstable.

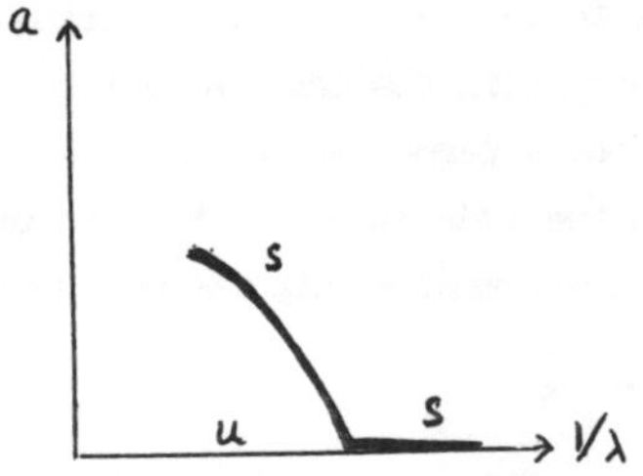

The familiar example of an iterative quadratic mapping

$$x_{n+1} = \lambda x_n (1 - x_n) \tag{2}$$

illustrates very well the bifurcation properties of the mapping. For λ in the range $0 < \lambda < 1$, there is only one fixed point, a, with a = 0. For λ however in the range $1 < \lambda < 3$, there is a new fixed point at

$$a = \frac{\lambda - 1}{\lambda} \qquad \text{STABLE} \tag{3}$$

while the other fixed point, a = 0 , is now UNSTABLE. The label, STABLE vs. UNSTABLE, is a mathematical one, reflecting the fact that upon iteration from any initial point in the neighborhood of the fixed point, a stable fixed point is one towards which the successive iterative mapping converges. While in general we have been conditioned to prefer what has been labelled as stable, there are however situations where physics dictates that we deal with the so-called unstable fixed points.

Incidentally, for the example given in eq. (2), for λ larger than 3, further bifurcations occur, until for λ greater than 3.57..., "chaos" sets in. For our field theory application, physics dictates that we shall be concerned only with the first bifurcation on the physical sheet.

In the study of bifurcation theory, the external parameter plays a very important role. For physics applications, that parameter may be the pressure, as in the case of the buckling Euler-Bernoulli rod, or it may be the temperature as in the case of the ferromagnetic phase transition or it may be the applied external magnetic field in the superconductor phase transition. However, in the relativistic field theory study that I shall be reporting, the role of the **external parameter** is taken over by the four momentum-squared.

Bifurcation phenomenon in renormalisation field theory thus takes place in **momentum space**. From the study of chiral symmetry breaking in QCD, we find that

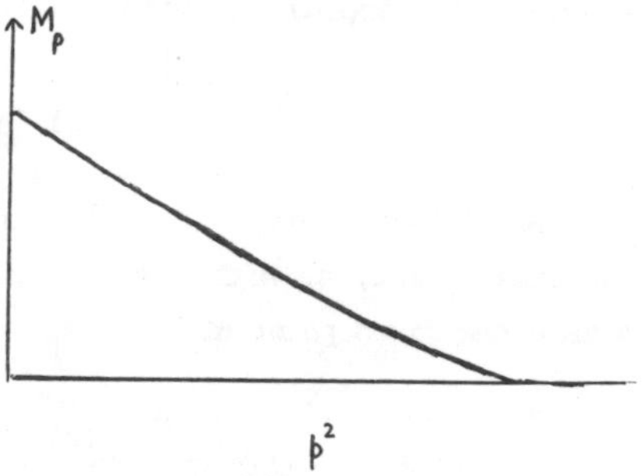

for spacelike p^2 GREATER than a threshold value, the QCD quantum system chooses the chiral symmetric state, while

for spacelike p^2 LESS than the threshold, the QCD quantum system chooses the chiral asymmetric state.

Dynamical Symmetry Breaking in field theory therefore takes on a space-time dimension that is very suggestive of the bag model, where chiral symmetry breaking is taken to be preserved **inside** the bag, but is broken **outside** the bag. The difference is, the bag model has a sharp spatial boundary, while in our result, the sharp boundary occurs in momentum space.

The problem of chiral symmetry breaking in QCD is of course intimately tied in with the fundamental problem of fermion masses in particle theory. Much progress has been made in the last decade on the study of strong, electromagnetic and weak forces. With the aid of Yang-Mills gauge fields[3], all these forces have now been unified into an elegant gauge theory, whose couplings are all determined by the local gauge principle.

The inelegant and indeed 'dirty' part of the unification theory has to do with how masses are generated. Usually, we rely upon the ubiquitous Higgs fields to 'spontaneously' break the gauge symmetry and produce huge masses for the X,Y and the W, Z gauge bosons. The same Higgs fields are supposed to have Yukawa interactions with the quarks and leptons, with coupling constants, h. These latter coupling constants, unlike the gauge ones, are arbitrary. Their magnitudes are dictated by the fermion masses, through the relation

$$-h\bar{\psi}\psi\phi$$

$$m = hV \tag{4}$$

where V denotes the vacuum expectation value of the Higgs field, determined here by the W and Z gauge boson masses. For the electron and the up and down quarks, the corresponding coupling constants, h_e, h_u, h_d, are respectively of the magnitudes

$$h_e \sim 10^{-5} \qquad h_{u,d} \sim 10^{-3} \tag{5}$$

to be compared with the gauge coupling constants of order 10^{-1}. A fundamental unified field theory with coupling constants that differ by so many orders of magnitude cannot be the ultimate fundamental theory.

It is in the hope of looking for alternates that Dynamical Symmetry Breaking mechanisms have been sought to break out of the bind that Higgs has put us in. The history of this line of research goes back to Nambu and Jona-Lasinio[4], who first demonstrated how a chiral invariant field theory with a Lagrangian that is strictly invariant under the chiral transformation can nevertheless describe massive physical fermions. Their key observation is that the physical ground state, just like in the ferromagnet analogy, may not be chirally symmetric. As a result of dynamical fermion pairing ('condensates"), the ground state has broken the

$$\psi(x) \to e^{i\alpha\gamma_5}\psi(x)$$

chiral invariance forbids

$$m\bar{\psi}\psi$$

chiral symmetry. In the ferromagnet analogy, the neighboring spins in the ground state have, as a result of the spin-spin interaction dynamics, all lined up, so that the total spin is N, rather than the rotationally symmetric J=0 state. Even though the Hamiltonian remains strictly rotationally invariant, it would have been naive to use a rotationally invariant ground state as the 'vacuum' state.

For the chiral symmetry breaking, Nambu and Jona-Lasinio(NJL) make the same observation and conclude that usual perturbation theory based on the chiral invariant Fock space vacuum is not correct. By developing perturbation theory around a massive fermion vacuum, and demanding that radiative corrections be self-consistent and not take them out of this massive fermion vacuum, they derived their famous Nambu gap equation, whose solution can give the dynamical fermion mass. Because the theory they considered was the relativistic generalisation of the BCS theory, their field theory was not renormalisable. In 1984, we applied the NJL mechanism to QCD and showed[5] that the corresponding Nambu gap equation is in fact a renormalisation group invariant.

But this is a long detour to tell you about a new development that finally connects Dynamical Symmetry Breaking with bifurcation theory. This development relies on a new understanding of renormalisation group analysis that will be discussed below. To introduce it, let me remind you of the steps that we go through in writing down a renormalised perturbation theory.

$$-\bar{\psi}\gamma\cdot D\psi - \frac{1}{4}G^a_{\mu\nu}G^a_{\mu\nu} - m_r\bar{\psi}\psi$$

Consider a chiral broken QCD theory so that in the tree Lagrangian the fermions are already massive, with a mass parameter, m_r. When you start computing with this Lagrangian, of course, you find infinities. To obtain finite, renormalised perturbative series, you must introduce counterterms in the Lagrangian such that the new total Lagrangian is expressible entirely in terms of the original Lagrangian with however bare parameters. Because of the counterterms, the resultant renormalised series now is not only a function of p, m_r, g_r, but also of μ , the renormalisation scale. This renormalisation scale is arbitrary, and renormalisation group (RG) analysis is simply the statement that the bare parameters are independent of μ , so that, for example, from

$$m_b = Z_m m_r \tag{6}$$

one easily derives the renormalisation group equations for m_r

$$\mu \frac{\partial}{\partial\mu} m_r = -6 C_f \lambda_r m_r \tag{7}$$

and similarly for $\lambda_r \; (\equiv g_r^2/16\pi^2)$

$$\mu \frac{\partial}{\partial\mu} \lambda_r = -b \lambda_r^2 \tag{8}$$

Because m_r and g_r are dependent on μ , they are not suitable as physical parameters. Instead, we can parametrize by the two RG-invariant mass scales, Λ_c and M_0, defined to one loop RG accuracy by the implicit relations

$$\frac{1}{\lambda_r} + \frac{b}{2} \log \frac{\Lambda_c^2}{\mu^2} = 0 \tag{9}$$

$$m_r = M_0 \left\{ 1 + \frac{\lambda_r b}{2} \log \frac{M_0^2}{\mu^2 e^{1/3}} \right\}^{\alpha} \tag{10}$$

$$\alpha = 6 C_f / b$$

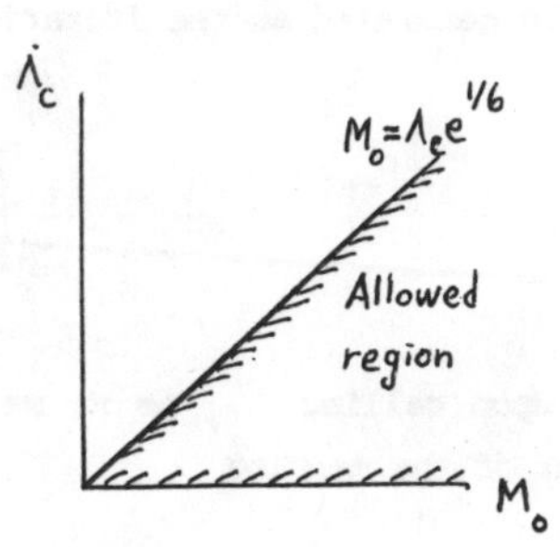

The physical requirement that m_r be real puts a constraint on the allowed region for the independent RG-invariant parameters,

$$M_0 \geq \Lambda_c e^{1/6}$$

The boundary of the allowed domain is described by

$$M_0 = \Lambda_c e^{1/6} \tag{11}$$

and corresponds to the chiral symmetric limit of vanishing m_r. This fact enables us to study the Dynamical Symmetry Breaking as the limit of a massive QCD theory.

ables us to study the Dynamical Symmetry Breaking as the **limit** of a massive QCD theory.

For every point in the allowed domain of parameter space in the figure above, we can develop the renormalised perturbative series for the n-point functions of the theory, accurate to, say, one loop renormalisation group analysis. Take in particular the chiral-flip part of the two point function for the function, M_p, with the representation ($\lambda_r \equiv g_r^2 / 16\pi^2$)

$$M_p = m_r \left(\frac{\lambda_r}{\lambda_p}\right)^{-\alpha} , \quad \alpha = 6C_f/b \tag{12}$$

Eq.(12) satisfies the requirements of renormalisation group analysis so long as $1/\lambda_p$ is an RG-invariant. The renormalised perturbation theory is here summarised by the series

$$\frac{\lambda_r}{\lambda_p} = 1 + \frac{b}{2}\int_0^1 dz \log\left[\frac{m_r^2 + zp^2}{\mu^2 e^{1/3}}\right] + \cdots \tag{13}$$

We claim that the series, accurate to one loop renormalisation group analysis, can be generated as the iterative solution to the following implicit equation

$$\frac{1}{\lambda_p} = \frac{1}{\lambda_r} + \frac{b}{2}\int_0^1 dz \; \log\left[\frac{m_r^2\left(\frac{\lambda_r}{\lambda_p}\right)^{-2\alpha} + zp^2}{\mu^2 e^{1/3}}\right] \tag{14}$$

or, upon calling $1/\lambda_p$ as x, we have the implicit equation for x as the fixed point of the mapping

$$x = \frac{1}{\lambda_r} + \frac{b}{2}\int_0^1 dz \; \log\left[\frac{m_r^2(\lambda_r x)^{-2\alpha} + zp^2}{\mu^2 e^{1/3}}\right] \tag{15}$$

and the connection with the bifurcation theory (see eq. (1)) becomes clear. From eq.(12), and using the definitions in eq. (9) & (10), the function M_p itself may be generated by the iterative equation

$$M_p = M_0\left\{1 + \frac{\lambda_0 b}{2}\log\left[\frac{M_p^2}{M_0^2 e}\left(1+\frac{p^2}{M_p^2}\right)^{\frac{M_p^2}{p^2}+1}\right]\right\}^{-\alpha} \tag{16}$$

$$\frac{1}{\lambda_0} = \frac{b}{2} \log\left[\frac{M_0^2}{\Lambda_c^2 e^{1/3}} \right] \tag{17}$$

Equation (16) is the crux of the matter as far as connection with bifurcation is concerned, for it is in the form of eq. (1) with M_0 and Λ_c as well as p^2 as **external parameters**. By an application of the Leray-Schauder fixed point theorem[6], it can be shown taht eq. (16) defines a complex function of p^2 that is analytic in a cut p^2 - plane with the cut along the time-like p^2 axis. This is true **for every set of parameters within the allowed domain**. This is of course reassuring since perturbative analyticity has long been established for renormalised field theory order by order.

When you go to the boundary of the allowed domain, i.e. at the critical limit of eq. (11), something remarkable happens. For spacelike p^2 greater than $M_0^2 e$, M_p vanishes in the critical limit. For spacelike p^2 less than $M_0^2 e$ however, M_p chooses the chiral asymmetric phase and does not vanish in the critical limit.

As a result, the vacuum expectation value of $\langle \bar{\psi} \psi \rangle$, defined to be the limit as x approaches y of the two-point function for the fermion, becomes finite and calculable. It is

$$\langle \bar{\psi} \psi \rangle = - 0.0398\, N_c\, \Lambda_c^3 \tag{18}$$

where N_c is the number of colors. To compare with current algebra determinations, it is necessary to multiply eq. (18) by M_0, the dynamically generated mass, to find

$$M_0 \langle \bar{\psi} \psi \rangle = - 0.0470\, N_c\, \Lambda_c^4 \tag{19}$$

The combination on the left hand side of eq.(19) is commonly believed to be RG-invariant. Using an accepted value of 0.15 GeV for Λ_c yields a value of 9.24×10^{-5} $(\mathrm{GeV})^4$, which is remarkably close to the current algebra estimate of 8.8×10^{-5} $(\mathrm{GeV})^4$ for the up quark.

The results that I have presented here are still new and not well digested. We have but scratched the surface of the full panoply of bifurcation theory. Questions of **chaos** spring to mind. In our investigation of the analyticity properties of M_p so far we have restricted ourselves to the physical sheet. What happens when we go to other unphysical sheets is still an open question. Presumably we will encounter further bifurcations typical of the iterative mappings. If **chaos** is found on the unphysical sheets, can Feigenvalues be far behind? And then, there is also the question of the temperature dependence of the bifurcations studied here. All these and more await the attention and development by all concerned, and I am pleased to have been given the opportunity to do so in front of such a distinguished audience.

REFEREMCES

1. L.N. Chang, N.P. Chang, Phys. Rev. Lett. **54**, 2407 (1985)
2. For an excellent exposition on the subejct, see Lecture Notes by S.N. Chow, **Bifurcation theory and Dynamical Systems**, Lecture NOtes No. 22, Math Dept, National University of Singapore, Kent Ridge, Singapore 0511.
3. C.N. Yang and R.L. Mills, Phys. Rev. **96**, 191 (1954)
4. Y. Nambu and G. Jona-Lasinio, Phys. Rev. 122, 345 (1961);
   Phys. Rev. 124, 246 (1961)
5. L.N. Chang and N.P. Chang, Phys. Rev. D29, 312 (1984);
   N.P. Chang and D.X. Li, Phys. Rev. D30, 790 (1984)
6. D.R. Smart, Fixed point theorems, Cambridge Univeristy Press (1974). See Theorem 10.3.10 on p. 82.

Status of Electroweak Theory for Heavy Quark Decays and CP Violation

Ling-Lie Chau

Department of Physics
Brookhaven National Laboratory
Upton, New York 11973

Abstract: The phenomena of quark mixing in weak interaction are described. The current experimental status of the mixing matrix, charm and beauty particle decays, and CP violations are reported. Future interesting experiments are pointed out, especially in the event that higher generations of quarks exist.

Table of Contents

Introduction: Some Historical Remarks

I. The Three-Generation Quark Mixing Matrix

II. CP Violation in the Kaon and Hyperon Decays

III. Mass-Matrix Mixing and CP Violation in the Heavy Quark System

III.a. Mass-Matrix Mixing and CP Violation -- Tagged-Neutral-Meson-Decay Experiments

III.b. Decay-Amplitude CP Violation -- Partial-Decay-Rate Differences

IV. Comparison of Models for CP Violation

V. Nonleptonic Decays

V.a Charm Mesons → Pseudoscalar-Vector Decays

V.b Charm Mesons → Pseudoscalar-Pseudoscalar Decays

VI. Beyond the Three Generations of Quarks

VII. Concluding Remarks and Outlook

References

Acknowledgments: The advances in the subject discussed here have been made by many theorists and experimentalists whose contributions are cited in the references. My contributions to the subject have been made in recent years in collaboration with H.-Y. Cheng (Indiana University, Bloomington), W.-Y. Keung (University of Illinois, Chicago), and also recently with F. Botella, a Fulbright Fellow from Valencia, Spain. I would like to thank Professor C.N. Yang for many enlightening discussions and for his encouragement.

I am also grateful to Professor Z. Maki and the organizers of the Meson 50 for inviting me to give this talk and to participate in this most memorable conference.

Introduction: Some Historical Remarks

It is 51 years since Fermi[1] proposed the Fermi coupling for β decay, 50 years since Yukawa[2] formulated his theory, and 35 years since Yang and Mills[3] constructed their theory. Nature's answer to these human endeavors has been most elegant and rewarding. The electroweak unified theory of Glashow, Salam, and Weinberg[4] has been confirmed by the recent discovery[5] of the long-awaited intermediate bosons $W^{\pm}$ and Z^0. Their masses,[6] and even their production and decay characteristics, were predicted. (The observed lepton-forward-backward asymmetry, i.e., ℓ^- from the W^- moves in the proton direction, and ℓ^+ from the W^+ moves in the antiproton direction, provides a truly remarkable confirmation of the quark-parton description of the reaction.)[7] This was the climax of the strikingly beautiful development of more than 50 years of history of weak interactions, as we have heard from many of the preceeding talks in this conference, and many of you in the audience have made major contributions. The list below demonstrates the beautiful interplay between theoretical ideas and experimental observations:

1930, Nuclear β decay ----------→ ν,[8]
discovered 1954,[9]

1934, Fermi coupling,[1]

1935, Nuclear force >> Fermi coupling[10] ------→ Yukawa theory,[2]
π meson discovered 1947,[11]

1954 Local gauge invariance ----------→ Yang-Mills theory,[3]

1953-56 θ, τ puzzle[12] -------------------→ parity violation,[13]
observed 1957,[14]

1960's Electroweak unification[4] -------→ neutral current observed 1973,[15]
$W^{\pm}$, Z^0 with mass predicted,[4,6]

1963 Universal suppression of hyperon decays -------→ Cabibbo mixing,[16]

1970 $K_L \to \mu^+\mu$ extremely small-[17]-----------→ the GIM mechanism,[18] and
charm, observed 1976,[19]

1973	CP violation from W coupling ------------------→	Kobayashi-Maskawa (KM) matrix[20] and (t), b quarks beauty particle observed 1982-1984[21]

In the light of this milestone of the great success of electroweak unified theory, I am going to review for you some particular areas of research[22-24] in weak interactions and their outlook in our ever ongoing pursuit to understand the constituents of matter and the forces governing their interactions.

I. Status of the Quark Mixing Matrix

The mixing matrix of the three left-handed generations of quarks,

$$\begin{pmatrix} d' \\ s' \\ b' \end{pmatrix} = \begin{pmatrix} V_{ud} & V_{us} & V_{ub} \\ V_{cd} & V_{cs} & V_{cb} \\ V_{td} & V_{ts} & V_{tb} \end{pmatrix} \begin{pmatrix} d \\ s \\ b \end{pmatrix}, \qquad (1.1)$$

can be determined by various experiments:

$|V_{ud}| = c_x\,(1 + 0(10^{-4})) = 0.9737$, from nuclear β decays;[25,26] (1.2)

$|V_{us}| = s_x\,(1 + 0(10^{-4})) = 0.225$, from strange particle decays;[25,26]

$|V_{cb}| = s_y\,(1 + 0(10^{-4})) = 0.059$, from $\tau_b = 10^{-12}$ sec;[27,29]

$|V_{ub}| = s_z \lesssim 0.0082$, from $\Gamma(b \to u)/\Gamma(b \to c) < 0.05$[28-29]

Besides the few percent errors for $|V_{ud}|$, ten to twenty percent are typical errors for other v_{ij}'s. Recent news on the ancient Cabibbo fit is that the old experimental discrepancy in the lepton-momentum asymmetry in Σ decays, which gave g_1/g_1 a wrong sign, has now disappeared. A recent high-statistics experiment[30] showed $g_1/f_1 = -0.29 \pm 0.07$, which is consistent with the Cabibbo theory. In addition, a somewhat smaller value of $|V_{us}|$, $|V_{us}| = 0.225$ is obtained, which is in better agreement with older fits.[25] One interesting experimental uncertainty is still the neutron lifetime.[31]

Our knowledge of the quark mixing matrix has dramatically improved recently owing to the b-lifetime measurement[25] and the $\Gamma(b \to u)/\Gamma(b \to c) < 0.05$[26] bound from the $b \to \ell^- X$ measurements, which indicated that X is mainly made out of states with charm. At present the errors on the b lifetime are still rather large. Here we use $\tau_b = 10^{-12}$ sec as an example to obtain Eq. (1.2). Using unitarity and a phase convention[32-34] such that the phase factor of the matrix elements appears with the smallest matrix elements $\lesssim 10^{-3}$, the values of the other matrix element can be obtained through unitarity

$$V = \begin{pmatrix} 1 & 0 & 0 \\ 0 & c_y & s_y \\ 0 & -s_y & c_y \end{pmatrix} \begin{pmatrix} c_z & 0 & s_z e^{-i\phi} \\ 0 & 1 & 0 \\ -s_z e^{-i\phi} & 0 & c_z \end{pmatrix} \begin{pmatrix} c_x & s_x & 0 \\ -s_x & c_x & 0 \\ 0 & 0 & 1 \end{pmatrix}$$

$$\begin{matrix} & & d & s & b & \\ s_x, s_y, s_z \backsim 0 & \Bigg(& c_x c_z & s_x c_z & s_z e^{-i\phi} & \Bigg) u \\ & & -s_x c_y - c_x s_y s_z e^{i\phi} & c_x c_y - s_x s_y s_z e^{i\phi} & s_y c_z & c \\ & & s_x s_y - c_x c_y s_z e^{i\phi} & -c_x s_y - s_x c_y s_z e^{i\phi} & c_y c_z & t \end{matrix} \qquad (1.3)$$

$$= \begin{pmatrix} 0.9737 & 0.225 & V_{ub} \lesssim 0.0082 e^{i\phi} \\ -0.23-0.059\, V_{ub} & 0.98 & 0.059 \\ 0.01-V_{ub} & -0.06+0.228\, V_{ub} & 1 \pm 0(10^{-2}) \end{pmatrix} . \qquad (1.4)$$

The interesting features of the mixing matrix are that the gap between the second and the third generations has been widened: $|V_{cb}|$ is about a quarter of $|V_{ud}|$. Because of such severe suppression, the 2 x 2 GIM matrix is almost unitary, i.e., $V^*_{cd}/V_{cd} \simeq -V^*_{us}/V_{ud}$. It is very important to check this relation independently. Significant deviation from it can have serious implications, e.g., result in breakdown of the model or the existence of the fourth generation. The nice feature of using the parameterization of Eq. (1.3) is that the imaginary parts appear at matrix elements with magnitudes $<10^{-3}$. It is also clear in this parametrization that V_{ub} cannot be zero in order to explain CP violation. Actually one can show that in order to give CP violation none of the matrix elements can be zero.[24] The analysis of CP violation in the K system can give an estimate of the lower bound on V_{ud}, which is not too far below the experimental upper bound, as we shall discuss later. In addition, each angle has a direct correspondence with experimental results, as shown in Eq. (1.2). Also, its relation to ideas in quark mass matrix can also be described in a more transparent way.[35]

The CP-violation effects from this 3x3 KM matrix have a quite striking feature. It was shown in Ref. (33) that if there are only three generations of quarks, the CP-violation phenomena in all different decays of three generations of quarks are characterized by one single parametrization-invariant parameter

$$X_{cp} = s_x s_y s_z s_\phi c_x c_y c_z^2 \sim 2.10^{-4} , \qquad (1.6)$$

as determined from ε in K decay.[33] Now this analysis has been carried out for more than three generations of quarks in Ref. (36), where we find that the number of invariant CP violation parameters $X_{cp,i}$ jumps to nine in the four-generation case, saturating the parameter space for generation numbers higher than three. Thus the unique X_{cp} in the three-generation case can serve as a base for searching for signals of the existence of higher generations. I shall come back to this in Section VI.

II. CP Violation in the Kaon and the Hyperon Decays

Weak interactions mediate $K^0 \rightleftarrows \bar{K}^0$, i.e., an off-diagonal term in the mass-matrix exists, so that the physical states are no more K^0, $\bar{K}^0$ but $|K_{S,L}\rangle =$

$N_{\bar{\varepsilon}}((1+\bar{\varepsilon})|K^0> \pm (1-\bar{\varepsilon})|\bar{K}^0>)$, where $N_{\bar{\varepsilon}}$ is a normalization factor. Such mixing gives unequal masses and decay width of the $K_{S,L}$ system, $\delta m \equiv m_S - m_L$, $\delta\Gamma \equiv \Gamma_S - \Gamma_L$, $\bar{\Gamma} \equiv (\Gamma_S + \Gamma_L)/2$. The mixing parameters δm, $\delta\Gamma$ can be measured in an interference experiment involving regeneration of K_S from a K_L beam. Because of $\Gamma_S >> \Gamma_L$, the K^0, $\bar{K}^0$ is a maximally mixed system, since no matter whether we have K^0 or $\bar{K}^0$ to begin with, soon K_S decays away and leaves only K_L which is almost an equal mixture of K^0, $\bar{K}^0$. The parameter $\eta \equiv |1-\bar{\varepsilon}|/|1+\bar{\varepsilon}| \neq 1$ is a phase-convention independent indicator of the mass-matrix CP violation, and was directly measured in K_L decay

$$\delta_\ell = \frac{\Gamma(K_L \to \pi^- \ell^+ \nu) - \Gamma(K_L \to \pi^+ \ell^- \bar{\nu})}{\text{"} + \text{"}}$$

$$= \frac{2\mathrm{Re}\bar{\varepsilon}}{1+|\bar{\varepsilon}|^2} = \frac{1-\eta^2}{1+\eta^2} = (3.3 \pm 0.12) \times 10^{-3} \quad . \qquad (2.1)$$

The mass-matrix CP violation can be calculated through the diagram, Fig. II.1. After calculating the box graph in the middle, which contains X_{CP}, we still need to calculate hadronic matrix element $B_K = <K^0|(\bar{d}\gamma_\mu(1-\gamma_5)s)^2|\bar{K}^0>(-4/3\ f_K^2 M_K)^{-1}$, which describes how the K^0, $\bar{K}^0$ are made of quarks. This is the main uncertainty in the calculation. There is still no reliable way to calculate this B_K parameter. For a discussion on the evaluation of the values of B_K, see Ref. (38). However, we shall see later, by fitting ε, the range of the value B_K is rather limited in the model of three generations of quarks.

The historically first measured CP violation[(39)] in K decays are

$$\eta_{+-} = \frac{A(K_L \to \pi^+\pi^-)}{A(K_S \to \pi^+\pi^-)} = \frac{(A_{+-}/\bar{A}_{+-}) - (1-\bar{\varepsilon})/(1+\bar{\varepsilon})}{\text{"} + \text{"}} = \varepsilon + \varepsilon' \ , \qquad (2.2a)$$

$$\eta_{+-} = \frac{A(K_L \to \pi^0\pi^0)}{A(K_S \to \pi^+\pi^-)} = \frac{(A_{00}/\bar{A}_{00}) - (1-\bar{\varepsilon})/(1+\bar{\varepsilon})}{\text{"} + \text{"}} = \varepsilon - 2\varepsilon' \ . \qquad (2.2b)$$

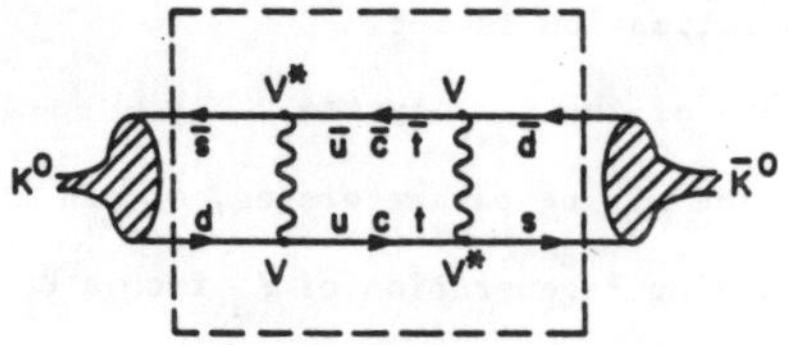

Fig. II.1 Box diagram for $K^0 \leftrightarrows \bar{K}^0$ transition, the B_K parameter describes the transition without the enclosed part; for details, see Ref. (38).

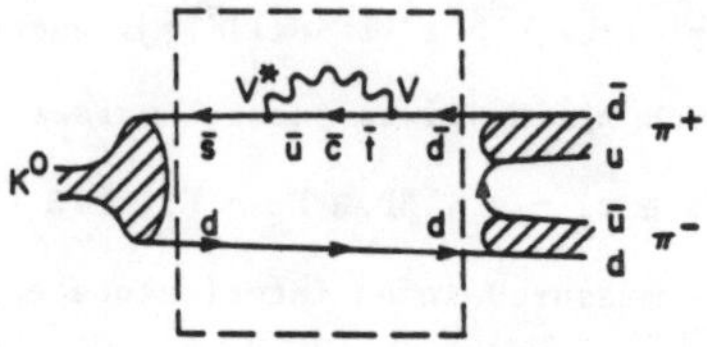

Fig. II.2 W-loop diagram for $K^0 \to \pi^+\pi^-$. Its one-gluon-exchange approximation is the "Penguin" diagram. The B_K' parameter describes the transition without the enclosed part; for details see Ref. (38,44).

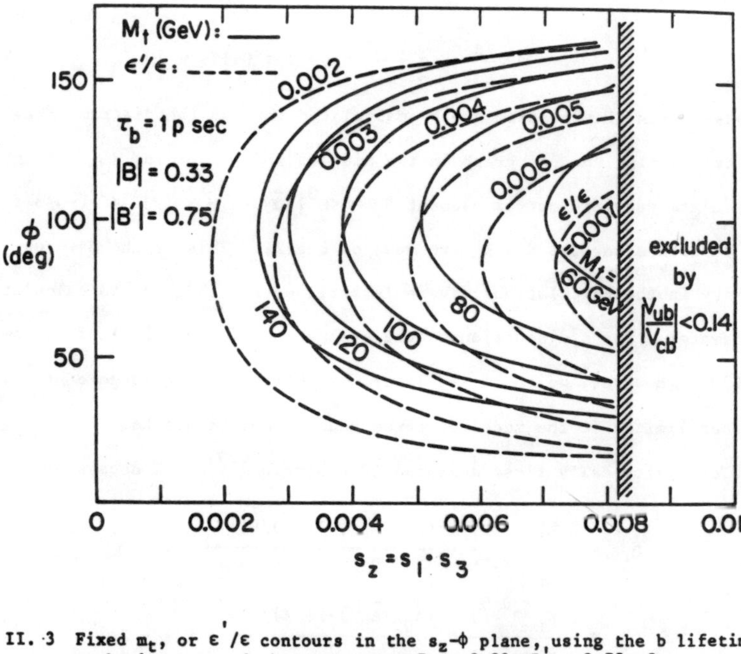

Fig. II. 3 Fixed m_t, or ϵ'/ϵ contours in the s_z-ϕ plane, using the b lifetime to be 1 psec, and the parameters $B_K = 0.33$, $B_K' = 0.75$; for discussions on long-range effects and details see Ref. (44).

One can easily show that[40,24] $\varepsilon' = 0$ only if $A_{+-}/\bar{A}_{+-}$ as well as $A_{00}/\bar{A}_{00}$ can simultaneously be made to be one, i.e., no decay-amplitude CP violation. In that case and phase convention, $\eta_{+-} = \eta_{00} = \varepsilon = \bar{\varepsilon}$. Thus $|\eta_{00}/\eta_{+-}|^2 - 1 \neq 0$ is a measurement of decay-amplitude CP violation. The current experimental data give

$$|\eta_{00}/\eta_{+-}|^2 - 1 = -6\,\mathrm{Re}(\varepsilon'/\varepsilon) = \begin{cases} -0.0046 \pm 0.0053 \pm 0.0024, & ^{(42)} \\ +0.0017 \pm 0.0082, & ^{(41)} \end{cases} \qquad (2.3)$$

which are still consistent with zero. It was noted by Gilman and Wise[43] that the KM model can give a nonvanishing number of ε'. The W-loop diagram shown in Fig. II.2 is essential in giving such an effect. The one-gluon exchange approximation (the "Penguin") has been used to estimate the effect. Again the main uncertainty comes from the hadronic amplitude of B_k', which describes how the K and 2π are related through quark fields, Fig. II.2. For a discussion of the values of B_K' from various calculation, see Refs. (38 and 44).

Since the KM matrix is designed to give CP violation, the CP-violation measurements ε, ε'/ε must give further constraint or the quark mixing matrix. After the determination of $s_y = 0.059$ from the b lifetime, and $s_z < 0.0082$ from $|V_{ub}/V_{cb}| < 0.14$, we still have s_z, s_ϕ, m_t as independent parameters. After fitting ε, assuming some values for B_K, B_K', we obtain fixed m_t, or fixed ε'/ε contours in the s_z, ϕ plane, Fig. II.3. From such analyses, we learn the following: 1) The mass of the t quark m_t cannot be too small, depending on the values of B_K', use of $m_t \lesssim 40$-GeV current data indicates $B_K > 0.33$. 2) other than all the other calculable factors, $|\varepsilon'/\varepsilon|$ depends upon B_K', and its sign depends upon the relative sign of B_K, B_K'; current data indicate $B_K' < 1$. 3) The lower limit on s_z from CP violation is already only a factor of 2 or 3 of the upper limit given by the experimental $|V_{ub}|$ bound.

We can see that the three-generation quark model is quite narrowed down by the current experiments.

There are now two experiments[45] being carried out measuring $|\eta_{00}/\eta_{+-}|$ of the K_S, K_L system with higher sensitivity. Another new type[46] of experiment is being planned to measure ε'/ε via the tagged $K^0,\bar{K}^0 \to 2\pi$ decays at LEAR, CERN, through the following reactions $p\bar{p} \to \pi^- K^+ \bar{K}^0$, or $\pi^- K^+ K^0$;[47] we eagerly await the results from these experiments, and most importantly the observation of the top quark and its mass.

Here I would like to point out an interesting historical irony. Because of the $\Delta = 1/2$ - dominance phenomenon from $\Gamma(K^+ \to \pi^+\pi^0)/\Gamma(K^0 \to \pi^+\pi^-) \simeq 1/670$, ε'/ε is automatically suppressed by a factor of 20. This same factor, which is now causing difficulty for experimentalists in their attempts to measure ε'/ε, actually made it possible for Professor Dalitz to have some fifteen τ events $(K^+ \to \pi^+\pi^+\pi^-)$ in the 1950's. If there is no such suppression of $\Delta I = 3/2$ decays of $K^+ \to \pi^+\pi^0$, the θ, τ puzzle would not have been established before 1956, and the discovery of parity violation would have been delayed for many years. Now this $\Delta I = 3/2$ suppression is making the measurement of ε'/ε difficult. So we began to ask the question whether there are decay channels of the kaon, such that this suppression rule does not apply. Indeed we found that in $K_L \to \gamma\gamma$, the decay amplitude CP-violation effect can be as big as the mass-matrix (superweak) CP-violation effect.[48] It is possible to observe the CP-violation effect in the K_L system only if it is not swamped by $K_S \to \gamma\gamma$. Interestingly, we found that $A(K_S \to \gamma\gamma) \simeq 2.4\ A(K_L \to \gamma\gamma)$, i.e., $Br\ (K_S \to \gamma\gamma) \simeq 5 \times 10^{-6}$. The CERN ε'/ε experiment NA31 can soon measure this branching ratio. If it turns out that $K_L \to \gamma\gamma$ is not overshadowed by $K_S \to \gamma\gamma$, as predicted by our calculation, it is possible to measure the CP violation in $K_L \to \gamma\gamma$. To measure such effects in

K_L, one <u>must</u> do the K^0, $\bar{K}^0$ tagging experiment as provided by the LEAR experiment.

Other decay-amplitude CP violation effects are the partial decay rate difference in $K^\pm \to (3\pi)^\pm$, and in Λ, $\bar{\Lambda}$ decays.(49) Convenient quantities to measure are

$$\frac{\Gamma(K^+\to\pi^+\pi^+\pi^-)/\Gamma(K^+\to\pi^+\pi^0\pi^0)}{\bar{\Gamma}(K^-\to\bar{\pi}\,\overset{+}{\bar{\pi}}\,\pi\,)/\Gamma(K^-\to\pi^-\pi^0\pi^0)} = 1 + 2\Delta_{K\to 3\pi}, \tag{2.4}$$

$$\frac{\Gamma(\Lambda\to\pi^- p)/\Gamma(\Lambda\to\pi^0 n)}{\Gamma(\bar{\Lambda}\to\pi^+\bar{p})/\Gamma(\bar{\Lambda}\to\pi^0\bar{n})} = 1 + 2\Delta_\Lambda\,. \tag{2.5}$$

Current estimates give $\Delta_{K\to 3\pi} \sim 0.74\,\varepsilon'$,(50) and $\Delta_\Lambda \lesssim 10^{-5}$. (51) The advantages of measuring such ratio of ratios rather then the percentage partial decay difference Δ itself are twofold: first the effects are two times as large, secondly some experimental errors tend to cancel. Another interesting CP-violation effect to look for is the difference of the momentum asymmetry parameter of the pion with respect to the polarization of Σ^+ in $\Sigma^+ \to \pi^+ n$ decay,

$$A_{\Sigma^+_+} \simeq 20X_{CP} \sim 4\times 10^{-3}\,, \tag{2.6}$$

according to an estimate by Chau and Cheng in Ref. (51). Such experiments will be done at BNL, LEAR, or another kaon, hyperon-rich laboratory.

III. <u>Mass-Matrix Mixing and CP Violation in Heavy Quark Decays</u>

III.a. <u>Mass-matrix mixing and CP violation</u> -- Tagged-Neutral-Particle-Decay Experiments:

We are fortunate with the kaon system to have the K_S, K_L system to study mass-matrix mixing (the neutral particle-antiparticle mixing) and the superweak CP-violation parameters $\eta \equiv |1-\bar{\varepsilon}|/|1+\bar{\varepsilon}|$ in that K_L indeed lives rather long because the three-pion mass is very close to the kaon mass. However, heavier quark systems like, D^0, $\bar{D}^0$; B^0, $\bar{B}^0$, both mass eigenstates D^0_S, D^0_L; B^0_S, B^0_L will be short lived. It is very difficult or impossible, to do K_L, K_S-type of intereference experiments.[51] So the question is whether we still can measure the mass-matrix mixing parameters δm, $\delta\Gamma$, and CP-violating parameter η separately. The answer is yes, as discussed in Refs. (24, 40). However, one has to do neutral-meson-tagged experiments. The D^0 can be tagged in $e^+e^- \to \psi$ (4030) $\to \pi^+D^-D^0$, $\pi^-D^+\bar{D}^0$. The three parameters δm, $\delta\Gamma$, η_D can be measured by the following three experiments: one to measure $D^0\bar{D}^0$ mixing parameter,

$$\gamma_D = \frac{"D^0 \to \bar{D}^0"}{"D^0 \to D^0"} = \eta_D^2 \frac{(\delta m/\tilde{\Gamma})^2 + (\frac{1}{2}\delta\Gamma/\tilde{\Gamma})^2}{(\delta m/\tilde{\Gamma})^2 + 2 - (\frac{1}{2}\delta\Gamma/\tilde{\Gamma})^2} , \tag{3.1}$$

which can be measured by a tagged D^0, $\bar{D}^0$ decays, $\gamma = "D^0 \to \ell^- x"/"D^0 \to \ell^+ x"$. Or it can manifest itself in the same sign dipletons productions. That $\delta m \neq 0$, $\delta\Gamma \neq 0$ are indications of mixing. Maximal mixing can happen in two ways: one way is when $\tilde{\Gamma} = (\Gamma_S + \Gamma_L)/2$ is dominated by either Γ_s or Γ_L so that $(1/2)\ \delta\Gamma/\tilde{\Gamma} \lesssim 1$, which is the case of neutral kaon decays; the other way is $\delta m/\tilde{\Gamma} >> 1$, and $\delta m/\tilde{\Gamma} >> (1/2)\ \delta\Gamma/\tilde{\Gamma}$. We shall see that the latter situation may happen in the heavy quark systems.

Two more experiments are to measure the $\ell^\pm$ partial decay rate differences:

$$\Delta(\ell^+) = \frac{"D^0 \to \ell^+ X" - "\bar{D}^0 \to \ell^+ X"}{" \quad " + " \quad "} = -\frac{2\mathrm{Re}\bar{\varepsilon}_D}{1 + |\bar{\varepsilon}_D|^2} + \frac{\Gamma_S\Gamma_L}{(\delta m)^2 + (\tilde{\Gamma})^2} , \tag{3.2}$$

and

$$\Delta(\ell^-) = \frac{"D^0 \to \ell^- X" - "\bar{D}^0 \to \ell^- X"}{" \quad " + " \quad "} = -\frac{2\mathrm{Re}\bar{\varepsilon}_D}{1 + |\bar{\varepsilon}_D|} + \frac{\Gamma_S\Gamma_L}{(\delta m)^2 + (\tilde{\Gamma})^2} ; \tag{3.3}$$

therefore,

$$\Delta(\ell^-) + \Delta(\ell^+) = -\frac{4\mathrm{Re}\bar{\varepsilon}_D}{1 + |\bar{\varepsilon}_D|^2} \equiv 2\,\frac{1 - \eta_D^2}{1 + \eta_D^2} , \tag{3.4}$$

is a pure CP violation effect, and

$$\Delta(\ell^-) - \Delta(\ell^+) = \frac{2\Gamma_S\Gamma_L}{(\delta m)^2 + (\tilde{\Gamma})^2} , \tag{3.5}$$

is a pure mass-matrix mixing effect. From Eqs. (3.1)-(3.5), knowing either Γ_S, or Γ_L, δm, $\delta\Gamma$, are measurable. The mass-matrix CP-violation effects $\eta_D \neq 1$ can be measured using Eq. (3.4).

It has been quite firmly established[29] that in the case of three generations of quarks, the only appreciable neutral meson $P^0\bar{P}^0$ (mass-matrix) mixing and CP-violation effects is in the B_S^0, $\bar{B}_S^0$ mixing. Observing deviation from this prediction can provide evidence for the new physics of higher-than-three generations of quarks.

III.b. <u>Partial-decay-rate differences</u>

Studying the partial-decay-rate differences is a convenient way to study the decay-amplitude CP violations.[23,49,53] As we mentioned previously, all CP-violation effects in the case of three generations of quarks are from a single CP-violating parameter X_{CP}, which with other quark-mixing matrix elements are rather well known. Given the same uncertainties in the hadronic decay amplitudes, we can search for the most likely channels where partial-decay-rate differences can be big. Consistently, we find in the B decays, charged[54] as well as neutral, that the partial[54,55] decay rates can be big (few x 10%), though the branching ratios of these exclusive decays are very small ($10^{-3} \sim 10^{-4}$). Here we have an interestingly reversed the situation, as in kaon decays, where the partial-decay-rate differences are very small ($\sim 10^{-3}$), but branching ratios large ($\sim$10%), since the number of events needed is inversely proportional to the <u>square</u> of partial-decay-rate difference, but inversely proportional to the branching ratio. Thus we have a more advantageous situation in the B decay.

Here we list some very encouraging examples,(54,55) e.g.

	Δ_{tree}	$\Delta_{W\text{-}loop}$	Br	# of events needed
$B_u^- \to D^- D^{0}*$	-1.6×10^{-2} (-0.86)	1×10^{-3} (7.3×10^{-4})	3×10^{-3} (4.1×10^{-3})	1.3×10^{6} (3.3×10^{2})
$B_c^- \to \pi^- D^0$	4×10^{-2} (-0.86)	3.7×10^{-3} (1×10^{-3})	2.3×10^{-4} (8.5×10^{-4})	2.7×10^{6} (1.6×10^{3})
$B_d^0 \to K^- \pi^+$	0	10×10^{-2}	1.6×10^{-4}	6.3×10^{5}
$B_S^0 \to D^- F^+$	0	4×10^{-2}	9.6×10^{-4}	5.6×10^{5}
$B_d^0 \to \pi^+ \pi^-$	0	25×10^{-2}	6.1×10^{-4}	2.6×10^{4}
$B_S \to D^+ D^-$	0	-27×10^{-2}	7.0×10^{-4}	2.0×10^{4}

The changed B partial rate differences are mainly from the tree graphs, e.g.,

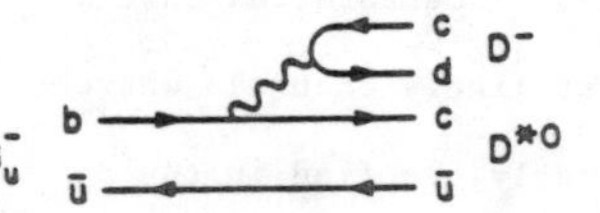

a

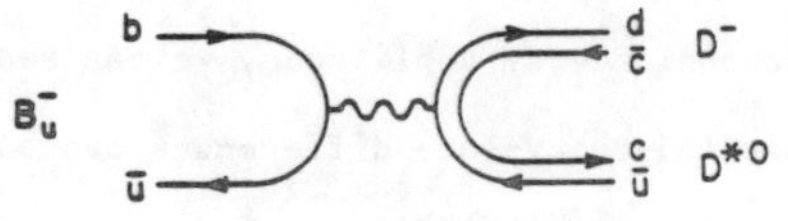

d

The neutral B^0 partial-rate differences can only come from interference between the tree graph and the loop group, e.g.,

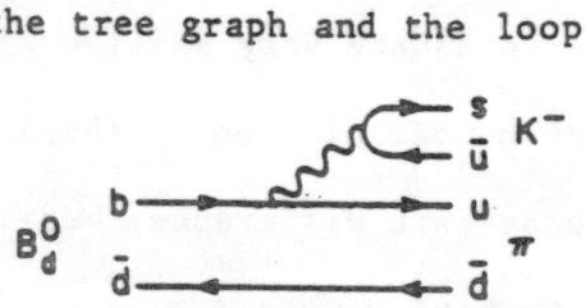

a

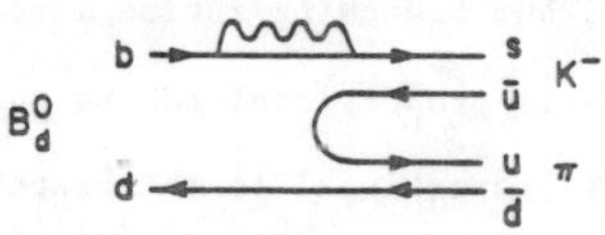

e

For final state with no definite CP, the partial rate differences are only decay-amplitude CP violation. However, for final states with definite CP, the partial rate differences are a combined result of mass-matrix mixing, and decay-amplitude CP violation. For details, see Refs. (54,55).

Such large partial-decay-rate differences can be searched for at CLEO II, Fermilab. LEP, SLC of current luminosity design gives barely enough B particles to do such experiments.

IV. Comparison of CP-Violating Mechanisms

Of course there is the left-right symmetric theories;[(56)] they have one more phase parameter than the KM model so that they give an easier fit to the current data.

CP violations can also come from the multiple Higgs fields. The more specific model of Weinberg[(57)] with three Higgs doublet at one point was thought to be ruled out,[(58)] i.e., $|\varepsilon'/\varepsilon| \gtrsim 1/20$ too large comparing even to the old experimental bound. Recently, after reexamining the calculations,[(59)] $|\varepsilon'/\varepsilon|$ becomes small enough to be compatible with present experimental measurement. But the neutron electric dipole moment d_n ironically is found to be much larger[(60)] than previous calculations. However, its violation of the current experimental bound of $d_n = (0.3 \pm 4.8) \times 10^{-25}$ cm[(61)] depends on an incalculable parameter. H.-Y. Cheng in Ref. (59) gave a very thorough analysis. For other CP-violation effects, see the talk by G. Segre in Ref. (22).

V. Nonleptonic Decays

Ever since the observation of the suppression of $K^+ \to \pi^+\pi^0$ in 1956,[(62)] $\Gamma(K^+ \to \pi^+\pi^0)/\Gamma(K^0 \to \pi^+\pi^-) = 1/670$, i.e., the $\Delta I = 1/2$ dominance rule theorists are still struggling to understand such phenomena in nonleptonic decays. The uncertainties in the calculation of the parameters of B_K and B_K' reflect the same difficulties. The "penguin" mechanism[(63)] is as elusive as ever. Many brave attempts in model calculations to estimate the nonleptonic charm decays are challenged again and again as more data become available, and the theory is forced to be amended and modified. Disappointing as it is, this is to be expected if we are to make progress. Here I would like to point out that this process in furthering understanding of nonleptonic decays can be guided by a model-independent analysis (i.e., the quark-diagram-formulation) of the available experimental data, in the same spirit as Fermi introduced his coupling where only the then-known model-independent information was used.

One fortunate feature of charm and heavier quark decays is that there are many channels open, unlike in the K decays which have only 2π, 3π decays. As has been known for some time now,[(64)(23)] all meson → 2 meson decays can be generally described in a model-independent way by six quark diagrams, as shown in Figs. V., multiplied by quark-mixing matrix elements, which are now quite well determined as discussed in Section I. There are twenty-plus channels of decays in D^+, D^0, F^+ to two-pseudoscalars, PP, decays, and double that number in pseudoscalar-vector, PV, decays. Thus here the model-independent quark-analysis can be very useful in such a quark-diagram approach. Here I shall demonstrate what we have learned from the present charm-meson-decay data, and what their implications are in this approach.

Recently, many two-body decays of charm particles have been beautifully measured,[67-71] which we list in Tables V.1 and V.2. Here we shall put the quark-diagram formalism to use, analyzing all existing charm two-body decay data and discussing their implications for various theoretical model calculations.

V.a Charm meson → pseudoscalar-vector decays

I begin with the PV decays because of the relative simplicity in presenting the discussion, though the data of PV decays are not yet as good as those for some of the PP decays. The simplicity in discussing the PV decays

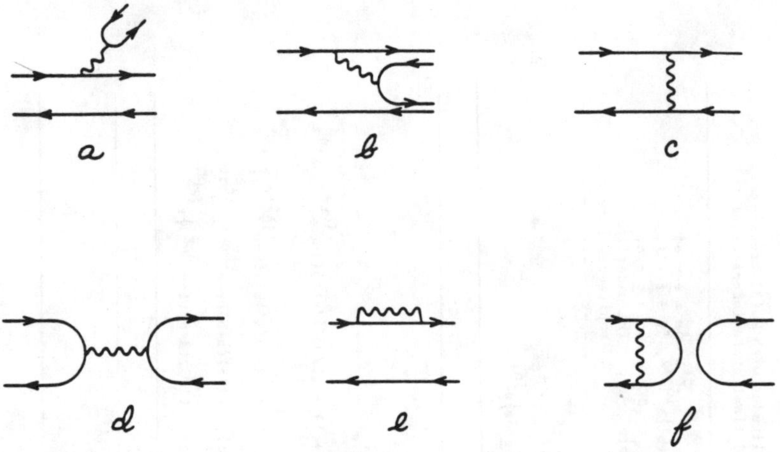

Fig. V.a The six quark diagrams for inclusive meson decay.

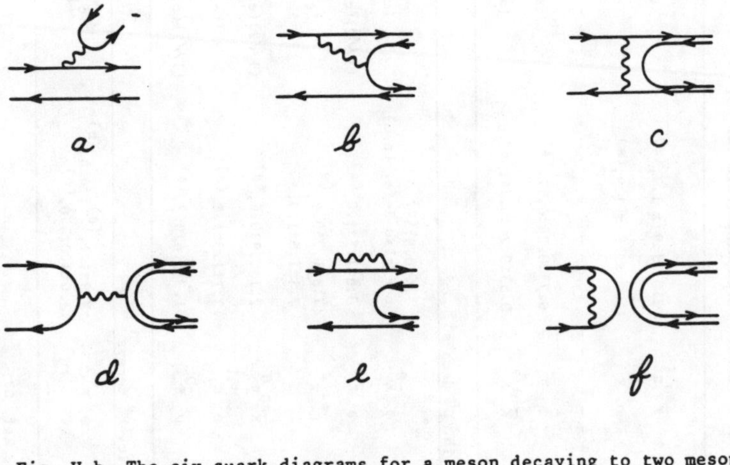

Fig. V.b The six quark diagrams for a meson decaying to two mesons.

Table V.1. Charm Meson Decays into a vector boson and a pseudoscalar meson

V.1a. D^+ decays	Experimental branching ratio (%)	Amplitudes with SU(3) Symmetry ($V_u V^*_{\ } \simeq - V_{ud}V_{cd}^* \simeq s_1c_1$ used)	Amplitudes with SU(3) breaking and final-state interactions
$\bar{K}^{*0}\pi^+$	3.0 ±1.9 ± 1.7,(a)	$(c_1)^2\ \{a'+b'\} \rightarrow$	$\{a'+b'\}e^{i\delta_{3/2}^{\bar{K}^*\pi}}$, $\delta e \equiv \underline{e} - e$; $\delta f \equiv \underline{f} - f$;
$\rho^+\ \bar{K}$	12.2 ±2.8 ± 1.9,(a)	$(c_1)^2\ \{a+b\} \rightarrow$	$\{a+b\}e^{i\delta_3^{\rho\bar{K}}}$, $\delta c \equiv \underline{c} - c$; The underlined amplitudes have $s\bar{s}$.
$\phi\ \pi^+$	0.93±0.26±0.17,(a)	$(s_1c_1)\ \{b'\} \rightarrow$	$\{b'\}e^{i\delta^{\phi\pi}}$,
$\bar{K}^{*0}K^+$	0.53±0.24±0.14,(a)	$(s_1c_1)\ \{a'-\underline{d}\} \rightarrow$	$\{a'-\underline{d}+\delta e\}e^{i\delta_1^{\bar{K}^*K}}$
V.1b. D^0 decays			
$\phi\ \bar{K}^0$	1.4±0.5;(c) see also Refs. (a,b)	$(c_1)^2\ \{c'\} \rightarrow$	$\{\underline{c}'\}e^{i\delta^{\phi\bar{K}}}$
$\omega\ \bar{K}^0$	3.8±1.5±1.0,(a)	$(1/\sqrt{2})(c_1)^2\ \{b+c\} \rightarrow$	$\{b+c\}e^{i\delta^{\omega\bar{K}}}$
$K^{*-}\pi^+$	7.8±1.2±0.9,(a) 7.1±1.6±1.3,	$(c_1)^2\ \{a'+c'\} \rightarrow$	$\{(a'+c') - (1/3)(a'+b')(1-e^{i\Delta_{\bar{K}^*\pi}})\}e^{i\delta_{1/2}^{\bar{K}^*\pi}}$
$\bar{K}^{*0}\pi^0$	2.1±0.9±0.6,(a)	$(1/\sqrt{2})(c_1)^2\ \{b'-c'\} \rightarrow$	$\{(b'-c') - (2/3)(a'+b')(1-e^{i\Delta_{\bar{K}^*\pi}})\}e^{i\delta_{1/2}^{\bar{K}^*\pi}}$
$\rho^+\ K^-$	13.7±1.3±1.5,(a)	$(c_1)^2\ \{a+c\} \rightarrow$	$\{(a+c) - (1/3)(a+b)(1-e^{i\Delta_{\rho\bar{K}}})\}e^{i\delta_{1/2}^{\rho\bar{K}}}$
$\rho^0\ \bar{K}^0$	1.3±0.4±0.3,(a)	$(1/\sqrt{2})(c_1)^2\ \{b-c\} \rightarrow$	$\{(b-c) - (2/3)(a+b)(1-e^{i\Delta_{\rho\bar{K}}})\}e^{i\delta_{1/2}^{\rho\bar{K}}}$
V.1c. F^+ decays			
$\phi\ \pi^+$	3.3±1.1,(d); 4.4;(b) 13.0±3.0±4.0,(e)	$(c_1)^2\ \{a'\} \rightarrow$	$\{a'\}e^{i\delta^{\phi\pi}}$

a) Ref. 67
b) Ref. 68
c) Ref. 69
d) Ref. 70
e) Ref. 71

Table V.2. Charm Meson Decays into Two Pseudoscalars (same notations a to f are used for amplitudes, but in general they have no relations to those in PV decays in Table I. The heading for the table is the same as Table I.)

V.2a D^+ decays (Data from Ref. (67))

$\bar{K}^0 \pi^+$	3.5 ±0.5 ±0.4,	$(c_1)^2 \{a+b\}$	$\to \{a+b\} e^{i\delta_{3/2}^{\bar{K}\pi}}$	
$\bar{K}^0 K^+$	1.11±0.34±0.21,	$(s_1c_1) \{a-d\}$	$\to \{a-\underline{d}+\delta e\} e^{i\delta^{\bar{K}K}}$,	$\delta e \equiv \underline{e} - e$; $\delta f \equiv \underline{f} - f$; $\delta c \equiv \underline{c} - c$; The underlined amplitudes have $s\bar{s}$.
$\pi^0 \pi^+$	≤ 0.53,(1)	$(1/\sqrt{2})(s_1c_1) \{a+b\}$	$\to \{a+b\} e^{i\delta_2^{\pi\pi}}$,	

V.2b D^0 decays

$K^- \pi^+$	4.9 ±0.4 ±0.4,	$(c_1)^2 \{a+c\}$	$\to \{(a+c) - (a+b)\ (1/3)\ (1-e^{i\Delta_{\bar{K}\pi}})\} e^{i\delta_{1/2}^{\bar{K}\pi}}$
$\bar{K}^0 \pi^0$	2.2 ±0.4 ±0.2,	$(1/\sqrt{2})(c_1)^2 \{b-c\}$	$\to \{(b-c) - (a+\underline{b})\ 2/3\ (1-e^{i\Delta_{\bar{K}\pi}})\} e^{i\delta_{1/2}^{\bar{K}\pi}}$
$\bar{K}^0 \eta$	1.8 ±0.8 ±0.3,	$\cos\theta\ A(D^0 \to \bar{K}^0 \eta_8) + \sin\theta\ A(D^0 \to \bar{K}^0 \eta_0)$	
$K^0 \bar{K}^0$	≤ 0.62,	$(s_1c_1)\quad 0$	$\to \{(-\delta e-2\delta f) + (a+\underline{c}-\delta e)\ (1/2)\ (1-e^{i\Delta_{\bar{K}K}})\} e^{i\delta_0^{K\bar{K}}}$
$K^- K^+$	0.60±0.10±0.08,	$(s_1c_1) \{a+c\}$	$\to \{(a+c) + (\delta e+2\delta f) - (a+\underline{c}-\delta e)\ (1/2)\ (1-e^{i\Delta_{\bar{K}K}})\} e^{i\delta_0^{\bar{K}K}}$
$\pi^+ \pi^-$	0.16±0.09±0.03,	$-(s_1c_1) \{a+c\}$	$\to \{(a+c) + (\delta e+28f) - (a+b)\ (1/3)\ (1-e^{i\Delta_{\pi\pi}})\} e^{i\delta_0^{\pi\pi}}$
$\pi^0 \pi^0$		$\{\sqrt{2}\}(1/2)(s_1c_1) \{b-c\}$	$\to \{(b-c) + (\delta e+2\delta f) - (a+b)\ (2/3)\ (1-e^{i\Delta_{\pi\pi}})\} e^{i\delta_0^{\pi\pi}}$

V.2c. F^+ decays

$\bar{K}^0 K^+$	$(c_1)^2 \{b+d\}$	$\to \{b+d\} e^{i\delta_1^{\bar{K}K}}$
$\pi^0 \pi^+$	0	$\to 0$
$\eta_8 \pi^+$	$-(\sqrt{2}/\sqrt{3})(c_1)^2 \{a-d\}$	$\to \{a-d\} e^{i\delta\pi\eta_8}$
$\eta_0 \pi^+$	$(1/\sqrt{3})(c_1)^2 \{a+2d\}$	$\to \{a+2d\} e^{i\delta\pi\eta_0}$

comes from the purity of the quark contents in ϕ and ω. Many PV decays are given by one tyoe of amplitude, as shown in Table I: e.g., $F^+ \to \phi\pi^+$ ($\propto$ a'), $D^+ \to \phi\pi^+$ ($\propto$ b'), $D^0 \to \phi\bar{K}^0$ ($\propto$ $\underline{c}$'), respectively. Thus from the decay rates, we can determine their absolute values:

$$|a'| = (2.50 \pm 0.42)10^{-6}, \; |b'| = (3.67 \pm 0.51)10^{-6}, \; |\underline{c}'| = (1.68 - 2.10)10^{-6} \;. \quad (5.1)$$

The only theoretical assumption used here is that $|e^{i\delta^{\phi\pi}}| = 1 = |e^{i\delta^{\phi\bar{K}}}|$. To obtain the decay widths from the measured branching ratios as given in Table I, the following charm-decay lifetimes are used:[72] $\tau(D^+) = (8.8^{+1.0}_{-0.8})\times 10^{-13}$ sec, $\tau(D^0) = (4.3^{+0.4})\times 10^{-13}$ sec, $\tau(F^+) = (2.8^{+1.4})\times 10^{-13}$ sec. Among the three measurements of $Br(F^+ \to \phi\pi^+)$ we used (3.3 ± 1.1)%. To obtain the amplitudes from the rates, we have made use of the the phase-space factor $p_c^{\,3}/(8\pi m_v^{\,2})$, where p_c is the center-of-mass momentum. Note here that neither amplitude b' nor amplitude $\underline{c}$' is negligible, as preferred by some model calculations.

$D^+ \to \bar{K}^{*0}\pi^+$ $[\propto(a' + b')]$ is an exotic channel which implies elastic and small $\delta^{\bar{K}^*\pi}_{3/2}$. Therefore, the rate gives

$$|a' + b'| = (1.16 \pm 0.37) \times 10^{-6} \;. \quad (5.2)$$

From Eqs. (5.1) and (5.2), it is evident that a' and b' are of opposite signs, and the |a'|, and |b'| obtained from $F^+ \to \phi\pi^+$, $D^+ \to \phi\pi^+$ are in excellent agreement with |a' + b'| determined from $D^+ \to \bar{K}^{*0}\pi^+$. Thus we have

$$a' = (2.50 \pm 0.42) \times 10^{-6}, \; b' = -(3.67 \pm 0.51) \times 10^{-6} \;. \quad (5.3)$$

Note that a' and b' are severely destructive, consistent with the decay width of D^+ being smaller than D^0. Since $D^0 \to K^{*-}\pi^+$, $\bar{K}^{*0}\pi^0$ are given in terms of (a' + b'), and (a' + c')/(a' + b') [= 1 - (b' - c')/(a' + b')], and the phase

shift $\Delta_{\bar{K}*\pi} = \delta_{1/2}^{\bar{K}*\pi} - \delta_{3/2}^{\bar{K}*\pi}$, and assuming $|e^{i\delta_{1/2}^{\bar{K}*\pi}}| = 1$,
we obtain the following two solutions,

$$(a' + c')/(a' + b') = 2.36 \pm 0.67, \quad \rightarrow c' = -(5.25 \pm 0.39)\times 10^{-6} ,$$
$$\Delta_{\bar{K}*\pi} = (52^{+30}_{-52})^\circ ; \qquad (5.4a)$$

$$(a' + c')/(a' + b') = -1.70 \pm 0.67, \quad \rightarrow c' = -(0.53 \pm 0.39)\times 10^{-6} ,$$
$$\Delta_{\bar{K}*\pi} = 180^\circ - (52^{+30}_{-52})^\circ . \qquad (5.4b)$$

We note that the errors on $\Delta_{\bar{K}*\pi}$ are so large that the data can be accommodated by real amplitudes without final-state interactions. The amplitude c' is quite different from <u>c</u>' in Eq. (5.1). (Underlined amplitudes involve strange quark and antiquark pair production.) To make definite conclusions, we need better measurements.

One nice prediction from this analysis is

$$Br(D^0 \rightarrow \phi\pi^0) = (1/2)Br(D^+ \rightarrow \phi\pi^+)\Gamma(D^+)/\Gamma(D^0) \simeq 0.21\% . \qquad (5.5)$$

This will be an important measurement if we are to test this scheme.

We next proceed to determine the unprimed amplitudes from $D \rightarrow \rho\bar{K}$ and $D^0 \rightarrow \omega\bar{K}$ decays. From Table V.1 it follows that $D^+ \rightarrow \rho^+\bar{K}^0$, $D^0 \rightarrow \omega\bar{K}^0$, determine

$$|a + b| = (2.18 \pm 0.25) \times 10^{-6} , \quad |b + c| = (2.57 \pm 0.51) \times 10^{-6} . \qquad (5.6)$$

From the measurements of $D^0 \rightarrow \rho^0\bar{K}^0$, ρ^+K^-, we find two solutions:

$$(a + c)/(a + b) = 1.55 \pm 0.16 , \quad \Delta\rho\bar{K} = (24^{+25}_{-24})^\circ , \qquad (5.7a)$$

$$(a + c)/(a + b) = -0.89 \pm 0.16 , \quad \Delta\rho\bar{K} = 180^\circ - (24^{+25}_{-24})^\circ . \qquad (5.7b)$$

These give the following three possible solutions for amplitudes a, b, c:

$$a = (4.06 \pm 0.38)10^{-6}\ ,\ b = -(1.88 \pm 0.28)10^{-6}\ ,$$

$$c = -(0.68 \pm 0.28)10^{-6}\ ,\ \Delta_{\rho\bar{K}} = (24^{+25}_{-24})^\circ\ ; \qquad (5.8a)$$

$$a = (1.50 \pm 0.38)10^{-6}\ ,\ b = (0.68 \pm 0.28)10^{-6}\ ,$$

$$c = (1.89 \pm 0.28)10^{-6},\ \Delta_{\rho\bar{K}} = (24^{+25}_{-24})^\circ\ ; \qquad (5.8b)$$

$$a = (1.41 \pm 0.38)\mathrm{x}10^{-6}\ ,\ b = (0.77 \pm 0.28)\mathrm{x}10^{-6},$$

$$c = -(3.34 \pm 0.28)\mathrm{x}10^{-6},\ \Delta_{\rho\bar{K}} = 180^\circ - (24^{+25}_{-24})^\circ\ . \qquad (5.8c)$$

Again, because of the large errors, the data are compatible with real amplitudes. The measurement of $D^+ \to \bar{K}^{*0}K^+$, Table V.1, gives us

$$|a' - \underline{d} + \delta e| = (2.70 \pm 0.61) \times 10^{-6}\ . \qquad (5.9)$$

Future measurements of $\bar{K}^{*0}\eta_8$ $\big(\propto(b' + c' - 2\underline{c})\big)$ and $\bar{K}^{*0}\eta_0$ $\big(\propto(b' + c' + \underline{c})\big)$, can help to determine amplitudes $\underline{c}$ and (b' + c'), and then c', since b' is known. From future measurements of $D^0 \to \bar{K}^{*0}K^0$, $K^{*0}\bar{K}^0$ $\big(\propto(c - c')\big)$, and the known information on c' we can determine the amplitude c. Then we can check which solution of equation (5.8) will be picked, and thus determine a and b individually. From $F^+ \to \rho^+\pi^0$ $\big(\propto(d - d')\big)$, $F^+ \to \omega\pi^+$ $\big(\propto(d + d')\big)$, we can determine d, d'. We then know all the amplitudes a, b, c, d, and a', b', c', d', and their relative signs. The rest of the PV decays are predictable, up to SU(3) breaking and final-state interactions. These results must be conformed to by any theoretical calculations.

Next we discuss the case of charm meson decay into two pseudoscalars, $P_c \to PP$. (Note that the amplitudes a to f here for PP decays have no relation to those for the PV decays. When needed for clarity, we use subscript PP to denote the distinction.) Here, the data are of greater accuracy than for the $P_c \to PV$ case, Table V.2. From $D^+ \to \pi^+\bar{K}^0$, $D^0 \to K^-\pi^+$, $\bar{K}^0\pi^0$, we can conclude definitely that real amplitudes $(a, b, c)_{PP}$, (without including effects like final-state interactions) cannot fit the data.[12,14] From $D^+ \to \pi^+\bar{K}^0$, we obtain

$$|a + b|_{PP} = (1.66 \pm 0.11)10^{-6} \text{ GeV} . \tag{5.10}$$

Then from $D^0 \to K^-\pi^+$, $\bar{K}^0\pi^0$, we obtain the following two solutions,

$$[(a + c)/(a + b)]_{PP} = 1.95 \pm 0.14 , \; \Delta_{\bar{K}\pi} = (79^{+10}_{-14})° ; \tag{5.11a}$$

or

$$[(a + c)/(a + b)]_{PP} = -1.28 \pm 0.14 , \; \Delta_{\bar{K}\pi} = 180 -(79^{+10}_{-14})° . \tag{5.11b}$$

We want to caution about the interpretation of the phase shift difference $\Delta_{\bar{K}\pi} \equiv \delta^{\bar{K}\pi}_{1/2} - \delta^{\bar{K}\pi}_{3/2}$ obtained here from charm decays. Its relation to the hadronic scattering phase shifts is complicated by the other competing channels, e.g. $\pi\pi\bar{K}$, $\pi\pi\pi\bar{K}$ (not including $P\bar{K}$, $\pi\bar{K}^*$, which do not communicate with $\bar{K}\pi$ through strong interactions). Only if all the strong-interaction communicating channels are negligible. $\delta^{\bar{K}\pi}$ here is the $\bar{K}\pi \to \bar{K}\pi$ scattering phase shift.

Unlike the PV decays, it is much harder here to determine individual amplitudes, since none of the decays is given by a single amplitude. It is interesting to point out that the nonspectator-diagram amplitudes c, $\underline{c}$ and d can be measured in a model-independent way by observing the following decay modes,

$$\frac{\Gamma(D^0 \to \bar{K}^0\eta_8)}{\Gamma(D^0 \to \bar{K}^0\eta_0)} = \frac{1}{2}\left|\frac{b+c-2\underline{c}}{b+c+\underline{c}}\right|^2_{PP} ; \quad \frac{\Gamma(F^+ \to \eta_8\pi^+)}{\Gamma(F^+ \to \eta_0\pi^+)} = 2\left|\frac{a-d}{a+2d}\right|^2_{PP} . \tag{5.12}$$

From the absolute rates of these decays we can determine $[(b+c), \underline{c}; \text{ and } a, d)_{PP}$. Combining with the solutions Eq. (5.10) and (5.11), we shall determine all amplitudes (a, b, c and d)$_{PP}$ and their relative signs.

Next we go to the mixing-matrix singly suppressed measurement of $\Gamma(D^0 \to K^+K^-)/\Gamma(D^0 \to \pi^+\pi^-) \neq 1$ first observedly the Mark II collaboration.[73] From Table V.2b, we see that such differences can be attributed to the SU(3) breaking effect of $(\delta e + 2\delta f)_{PP}$, which contri-butes with opposite sign to $D^0 \to K^+K^-$, $\pi^+\pi^-$ and/or to the final-state-inter-action effect (e.g., $\delta_0^{\pi\pi}$ has a larger absorptive part than $\delta_0^{K\bar{K}}$). To clarify these mechanisms it is of paramount importance to measure $D^0 \to \pi^0\pi^0$ (see Table II.b), since the same unknown $(\delta e + 2\delta f)_{PP}$, $\delta_0^{\pi\pi}$ are present, but the rest of the amplitudes $(b-c)_{PP}$, $(a+b)_{PP}$ are known [from Eqs. (5.10) and (5.11)].

Using the known relation between η, η' and η_8, η_0 (here the mixing angle of (-10°) is used. In the analysis on future measurements of decays involving η', care should be taken to substract any component in η' that is not η_0 or η_8), the current measurement of $D^0 \to \bar{K}^0\eta$, Table V.2b, gives $|1.23b - 0.49c|_{PP} = (4.57 \pm 1.01) \times 10^{-6}$ GeV, if there is no SU(3) breaking, i.e. $\underline{c} = c$; or it gives $|b + c|_{PP} = (3.71 \pm 0.83) \times 10^{-6}$ GeV, if SU(3) breaking is maximal, i.e., $\underline{c} = 0$. Future measurements of $D^0 \to \bar{K}^0\eta_8$ $[\propto(b + c - 2\underline{c})]$, or $\bar{K}^0\eta'$, will help to determine amplitudes $[(b+c)$, and $\underline{c}$, thus $b, c)_{PP}$ individually when combining with the results of Eqs. (5.10) and (5.11). The measurement of $D^+ \to \bar{K}^0K^+$ gives $|a - \underline{d} + \delta e|_{PP} = (4.29 \pm 0.66) \times 10^{-6}$ GeV. Future measurements of $F^+ \to \pi^+\eta_8[\propto(a-d)_{PP}]$, and $\pi^+\eta_0[\propto(a+2d)_{PP}]$, can give $[a$, and $d)_{PP}$; $\bar{\Gamma}(D^+ \to \bar{K}^0K^+)$ is predicted to be equal to $(s_1/c_1)^2\, \bar{\Gamma}(F^+ \to \pi^+\eta_8)$ if there is no SU(3) breaking. The measurements of $D^0 \to K^0\bar{K}^0$, $\eta_0\eta_0$, which are nonzero only from SU(3) breaking, can give a direct modification of SU(3) breaking effects. The long-predicted relation $\bar{\Gamma}(D^+\to\pi^+\pi^0)/\bar{\Gamma}(D^+\to\bar{K}^0\pi^+) = 1/2\ |V_{cd}/V_{cs}|^2$ should be checked by experiments.

As discussed in the previous sections future measurements of $\bar{K}^{*0}\eta_8$, $\bar{K}^{*0}\eta_0$, $D^0\to\bar{K}^{*0}K^0$, $K^{*0}\bar{K}^0$, $F^+\to\rho^+\pi^0$, $\omega\pi^+$; and $D^0\to\bar{K}^0\eta_8$, $\bar{K}^0\eta_0$, $F^+\to\pi^+\eta_8$, $\pi^+\eta_0$, will give definite and model-independent results about individual amplitudes, to which theoretical calculations must conform.

I hope that I have demonstrated to you that the general quark-diagram approach provides a framework in which experimental results can be analyzed in a model-independent way, and new experiments can be pointed out to further test certain specific model calculations.(74) The current experimental results have already provided much information on non-leptonic decay mechanism in term of the quark diagram amplitudes, which can be used for comparison with theoretical calculations. Many interesting predictions have resulted. The story of non-leptonic decay is very complex, but also very interesting. It will take our persistent effort, both theoretically and experimentally, to find the conclusion of the story.

VI. Beyond the Three Generations of Quarks

As we see the experiments measuring the b lifetime, $|V_{ub}/V_{cb}|$, ε, ε'/ε, and the hint of m_t from UA1 experiment at CERN have really pushed the three-generation model to a corner. Future measurements of ε'/ε, $|V_{ub}/V_{cb}|$ with improved error bars, and with m_t pinpointed can really imply very specifically about B_K, B_K', and the likelihood of the three generation model. So it's a good time to think ahead if there are four generations of quarks.

If there are fourth generation of quarks b', t', the number of parameters increase from the three generation 4 to 9, six angles and three phases. We generalize the parameterization of Eq. (1.3) as follows,[36]

$$V_4 = \begin{bmatrix} 1 & 0 & 0 & 0 \\ 0 & 1 & 0 & 0 \\ 0 & 0 & c_u & s_u \\ 0 & 0 & -s_u & c_u \end{bmatrix} \begin{bmatrix} 1 & 0 & 0 & 0 \\ 0 & c_v & 0 & s_v e^{-i\phi_3} \\ 0 & 0 & 1 & 0 \\ 0 & -s_v e^{i\phi_3} & 0 & c_v \end{bmatrix} \begin{bmatrix} c_w & 0 & 0 & s_w e^{-i\phi_2} \\ 0 & 1 & 0 & 0 \\ 0 & 0 & 1 & 0 \\ -s_w e^{i\phi_2} & 0 & 0 & c_w \end{bmatrix} \left[\begin{array}{ccc|c} & & & 0 \\ & V_3 & & 0 \\ & & & 0 \\ \hline 0 & 0 & 0 & 1 \end{array}\right] \quad (6.1)$$

where V_3 is the 3x3 matrix given in Eq. (1.3). The reason we put the additional phases ϕ_2, ϕ_3 where there are in Eq. (6.1) is to anticipate that the widening of the generation gap will keep its pace. Of course physical consequences are independent of ways of parameterization. What are the different physical phenomena if fourth generation do exist?

(1) If the masses are low enough, we should search for them in e^+e^-, $p\bar{p}$, pp and cosmic rays.

(2) $|V^*_{cd}/V_{cs}| \simeq |V^*_{us}/V_{ud}|$ does not have to be true. More accurate measurements of $|V_{cd}/V_{cs}|$ should be studies from semileptonic decays of $D \to \ell^{\pm}X$ containing s or not, and from the long predicted,[79] $\Gamma(D^+ \to \pi^+\pi^0)/\Gamma(D^+ \to \bar{K}^0\pi^+) = 1/2\,|V_{cd}/V_{cs}|^2$.

(3) As mentioned in Sect. I, we showed a suprising result in Ref. (36), that rather than the unique phase-convention-invariant CP violation parameter X_{CP} in the three-generation case, there are now nine of them $X_{CP,i}$, $i = 1, \ldots 9$, for generation number $n = 4$, and always saturating the whole parameter space for $n \geq 4$. So the consequences of CP violations are much more varied then the three generation case, e.g. ε'/ε can be zero without making B_K'

$= 0$; V_{ub} can be zero without annihilating the CP-violation effect ϵ. Due to the presence of more $X_{CP,i}$ variables, many more appreciable partial rate differences can happen. Those effects from the tree graphs are actually only dependent on the presence of the higher generation, not on their masses. For example the partial-decay-rate difference Δ_F in mixing-matrix singly suppressed decay $F \to K^0\pi$ can be increased to 10^{-1}, and Δ_T for t quark decays can be very large too. This makes the improvement of SPEAR experiment, and BEPC extremely important. For details, see Ref. (36).

(4) There are new contributions in the mass matrix from the new generation of b', t', so that the mass-matrix mixing (i.e. the neutral meson-antimeson mixing results from the three-generation model can be drastically changed. However such effects are from loop diagrams and very sensitive to the b',t' quark masses. As shown in Ref. (75). If one picks suitable values of the quark-mixing matrix and masses of the new quarks, e.g. $D^0, \bar{D}^0$ mixing can be increased from $\ll 10^{-4}$ to 10^{-2} if $m_{b'} \gtrsim 200$ Gev; $B_d^0, \bar{B}_d^0$ can also be appreciable; the mass-matrix CP violation in D, B mesons can also be large so that $("\ell^+\ell^+"-"\ell^-\ell^-")/("\ell^+\ell^+"+"\ell^-\ell^-")$, $(KK-\bar{K}\bar{K})/(KK+\bar{K}\bar{K})$ are reasonably large. Such possibilities are extremely interesting to search for experimentally.

VII. Quark Mass Matrix

So far in the electroweak unified gauge theory, the most obscure part is the mass generating mechanism for gauge particles and for quarks. Here I would like to reiterate, agreeing with what Dr. Kobayashi had emphasized in his talk, the intimate relations between the quark mass matrix M' in the bases of quark eigen-state in weak interaction and the mixing matrix V. The mass term in the interaction Lagrangian can be written in the strong-interaction eigen quark state u, d, or equivalently the weak-interaction eigenstate u', d',

$$\mathscr{L}\text{mass} = \bar{u}_L M(2/3)\, u_R + \bar{d}_L\, M(-1/3)\, d_R + \text{H.C.},$$

$$= \bar{u}_L M(2/3)\, u_R + \bar{d}'_L\, M'(-1/3)\, d'_R + \text{H.C.}, \qquad (7.1)$$

where $V_L M'(-1/3)\, V_R^+ = M(-1/3)$, and $V_R^+ V_R = 1 = V_L^+ V_L$.

We can show that M'(-1/3) can always either chosen to be Hermitean or symmetric:[76]

From $$M'(-1/3) = V_L^{\ +} M(-1/3)\, V_R\ , \qquad (7.2)$$

we multiply an unitary matrix W on the right hand side of Eq. (7.2),

$$M'' \equiv M'\, W = V_L^{\ +} M\, V_R W\ . \qquad (7.3)$$

If we require M'' to be Hermitean $M''^{+} = M''$ i.e. $W^+ V_R^+ M V_L = V_L^{\ +} M V_R W$ or $V_L W^+ V_R^+ M = M V_R\, W V_L^{\ +}$. A sufficient solution for W is $V_R W V_L^+ = 1$, i.e. $W = V_R^+ V_L$, and $M'' = V_L^+\, M V_L$; or if we want M'' to be symmetric $M''^T = M''$, i.e. $W^T V_R^T M V_L^{\ *} = V_L^{\ +} M V_R W$. A sufficient solution for W is $V_R W V_L^{\ T} = 1$, i.e. $W = V_R^+ V_L^{\ *}$, and $M'' = V_L^+ M\, V_L^* = V_L^+ M V_L^{+T}$. All this says that we can choose a base such that M'(-1/3) is either Hermitean or symmetric. Now the charged weak current is

$$J_\mu = \bar{u}_L\, \Gamma_\mu\, d_L{}' = \bar{u}_L\, \Gamma_\mu\, V\, d_L\ . \qquad (7.4)$$

This V is the quark mixing matrix. So in the d'_L base the charge - (-1/3) mass-matrix is

$$M'_H\ (-1/3) = V^+\ M(-1/3)\ V, \quad \text{if in the Hermitial base} \qquad (7.5)$$

$$M'_S\ (-1/3) = V^T M(-1/3)\ V, \quad \text{if in the symmetric base;} \qquad (7.6)$$

using the V matrix given in Eq. (1.3), we obtain M'_H (-1/3) and $M'_S(-1/3)$. Picking the hermitian case, the mass matrix becomes:

$$M_H \equiv V\, M^D\,(\text{diag.})\, V^+ =$$

$$= \left[\begin{array}{c|c|c} m_d + m_s s_x^2 & m_s s_x + m_b s_y s_z e^{-i\phi_1} & m_b s_z e^{-i\phi_1} - m_s s_x s_y \\ \hline m_s s_x + m_b s_y s_z\, e^{i\phi_1} & m_s c_x^2 + m_b s_y^2 & \begin{array}{c}(m_b - m_s) s_y \\ - m_s s_x s_z e^{-i\phi_1}\end{array} \\ \hline m_b s_z e^{i\phi_1} - m_s s_x s_y & \begin{array}{c}(m_b - m_s) s_y \\ - m_s s_x s_z e^{i\phi_1}\end{array} & m_b c_y^2 \end{array}\right] \tag{7.7}$$

To get (7.7) we have used the smallness of s_y, s_z and $m_d/m_b \lesssim \lambda^4$, $m_s/m_b \lesssim \lambda^2$. Similar we can get the symmetric mass matrix VM^D (diag.) V. Equation (7.7) is accurate up to order $m_b\lambda^4$. These are the mass matrices the weak decay phenomenology has told us. They should be conformed by model building.

We can see that the quark mixing matrix and its complex phase, the source of CP violation, is intimately related to the quark mass matrix in the base of weak-interaction quarks. Thus the mass-generating sector, i.e. the Higgs sector, is truely now the reservoir of our ignorance in the gauge theory description of the particle world. It is of great importance to make progress on it.

VIII. Concluding Remarks and Outlook

As we can see that there are many interesting experiments can be done to shed light on our understanding of the physics of heavy-quark decays and CP violation. Here I list them briefly.

VIII.a. The Measurements of the b Particle Lifetime and $\Gamma(b\to u)/\Gamma(b\to c)$ have well constrained the quark mixing matrix. It is important to further study the quark mixing matrix elements: V_{ub} via $B\to\tau\nu_\tau$, and the charmless b decays; V_{cs}, V_{cd} via $D\to\ell\, X_{s,d}$. The fitting of the CP-violation effects in K decays,

ε and ε'/ε, has cornered the model with three generations of quarks, and has constrained the nonleptonic dynamical parameter B_K, B_k', and the t quark mass. If V_{ub} becomes smaller than needed for CP violating effects, it's an indication of trouble with the KM scheme for CP violation, or of the exciting possibility of the existence of higher than three generations of quarks.

VIII.b. There are still many important and interesting CP violating effects to be measured in the kaon and hyperon decays: ε'/ε, ϕ_{00}; η_{+-0}; $Br(K_S^0 \to \gamma\gamma)$; R_+/R_-, where $R_\pm = \Gamma(K^\pm \to \pi^\pm\pi^\pm\pi^\mp)/\Gamma(K^\pm \to \pi^\pm\pi^0\pi^0)$; $R_\Lambda/R_{\bar\Lambda}$, where $R_\Lambda \equiv \Gamma(\Lambda \to \pi^- p)/\Gamma(\Lambda \to \pi^0 n)$, $R_{\bar\Lambda} = \Gamma(\bar\Lambda \to \pi^+ \bar p)/\Gamma(\bar\Lambda \to \pi^0 \bar n)$; Pion momentum asymmetry difference in $\Sigma^\pm \to \pi^\pm n$ decays. The tagged $K^0, \bar K^0$ experiments for partial rate difference $\Delta(K^0, \bar K^0 \to 2\pi)$, $\Delta(K^0, \bar K^0 \to \gamma\gamma)$ are very interesting and will mark the beginning of this new type of CP experiments.

VIII.c. Some partial decay rate differences in beauty particle decays can be very large, such that only some tens of restructed beauty decays are needed:

$$\Delta(B_u^\pm \to D^\pm \overset{(-)}{D}{}^{*0}),\ \Delta(\overset{(-)}{B}{}_d^0 \to K^\mp \pi^\pm),\ \Delta(\overset{(-)}{B}{}_s^0 \to D^+ D^-),\ \Delta(\overset{(-)}{B}{}_d^0 \pi^+ \pi^-),$$

$\Delta(B_c^\pm \to \pi^\pm \overset{(-)}{D}{}^0)$, etc. See Table in Section IIIb, and Refs. (54, 55).

In the three-generation case, such partial decay rate differences for charm and top quarks are in general an order of magnitude smaller. This information can serve as a base for looking for hints of the existence of higher generations of quarks.

VIII.d Look for the B_S^0, $\bar B_S^0$ mixing effects, e.g. same-sign dileptons, which has been consistantly predicted to be substantial in the case of three generations of quarks while mixing in B_d^0, $\bar B_d^0$ is small and sensitive to parameters, and mixing in $D^0, \bar D^0$ and $T^0, \bar T^0$ are extremely small.

VIII.e. Experiments tagging neutral mesons can provide new ways to study CP-violation effects that the conventional interference experiments can not do, e.g. measuring decay-amplitude CP violation effects in $K_L \to \gamma\gamma$; by measuring $\Delta(K^0, \bar{K}^0 \to \gamma\gamma)$; separately measuring the neutral meson $P^0, \bar{P}^0$ (mass-matrix) mixing parameters δm, $\delta\Gamma$, and the mass-matrix CP violation $\eta_p \equiv |1-\bar{\epsilon}_p|/|1-\bar{\epsilon}_p|$ for short lived heavy quark systems.

VIII.f. Experiments should be planned to look for indications of higher then three generations of quark: the most important are the direct observations of the top particles, and the heavier ones; Since the phenomenalogical consequences in the case of three generations of quarks have been quite well determined, any observation of substantial deviations from them are possible indication of the existence of higher generations, e.g. large partial decay rates in charm and top quark decays, in addition to the beauty particle decays as which are the only ones calculated to be substantial in the three-generation case; large D^0, $\bar{D}^0$, B_d^0, $\bar{B}_d^0$ mixing in addition to B_s^0, $\bar{B}_s^0$ which is the only one calculated to be substantial in the three-generation case; large mass-matrix CP-violation in B^0, D^0 decays, e.g. $\Delta(\ell^+\ell^+ - \ell^-\ell^-)=0$; "long" lived top quark ($\tau_T \gtrsim 10^{-18}$ sec).

VIII.g. The well determination of the quark mixing matrix also helps for a systematic study on nonleptonic decays via the many channels available in charm and beauty particle decays. Recent charm decay data analyzed via the quark-diagram scheme already have shed some light on the structure of charm decays. The quark-diagram amplitudes for charm meson $\to$ pseudoscalar-vector-meson decays are quite well determined in magnitudes and relative signs. This leads to many predictions, e.g. $Br(D^0 \to \phi\pi^0) \simeq 0.21\%$; future measurements of $D^0 \to \bar{K}^{*0}K^0$, $K^{*0}\bar{K}^0$ can shed light on amplitude c'$-$c; $F^+ \to \rho^+\pi^0$, $\rho^0\pi^+$ can give information on amplitude d-d'; and $F^+ \to \omega\pi^+$ can give information on d+d'; and many more. For charm meson → pseudoscalar-pseudoscalar meson decays, measurements of the following few charm decays will be most helpul: $(D^+ \to \pi^+\pi^0)$, so that the long predicted relation $\Gamma(D^+ \to \pi^+\pi^0)/\Gamma(D^+ \to \bar{K}^0\pi^+) = 1/2\,|V_{cd}/V_{cs}|^2$ can be checked Measurements on $D^0 \to \bar{K}^0\eta'$ in addition to the recently measured $D^0 \to \bar{K}^0\eta$ can test the importance of W-exchange amplitude c; the measurement of $D^0 \to \pi^0\pi^0$ can help us to understand the details of the mechanism for the ratio $\Gamma(D^0 \to K^+K^-)/\Gamma(D^0 \to \pi^+\pi^-) \neq 1$; the observation of $D^0 \to K^0K^0$, $\eta_0\eta_0$ can give a clear indication and measurements of SU(3) breaking.

VIII.h. Measurement of F, $D \to \tau\nu_\tau$, $\mu\nu_\mu$ can help to determine g_τ/g_μ and the ν_τ mass.[77]

VIII.i. Measurements of rare decays can put standard models to stringent tests, and provide windows for a glimpse of possible new physics. The case of $K_L \to \mu^+\mu^-$ (Br = 9.1×10^{-9}) was such an example. Now there are four rare decay experiments being carried out at Brookhaven National Laboratory:[78]

Exp. 777, BNL - U. of Washington-Yale Collaboration,

$K^+ \to \pi^+ e^- \mu^+$, sensitivity Br $\sim 10^{-11}$,

Exp. 780, BNL-Yale Collaboration,

$K_L \to \mu e$, sensitivity Br $\lesssim 10^{-10}$,

Exp. 787, BNL-Carnegie-Mellon-Columbia-Princeton-TRILIMF Collaboration,

$K^+ \to \pi^+ \nu\bar{\nu}$, π^+ "?", sensitivity Br $\sim 10^{-10}$,

Exp. 791, UCLA-Los Alamos-U. of Pennsylvania-Princeton-Stanford-Temple Collaboration,

$K_L \to \mu e$, Sensitivity 10^{-12},

$K_L \to \pi^0 e^+ e^-$, Sensitivity 10^{-12},

μ polarization $K_L \to \mu^+\mu^-$, sensitivity 10 $\sim$ 20%.

Further tightening of the already very impressive bound on the neutron electric dipole moment will help to test various CP-violating models.

We can see that despite the milestone of the observation of the long awaited intermediate bosons, many fundamental questions in electroweak interactions still remain to be answered: how are the masses of gauge bosons and quarks generated? What is the reason for the V-A nature of the weak current? Where is the origin of CP noninvariance? Why do the quarks and leptons mix and repeat themselves? How many more are there? The apparent complexity of my discussions presented here reflect the immaturity of the field. However, this is the very nature of our pursuit: once a good question is answered, more new meaningful questions can be asked. There will be frontier research as long as there is life itself. I look forward to Meson 60 to see what new milestones have been achieved and what the new frontiers are to be conquered.

REFERENCES

1. E. Fermi, Zeit. f. Phys. 88, (1934) 161.

2. H. Yukawa, Proc. Phys. Math. Soc., Japan 17, (1935) 48.

3. C.N. Yang and R.L. Mills, Phys. Rev. 96, (1954) 191.

4. S.L. Glashow, Nucl. Phys. 22, (1961) 579; S. Weinberg, Phys. Rev. Lett. 19, (1967) 1264; A. Salam in: "Elementary Particle Theory", N. Svartholm, ed., Almquist and Wiksell, Stockholm (1968); see also A. Salam and J.C. Ward, Phys. Lett. 13, (1964) 168. For renormalization of the theory see G.'t Hooft, Nucl. Phys. B33, (1971) 173.

5. UA1 collaboration, CERN, G. Arnison, et al., Phys. Lett. 122B, (1983) 103, ibid. 126B, (1983) 398; UA2 collaboration, CERN, M. Banner, et al., ibid. 122B, (1983) 476. Talk by C. Rubbia and L. Dilella at 1985 Int'l. Conf. on Lepton Photon Conf. at Kyoto, Japan, Aug. 1985.

6. See review talk by W.J. Marciano, Proceedings of the 1983 Int'l. Sym. on Lepton and Photon Interactions at High Energies, Cornell U., Aug. 4-9, 1983.

7. R.F. Peierls, T.L. Trueman and, L-L. (Chau) Wang, Phys. Rev. D16, (1977) 1397. See also L-L. (Chau) Wang Theoretical Implications of ISABELLE Physics, in Proc. Int'l. School, The High Energy Limit, Erice, Italy, July 31-August 13, 1980; Plenum, Ed. A. Zichichi.

8. W. Pauli, in open letter dated Dec..4, 1930, now reproduced in: "Collected Scientific Papers by W. Pauli", R. Kronig and V.F. Weisskopf, eds., Interscience, New York (1964).

9. Neutrino observation, C.L. Cowan, F. Reines, and F.B. Harrison, Phys. Rev. 96, (1954) 1294.

10. Ig. Tamm, Nature 133, (1934) 931; D. Iwanenko, ibid. (1934) 981.

11. The discovery of the muon, C.D. Anderson, S.H. Neddermeyer, Phys. Rev. 51, (1937) 894. The discovery of charged pions, C.M.J. Lattes, H. Muirhead, J.P.S. Occhialini, C.F. Powell, Nature 159 (1947) 694.

12. R. Dalitz, Phil. Mag. 44, (1953) 1068, and Phys. Rev. 94 (1954) 1046. e. Fabri, Nuovo Cimento 11, (1954), 479.

13. T.D. Lee and C.N. Yang, Phys. Rev. 104, (1956) 254.

14. Wu, C.S., Ambler, E., Hayward, R.W., Hoppes, D.D., and Hudson, R.P., Phys. Rev. 105, (1957) 1413; Garwin, R.L., Lederman, L.M., and Weinrich, M., Phys. Rev. 105, (1957) 1415; Friedman, J.I., and Telegdi, V.L., Phys. Rev. 105, (1957) 1681; M. Goldhaber, L. Grodzins, and A.W. Sunyar, Phys. Rev. 109, (1958) 1015.

15. Neutral current observation, F. Hasert et al., Phys. Lett. 46B, (1973) 138; Nucl. Phys. B73, (1974) 1.

16. N. Cabibbo, Phys. Rev. Lett. 10 (1963) 531; for earlier work on the subject, see M. Gell-Mann and M. Levy, Nuovo Cim. 16 (1960) 705.

17. Particle Data Group, Rev. Mod. Phys. 52 (1980) 5. Shocket et. al. Phys. Rev. Lett. 39 (1979) 59.

18. S. Glashow, J. Iliopoulos and L. Maiani, Phys. Rev. D2, (1970) 1285.

19. For implicit charm J/ψ, J.J. Aubert, et al., Phys. Rev. Lett 33, (1974), 235; J.E. Augustin, et al., Phys. Rev. Lett. 33, (1974) 1406. For charm particle observation in $\nu_\mu p$ reaction in the BNL 7-foot bubble chamber, see E.G. Cazzoli, et al., Phys. Rev. Lett. 34, (1975) 1125; in e^+e^- reaction, see G. Goldhaber, et al., Phys. Rev. Lett. 37, (1976) 255, in pp reaction, see M. Basile, et al., Nuovo Cim. 63A, (1981) 230; for hint from earlier cosmic ray experiment see, K. Niu, E. Mikumo, Y. Maeda, Prog. Theor. Phys., 46 (1971) 1644.

20. M. Kobayashi and T. Maskawa, Prog. Theor. Phys. 49, (1973) 652.

21. For implicit beauty Υ, S.W. Herb et al., Phys. Rev. Lett. 39, (1977) 252, for explicit beauty particle observation, S. Behrends et al., Phys. Rev. Lett. 50 (1983), 881, and talk by P. Avery in Ref. 22.

22. For recent development on the subject, see talks in Proceedings of "Flavor Mixing in Weak Interactions," Europhysics Conference, March 5-10, 1984, Erice, Italy, Ed. L.-L. Chau, Plenum 1985.

23. For a general survey, see L.L. Chau, Phys. Rept. 95 (1983) 1.

24. L.L. Chau, "Comments on Heavy Quark Decays and CP Violation," Proceedings of the G.F. Chew Jubilee (Sept. 29, 1984), to be published by the World Scientific Pub. Co.

25. For earlier fits to find V_{ud}, V_{us}, R. Shrock, and L.L. (Chau) Wang, Phys. Rev. Lett. 41 (1978) 1692; for more recent fits, see the next reference.

26. J.F. Donoghue and B.R. Holstein, Phys. Rev. D25 (1982) 2015; A. Garcia and P. Kielanowski, Phys. Lett. 110B (1982) 498; and the most recent fits, WA2 experiment at CERN, M. Bourquin et al., "IV. Tests of the Cabbibbo Model," CERN preprint (1983). See talk by H.W. Siebert in Ref. 22; J.F. Donoghue, B.R. Holstein, Phys. Lett. 160B (1985) 173; W.J. Marciano and A. Sirlin, Phys. Rev. Lett. 56 (1986) 22, found $V_{ud} = 0.9729 \pm 0.0012$; and A. Bohm, M. Kmiecik, Phys. Rev. D31, (1985) 3005, include new Fermilab hyperon data, Ref. 30, and found $|V_{us}| = 0.225 \pm 0.002$ to be in better agreement with the older Ke3 fit of Ref. 25.

27. For the b lifetime, E. Fernandez, Phys. Rev. Lett. 51 (1983) 1022; N.S. Lockyer et al., Phys. Rev. Lett. 51 (1983) 1316; for more up to date information, see talks by W.T. Ford, and G.H. Trilling in Ref. 22. see P. Ginsparg, S. Glashow, M. Wise, Phys. Rev. Lett. 50 (1983), 1415.

28. For $\Gamma(b \to u)/\Gamma(b \to c)$, C. Klopfenstein et al., Phys. Lett. 103B, (1983) 444; A. Chen et al., Phys. Rev. Lett. 111B, (1984) 1084; and talks by J. Lee-Franzini, and P. Avery in Ref. 22.

29. L.-L. Chau, W.-Y. Keung, and M.D. Tran, Phys. Rev. D27, (1983) 2145, L.-L. Chau and W.-Y. Keung, Phys. Rev. D29, (1984) 592.

30. Fermilab Polarized Σ^- experiment, S.Y. Hsueh et al., Phys. Rev. Lett. 54, (1985), 2399.

31. See H.W. Siebert's talk in Ref. (22).

32. L. Wolfenstein, Phys. Rev. Lett. 51, (1984) 1945.

33. L.-L. Chau and W.-Y. Keung Phys. Rev. Lett. 53, (1984) 1802.

34. L. Maiani, Phys. Lett. 62B (1976) 183; R. Mignami Lett. Al Nuovo Cimento, 28 (1980), 529.

35. H. Fritzsch, Phys. Rev. D32 (1985), 3058.

36. F.J. Botella and L.-L. Chau, "Anticipating the Higher Generations of Quarks from Rephasing Invariance of the Mixing Matrix," Brookhaven preprint (1985).

37. M.K. Gaillard and B.W. Lee, Phys. Rev. D10 (1974) 897.

38. See L.-L. Chau, H.-Y. Cheng and W.-Y. Keung, "A KM Model Study for ε, ε/ε, and M_t", BNL preprint BNL-35163, July '84 (unpublished); and Phys. Rev. 32, (1985) 1837; the uncertainties in the B_K parameter and the numerical findings of these papers were presented by L.-L. Chau in a talk immediately following B. Winstein at the APS Meetings at Washington, DC, 23-26, April '84. J. Bijnens, H. Sonoda and M.B. Wise, Phys. Rev. Lett. 53, (1984) 2367. A. Pich and E. de Rafael, Phys. Lett. 158B, (1985) 477.

39. J.H. Christenson, J.W. Cronin, V.L. Fitch and R. Turlay, Phys. Rev. Lett. 13, (1964) 138; and the classical theoretical discussions on CP violation. T.D. Lee, R. Oehme and C.N. Yang, Phys. Rev. 106, (1957) 340; T.T. Wu and C.N. Yang, Phys. Lett. 13, (1964) 380.

40. L.-L. Chau, "Comments on CP Violation" Brookhaven preprint, (1984).

41. J.K. Black et al., Phys. Rev. Lett. 54, (1985) 1628.

42. R.H. Bernstein et al. ibid, 54, (1985) 1631.

43. F.J. Gilman and M.B. Wise, Phys. Lett. 83B, (1979) 83.

44. L.-L. Chau, H.-Y. Cheng, and W.-Y. Keung, "CP Violation in the Kaon Systems", Brookhaven preprint 1985.

45. The current experiments on ε'/ε: Chicago, -Fermilab, -Orsay-Princeton collaboration, Fermilab experiment #731; CERN-Dortmund-Edinburgh-Orsay-Pisa-Siegen collaborations, CERN NA31.

46. For discussion partial rate differences in K decays see Sect. 3.2.2 of Ref. (6); L.-L. Chau, talk at Erice Conference on "Electroweak Effects at High Enrgies," Feb. 1-12 '83; C. Kounnas, A.B. Lahanas and P. Pavlopoulos, Phys. Lett. 127B, 381 (1983):

47. "Physics at LEAR with Low-Energy Cooled Antiprotons," edited by U. Gastaldi and R. Klapisch, Ettore Majorana Int. Sci. Series, ed. A. Zichichi (Plenum); P. Pavlopoulos, Talk in Ref. (22); tagged K^0 experiment: Athen-Basel-ETH-Zurich-Liverpool-Saclay-Sin collaboration, CEAR, LEAR, p82.

48. L.-L. Chau, H.-Y. Cheng, Phys. Rev. Lett. 54, (1985) 176B.

49. L.-L. Chau Wang, AIP Conf. Proc. No. 72, Particle and Fields, Subseries No. 23, Virginia Polytechnic Inst. 1980, eds. G.B. Collins. L.N. Chang and J.R. Ficenec; L.-L. Chau, Proceedings of Workshop on Weak Interactions and Neutrinos, Javea, Spain, Sept. 5-11, 1983.

50. C. Avilez, Phys. Rev. D23, (1981) 1124. See also B. Grinstein, S.-J. Rey and M.B. Wise, CALT-68-1286 (1985).

51. L.-L. Chau and H.Y. Cheng, Phys. Lett. 131B, (1983) 202; T. Brown, S.F. Tuan and S. Pakvasa, Phys. Rev. Lett. 51, (1983) 1823. I would like to thank Dr. K. Kilian for illuminating discussion on the possibility of doing this experiment at LEAR.

52. For a review of CP violation in the K system see, K. Kleinknecht, Ann. Rev. Nucl. Sci. 26, 26 (1976). R.G. Sachs, Ann. Phys. 22 (1963), 239. T.D. Lee, and C.S. Wu, Ann. Rev. Nucl. Sci. 15, 381 (1966); "Theory of Weak Interactions in Particle Physics", R.E. Marshak, Riazuddin, C.P. Ryan, Publisher Wiley-Insterscience (1969).

53. L.M. Sehgal, and L. Wolfenstein, Phys. Rev. 162 (1976) 1362, O.E. Overseth and S. Pakvasa, Phys. Rev. 189 (1969) 1663. A. Pais and S.B. Treiman, Phys. Rev. D12, 2744 (1975); L.B. Okun, V.I. Zakharov and B.M. Pontecorvo. Lett. Nuovo Cim. 13, 218 (1975). J. Barshay and J. Geris, Phys. Lett. 84B, (1979), 319; M. Bander, D. Silverman and A. Soni, Phys. Rev. Lett. 43, (1979) 242; J. Barnabeu and C. Jarlskog, Z. Phys. C8, (1981) 233. I.I. Bigi and A.I. Sanda, Nucl. Phys. B193, (1981) 81. L. Wofenstein, Nucl. Phys. B200 (1984), 45 and talk by K.-C. Chou in Ref. 22.

54. L.-L. Chau H.-Y. Cheng, Phys. Rev. Lett. 53, (1984) 1037.

55. L.-L. Chau, H.-Y. Cheng, "CP Violation in Decay Rates of Neutral B Mesons", Brookhaven preprint, BNL 36915 (1985), Phys. Lett. 165B (1986).

56. For left-right symmetric theory, see talk by R.N. Mohapatra in Ref. (22). J.C. Pati and A. Salam, Phys. Rev. D10, (1974) 275; R.N. Mohapatra and J.C. Pati, Phys. Rev. D11, (1975) 566, 2558.

57. See, e.g. S. Weinberg, Phys. Rev. Lett. 37, (1976) 657, Physica 96A, (1979) 327.

58. A.I. Sanda, Phys. Rev. D23 (1981) 2647; N.G. Deshpande, ibid D23 (1981) 2654.

59. Y. Dupont and T.N. Pham., Phys. Rev. D28, (1983) 2169. J.F. Donoghue and B.R. Holstein, UMHEP-213 (1984). H.-Y. Cheng, "Weinberg CP Violation Model Revisited", Brandeis preprint (1984).

60. A.A. Anselm, V.E. Bunakov. V.P. Gudkov, and N.G. Uraltsev, Leningrad preprint (1984).

61. N.F. Ramsey, Rep. Prog. Phys. 45, (1982) 95; J.M. Pendlebury et al., Phys. Lett. 136B, (1984) 327; I.S. Alterev et al., ibid 102B, (1981) 13; and B. Heckel's talk in Ref. (22).

62. Observation of $K^+ \to \pi^+ \pi^0$ suppression R.W. Birge et al., Nuovo cimento 4 (1956) 834; G. Alexander et al., Nuovo cimento 6 (1957) 478, for theoretical discussions, see M. Gell-Mann and A. Pais, Proc. Int'L Conf. of High Energy Physics, Glasgow (Pergamon Press, London 1955); Proc. Fifth Annual Rochester Conf. on High Energy Physics (NY 1955) p. 136;

63. M.A. Shifman, A.I. Vainshtein, V.I. Zakharov and Nucl. Phys. B120, 316 (1977); Sov. Phys. JETP 45 (1977) 670.

64. L.-L. Chau, talk in the Proceedings of the 1980 Quangzhou Conference (Jan. 5-10, 1980), Science Press, Beijing, Reinhold Comp., China; Van Nostr. and Proc. of VI Int'l. Conf. on Meson Spectroscopy, BNL, April 24-25, 1980; AIP.

65. For charm → PP, see Table 2.5, Ref. 23), and footnote (2) of Ref. 24); For Charm → PV see Table 2.6, Ref. (23) and M. Gorn, Nucl. Phys. B191, (1981). X.-Y. Li, S.-F. Tuan (to be submitted to Zeits für Physik C). For treatment of SUB(3) breaking and final-state-interaction, see next reference.

66. For charμ → PP, PV including SU(3) breaking and final-state-interaction effects, see L.-L. Chau and H.-Y. Cheng "Exclusive Two Body Decays of Charm Meson", Brookhaven-Brandes preprint 1985; and L.-L. Chau and H.-Y. Cheng, "Quark Diagram Analysis of Charm Meson Decays", Brookhaven-Indiana preprint 1985.

67. MARK III collaboration, R. Baltrusaitis, Phys. Rev. Lett. 55, (1985) 150. R.H. Schindler, "New Results on Charmed D Meson Decay", invited talk at the 1985 SLAC Summer Institute for Particle Physics.

68. CLEO collaboration, A. Chen et al., Phys. Rev. Lett. 51, (1983) 634; see also talk by S. Stone, Int. Sym. or Lepton and Photon, Interaction at High Energies, Cornell July 83; and paper contributed to the 1985 Int'l. Lepton and photon conference at Kyoto, Japan, Aug. '85.

69. ARGUS collaboration C. Darden et al., DESY preprint 84-04-3, H. Albrecht et. al., Phys. Lett. 158B (1985) 525.

70. TASSO collaboration, M. Althoff et al., Phys. Lett. 136B, (1984) 130.

71. HRS collaboration, M. Derrick et al., Phys. Rev. Lett. 54, (1985) 2568.

72. For a review on charm lifetime see talk by K. Niu in Ref. (22). I would like to thank Professor N. Reay for very informative discussions on the subject.

73. R. Schindler et al., Phys. Rev. D24, (1981) 78. Talk by G. Goldhaber at 18th Moriond Conference, March 13-19, 1983.

74. For current comparisons with various model calculations, see discussions in Ref. (66); and for other ways of analyzing the data and comparison with calculations. See the rather exhaustive references given there in, and talk by R. Rückel in Ref. (22), and talk by B. Stech at Moriond Conference, Spring, 1985.

75. X.-G. He and S. Pakvasa Phys. Lett. 156B, (1985) 236. A.A. Anselm et al., Phys. Lett. 156B, (1985) 102. M. Gronau and J. Schechter Phys. Rev. D31, (1985) 1668. U. Türke et. al., DO-TH 84/26. I.I. Bigi PITHA 84/19.

76. P.H. Frampton and C. Jarlslog Phys. Lett. 154B (1985) 421, R.N. Mohapatra, Private communication.

77. Chao Kuang-ta, Chau Ling-Lie, Huang Tao, Du Dong-Sheng, Wu Dan-di, Chinese Phys. Lett. 2, (1985) 117.

78. See Talk by L. Littenberg in Ref. (22).

79. L.-L. chau Wang, and F. Wilczek, Phys. Rev. Lett. 43 (1979), 816.

Geometrical Integrability and Equations of Motion: A Unifying Description of Equations of Motion

Ling-Lie Chau
Physics Department
Brookhaven National Laboratory
Upton, NY 11973

Abstract

It is pointed out that many equations of motion in physics, including gravitational and Yang-Mills equations, have a common origin: i.e. they are the results of certain geometrical integrability conditions. These integrability conditions lead to linear systems that are important in integrating these equations of motion.

Introduction

Recently it had been shown[(1)] that light-like integrability in curved extended superspace leads to equations of motion. This completes the picture of a unifying description of equations of motion from the point of view of geometrical integrability, which had its origin in the study of many two-dimensional nonlinear systems,[(2)] and in the study of Yang-Mills equations.[(3)] Here I would like to describe these geometrical integrability schematically. I shall first begin with the self-dual Yang-Mills equation.

I. Self-Dual Yang-Mills Equations

I.a. Null plane Integrability and Equation of Motion

It has become clear, as originally advocated by Penrose and collaborators,[(4)] that the spinor formulation of the space time variable can be conveniently described by writing x^μ by a two-by-two matrix

$$x_{\alpha\dot{\beta}} \equiv (x_\mu \sigma^\mu)_{\alpha\dot{\beta}} ,$$

where

$$\sigma^\mu = (I, \vec{\sigma}) ,$$

and

$$\sigma^0 = \begin{pmatrix} 1 & 0 \\ 0 & 1 \end{pmatrix} , \quad \sigma^1 = \begin{pmatrix} 0 & 1 \\ 1 & 0 \end{pmatrix} , \quad \sigma^2 = \begin{pmatrix} 0 & -i \\ i & 0 \end{pmatrix} , \quad \sigma^3 = \begin{pmatrix} 1 & 0 \\ 0 & -1 \end{pmatrix} ,$$

for Minkowski space time, and $\sigma^\mu = (I, i\vec{\sigma})$ for Euclidean spacetime. The length of the $x_{\alpha\dot{\beta}}$ is given by its determinant, $|x|^2 = \det(x_{\alpha\dot{\beta}})$. For given pairs of complex numbers $\lambda^{\dot{\alpha}}$, and ω_α,

$$x_{\alpha\dot{\alpha}} \lambda^{\dot{\alpha}} = \omega_\alpha \qquad (1a.1)$$

defines a null plane, since $\det(x_{\alpha\dot{\alpha}} - x'_{\alpha\dot{\alpha}}) = 0$ for nontrivial $\lambda^{\dot{\alpha}}$.

It is obvious that when there are no fields

$$(\partial_{1\dot{\alpha}}, \partial_{2\dot{\alpha}}) = 0 \ , \text{ where } \partial_{\alpha\dot{\alpha}} \equiv \frac{d}{dx^{\alpha\dot{\alpha}}} \lambda^{\dot{\alpha}} \ , \tag{1a.2}$$

i.e. ordinary differentiation commutes. Now let us see if we require Eq. (1a.2) to be still true in the presence of a gauge field $A_{\alpha\dot{\beta}}$, i.e., $\partial_{\alpha\dot{\beta}} \to \nabla_{\alpha\dot{\beta}} \equiv \partial_{a\dot{b}} + A_{\alpha\dot{\beta}}$.

$$(\nabla_{1\dot{\beta}} \ , \nabla_{2\dot{\beta}}) = 0 \ . \tag{1a.3}$$

Without losing generality and to be more specific we take $\lambda^{\dot{\alpha}} = \binom{1}{\lambda}$, where λ is an arbitrary complex number. More explicitly, Eq. (1a.3) is

$$(\partial_{1\dot{1}} + A_{1\dot{1}} + \lambda\partial_{1\dot{2}} + \lambda A_{1\dot{2}}, \partial_{2\dot{1}} + A_{2\dot{2}} + \lambda\partial_{2\dot{2}} + \lambda A_{2\dot{2}}) = 0 \ . \tag{1a.4}$$

For the λ^0 term:

$$(\partial_{1\dot{1}} + A_{1\dot{1}}, \partial_{2\dot{1}} + A_{2\dot{1}}) = 0 \ , \quad \to f_{1\dot{1},2\dot{1}} = 0 \ . \tag{1a.5}$$

For the λ^1 term:

$$(\partial_{1\dot{2}} + A_{1\dot{2}}, \partial_{2\dot{1}} + A_{2\dot{1}}) + (\partial_{1\dot{1}} + A_{1\dot{1}}, \partial_{2\dot{2}} + A_{2\dot{2}}) = 0 \ ,$$

$$\to F_{1\dot{2},2\dot{1}} + f_{1\dot{1},2\dot{2}} = 0 \ . \tag{1a.6}$$

For the λ^2 term:

$$(\partial_{1\dot{2}} + A_{1\dot{2}}, \partial_{2\dot{2}} + A_{2\dot{2}}) = 0 \ , \quad \to f_{1\dot{2},2\dot{2}} = 0 \ . \tag{1a.7}$$

All these mean that

$$f_{\alpha\dot{\beta}}{}^{\alpha}{}_{\dot{\gamma}} = 0 \;, \tag{1a.8}$$

i.e., the field is self-dual, which is obvious from the theorem that any skew-symmetric tensor written in spinor form can always decompose into the sum of self-dual and anti-self-dual terms:

$$F_{\alpha\dot{\beta}\,\gamma\dot{\delta}} = 1/2\,(\varepsilon_{\alpha\gamma}\,F_{\alpha\beta}{}^{\alpha}{}_{\dot{\delta}} + \varepsilon_{\dot{\beta}\dot{\delta}}\,F_{\alpha\dot{\beta}\gamma}{}^{\dot{\beta}}) \;.$$

So the integrability conditions on the self-dual null-plane, shown in Eq. (1a.3), imply that the field is anti-self-dual.

As in Eq. (1a.1), we can also define an anti-self-dual plane, for given two pairs of complex number λ^{α}, $\omega_{\dot{\alpha}}$,

$$\lambda^{\alpha}\chi_{\alpha\dot{\alpha}} = \omega_{\dot{\alpha}} \;. \tag{1a.9}$$

Then the integrability condition:

$$[\nabla_{\alpha\dot{1}}, \nabla_{\beta\dot{2}}] = 0 \;, \tag{1a.10}$$

implies that the curvature field is anti-self-dual, i.e.

$$f_{\alpha\dot{\beta}\gamma}{}^{\dot{\beta}} = 0 \;. \tag{1a.11}$$

In tensor notation Eqs. (1a.8) and (1a.11) imply

$$F_{\mu\nu} = \pm{}^{*}F_{\mu\nu} \;. \tag{1a.12}$$

From the Bianchi identity,

$$\nabla_{\mu}{}^{*}F^{\mu\nu} = 0 \;, \tag{1a.13}$$

the consequence of the geometrical integrability Eq. (1a.12) lead to the tensor field $F_{\mu\nu}$ satisfy the Maxwell-Yang-Mills equation

$$\nabla_{\mu} F^{\mu\nu} = 0. \tag{1a.14}$$

It was in this spirit that the authors of Ref. (3) attempted to find geometrical integrability conditions in higher dimensions that lead to the full equations of motion, instead of just to the self-dual or anti-self-dual part. Further work had been carried out in higher-than-four ordinary space.[(5)] Here I shall elaborate the development in generalizing to superspaces, as originally proposed by Witten,[(3)] and Ferber[(6)] in their attempt to generalize the twistor formulation in superspace.

I.b. The Linear Systems and Integrability Properties

Equation (1a.3) is precisely the integrability condition of the existence of the wave function (analytic bundle in the parameter λ) of the following two linear systems:

$$\nabla_{\alpha\dot{\beta}}\, \psi(x,\lambda) = 0, \text{ for the self-dual case;} \tag{1b.1}$$

$$\nabla_{\dot{\alpha}\beta}\, \bar{\psi}(x,\bar{\lambda}) = 0, \text{ for the anti-self-dual case;} \tag{1b.2}$$

or more explicitly

$$\left(\partial_{\alpha\dot{1}} + A_{\alpha\dot{1}} + \lambda\, (\partial_{\alpha\dot{1}} + A_{\alpha\dot{2}})\right) \psi = 0 \ , \tag{1b.1a}$$

$$\left(\partial_{\dot{1}\beta} + A_{\dot{1}\beta} + \bar{\lambda}\, (\partial_{\dot{2}\beta} + A_{\dot{2}\beta})\right) \bar{\psi} = 0 \ . \tag{1b.2a}$$

Another way of writing Eq. (1b.1) is

$$A_{\alpha\dot{\beta}} = (\partial_{\alpha\dot{\beta}}\psi)\psi^{-1} \ , \tag{1b.3}$$

i.e., the gauge connections on the light-like plane are pure gauge.

Note that $A_{\alpha\dot\beta} = A_{\alpha\dot 1}(x) + \lambda\, A_{\alpha\dot 2}(x)$, where $A_{\alpha\dot\beta}\ (x)$ is independent of λ. The main problem in obtaining solutions for the (anti) self-dual part of the solutions Eqs. (1a.8) (1a.11) to Eq. (1a.14) is to find $\psi\ (x,\psi,\lambda)$ which are nontrivial functions both for the space coordinate as well as for the parameter λ such that the resultant $A_{\alpha\dot\beta}$ in Eq. (1b.3) is only a function of space coordinate, independent of λ.

The linear systems, Eq. (1b.1) were first obtained from the twister program,(4) and by Belavian and Zhakharov(4) from Lax-pair point of view. There is another form of the linear systems, which was derived from quite a different point of view.(7,8) This latter formulation has the advantage of being gauge invariant, being possible to give infinite non-local conservation laws, and, in the author's opinion, of being more appropriate for future quantum-field-theory formulation. Though historically the gauge-independant formulation, which has customarily been called the J-formulation, was derived in quite a different way;(9,10) here I shall derive them starting from Eqs. (1b.1). Putting $\lambda = 0$ in Eq. (1b.1), we obtain

$$A_{\alpha\dot 1}(x) = -\left(\partial_{\alpha\dot 1}\psi(\lambda=0)\right)\psi^{-1}(\lambda=0) \equiv -\left(\partial_{\alpha\dot 1}D^{-1}(x)\right)D(x) = D^{-1}(x)\,\partial_{\alpha\dot 1}\,D(x)\ . \tag{1b.4a}$$

Putting $\lambda = \infty$, we obtain

$$A_{\alpha\dot 2}(x) = -\left(\partial_{\alpha\dot 2}\,\psi(\lambda=\infty)\right)\psi^{-1}(\lambda=\infty) \equiv -\left(\partial_{\alpha\dot 2}\bar D^{-1}(x)\right)\bar D(x) = \bar D^{-1}(x)\,\partial_{\alpha\dot 2}\bar D(x)\ . \tag{1b.4b}$$

Since $A_{\alpha\dot 1}(x)$, and $A_{\alpha\dot 2}(x)$ are pure gauge, we can choose a gauge such that, say, $A_{\alpha\dot 1}(x) \to 0$:

$$g(x) \times \text{Eq (1b.1b)} \times g^{-1}(x) \to \left\{\partial_{\alpha\dot 1} + \lambda\left(\partial_{\alpha\dot 2} + J(x)^{-1}\,\partial_{\alpha\dot 2}\,J(x)\right)\right\}\chi(x,\lambda) = 0, \tag{1b.5}$$

where

$$J(x) = D(x)\,\bar{D}^{-1}(x), \text{ and } \chi(x,\lambda) \equiv D(x)\psi(x,\lambda)D^{-1} \qquad (1b.5a)$$

This is the linear system in the J-formulation, which we can write in a concise way:

$$\nabla'_{\alpha\dot{\beta}}\,\chi(x,\lambda) = 0 \,, \qquad (1b.6)$$

where $\nabla'_{\alpha\dot{\beta}} \equiv \partial_{\alpha\dot{\beta}} + B_{\alpha\dot{\beta}}$, $B_{\alpha\dot{1}} \equiv 0$, $B_{\alpha\dot{2}} \equiv J(x)^{-1}\,\partial_{\alpha\dot{2}}\,J(x)$.

Now we can ask what the integrability equations for the linear system Eq. (1b.5) are:

$$\left[\nabla'_{1\dot{\beta}}\,,\,\nabla'_{2\dot{\beta}}\right] = 0 \,, \qquad (1b.7)$$

$$\rightarrow \lambda^{o} \text{ term: } \left[\partial_{1\dot{i}},\,\partial_{2\dot{i}}\right] = 0, \text{ automatically satisfied;} \qquad (1b.7a)$$

$$\rightarrow \lambda^{2} \text{ term: } \partial_{1\dot{2}}\,B_{2\dot{2}} - \partial_{2\dot{2}}\,B_{1\dot{2}} + \left[B_{1\dot{2}},\,B_{2\dot{2}}\right] = 0 \,, \qquad (1b.7b)$$

i.e., $B_{\alpha\dot{2}}$ are curvatureless and a $J(x)$ exists so that $B_{\alpha\dot{2}} = J^{-1}(x)\,\partial_{\alpha\dot{2}}J(x)$;

$$\rightarrow \lambda^{1} \text{ term: } \partial_{2\dot{1}}\,B_{1\dot{2}} + D_{1\dot{1}}\,B_{2\dot{2}} = 0 \,,$$

or equivallently

$$\partial_{2\dot{1}}\left(J^{-1}(x)\partial_{1\dot{2}}J(x)\right) + \partial_{1\dot{1}}\left(J^{-1}(x)\,\partial_{2\dot{2}}J(x)\right) = 0, \qquad (1b.7c)$$

which is the nonlinear second-order differential equation for $J(x)$. Thus solving the self-dual Yang-Mills equation (1a.8) becomes solving for the matrices $J(x)$ which satisfy Eq. (1b.7c).

Under gauge transformation,

$$A_{\alpha\dot{\beta}} \rightarrow G^{-1}\,A_{\alpha\dot{\beta}}\,G + G^{-1}\,\partial_{\alpha\dot{\beta}}\,G,$$

or

$$D,\ \bar{D} \rightarrow DG,\ \bar{D}G, \text{ respectively;}$$

since $J \equiv D \bar{D}^{-1}$. Therefore J is gauge invariant.

Futher from the linear system in the J-formulation, an infinite number of non-local observation laws can be derived; a finite Riemann-Hilbert transformation can be formulated; then the infinitesimal R-H transformation leads to the Affine algebra of Kac and Moody. The readers are referred to Ref. (10) for detailed discussions and for further original references, in addition to Refs. (9) and (11).

II. Supersymmetric Yang-Mills Equations

a. Light-Like Integrability in Superspace and Equations of Motion

Now our space is enlarged from the ordinary space $x^{\alpha\dot{\beta}}$ to $x \equiv (x^{\alpha\dot{\beta}}, \theta^s_\alpha, \bar{\theta}^{\dot{\beta}t})$, where $\alpha,\beta = 1,2$ the spinor indices; $s,t = 1,\ldots,N$ the superspace dimension. The differentiations are[12,13]

$$\partial_{\alpha\dot{\beta}} = \frac{d}{dx^{\alpha\dot{\beta}}} \; ; \; D^s_\alpha = \frac{d}{d\theta^\alpha_s} + i\bar{\theta}^{\dot{\beta}s}\,\partial_{\alpha\dot{\beta}} \; ; \; \bar{D}_{\dot{\beta}t} = \frac{d}{d\bar{\theta}^{\dot{\beta}t}} - i\theta^\alpha_t\,\partial_{\alpha\dot{\beta}} \tag{2a.1a}$$

which satisfy the following anticommutation relations, contrasting with the commutation relations $[\partial_{\alpha\dot{\beta}}, \partial_{\gamma\dot{\delta}}] = 0$ satisfied by the ordinary differentiation,

$$\{D^s_\alpha, D^t_\beta\} = 0 = \{\bar{D}_{\dot{\beta}s}, \bar{D}_{\dot{\alpha}t}\} \; ; \; \{D^s_\alpha, \bar{D}_{\dot{\beta}t}\} = -2i\delta^s_t\partial_{\alpha\dot{\beta}} \; . \tag{2a.1b}$$

In compactified notation,

$$[D_A, D_B) = 0 - t_{AB}{}^C D_C \; , \tag{2a.2}$$

where $A,B = \{\mu \text{ (or } \alpha\dot{\beta}), \overset{s}{\alpha}, \dot{\beta}t\}$, and the flat superspace torsion term $t_{AB}{}^C = 0$ except for $A = \overset{s}{\alpha}$, $B = \dot{\beta}t$. Then $t_{AB}{}^C D_C = 2i\delta^s_t\partial_{\alpha\dot{\beta}}$.

In Minkowski space, a light-like vector in spinor notation takes the form[4] $V^{\alpha\dot{\alpha}} = \lambda^\alpha\bar{\lambda}^{\dot{\alpha}}$, where λ^α, $\bar{\lambda}^{\dot{\alpha}}$ are pairs of independent complex numbers. In the extended superspace, the differentiation operators along spinorial light-like lines[3,6] are

$$\lambda^\alpha D^s_\alpha + \bar{\lambda}^{\dot{\alpha}}\bar{D}_{\dot{\alpha}s} \equiv D^s_{\mathcal{A}} + \bar{D}_{\dot{\mathcal{A}}s} \equiv D_{\underline{\mathcal{A}}} + D_{\underline{\dot{\mathcal{A}}}} \; , \tag{2a.3}$$

whose square becomes the differentiation along the light-like direction

$$V^{\alpha\dot{\alpha}}\,\partial_{\alpha\dot{\alpha}} \equiv \partial_{\mathcal{A}\dot{\mathcal{A}}} \; . \tag{2a.4}$$

In a compact notation we denote differentiation along light-like directions by $D_{\underset{\sim}{A}}$, where $\underset{\sim}{A} = (\underset{\sim}{\alpha\dot{\alpha}}, \underset{\sim}{\alpha}, \underset{\sim}{\alpha})$. Without gauge fields, the integrability conditions, such as Eq. (2a.2), of course hold along all these light-like directions, i.e.

$$\{D_{\underset{\sim}{A}}, D_{\underset{\sim}{B}}\} = 0 - t_{\underset{\sim}{AB}}{}^{C} D_C \ . \tag{2a.5}$$

When there are gauge fields, the commutation relations give

$$\{\nabla_A, \nabla_B\} = F_{AB} - t_{AB}{}^{C}\nabla_C \ , \tag{2a.6}$$

where $\nabla_B = D_B + A_B$, A_B being the gauge potential F_{AB} are the curvature, and $T_{AB}{}^{C}\nabla_C$ is the kinematical torsion term as given in Eq. (2a.2); (in Section III, when we discuss the gravity case, we will use the true torsion terms.) In the light-like directions,

$$\{\nabla_{\underset{\sim}{A}}, \nabla_{\underset{\sim}{B}}\} = F_{\underset{\sim}{AB}} - t_{\underset{\sim}{AB}}{}^{C}\nabla_C \ . \tag{2a.7}$$

It was observed by Witten[3] and Sohnius[13] that if we require the integrability condition, Eq. (2a.5) still holds in the presence of gauge fields, i.e.,

$$\{\nabla_{\underset{\sim}{A}}, \nabla_{\underset{\sim}{B}}\} = 0 - t_{\underset{\sim}{AB}}{}^{C}\nabla_C \ , \tag{2a.8}$$

then for N = 3, through Bianchi identity,

$$\oint_{ABC} \{\nabla_A, \{\nabla_B, \nabla_C\}\} = 0 \ , \tag{2a.9}$$

where $\oint_{ABC}$ stands for cyclic permutation in ABC: as with Eq. (1a.13) for the self-dual Yang-Mills case, the gauge fields F_{AB} can be shown to satisfy equations of motion. The original work done by Witten was in the linearized version (i.e., terms higher than linear in A_B were neglected). Recently the resulting equations of motion have been done also for the complete nonlinear case.[14]

Now let us spell out in more detail the integrability conditions Eq. (2a.8):

For $\underset{\sim}{A},\underset{\sim}{B} = \underset{\sim}{\alpha\dot{\alpha}}$ condition Eq. (2a.8) is automatically satisfied. For $\underset{\sim}{A},\underset{\sim}{B} = \underset{\sim}{\alpha}$, condition (2a.8) implies

$$\lambda^{\alpha}\lambda^{\beta}\{\nabla^s_\alpha, \nabla^t_\beta\} = 0, \qquad \rightarrow F^{st}_{\alpha\beta} + F^{st}_{\beta\alpha} = 0 \; ; \tag{2a.10}$$

$$\bar{\lambda}^{\dot{\alpha}}\bar{\lambda}^{\dot{\beta}}\{\nabla_{\dot{\alpha}s}, \nabla_{\dot{\beta}t}\} = 0, \qquad \rightarrow F_{\dot{\alpha}s\dot{\beta}t} + F_{\dot{\beta}s\dot{\alpha}t} = 0 \; ; \tag{2a.11}$$

$$\lambda^{\alpha}\bar{\lambda}^{\dot{\beta}}\{\nabla^s_\alpha, \bar{\nabla}_{\dot{\beta}t}\} = -2i\delta^s_t\nabla_{\alpha\dot{\beta}} \qquad \rightarrow F^s_{\alpha\dot{\beta}t} = 0 \; . \tag{2a.12}$$

It turned out that these are the same conditions one has to use to eliminate unphysical fields.[13-15]

Essentially, the geometrical constraints Eq. (2a.10) define the spin-zero fields $\bar{W}^{st}$, W_{st}

$$F^{st}_{\alpha\beta} = \epsilon_{\alpha\beta}\bar{W}^{st} \; , \qquad F_{\dot{\alpha}s\dot{\beta}t} = \epsilon_{\dot{\alpha}\dot{\beta}} W_{st} \; , \tag{2a.13}$$

where the $\bar{W}$, W are antisymmetric in s and t. From the Bianchi identities, one can show that the only other physical fields are the spin-half fields

$$\Lambda^{\alpha}_s = \epsilon^{\alpha\beta}\nabla^t_\beta W_{ts} \; , \qquad \bar{\Lambda}^{\dot{\alpha}s} = \epsilon^{\dot{\alpha}\dot{\beta}}\bar{\nabla}_{\dot{\beta}t}\bar{W}^{ts} \; , \tag{2a.14}$$

and the spin-1 fields

$$F_{\mu\nu} \equiv \frac{i}{16N(N-1)} \{(\nabla\varepsilon\sigma_{\mu\nu})^{st} W_{st} + (\bar{\nabla}\bar{\sigma}_{\mu\nu}\varepsilon\bar{\nabla})_{st} \bar{W}^{st}\} , \tag{2a.15}$$

and they satisfy the following equations of motion:

$$\nabla_{\alpha\dot{\beta}} \bar{\Lambda}^{\dot{\beta}s} = \nabla_{\alpha\dot{\beta}} \varepsilon^{\dot{\beta}\dot{\gamma}} \bar{\nabla}^{\bullet}_{\dot{\gamma}k} \bar{W}^{ks} = -\frac{1}{2} (\nabla^{k}_{\alpha} W_{kt}, \bar{W}^{st}) . \tag{2a.16}$$

$$\begin{aligned} \bar{W}^{st} = &- \frac{i}{36} \varepsilon^{\dot{\alpha}\dot{\beta}} \{\bar{\nabla}^{\bullet}_{\alpha k} \bar{W}^{ks}, \bar{\nabla}^{\bullet}_{\dot{\beta}\ell} \bar{W}^{\ell t}\} \\ &+ \frac{i}{72} \varepsilon^{stk\ell}\varepsilon^{\alpha\beta}\{\nabla^{m}_{\alpha} W_{mk}, \nabla^{n}_{\beta} W_{n\ell}\} + \frac{1}{8} (\bar{W}^{tk}, W_{k\ell}) \bar{W}^{s\ell} \quad ; \end{aligned} \tag{2a.17}$$

and W_{ij} satisfied a conjugated equation;

$$\begin{aligned} \nabla^{\nu} F_{\mu\nu} = &\frac{i}{96} (3((\nabla_{\mu}W_{ij}, \bar{W}^{st}) + (\nabla_{\mu} \bar{W}^{st}, W_{st})) \\ &- \frac{2}{3} i (\bar{\sigma}_{\mu})^{\dot{\alpha}\alpha} \{\nabla^{t}_{\alpha} W_{st}, \bar{\nabla}^{\bullet}_{\dot{\alpha}k} \bar{W}^{ks}\}) \quad . \end{aligned} \tag{2a.18}$$

II.b The Linear Systems and Integrability Properties

Similar to the discussions in the last section, the integrability along the light-like lines Eq. (2a.8) are precisely the integrability condition of the wave function of the following linear systems.(16)

$$\{\nabla^s_{\mathcal{A}}, \nabla^t_{\mathcal{B}}\} = 0 \qquad \rightarrow \nabla^s_{\mathcal{A}}\underline{\psi}(x,\theta,\bar{\theta},\lambda) = 0 \text{ integrable,} \tag{2b.1}$$

$$\{\nabla_{\dot{\mathcal{A}}s}, \nabla_{\dot{\mathcal{B}}t}\} = 0 \qquad \rightarrow \nabla_{\dot{\mathcal{A}}}\underline{\psi}'(x,\theta,\bar{\theta},\lambda) = 0 \text{ integrable,} \tag{2b.2}$$

$$\{\nabla^s_{\mathcal{A}}, \nabla_{\dot{\mathcal{B}}t}\} = -2\, i\delta^s_{\underline{t}}\nabla_{\mathcal{A}\dot{\mathcal{B}}} \qquad \rightarrow \underline{\psi}' = \underline{\psi}\ . \tag{2b.3}$$

In more explicit expression, we have

$$\left(D^s_1 + A^s_1 + \lambda\left(D^s_2 + A^s_2\right)\right)\underline{\psi}\,(x,\theta,\bar{\theta},\lambda) = 0\ , \tag{2b.1a}$$

$$\left(D_{\dot{1}t} + A_{\dot{1}t} + \bar{\lambda}\,(D_{\dot{2}t} + A_{\dot{2}t})\right)\underline{\psi}\,(x,\theta,\bar{\theta},\lambda) = 0\ . \tag{2b.2a}$$

As in previous discussions the third integrability condition implies

$$\begin{aligned} f^s_{\alpha\dot{\alpha}}{}^t &= 0 \\ &= \bar{D}_{\dot{\alpha}t}\, A^s_\alpha + D^s_\alpha\, A_{\dot{\alpha}t} + \{A_{\dot{\alpha}t}, A^s_\alpha\} + A_{\alpha\dot{\alpha}}\ . \end{aligned} \tag{2b.4}$$

With this definition of $A_{\alpha\dot{\alpha}}$, the third requirement of $\underline{\psi}$, i.e., $\nabla_{\alpha\dot{\beta}}\,\underline{\psi} = 0$, is automatically satisfied. Also, if we always take the definition Eq. (2b.4) of $A_{\alpha\dot{\alpha}}$, λ and $\bar{\lambda}$ can be related as $\lambda = \bar{\lambda}$ or $\lambda = 1/\bar{\lambda}$. Therefore we can treat this case as a one-parameter case. With this definition of $A_{\alpha\dot{\alpha}}$, we need only to deal with the two linear systems Eq. (2b.1) and Eq. (2b.2) or Eq. (2b.1a) and Eq. (2b.2a). Observing the similarity between Eq. (2b.1) and Eq. (2b.2) with Eq. (1b.2), and Eq. (1b.1), respectively, we can follow the same procedure to find the super J-formulation.

For $\lambda = 0$

$$A_1^s = -\left(D_1^s \psi(\lambda=0)\right)\psi^{-1}(\lambda=0) \equiv -\left(D_1^s g^{-1}(x)\right\}' g(x) = g^{-1}(x) D_1^s g(x) \; ; \tag{2b.5}$$

For $\lambda = \infty$

$$A_2^s = -\left(D_2^s \psi(\lambda = \infty)\right)\psi^{-1}(\lambda = \infty) \equiv -\left(D_2^s h^{-1}(x)\right\} h(x) = h^{-1}(x) D_2^s h(x) \; ; \tag{2b.6}$$

From Eq. (2b.2a), for $\bar{\lambda} = 0$

$$\begin{aligned} A_{1t}^{\bullet} &= -\left(D_{1t}^{\bullet} \underline{\bar{\psi}} (\lambda = 0)\right\} \underline{\bar{\psi}}^{-1}(\bar{\lambda} = 0) = -\left(D_{1t}^{\bullet} g'^{-1}(x)\right) g'(x) \\ &= g'^{-1}(x) D_{1t}^{\bullet} g'(x); \end{aligned} \tag{2b.7}$$

for $\bar{\lambda} = \infty$

$$\begin{aligned} A_{2t}^{\bullet} &= -\left(D_{2t}^{\bullet} \bar{\psi}(\bar{\lambda} = \infty)\right\} \bar{\psi}^{-1}(\bar{\lambda} = \infty) \equiv -\left(D_{2t}^{\bullet} h'^{-1}(x)\right\} h'(x) \\ &= h'^{-1}(x) D_{2t}^{\bullet} h'(x) \, . \end{aligned} \tag{2b.8}$$

From Eq (2b.3), we can again show that either $g'(x) = g(x)$, $h'(x) = h(x)$, for $\lambda \neq 1/\bar{\lambda}$, (for this case we can take $\lambda = \bar{\lambda}$); or $g'(x) = h(x)$, $h' = g(x)$, for $\lambda \neq \bar{\lambda}$, (for this case we can take $\lambda = 1/\bar{\lambda}$).

Considering the first case, we have

$$\left\{D_1^s + g^{-1} D_1^s g + \lambda \left(D_2^s + (h^{-1} D_2^s h)\right)\right\} \underline{\bar{\psi}} = 0 \; , \tag{2b.9}$$

$$\left\{D_{1t}^{\bullet} + g^{-1} D_{1t}^{\bullet} g + \bar{\lambda}\left(D_{2t}^{\bullet} + h^{-1} D_{2t}^{\bullet} h\right)\right\} \underline{\bar{\psi}} = 0 \; . \tag{2b.10}$$

Now we make a gauge transformation:

$$g \times \text{Eq. (2b.9)} \times g^{-1} \rightarrow \{D_1^s + \lambda\,(D_2^s + J^{-1}(D_2^s J))\}\,\psi = 0\ , \qquad (2b.10)$$

where $J = h\,g^{-1}$

$$g \times \text{Eq. (2b.10)} \times g^{-1} \rightarrow \{D_{\dot{1}t} + \bar{\lambda}\,(D_{\dot{2}t} + J^{-1}(D_{\dot{2}}tJ))\}\,\bar{\psi} = 0 \qquad (2b.11)$$

here $\lambda \neq 1/\bar{\lambda}$, but we can take $\lambda = \bar{\lambda}$.

Now we rewrite Eqs. (2b.10) and (2b.11) in a more compact form:

$$\nabla'^{s}_{\alpha}\,\psi = 0\ , \qquad (2b.12)$$

$$\nabla'_{\dot{\beta}t} = \psi = 0\ , \qquad (2b.13)$$

where

$$\nabla'^{s}_{\alpha} = D^{s}_{\alpha} + B^{s}_{\alpha};\ \nabla'_{\dot{\beta}t} = D_{\dot{\beta}t} + B_{\dot{\beta}t};$$

$$B_1^s = 0,\ B_2^s = J^{-1}(D_2^s J);\ B_{\dot{1}t} = 0,\ B_{\dot{2}t} = J^{-1}D_{\dot{2}t}J.$$

The integrability of Eqs. (2b.12) and (2b.13) imply the nonlinear equations of motion for J

$$D_1^s\,(J^{-1}\,D_2^t\,J) + D_2^t\,(J^{-1}\,D_1^s J) = 0\ , \qquad (2b.14)$$

$$D_{\dot{1}s}\,(J^{-1}\,D_{\dot{2}t}\,J) + D_{\dot{2}t}\,(J^{-1}D_{\dot{1}s}J) = 0\ . \qquad (2b.15)$$

Therefore, solving supersymmetric Yang-Mills equations have now been reduced down to solving these J equations. After solving J, to solve the potentials, we first split $J = hg^{-1}$, then we obtain the potentials from

$$A_1^s = g^{-1} D_1^s g, \quad A_2^s = h^{-1} D_2^s h ,$$

$$A_{\dot{1}t} = g^{-1} D_{\dot{1}t} g, \quad A_{\dot{2}t} = h^{-1} D_{\dot{2}t} h,$$

and $A_{\alpha\dot{\beta}}$ from Eq. (2b.4b).

The corresponding integrability properties have all been worked out: the Bäckland transformation,[17-19] the finite Riemann-Hilbert transformation, the infinitesimal R-H transformation, and its corresponding infinite dimensional algebra of Kac-Moody.

III. Extended Supergravity Equations

Here the basic fields are the vielbeins $E_M{}^A$, in addition to the connection field $(\Omega_M)^A_B$, where M is the superspace world indices, and A, B are the tangent space indices. When there is no field, the discussion is the same as in Section II. In the presence of gravity field, Eq. (2a.b) become

$$\{\nabla_A, \nabla_B\} = R_{AB} - T_{AB}{}^C \quad , \tag{3.1}$$

where R_{AB} are the curvature tensor, and $T_{AB}{}^C$ are the torsion fields. The light-like geometrical integrability is the same as in the supersymmetric Yang-Mills field

$$\{\nabla_{\underset{\sim}{A}}, \nabla_{\underset{\sim}{B}}\} = 0 - t_{AB}{}^C \, \nabla_C ,$$

i.e.,

$$R_{\underset{\sim}{AB}} = 0, \; T_{\underset{\sim}{AB}}{}^C \nabla_C = 0 \quad , \tag{3.2}$$

or more explicitly

$$R^{ij}_{\alpha\beta} + R^{ij}_{\beta\alpha} = 0; \quad R_{\dot\alpha i \dot\beta j} + R_{\dot\beta i \dot\alpha j} = 0; \tag{3.3}$$

$$T^{ij}_{\alpha\beta} + T^{ij}_{\beta\alpha} = 0; \quad T_{\dot\alpha i \dot\beta j} + T_{\dot\beta i \dot\alpha j} = 0; \tag{3.4}$$

$$T^{i}_{\alpha\dot\beta j} = -2i\delta^i_j \, \partial_{\alpha\dot\beta} \; . \tag{3.5}$$

In an extremely long and tedious calculation,(1) we first find all the physical fields from the constraints and Bianchi: the spin-1/2 fields $\bar\Lambda^{(ijk)}{}_{\dot\alpha}$, the spin-one fields $\bar F^{(ij)}_{\{\dot\alpha\dot\beta\}}$, the spin-3/2 field $\bar\Sigma^{i}_{\{\dot\alpha\dot\beta\dot\gamma\}}$, and the gravitational field $\bar V_{\{\dot\delta\dot\gamma\dot\beta\dot\gamma\}}$, we then establish that they satisfy, in the linearized case, the following equations of motion:

$$(N-2)\,(N-3)\,(N-4)\ \Box\partial^{\dot{\alpha}}_{\alpha}\ \bar{\lambda}^{(ijk)}{}_{\dot{\alpha}} = 0\ , \tag{3.6}$$

$$(N-2)\,(N-3)\,(N-4)\ \partial^{\dot{\alpha}}_{\alpha}\ \partial^{\dot{\beta}}_{\beta}\ \bar{F}^{(ij)}_{\{\dot{\alpha}\dot{\beta}\}} = 0\ , \tag{3.7}$$

$$(N-2)\,(N-3)\,(N-4)\ \partial^{\dot{\alpha}}_{\alpha}\ \partial^{\dot{\beta}}_{\beta}\ \bar{\Sigma}^{i}{}_{\{\dot{\alpha}\dot{\beta}\dot{\gamma}\}} = 0\ , \tag{3.8}$$

$$(N-2)\,(N-3)\,(N-4)\ \partial^{\dot{\alpha}}_{\alpha}\ \partial^{\dot{\beta}}_{\beta}\ \bar{V}_{\{\dot{\delta}\dot{\gamma}\dot{\beta}\dot{\alpha}\}} = 0\ . \tag{3.9}$$

Note that these equations are conformal-like and are equations in higher order in differential. However if we find solutions such that an intermediate field[(1)] $X^{(ij,k)}_{\dot{i}}$ becomes zero, we can show that all these equations of motion Eqs. (3.6-9) become first order differential equations. Just as the N = 4 supersymmetric Yang-Mills equations contain solutions of the N = 0, N = 1, and N = 2 Yang-Mills fields, the extended supergravity equations do contain solutions to gravitational fields of lower N.

These results put extended supergravity in four-dimensional space with N > 4 on the same footing as supersymmetric Yang-Mills in four dimensions with N > 2.

The corresponding integrability properties for this case will be worked out in my next lecture.

IV. Two-Dimensional Models

a. Chiral fields

It is well known that the principal chiral field equations

$$\partial_\xi A_\eta + \partial_\eta A_\xi = 0 , \tag{4a.1a}$$

and

$$A_\xi \equiv g^{-1}\partial_\xi g, \; A_\eta = g^{-1}\partial_\eta g , \tag{4a.1b}$$

where g is a group element, are consequence of the integrability requirement

$$[\nabla_\xi, \nabla_\eta] = 0 \tag{4a.2}$$

of the following two linear systems

$$\nabla_\xi \psi(\ell,\xi,\eta) = 0 , \quad \nabla_\eta \psi(\ell,\xi,\eta) = 0 , \tag{4a.3}$$

where $\nabla_\xi = (1-\ell)\partial_\xi - \ell A_\xi$, $\nabla_\eta = (1+\ell)\partial_\eta + \ell A_\eta$, and ℓ is an arbitrary parameter.

b. Other two-dimensional models: KdV, S-G, and Liouville equations

In 1978 it was pointed out in a beautiful paper by Pirani et al.[2] that many integrable two-dimensional non-linear equations are results of similar geometrical integrability. Here I shall list them. Consider the linear systems:

$$\nabla_x \psi = 0 , \; \nabla_t \psi = 0 , \tag{4b.1}$$

where

$$\nabla_x = \partial_x + B_x , \; \nabla_t = \partial_t + B_t ,$$

$$B_x = \begin{pmatrix} -i\lambda & -u(x,t) \\ -1 & i\lambda \end{pmatrix}, \; B_t = \begin{pmatrix} A & B \\ C & -A \end{pmatrix} ,$$

and λ is a parameter, U, A, B, C are functions of x and t,

when

$$A = -4i\lambda^3 - 2i\lambda U + U_x ,$$

$$B = -4\lambda^2 U - 2i\lambda U_x - 2U^2 + U_{xx} ,$$

$$C = -4\lambda^2 - 2U . \tag{4b.2}$$

The integrability condition

$$(\nabla_x, \nabla_t) = 0 \quad \rightarrow \text{KdV equation } U_t + 6UU_x - U_{xxx} = 0 , \tag{4b.3}$$

where the subscripts denote differentiation;

when

$$B_x = \begin{pmatrix} \frac{i}{2} U_x & \frac{1}{2}\lambda \\ \frac{1}{2}\lambda_x & -\frac{i}{2} U_x \end{pmatrix} , \quad B_t = \begin{pmatrix} 0 & \frac{1}{2\lambda} e^{iU} \\ \frac{1}{2\lambda} e^{-iU} & 0 \end{pmatrix} ,$$

$$\text{then } (\nabla_x, \nabla_t) = 0 \quad \rightarrow \text{the S-G equation, } U_{xt} - \text{Sin}U = 0 ; \tag{4b.4}$$

when

$$B_x = \begin{pmatrix} \frac{i}{2} U_x & \frac{1}{2}\lambda \\ \frac{1}{2}\lambda & \frac{-i}{2} U_x \end{pmatrix} , \quad B_t = \begin{pmatrix} 0 & \frac{1}{2\lambda} e^{U} \\ 0 & 0 \end{pmatrix}$$

$$\text{then } (\nabla_x, \nabla_t) = 0 \quad \rightarrow \text{Liouville equation, } U_{xt} - 2e^U = 0 . \tag{4b.5}$$

Concluding Remarks

It is indeed remarkable that so many equations of motion in physics have a geometrical-integrability origin. Such geometrical integrability provides a way of linearization of the original nonlinear equations. So far this approach has given a very fruitful understanding of the two-dimensional nonlinear equations.[2] It will be most interesting to find out what these geometrically-integrable linear systems will contribute to the understanding of the Yang-Mills and the Einstein equations.

Acknowledgement

I would like to thank Chong-Sa Lim, and Huan-Cheh Yen for their collaborations and enlightening discussions.

References

1. L.L. Chau, and C.-S. Lim, "Geometrical Constraints and Equations of Motion in Extended Supergravity", Phys. Rev. Lett. 56, 294 (1986).

2. F.A.E. Pirani, D.C. Robinson, "The Soliton Connection", Lett. in Math. Phys. 2, 15 (1977), G.L. Lam Jr. "Elements of Soliton Theory" (Wiley, New York 1980); M.J. Ablowitz and H. Segur, "Solitons and the Inverse Scattering Transform" (SIAM, Philadelphia 1981); F. Calogero and A. Degasperis, "Spectral Transform and Solitons I" (North-Holland, Amsterdam 1982).

3. E. Witten, "An Interpretation of Classical Yang-Mills Theory", Phys. Lett. 77B (1978) 394, J. Isenberg, P.B. Yasskin and P.S. Green, Non-Self-Dual Gauge Fields Phys. Lett. 78B (1978) 462.

4. R. Penrose, W. Rindler, "Spinors and Space-Time", Cambridge University Press. S. Ward, Phys. Lott, 61A (1977) 81; A.A. Bolavian and V.E. Zakharov, Phys. Lett. 73B, 2528 (1978).

5. R.S. Ward, "Completely Soluable Gauge-Field Equations in Dimensions Greater than Four," Nucl. Phys. B236, 381 (1984), E. Corrigan, C. Devchand, D.B. Fairlie, J. Nuyts, "First Order Equations for Gauge Fields in Spaces of Dimension Greater than Four," Durham e U. de l'Ethal å Mons preprint 1982.

6. A. Ferber, "Supertwistors and Conformal Supersymmetry" Nucl. Phys. B132, 55 (1978).

7. L.-L. Chau Wang, "Bäcklund Transformations, Conservation Laws and Linearization of the Self-dual Yang Mills and Chiral Fields," Proc. Guanzhou (Canton) Conf. on Theoretical Partical Physics p. 1082, 1980.

8. K. Pohlmeyer, Comm. Math. Phys. 72, 37 (1980).

9. C.N. Yang, "Condition of Self-duality for SU(2) Gauge Fields on Euclidean Four-dimensional Space", Phys. Rev. Lett 38 (1977) 1377; Y. Brihaye, D.B. Fairlie, J. Nuyts, R.F. Yates, JMP 19 (1978) 2528.

10. For the affine Lie algebra of Kac-Moody in the SDYM field, see L.-L. Chau, M.-L. Ge and Y.-S. Wu, "Kac-Moody Algebra in the Self-dual Yang Mills Equations", Phys. Rev. D25 (1982) 1086; and Ref. 24-25. K. Ueno and Y. Nakamura, "The Hidden Symmetry of Chiral Fields and Riehmann-Hilbert Problem", Phys. Lett. 117B (1982) 208. L.-L. Chau, Y.-S. Wu, "More about Hidden Symmetry of the Self-dual Yang Mills Fields", Phys. Rev. D26 (1983) 3581; L.-L. Chau, M.-L. Ge, A. Sinha, Y.-S. Wu, "Hidden Symmetry Algebra for the Self-dual Yang Mills Equation", Phys. Lett. 121B (1983) 391.

11. Proceedings of "Nonlinear Phenomena", L.-L. Chau, "Chiral Fields, Self-dual Yang Mills Fields as Integrable Systems and the Role of Kac-Moody Algegra" Oaxtepec, Mexico, Nov. 19-Dec. 17, 1982, Lecture Notes in Physics #189, edited by K.B. Wolf.

12. J. Wess and B. Zumino "Super Gauge Invariant Extension of Quantum Electrodynamics", Nucl. Phys. B78, 1 (1974); Phys. Lett. 66B, 361 (1977); S. Ferrara and B. Zumino, "Super Gauge Invariant Yange-Mills Therory" Nucl. Phys. B79, 413 (1974); A. Salam and J. Strathdee, Supersymmetry and Non-Abelian Gauges" Phys. Lett. 51B, 353 (1974); R. Grimm, M. Sohnius, and J. Wess, "Extended Supersymmetry and Gauge Theories" Nucl. Phys. B133, 275 (1978). J. Wess and J. Bagger, Supersymmetry and Supergravity (Princeton Univ. Press, Princeton, N.J., 1983).

13. M. Sohnius, Nucl. Phys. B136, 461 (1978).

14. L.-L. Chau, M.-L. Ge, C.-S. Lim, "Constraints and Equation of Motion in Supersymmetric Yang-Mills Theories", Phys. Rev. 33 1056 (1986).

15. M. Rocek and W. Siegel, "On Off-shell Supermultiplets[4], Phys. Lett. 105B, 275 (1981); W. Siegel, "On-shell O(N) Super Gracity in Superspace", Nucl. Phys. B177, 325 (1981); V. Rivelles and J.G. Taylor, J. Phys. A 15, 163 (1982); B.E.W. Nilsson, "Simple 10-Dimensional Supergravity in Superspace", Nucl. Phys. B188, 176 (1981).

16. I.V. Volovich, "Supersymmetric Yang-Mills Equations and Twistors," Phys. Lett. 129B, 429 (1983); Teor. Mat. Fiz. 54, 39 (1983).

17. L.-L. Chau, "Supersymmetric Yang-Mills Fields as an Integrable System and Connections with Other Linear Systems," Proceedings of the Workshop on Vertex Operations in Math Physics, Berkeley, 1983, edited by J. Lepowsky, S. Mandelstam, and I.M. Singer (springer, New York, 1984).

18. C. Devchand, "An Infinite Number of Continuity Equations and Hidden Symmetries in Supersymmetric Gauge Theories", Nucl. Phys. B238, 333 (1984).

19. L.-L. Chau, M.-L. Ge, and Z. Popwicz, "Riemann-Hilbert Transforms and Bianchi-Bäcklund Transformations for the Supersymmetric Yang-Mills Fields.," Phys. Rev. Lett. 52, 1940 (1984).

THE CHARGED PARTICLE MULTIPLICITY IN $P\bar{P}$ ANNIHILATION

Chi-jiang CHEN and Jing-fa LU
Department of Physics, Nankai University, Tianjin, China

ABSTRACT

Using the three chains diagram based on the Dual Partonic Scheme, combining with the FF quark fragmentation mechanism, the charged particle multiplicity, the yields of π, K mesons, the single particle inclusive distribution and the central plateau particle density of soft high energy $P\bar{P}$ annihilation are calculated and compared with $P\bar{P}$ experiments on CERN SPS.

The results supporte the view that

1) The small $P_\perp$ multiparticle production in $P\bar{P}$ interaction consists of annihilation and collision processes. The annihilation process corresponding to the three chains diagram, the collision process to two chains diagram.

2) Annihilation has a richer charged particle multiplicity, more abundant amount of strange particles and higher inclusive distribution than PP collision. The difference between the two cases grows clearer with energy.

3) The annihilation plateau density grows with energy faster than collision, the Feynman scaling is violated sharper.

NEGATIVE BINOMIAL DISTRIBUTION FOR MULTIPLICITY DISTRIBUTIONS IN e^+e^- ANNIHILATION

C.K. CHEW and Y.K. LIM

Physics Department, National University of Singapore 0511

We show that the negative binomial distribution fits excellently the available charged-particle multiplicity distributions of e^+e^- annihilation into hadrons at three different energies $\sqrt{s}$ = 14, 22 and 34 GeV.

A clear evidence of KNO[1] scaling violation in the high multiplicity region has recently been reported by the UA5 Collaboration[2] from the CERN $p\bar{p}$ collider experiment. We have subsequently shown that the negative bionomial distribution[3] can be used to give an excellent fit to the charged-particle multiplicity distributions of the e^+e^- annihilation at the available energies of $\sqrt{s}$ = 14, 22, and 34 GeV from Tesso[4] with only one free parameter k.

The negative binomial distribution is given by

$$P_n = (-1)^n \binom{-k}{n} p^n q^k \quad , \tag{1}$$

where

$$p = (\langle n\rangle/k)/(1 + \langle n\rangle/k) \quad ,$$
$$q = 1 - p = 1/(1 + \langle n\rangle/k) \; .$$

From the characteristic functions

$$\phi(t) = q^k \; (1 - pe^{it})^{-k} \tag{2}$$

the moment $m_\ell = \langle n^\ell\rangle$ may be calculated by

$$m_\ell = \phi^\ell(0)/i^\ell \tag{3}$$

where $\phi^\ell(0)$ is the ℓth derivative of the characteristic function at t = 0.

The first five scaled moment $C_\ell = m_\ell/\langle n\rangle^\ell$ are given as follows:

$$C_1 = 1 \; ,$$
$$C_2 = 1/\langle n\rangle + 1/k + 1 \quad , \tag{4}$$
$$C_3 = 1/\langle n\rangle^2 + (3/\langle n\rangle)(1 + 1/k) + (1 + 1/k)(1 + 2/k) \quad , \tag{5}$$
$$C_4 = 1/\langle n\rangle^3 + (7/\langle n\rangle^2)(1 + 1/k) + (6/\langle n\rangle)(1 + 1/k)(1 + 2/k) + (1 + 1/k)(1 + 2/k)(1 + 3/k) \; , \tag{6}$$
$$C_5 = 1/\langle n\rangle^4 + (15/\langle n\rangle^3)(1 + 1/k) + (25/\langle n\rangle^2)(1 + 1/k)(1 + 2/k) + (10/\langle n\rangle)(1 + 1/k)(1 + 2/k)(1 + 3/k) + (1 + 1/k)(1 + 2/k)(1 + 3/k)(1 + 4/k) \; . \tag{7}$$

The average charged multiplicity <n> from experiment is taken to be the input for the negative binomial distribution and k is the only parameter to be determined by χ^2_{min} fitting. We have found that k = 35, 44, 62 for CMS

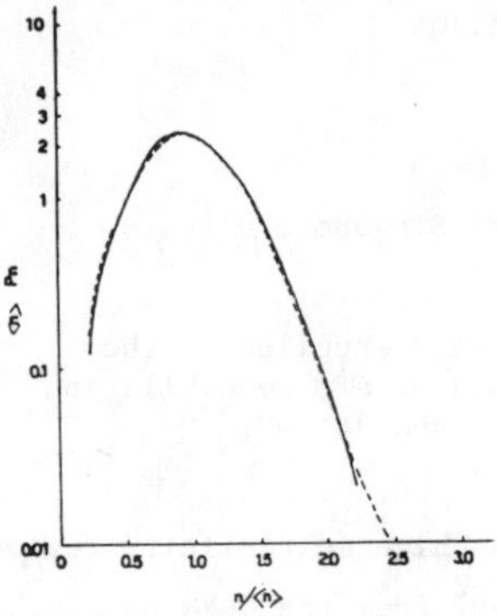

Fig. 1. Comparison of the multiplicity distributions of the e^+e^- annihilation at the CMS energy $\sqrt{s}$ = 14 GeV between the experimental data (dashed curve) taken from Tasso and the negative binomial distribution (solid curve) with k = 62.

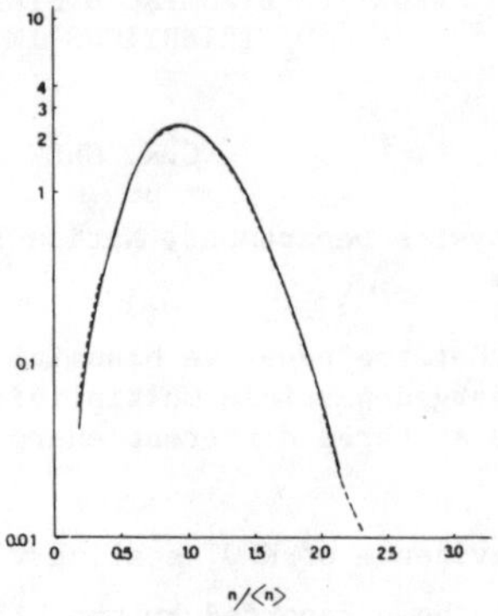

Fig. 2. Comparison of the multiplicity distributions of the e^+e^- annihilation at CMS energy $\sqrt{s}$ = 22 GeV between the experimental data (dashed curve) taken from Tasso and the negative binomial distribution (solid curve) with k = 44.

energy $\sqrt{s}$ = 34, 22, 14 (GeV), respectively. The fits of the experimental and theoretical distributions are remarkably excellent as shown in figs. 1, 2 and 3. Higher scaled moments are calculated from eqs. (4) - (7) and are compared to those calculated from the data and are presented in table 1. The agreement is so excellent that we believe that the negative binomial distribution could be considered as the best theoretical distribution for the charged multiplicity distribution for e^+e^- annihilation.

From the above k, s relation it is easy to see that k^{-1} is linearly proportional to ℓns as below

$$k^{-1} = a + b\ell ns \tag{8}$$

where $a = -2.09 \times 10^{-3}$ and $b = 7.01 \times 10^{-3}$.

This equation may be used to predict the parameter k and hence the multiplicity distribution for e^+e^- annihilation in the higher energy region.

At very high energy, the average multiplicity is much larger than k, the asymptotic form of the negative binomial distribution multiplied by <n> can be expressed in terms of the scaled variable z = n/<n> by

$$\langle n\rangle P_n = [k^k/\Gamma(k)]\, z^{k-1} e^{-kz} \quad . \tag{9}$$

The higher scaled moments become

$$C_2 = 1 + 1/k \quad , \tag{10}$$

$$C_3 = 1 + 3/k + 2/k^2 \, , \tag{11}$$

$$C_4 = 1 + 6/k + 11/k^2 + 6/k^3 \, , \tag{12}$$

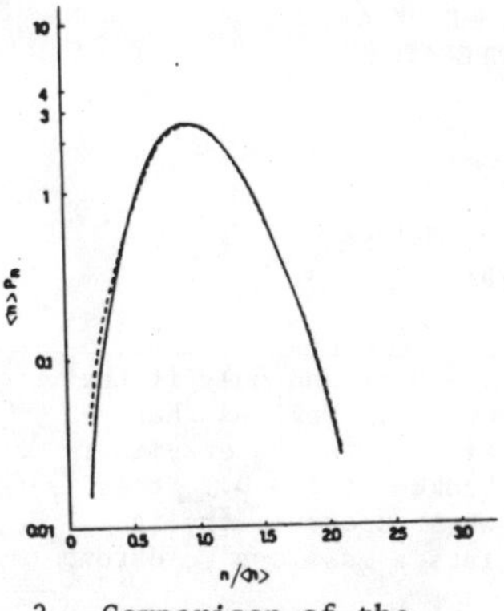

Fig. 3. Comparison of the multiplicity distributions of the e^+e^- annihilation at CMS energy $\sqrt{s}$ = 34 GeV between the experimental data (dashed curve) taken from Tasso and the negative binomial distribution (solid curve) with k = 35.

$$C_5 = 1 + 10/k + 35/k^2 + 50/k^3 + 24/k^4 \qquad (13)$$

These higher moments must remain constant for the validity of KNO scaling and this requires a constant k. However, k decreases linearly with ℓns and hence KNO scaling becomes increasingly violated with increasing energy and this is also supported by the UA5 report on $p\bar{p}$ collisions at 540 GeV. As pointed out by the UA5 Collaboration the mechanism of secondary particle production by k independent cells or by k clusters which leads to the negative binomial distribution with the assumption that k increases with energy is also in contradiction to our findings in e^+e^- annihilation.

TABLE 1

Parameter k of negative binomial distribution fitted to χ^2_{min}/point, the scaled moments $C_\ell = \langle n^\ell \rangle / \langle n \rangle^\ell$ calculated from negative binomial distribution and within brackets the corresponding quantities calculated from the data samples.

$\sqrt{s}$(GeV)	<n>	k	C_2	C_3	C_4	C_5	χ^2/point
14	9.0740	62	1.1263 (1.128 ± 0.057)	1.3970 (1.400 ± 0.091)	1.8810 (1.885 ± 0.156)	2.7215 (2.730 ± 0.283)	1.24/11
22	11.2285	44	1.1118 (1.112 ± 0.071)	1.3504 (1.351 ± 0.108)	1.7709 (1.771 ± 0.175)	2.4860 (2.483 ± 0.297)	0.99/13
34	13.4243	35	1.1031 (1.106 ± 0.021)	1.3228 (1.328 ± 0.034)	1.7069 (1.712 ± 0.0571)	2.3521 (2.345 ± 0.101)	18.15/14

References

1. Z. Koba, H.B. Nielson and P. Olesen, Nucl. Phys. B40 (1972) 317.
2. G.J. Alner et al., Phys. Lett. 160B (1985) 199.
3. C.K. Chew and Y.K. Lim, Phys. Lett. 163B (1985) 257.
4. TASSO Collab., M. Althoff et al., Z. Phys. C22 (1984) 307.

SPONTANEOUS BREAKDOWN OF SUPERSYMMETRY AND GOLDSTONE FERMION AT NONZERO TEMPERATURES

Swee-Ping Chia

Physics Department, University of Malaya
59100 Kuala Lumpur, Malaysia

It is shown that supersymmetry is manifest at $T \neq 0$ if and only if the thermal average of every auxiliary field vanishes. It follows that broken supersymmetry is never restored at higher T. When supersymmetry is manifest at $T = 0$, it is not automatically broken at $T \neq 0$. It is also demonstrated, by explicit examples, that when supersymmetry is spontaneously broken at $T \neq 0$, there always exists a massless Goldstone fermion.

In a supersymmetric theory, fermions and bosons belonging to the same supermultiplet have identical mass at $T = 0$. But at nonzero T, this fermion-boson mass degeneracy seems to be lost. Does it necessarily implies that supersymmetry is automatically broken at $T \neq 0$?[1,2] To answer this question, we must first find the proper criterion for spontaneous breaking of supersymmetry at $T \neq 0$.

Supersymmetry is maifest at a certain temperature T if and only if for every operator R, the change δR under a supersymmetry transformation satisfies[2-7]

$$\langle \delta R \rangle_T = 0, \tag{1}$$

where $\langle\ \rangle_T$ denotes ordinary thermal average at temperature T. Let us take R in (1) to be a scalar superfield $S \equiv \{\phi, \psi, F\}$,

$$S = \phi - i\sqrt{2}\bar{\theta}_L\psi_L - i\bar{\theta}_L\theta_L F + \frac{1}{2}\bar{\theta}\not{\partial}\phi\theta + i\bar{\theta}\theta\bar{\theta}\not{\partial}\psi_L - \frac{1}{8}(\bar{\theta}\theta)^2\Box\phi, \tag{2}$$

where θ is an anti-commuting variable. Criterion (1) then implies that $\langle\bar{\theta}_L\delta\psi_L\rangle_T = 0$. But under a supersymmetry transformation,

$$\delta\psi_L = -\sqrt{2}\alpha_L F - i\sqrt{2}\not{\partial}\phi\alpha_R. \tag{3}$$

Now since $\langle\not{\partial}\phi\rangle_T = 0$, the criterion therefore becomes $\langle\bar{\theta}_L\alpha_L F\rangle_T = 0$. As $\bar{\theta}_L\alpha_L$ is bosonic, we finally arrive at the following criterion for supersymmetry to be manifest at $T \neq 0$,[8]

$$\langle F \rangle_T = 0\ . \tag{4}$$

This criterion must be satisfied by each auxiliary field in the theory. Thus supersymmetry is spontaneously broken at $T \neq 0$ if the thermal average of at least one of the auxiliary fields does not vanish. Our criterion is distinctly different from that derived by Fuchs,[6] who uses superthermal average instead of ordinary thermal average.

Having derived the criterion for supersymmetry to be manifest at nonzero T, we can now proceed to find out whether supersymmetry is automatically broken spontaneously at nonzero T. For a system of N scalar superfields S_a ($a = 1, 2, \ldots, N$) interacting via a polynomial superpotential $f(S)$, the auxiliary fields F_a are given in terms of the scalar fields ϕ_a by

$$F_a^* = -i\partial f(\phi)/\partial\phi_a , \tag{5}$$

and the effective potential, at the one-loop level, is given by

$$V_{eff}^{(T)} = \sum_a |\partial f(\phi)/\partial\phi_a|^2 + \frac{1}{8}T^2 \sum_{a,b} |\partial^2 f(\phi)/\partial\phi_a\partial\phi_b|^2 , \tag{6}$$

The ground states of the system at $T \neq 0$ is obtained by finding the minima of $V_{eff}^{(T)}$ in the ϕ-field space. Let us denote the positions of such ground states by $\phi_a = \hat{\phi}_a^{(T)}$. Supersymmetry is not broken if, by criterion (4), $\phi_a = \hat{\phi}_a^{(T)}$ also satisfies

$$\partial f(\phi)/\partial\phi_a = 0 . \tag{7}$$

Now if supersymmetry is spontaneously broken at $T = 0$, then Eq. (7) will have no solution. This means that Eq. (7) can never have a solution at any T. Thus spontaneously broken supersymmetry can never be restored at all.[8] On the other hand, if supersymmetry is manifest at $T = 0$, then there exists at least one solution to Eq. (7), which we denote by $\phi_a = \hat{\phi}_a^{(0)}$. Each solution gives a supersymmetric ground state of the system at $T = 0$. As T increases, the ground state, denoted by $\hat{\phi}_a^{(T)}$, may, in general, move away from $\hat{\phi}_a^{(0)}$, and Eq. (7) would not be satisfied. Supersymmetry would then be spontaneously broken. However, exceptions can occur. There are two possibilities:[8]

(i) The ground state may be fixed at the same point in ϕ-field space at all T, i.e. $\phi_a^{(T)}$ = constant.

(ii) There may exist a continuous set of ground states at $T = 0$, but at $T \neq 0$, this degeneracy is lifted, and the ground state is fixed at one point. Here $\phi_a^{(T)}$ = constant is a subset of $\phi_a^{(0)}$.

In both cases, $\phi_a^{(T)}$ has no temperature variation, and supersymmetry will be manifest at all T. Our conclusion therefore is: Supersymmetry is not necessarily broken automatically at $T \neq 0$. Simple systems can easily be constructed in which supersymmetry is manifest at all T.[8]

When supersymmetry is spontaneously broken at $T = 0$, there exists a massless Goldstone fermion.[9-12] Is it still true for $T \neq 0$? We shall investigate this through three explicit examples.[13] The first system is the massive Wess-Zumino model[14] given by the superpotential

$$f(S) = -\lambda S + \frac{1}{2}mS^2 + \frac{1}{6}gS^3 . \tag{8}$$

For this system, the spin-zero field ϕ develops a nonzero thermal average at $T \neq 0$ given by[8]

$$\begin{aligned} \langle\phi\rangle_T = \frac{1}{\sqrt{2}}v &= -m/g \pm \frac{1}{2}(8\lambda g+4m^2-g^2T^2)^{\frac{1}{2}}/g , \quad T < T_c = (8\lambda g+4m^2)^{\frac{1}{2}}/g, \\ &= -m/g , \qquad\qquad T > T_c . \end{aligned} \tag{9}$$

Since $\langle F\rangle_T \neq 0$, supersymmetry is broken at $T \neq 0$.[8] We shift ϕ by

$$\phi = \frac{1}{\sqrt{2}}(v + A + iB) . \tag{10}$$

Perturbative calculations can then be performed in terms of the fields A, B and ψ, whose tree-level masses are

$$\begin{aligned} \mu^2_{A,B} &= \frac{1}{2}g^2(v + \frac{\sqrt{2}m}{g})^2 \pm \{\frac{1}{4}g^2(v + \frac{\sqrt{2}m}{g})^2 - (\lambda g + \frac{1}{2}m^2)\} , \\ \mu^2_\psi &= \frac{1}{2}g^2(v + \frac{\sqrt{2}m}{g})^2 . \end{aligned} \tag{11}$$

The fermion self-energy Σ_ψ at $T \neq 0$ is easily calculated in the one-loop approximation.[13] The relevant interaction term is

$$\mathcal{L}_I = - \frac{g}{2\sqrt{2}} \bar{\psi}(A + iB\gamma_5)\psi . \tag{12}$$

We obtain[13]

$$\Sigma_\psi = \frac{1}{2}g\mu_\psi T^2/(gv^2 + 2\sqrt{2}mv - 4\lambda) . \tag{13}$$

Thus the full fermion mass at $T \neq 0$ is given by

$$m_\psi(T) = \mu_\psi(gv^2+2\sqrt{2}mv-4\lambda+ \frac{1}{2}gT^2)/(gv^2+2\sqrt{2}mv-4\lambda) , \tag{14}$$

which vanishes because the numerator vanishes at all T, both below and above T_c.[13] That this massless fermion is indeed the Goldstone fermion is easily established by looking at its transformation property under a supersymmetric transformation,

$$\langle\delta\psi_L\rangle_T = -\sqrt{2}\alpha_L\langle F\rangle_T \neq 0 . \tag{15}$$

It is significant to note that ψ which is massive at $T = 0$, turns into the massless Goldstone fermion when $T \neq 0$, as supersymmetry is spontaneously broken.

Our next system is the well-known O'Raifeartaigh model[15)]

$$f(S) = -\lambda S_0 + mS_1S_2 + gS_0S_2^2, \tag{16}$$

with $2\lambda g > m^2$. For this system, supersymmetry is spontaneously broken at T = 0 with the existence of a massless Goldstone fermion. We shall show that the massless Goldstone fermion continues to exist at $T \neq 0$. Here the thermal averages of the spinless fields are

$$\begin{aligned} \langle\phi_0\rangle_T &= \langle\phi_1\rangle_T = 0\,, \\ \langle\phi_2\rangle_T &= \frac{1}{\sqrt{2}}v = (\lambda g - \tfrac{1}{2}m^2 - \tfrac{1}{2}g^2T^2)^{\frac{1}{2}}/g, \qquad T < T_c = (2\lambda g - m^2)^{\frac{1}{2}}/g, \\ &= 0\,, \qquad\qquad T > T_c\,. \end{aligned} \tag{17}$$

Again, we express the the ϕ_a fields in terms of the scalar and the pseudoscalar components, with ϕ_2 shifted,

$$\phi_{0,1} = \frac{1}{\sqrt{2}}(A_{0,1} + iB_{0,1})\,, \qquad \phi_2 = \frac{1}{\sqrt{2}}(v + A_2 + iB_2)\,. \tag{18}$$

The Lagrangian, after rearrangement, now contains the following mass eigenstates:

$$\begin{aligned} &A_0',\ B_0',\ \psi_0' &&: \mu^2 = 0\,; \\ &A_1',\ B_1',\ \psi_1',\ \psi_2' &&: \mu^2 = m^2 + 2g^2v^2 = \mu_1^2\,; \\ &A_2,\ B_2 &&: \mu^2 = m^2 + 2g^2v^2 \pm (g^2v^2 - 2\lambda g)\,. \end{aligned} \tag{19}$$

The primed fields are related to the unprimed fields by

$$\begin{bmatrix} A_0',\ B_0' \\ A_1',\ B_1' \end{bmatrix} = \frac{1}{\mu_1}\begin{bmatrix} m & -\sqrt{2}gv \\ \sqrt{2}gv & m \end{bmatrix}\begin{bmatrix} A_0,\ B_0 \\ A_1,\ B_1 \end{bmatrix} \tag{20}$$

$$\begin{pmatrix} \psi_{0L}' \\ \psi_{1L}' \\ \psi_{2L}' \end{pmatrix} = \frac{1}{\sqrt{2}\mu_1}\begin{pmatrix} \sqrt{2}m & -2gv & 0 \\ -\sqrt{2}gv & -m & -\mu_1 \\ i\sqrt{2}gv & im & -i\mu_1 \end{pmatrix}\begin{pmatrix} \psi_{0L} \\ \psi_{1L} \\ \psi_{2L} \end{pmatrix} \tag{21}$$

The fermion self-energies are straight-forwardly calculated in the one-loop approximation.[13)] To leading order in T, the resulting fermion mass matrix $\mathcal{M}$ is[13)]

$$\mathcal{M} = \gamma_+ \begin{bmatrix} 0 & b & -ib \\ b & \mu_1 - a & 0 \\ -ib & 0 & \mu_1 - a \end{bmatrix} + \gamma_- \begin{bmatrix} 0 & b & ib \\ b & \mu_1 - a & 0 \\ ib & 0 & \mu_1 - a \end{bmatrix} \tag{22}$$

$$a = g^3 v^2 T^2/\{\mu_1(2\lambda - gv^2)\}, \qquad b = mg^2 vT^2/\{2\mu_1(2\lambda - gv^2)\}. \tag{23}$$

The mass matrix is easily diagonalized, and we obtain the following physical fermion masses at $T \neq 0$:[13)]

$$0, \quad \{(\mu_1 - a)^2 + 2b^2\}^{\frac{1}{2}}, \quad \{(\mu_1 - a)^2 + 2b^2\}^{\frac{1}{2}}. \tag{24}$$

The wave-function corresponding to the massless fermion state is

$$\chi_L = c\{\psi_{0L} - \sqrt{2}mgv(m^2 + g^2T^2)^{-1}\psi_{1L}\}. \tag{25}$$

It is trivial to show that χ transforms inhomogeneously under a supersymmetry transformation. The above calculation is for $T < T_c$. For $T > T_c$, since $v = 0$, there is no one-loop corrections to the fermion masses, and ψ_0 is the massless Goldstone fermion.

Our last example is the supersymmetric O(N) model[8, 16)] given by

$$f(S) = -\lambda S_0 + \frac{1}{6}gS_0^3 + \frac{1}{2}hS_0(S_1^2 + S_2^2 + \ldots + S_N^2). \tag{26}$$

This system admits of two phases, the O(N)-violating phase and the O(N)-conserving phase. In both phases, supersymmetry is spontaneously broken at $T \neq 0$.[8,16)] The O(N)-violating phase is the dominant phase at $T \neq 0$, and is characterised by[8, 16)]

$$\begin{aligned} &\langle\phi_0\rangle_T = \langle\phi_1\rangle_T = \ldots = \langle\phi_{N-1}\rangle_T = 0, \\ &\langle\phi_N\rangle_T = \frac{1}{\sqrt{2}}v = \left(\frac{2\lambda}{h} - \frac{1}{2}T^2\right)^{\frac{1}{2}}, \qquad T < T_c = (4\lambda/h)^{\frac{1}{2}}, \\ &\qquad\quad = 0, \qquad T > T_c. \end{aligned} \tag{27}$$

Again, expressing the ϕ fields in terms of A and B fields

$$\phi_N = \frac{1}{\sqrt{2}}(v + A_N + B_N), \qquad \phi_a = \frac{1}{\sqrt{2}}(A_a + B_a), \qquad a = 0,1,2,\ldots,N-1, \tag{28}$$

we find that the tree-level mass eigenstates are

$$A_0,\ B_0 \ :\ \mu^2 = \pm g(-\lambda + \tfrac{1}{4}hv^2) + \tfrac{1}{2}h^2v^2\ ;$$

$$A_N,\ B_N \ :\ \mu^2 = \pm h(-\lambda + \tfrac{1}{4}hv^2) + \tfrac{1}{2}h^2v^2\ ;$$

$$A_k,\ B_k \ :\ \mu^2 = \pm h(-\lambda + \tfrac{1}{4}hv^2)\ , \qquad k = 1,2,\ldots,N-1\ ;$$

$$\psi'_0,\ \psi'_N \ :\ \mu^2 = \tfrac{1}{2}h^2v^2 = \mu_\psi{}^2\ ;$$

$$\psi_k \ :\ \mu^2 = 0\ , \qquad k = 1,2,\ldots,N-1\ . \tag{29}$$

The states ψ'_0 and ψ'_N are linear combinations of ψ_0 and ψ_N,

$$\psi'_{OL} = \frac{1}{\sqrt{2}}(\psi_{OL} + \psi_{NL})\ , \qquad \psi'_{NL} = \frac{i}{\sqrt{2}}(\psi_{OL} - \psi_{NL})\ . \tag{30}$$

We see that at $T = 0$, ψ'_0 and ψ'_N are both massive. At $T \neq 0$, the one-loop corrections to the ψ'_0 and ψ'_N masses are[13)]

$$\Sigma(\psi'_0) = \Sigma(\psi'_N) = \mu_\psi hT^2/(hv^2 - 4\lambda)\ , \quad \mu_\psi = \frac{1}{\sqrt{2}}hv\ . \tag{31}$$

The one-loop corrected mass for ψ'_0 and ψ'_N is therefore

$$m_\psi(T) = \mu_\psi(-4\lambda + hv^2 + hT^2)/(hv^2 - 4\lambda)\ , \tag{32}$$

which vanishes because of (27) for both T below and above T_c. The Goldstone fermion is actually ψ_0 on account of its inhomogeneous transformation property.[13)]

The O(N)-converving phase of the supersymmetric O(N) model is the less favoured phase, and is characterized by[8, 16)]

$$\langle\phi_i\rangle_T = 0, \qquad i = 1,2,\ldots,N\ ,$$

$$\langle\phi_0\rangle_T = \frac{1}{\sqrt{2}}v' = \frac{1}{2}\{8\lambda g - (g^2 + Nh^2)T^2\}^{\frac{1}{2}}/g\ . \tag{33}$$

This phase exists only for

$$T < \{16\lambda gh/(g^3 + Ngh^2 + 2Nh^3)\}^{\frac{1}{2}}\ . \tag{34}$$

Shifting the ϕ_0 field, we find that the ψ_0, which has a tree-level mass of

$$\mu(\psi_0) = \frac{1}{\sqrt{2}}gv'\ , \tag{35}$$

receives an one-loop correction given by[13)]

$$\Sigma(\psi_0) = -(g^2 + Nh^2)v'T^2/\{2\sqrt{2}(4\lambda - gv'^2)\}\ . \tag{36}$$

The mass of ψ_0 is therefore zero at $T \neq 0$ on account of (33), establishing that ψ_0 is the massless Goldstone fermion.[13)]

From the three explicit examples investigated above, we have demonstrated that the when supersymmetry is spontaneously broken at $T \neq 0$, there always exists a massless Goldstone fermion.

References:

1. L. Girardello, M.T. Grisaru and P. Salomonson, Nucl. Phys. B178, 331 (1981).
2. L. Van Hove, Nucl. Phys. B207, 15 (1982).
3. D.A. Dicus and X.R. Tata, Nucl. Phys. B239, 237 (1984).
4. T.E. Clark and S.T. Love, Nucl. Phys. B217, 349 (1983).
5. M.B. Paranjape, A. Taormina and L.C.R. Wijewardhana, Phys. Rev. Lett. 50, 1350 (1983).
6. J. Fuchs, Nucl. Phys. B246, 279 (1984).
7. D. Boyanovsky, Phys. Rev. D29, 743 (1984).
8. S.P. Chia, University of Malaya Preprint UMKL-85-3 (1985).
9. P. Fayet and J. Iliopoulos, Phys. Lett. 51B, 461 (1974).
10. A. Salam and J. Strathdee, Phys. Lett. 49B, 465 (1974).
11. J. Iliopoulos and B. Zumino, Nucl. Phys. B76, 310 (1974).
12. S.P. Chia, J. Fiz. Mal. 5, 197 (1984).
13. S.P. Chia, University of Malaya Preprint UMKL-85-4 (1985).
14. J. Wess and B. Zumino, Nucl. Phys. B70, 39 (1974).
15. L. O'Raifeartaigh, Nucl. Phys. B96, 331 (1975).
16. S.P. Chia, J. Fiz. Mal. 6, 131 (1985).

THE β-FUNCTIONS OF THE SU(2) LATTICE GAUGE THEORIES WITH SIN AND EXP ACTIONS

Dong Shao-jing and Li Wen-zhou

Physics Department, Zhejiang University, Hangzhou

The β-functions of the SU(2) lattice gauge theories with SIN and EXP actions are investigated by means of Monte Carlo renormalization group method (MCRG). Only one ultraviolet fixed point is observed. The finite size effect and the connection with continuum limit are discussed. We get some informations that the crossover is not equal to the scaling.

RELATIVISTIC EFFECTS, E1 AND M1 TRANSITIONS FOR HEAVY QUARKONIA

Yi-Bing Ding
Graduate School, Academy of Sciences of China

Ju He
High Energy Physics Institute, Academy of Sciences of China

Shen-Ou Cai, Dan-Hua Qin, Kuang-Ta Chao
Peking University

Abstract

We suggest a potential model with relativistic corrections which contains a scalar potential $S(r)=8\pi\Lambda^2 r/(33-2n_f)$ and a vector potential $V(r)=(-4/3)\alpha_s(r)/r$, $\alpha_s(r)=[(-6\pi)(1-\Lambda r)]/[(33-2n_f)(1+\Lambda r)\ln\Lambda r]$ involving only one parameter Λ corresponding to the QCD scale parameter. Our results of the mass spectrum for $c\bar{c}$ and $b\bar{b}$ are compatible with experiment. Especially, a good agreement of the fine splitting of the $b\bar{b}$ P-wave states with data is obtained. The relativistic corrections to the wave functions are found to be particularly important for some $c\bar{c}$ E1 transitions.

QCD-inspired potential models for heavy quarkonia have been extensively studied and turned out to be very successful in understanding many essential features of heavy quark systems.* However, it is still needed to study problems such as relativistic effects in a more consistent and systematical manner. In this note we would like to suggest a model which contains only one parameter in the potentials and to study various relativistic effects such as corrections to the mass spectrum, E1, and M1 transitions.

Our potential model contains a vector potential V(r) and a scalar potential S(r). The vector potential representing one gluon exchange is taken to be

$$V(r) = -\frac{4}{3}\frac{\alpha_s(r)}{r} \tag{1}$$

* For a comprehensive review on related theoretical and experimental work, see J. L. Rosner, talk presented at International Symposium on Lepton and Photon Interactions, Kyoto, Japan, Aug. 19-24, 1985 and references therein, see also Particle Data Group, Particle Properties Data Booklet, 1984.

where

$$\alpha_s(r)=\frac{-6\pi}{33-2n_f}\frac{1}{\ln\Lambda r}\frac{1-\Lambda r}{1+\Lambda r} \tag{2}$$

where Λ is the QCD scale parameter which is to be determined by fitting the spectrum and the flavor number is taken to be $n_f = 4$. The characteristic of this potential is that $\alpha_s(r)$ has asymptotic-free behavior at very short distances i.e.

$$\alpha_s(r)\longrightarrow\frac{-6\pi}{33-2n_f}\frac{1}{\ln\Lambda r}\qquad \text{as } r\to 0 \tag{3}$$

and $\alpha_s(r)$ increases gradually as r increases, and then it is almost saturated in a rather wide range where most of the low lying $c\bar{c}$ and $b\bar{b}$ states lie, i.e.

$$\alpha_s(r)\sim 0.36-0.38 \qquad \text{for } r\sim 1-4\ \text{GeV}^{-1} \tag{4}$$

Here a tentative value $\Lambda = 0.470$ GeV is used to estimate $\alpha_s(r)$ from (2). The long range (say, $r \gtrsim$ 1fm) behavior of $\alpha_s(r)$ is disregarded since one gluon exchange is no longer meaningful there. The confining potential is assumed to be a scalar linear potential which is related to the scale parameter Λ :

$$S(r)=\frac{8\pi}{33-2n_f}\Lambda^2 r \tag{5}$$

If both vector and scalar potentials are treated as one particle exchange interactions, the standard nonrelativistic reduction for a heavy $Q\bar{Q}$ system (quark mass $m_Q = m$) will give rise to a perturbatively expanded Hamiltonian, which reads up to the first order:

$$H=H_0+H_1, \tag{6}$$

$$H_0=\frac{\vec{p}^2}{m}+S+V, \tag{7}$$

$$H_1=-\frac{\vec{p}^4}{4m^3}+V_{SD}+V_{SI}, \tag{8}$$

$$V_{SD}=\frac{1}{2m^2}\left(\frac{3}{r}V'-\frac{1}{r}S'\right)(\vec{S}_1+\vec{S}_2)\cdot\vec{L}+\frac{2}{3m^2}\vec{S}_1\cdot\vec{S}_2\nabla^2V$$
$$-\frac{1}{3m^2}\left[3(\vec{S}_1\cdot\hat{r})(\vec{S}_2\cdot\hat{r})-(\vec{S}_1\cdot\vec{S}_2)\right]\left(V''-\frac{1}{r}V'\right), \tag{9}$$

$$V_{SI}=\frac{1}{4m^2}\left\{[\vec{p}^2, V-rV']+2(V-rV')\vec{p}^2+\frac{1}{2}\left(\frac{8}{r}V'+V''-rV'''\right)+\frac{2}{r}V'\vec{L}^2\right\}$$
$$-\frac{1}{4m^2}\left\{2[\vec{p}^2,S]+4S\vec{p}^2+\frac{2}{r}S'+S''\right\}. \tag{10}$$

However, we must bear in mind that the scalar confinement potential is probably due to multi-gluon exchange and therefore the one particle exchange picture for it may not be justified. While the Thomas precession term $-\frac{1}{2m^2}\frac{1}{r}S'(\vec{S}_1+\vec{S}_2)\cdot\vec{L}$ in V_{SD} may be explained in terms of a rotating flux tube or of a QCD electric confinement picture, the spin-independent (SSI) terms induced by scalar confinement in V_{SI} are truly questionable.

We first use (1),(2),(5), and (7) to solve the zeroth order Schrodinger equation and then treat H_1 as a perturbation and use (8)-(10) to calculate first order relativistic corrections. The forementioned SSI terms are found to be unreasonably large and not justified theoretically, and therefore they have been ignored in all calculations. The parameters determined by fitting the mass spectra are

$$\Lambda = 0.470\ \text{GeV}, \quad m_c = 1.833\ \text{GeV}, \quad m_b = 5.170\ \text{GeV} \tag{11}$$

The main results and conclusions are as follows:

(1) The obtained mass spectra for low lying $c\bar{c}$ and $b\bar{b}$ states are compatible with data (experimental values are shown in parentheses in unit of MeV):

$c\bar{c}$:	1S(3097),	$1P_{c.o.g.}$(3525),	2S(3686),	3S(4030?)	
	input	3547	3682	4112	
$b\bar{b}$:	1S(9460),	$1P_{c.o.g.}$(9903),	2S(10023),	$2P_{c.o.g.}$(10261),	3S(10356)
	input	9901	10016	10282	10381

The relative values of leptonic decay widths (compared with the width for the 1S state) are also compatible with data.

(2) The fine splittings of the $b\bar{b}$ P-wave states are found to be in fairly good agreement with data (experimental values shown in parentheses):

$M(1P_2) - M(1P_0) = 38\ (\sim 40)$ MeV, $\quad R_1 = 0.71(\sim 0.93)$;

$M(2P_2) - M(2P_0) = 30\ (\sim 36)$ MeV, $\quad R_2 = 0.72\ (\sim 0.89)$,

where $R_n = [M(nP_2) - M(nP_1)]/[M(nP_1) - M(nP_0)]$. However, for the $c\bar{c}$ states the calculated splitting $M(1P_2) - M(1P_0) = 62\ (\sim 141)$ MeV is too small while the ratio $R_1 = 0.47\ (\sim 0.48)$ seems to be satisfactory.

The calculated hyperfine splittings are 53(~ 114)MeV for $\psi - \eta_c$ and 29 (no experiment)MeV for $\Upsilon - \eta_b$

However, if we use a fixed value for α_s [= 0.38, see (4)], instead of using a running to calculate first order corrections, the obtained $\psi - \eta_c$ splitting would be $\sim$ 110 MeV in good agreement with data, and the fine splittings for $c\bar{c}$ P-wave states would also be much improved. Since terms related to the derivatives of $\alpha_s(r)$ with respect to r are responsible for reducing the fine and hyperfine splittings, and they are all of higher orders of α_s it might be justified to neglect these terms within the first order approximation. Nevertheless, our results imply that corrections due to higher orders of α_s are indeed important for $c\bar{c}$ states.

(3) Various E1 transition rates are calculated and relativistic corrections to the wave functions are found to be particularly important for some $c\bar{c}$ transitions. Following examples are given:

Transition	J	Nonrelativistic Widths(keV)	Relativistic corrected Widths(keV)	Experiment (keV)
$\psi' \to \chi_c(J) + \gamma$	0	42	25	21 ± 6
$(2S \to 1P + \gamma)$	1	36	27	16 ± 5
	2	25	22	17 ± 5
$\Upsilon' \to \chi_b(J) + \gamma$	0	0.93	0.73	1.0
$(2S \to 1P + \gamma)$	1	1.76	1.58	1.7
	2	1.77	1.74	1.8

(4) The M1 transition rates may be suppressed by including scalar confinement potential and other terms into relativistic corrections. For example, the observed branching ratio of $\psi \to \eta_c + \gamma$ is (1.27 + 0.36)%, the uncorrected value is $\sim$ 2.6%, whereas the relativistic corrected value is $\sim$1.1%. This suppresssion is mainly due to the fact that the scalar potential increases the effective quark mass and therefore reduces the quark magnetic moment.

To sum up, our potential model provides a better understanding for many important features related to relativistic corrections for heavy quarkonia. However, problems connected with higher order corrections are still open and worth studying.

BASIC FEATURES OF STRING THEORIES

Jean-Loup Gervais
Laboratoire de Physique Théorique de l'Ecole Normale Supérieure
24, rue Lhomond, 75231 Paris cedex 05, France

At the present time it is hardly necessary to emphasize the fundamental importance of string models since superstring theories are the most promising candidates for a completely unified theory of all interactions. Moreover string concepts have played an important role in the recent developments of theoretical physics and mathematics by suggesting many new important ideas, such as in particular supersymmetry[1], and have led to very interesting progress in the related critical models in two dimensions.

It is hopeless to try and cover the whole subject of string theories. These lectures will mostly concentrate on the basic features of free strings and tree string scattering amplitudes. The bosonic string will be used in order to illustrate, without too much technical complications, the common features of all string models.

The old covariant operator formalism is presented in section (I) as a warming up. It is based on clever manipulations of trivial free fields in the two dimensional space of the string parameters. Our discussion will allow to see the fundamental role of the conformal algebra in two dimensions explicitly, and to discuss in a simple fashion the ghost-killing mechanism and the emergence of the critical space time dimension. In section (II), we shall show how the above conformal invariance appears as a remnant of the reparametrization invariance of the world sheet swept out by the string. We shall review the functional approach to quantization of the string modes, from the viewpoint of gauge theory in two dimensions. The relationship between the path integrals based on the new and the old reparametrization invariant actions will be discussed. The connection of the latter with the string scattering amplitudes in the light cone gauge will be summarized. Finally we shall review, in section (III), the possibility of building new string models in the Lorentz covariant formalism from interacting field theories in two dimensions. This brings in the question of conformally invariant field theories in general, and of the related critical systems.

There already exists a large number of review articles[2)3)4)5)6)], and two reprint volumes are about to appear[7)]. The viewpoint chosen here is somewhat different.

1. The operator formalism.

Before discussing the more sophisticated viewpoints which were later developed, it is useful to review the operator formalism where the relevant features are derived from simple harmonic oscillators in a pedestrian way.

a) Free bosonic modes.

The relevant quantity is a field $X^\mu(\sigma,\tau)$ describing the space-time position of the point of the string, which is characterized by the parameter σ, at time τ. μ is thus a Lorentz index ($\mu = 0, 1, \ldots, \mathcal{D}-1$) the space-time dimension $\mathcal{D}$ is kept as a free parameter. Since the string has a finite extension, σ varies over a finite range. For closed strings, which we shall discuss first, X^μ is a periodic function of σ. We can take, by convention, the period to be equal to 2π and introduce the simple action

$$S = \frac{1}{4\pi}\int_0^{2\pi} d\sigma d\tau \left((\dot{X}^\mu)^2 - (X'^\mu)^2\right); \quad \dot{X} = \frac{\partial X}{\partial \tau}; \; X' = \frac{\partial X}{\partial \sigma} \tag{1.1}$$

The factor in front of S can be modified by a simple rescaling of X^μ. Imposing the periodicity condition

$$X^\mu(\sigma + 2\pi, \tau) = X^\mu(\sigma,\tau) \tag{1.2}$$

gives a simple set of harmonic oscillators with $\omega_m = |m|$, m integer $\neq 0$, and we can write the standard free field decomposition in a box

$$X^\mu(\sigma,\tau) = q^\mu + p^\mu\tau + \sum_{m\neq 0} \frac{1}{\sqrt{2|m|}}\left[\alpha_m^\mu\, e^{i(m\sigma - |m|\tau)} + \alpha_m^{\mu\dagger}\, e^{-i(m\sigma-|m|\tau)}\right] \tag{1.3}$$

$$[\alpha_m^\mu, \alpha_m^\nu] = [q^\mu, q^\nu] = [p^\mu, p^\nu] = 0 \tag{1.4}$$

$$[\alpha_m^\mu, \alpha_m^{\nu\dagger}] = \delta_{m,m}\,\eta^{\mu\nu} \tag{1.5}$$

$$[q^\mu, p^\nu] = i\eta^{\mu\nu} \tag{1.6}$$

q^μ and p^μ are the center of mass position and total momentum, respectively. The space-time flat metric, $\eta^{\mu\nu}$, is equal to +1 for the diagonal space like components. On (1.5) one sees that, as usual, the time components α_m^0, α_m^{0+} generate a Fock space with a non positive definite metric. We shall come back to this later on. For $m > 0$, $\alpha_m^\mu (\alpha_{-m}^\mu)$ annihilate right movers (left movers). It is convenient to separate these modes and rewrite, for $m > 0$,

$$a_m^\mu \equiv i\sqrt{m}\,\alpha_{-m}^\mu \quad ; \quad a_{-m}^\mu \equiv -i\sqrt{m}\,\alpha_{-m}^{\mu +}$$

$$\bar{a}_m^\mu \equiv i\sqrt{m}\,\alpha_m^\mu \quad ; \quad \bar{a}_{-m}^\mu \equiv -i\sqrt{m}\,\alpha_m^{\mu +} \tag{1.7}$$

in such a way that

$$X^\mu = q^\mu + p^\mu \tau + \frac{i}{\sqrt{2}} \sum_{m \neq 0} \left(a_m^\mu e^{-imu} + \bar{a}_m^\mu e^{-imv} \right) \frac{1}{m}$$

$$u \equiv \tau + \sigma \; ; \; v \equiv \tau - \sigma \tag{1.8}$$

$$[a_m^\mu, a_m^\nu] = [\bar{a}_m^\mu, \bar{a}_m^\nu] = m \eta^{\mu\nu} \delta_{m,-m}$$

$$[a_m^\mu, \bar{a}_m^\nu] = 0 \quad ; \quad a_m^{\mu +} = a_{-m}^\mu \; ; \; \bar{a}_m^{\mu +} = \bar{a}_{-m}^\mu \tag{1.9}$$

The point of (1.7) (1.8) is to introduce the following simple expansions

$$P^\mu(z) \equiv \frac{1}{\sqrt{2}} (\dot{X}^\mu + X'^\mu) = \sum_m a_m^\mu z^{-m}$$

$$\bar{P}^\mu(z) \equiv \frac{1}{\sqrt{2}} (\dot{X}^\mu - X'^\mu) = \sum_m \bar{a}_m^\mu \bar{z}^{-m}$$

$$z = e^{iu} \; ; \; \bar{z} = e^{iv} \; ; \quad a_0^\mu = \bar{a}_0^\mu \equiv \frac{p^\mu}{\sqrt{2}} \tag{1.10}$$

The energy-momentum tensor is such that

$$T_0^0 \pm T_0^1 = \frac{1}{4\pi} :(\dot{X} \pm X')^2 \tag{1.11}$$

According to (1.10) its Fourier modes are such that

$$\frac{1}{2}\left(T_0^0 + T_0^1 \right) = \frac{1}{2\pi} \sum_m L_m z^{-m}$$

$$\frac{1}{2}\left(T_0^0 - T_0^1 \right) = \frac{1}{2\pi} \sum_m \bar{L}_m \bar{z}^{-m} \tag{1.12}$$

$$L_m = \frac{1}{2}\sum_n' : a_n^\mu a_{m-n}^\mu : \quad ; \quad \bar{L}_m = \frac{1}{2}\sum_n' : a_n^\mu a_{m-n}^\mu : \tag{1.13}$$

From (1.8), (1.9), one deduces that

$$[L_m, X^\mu] = z^{m+1}\frac{\partial}{\partial z}X^\mu$$

$$[\bar{L}_m, X^\mu] = \bar{z}^{m+1}\frac{\partial}{\partial \bar{z}}X^\mu \tag{1.14}$$

and L_m, $\bar{L}_m$ respectively generate the infinitesimal transformations

$$z \to z(1+\varepsilon z^m) \quad ; \quad \bar{z} \to \bar{z}$$

$$z \to z \quad ; \quad \bar{z} \to \bar{z}(1+\varepsilon \bar{z}^m) \tag{1.15}$$

They are all the transformations which preserve the angles with the metric $g_{\sigma\sigma} = -g_{\tau\tau} = 1$, $g_{\sigma\tau} = 0$, and are compatible with the periodicity condition (1.2). This last fact is an immediate consequence of the fact that the infinitesimal change being only a function of

$$z = e^{i(\sigma+\tau)} \quad ; \quad \bar{z} = e^{i(\tau-\sigma)} \tag{1.16}$$

is a periodic function of σ with period 2π. The subset of the operators $L_{\pm 1}, L_0$ ($\bar{L}_{\pm 1}, \bar{L}_0$) are the infinitesimal generators of the Möbius transformations of $z(\bar{z})$: $z \to (az+b)/(cz+d)$, $(\bar{z} \to (a\bar{z}+b)/(c\bar{z}+d))$ In particular, it is clear from (1.15) and (1.16) that $L_0 + \bar{L}_0$ and $L_0 - \bar{L}_0$ respectively generate τ and σ translations. The algebra of the L_m or $\bar{L}_m$ operators (the Virasoro algebra) closes apart from a c-number term (central charge) which plays a crucial role and we now discuss its calculation in a way which will easily generalize to different cases. This is best done by going to the Euclidean 2-D space such that one replaces τ by $-i\nu$, ν real. Equations (1.16) become

$$z = e^{\nu + i\sigma} \quad ; \quad \bar{z} = e^{\nu - i\sigma} \tag{1.17}$$

z, $\bar{z}$ are now the two complex conjugate variables and the transformations (1.15) simply become analytic transformations. We shall only consider the L_m algebra since the discussion of $\bar{L}_m$ is exactly the same. Clearly, equation (1.13) leads to

$$L_n = \frac{1}{4\pi i}\oint \frac{dz}{z} z^n :P^2(z): \tag{1.18}$$

where the integral is in the complex z plane around an arbitrary contour surrounding the origin. We start the evaluation of $[L_n, L_m]$ by writing

$$L_n L_m = \frac{1}{(4\pi i)^2}\oint \frac{dx}{x}\oint \frac{dy}{y} x^n y^m :P^2(x)::P^2(y): \tag{1.19}$$

$$L_m L_n = \frac{1}{(4\pi i)^2}\oint \frac{dx}{x}\oint \frac{dy}{y} x^n y^m :P^2(y)::P^2(x): \tag{1.20}$$

Apply Wick's theorem to each expression. The contraction formula

$$\langle 0| P^\mu(z_1) P^\nu(z_2)|0\rangle = \eta^{\mu\nu}\,\frac{z_1 z_2}{(z_1-z_2)^2} \tag{1.21}$$

makes sense only if $|z_1|>|z_2|$ (otherwise the series diverges). Hence we have to choose $|x|>|y|$ in (1.19) and $|y|>|x|$ in (1.20). In the difference between (1.19) and (1.20) we thus pick up the poles of the Wick ordered expansion at x=y. This leads to[5)]

$$[L_n, L_m] = (n-m)L_{n+m} + \frac{D}{12}(n^3-n)\,\delta_{n,-m} \tag{1.22}$$

The first term, which is linear in the L_n is the standard Lie algebra contribution. The second inhomogeneous term is the central charge which is proportional to D.

We now turn to open strings. It is convenient to choose $0 \le \sigma \le \pi$. The boundary conditions are

$$X'^\mu = 0 \quad , \quad \text{at } \sigma = 0, \pi \tag{1.23}$$

The action

$$S = \frac{1}{2\pi}\int_0^\pi d\sigma \int d\tau \left((\dot{X}^\mu)^2 - (X'^\mu)^2 \right) \tag{1.24}$$

only describes one set of modes since right and left movers mix at the end points. One now has

$$X^\mu = q^\mu + p^\mu\tau + i\sum_{m\neq 0} \frac{a_m^\mu}{m} e^{-im\tau}\cos(m\sigma) \tag{1.25}$$

$$\frac{1}{2}(T_0^0 \pm T_0^1) = \frac{1}{4\pi} :\left(\sum_m a_m^\mu e^{-im(\tau\pm\sigma)}\right)^2: \qquad a_0^\mu \equiv p^\mu \tag{1.26}$$

Obviously there is a single set of Virasoro generators. It satisfies the algebra (1.22).

b) The dual tree amplitude between tachyons.

We shall mostly consider open strings which are simpler. For an arbitrary momentum vector k^μ, one introduces the basic vertex operator

$$V_k(z) = :e^{ik^\mu X^\mu}: \equiv e^{ik.q}\, z^{k.p}\, e^{-\sum_{m<0} \frac{k.a_m}{m} z^{-m}}\, e^{-\sum_{m>0} \frac{k.a_m}{m} z^{-m}} \tag{1.27}$$

Consider the ground state $|0\rangle$ such that

$$a_m^\mu |0\rangle = 0 \quad m>0 \;;\; p^\mu |0\rangle = 0 \tag{1.28}$$

One easily obtains

$$\langle 0| V_{k_N}(z_N)\, V_{k_{N-1}}(z_{N-1}) \cdots V_{k_1}(z_1)|0\rangle = \prod_{\ell>m} (z_\ell - z_m)^{k_\ell . k_m} \tag{1.29}$$

if the k_ℓ are conserved

$$\sum_{\ell=1,N} k_\mu^\ell = 0 \,. \tag{1.30}$$

Equation (1.29), however, is derived by a summing series which converges only if $|z_N| > |z_{N-1}| > \cdots > |z_1|$. As we shall immediately see, the Möbius transformations of all z's play a crucial role. To investigate this feature we introduce the cross ratios $(m>\ell)$

$$u_{m\ell} \equiv \frac{(z_m - z_{\ell-1})(z_{m-1} - z_\ell)}{(z_m - z_\ell)(z_{m-1} - z_{\ell-1})} \tag{1.31}$$

One obtains, assuming $(k^\ell)^2 = k^2$ independent of ℓ,

$$\prod_{m>\ell}(z_m-z_\ell)^{k_\ell\cdot k_m}=\prod_{m>\ell}u_{m\ell}^{-\alpha_{m\ell}-1}\prod_\ell(z_{\ell+1}-z_\ell)^{(k^2-\alpha_0-1)}(z_{\ell+2}-z_\ell)^{(\frac{k^2}{2}-\alpha_0-1)} \tag{1.32}$$

where

$$\alpha_{m\ell}=\alpha_0-\frac{1}{2}\Big(\sum_\ell^{m-1}k_\ell\Big)^2 \tag{1.33}$$

α_0 and k^2 are arbitrary so far, but if we choose

$$\alpha_0 = 1 \tag{1.34}$$

$$k^2 = 2 \tag{1.35}$$

we arrive at a Möbius invariant integrand

$$\prod_\ell dz_\ell\prod_{m>\ell}(z_m-z_\ell)^{k_\ell\cdot k_m}=\prod_\ell\frac{dz_\ell}{z_{\ell+2}-z_\ell}\prod_{m>\ell}u_{m\ell}^{-\alpha_{m\ell}-1} \tag{1.36}$$

The multi-Veneziano formula is basically the corresponding integral. Some more care is needed however. First, the vertex (1.27) describes a source k^μ located at the point $z=e^{i(\sigma+\tau)}$. As will become clearer in the next section, open string scattering amplitudes are described by external sources located at the end points $\sigma = 0$ or π . In the Euclidean, $z_\ell=e^{\nu_\ell+i\sigma_\ell}$, we choose $\sigma_\ell=0$ and all z_ℓ are on the real axis. Remembering the above condition of validity of (1.29), one integrates for $z_N > z_{N-1} > \cdots > z_1$. This is still not all, however, since (1.36) is Möbius invariant and we must divide out the integration over all Möbius transformations which preserve the real axis. This is done by giving arbitrary fixed values to three z variables (the choice is irrelevant). The multi-Veneziano formula reads, finally,

$$\mathcal{S}_N=\int d\mu(z)\,\langle 0|\,V_{k_N}(z_N)\,V_{k_{N-1}}(z_{N-1})\cdots V_{k_1}(z_1)|0\rangle$$

$$d\mu(z)=\prod_{\ell\neq a,b,c}(dz_\ell)\,|z_a-z_b||z_b-z_c||z_c-z_a|\prod_\pi\theta(z_{\pi+1}-z_\pi) \tag{1.37}$$

By Möbius transformation we can map the real line onto the unit circle.

This shows that (1.37) is invariant under cyclic permutations of the z 's. The complete, crossing symmetric S-matrix is obtained by summing over non cyclic permutations. The pole structure of (1.37) arises from the coincidence of z variables. From (1.31)(1.36) one sees that all poles are located at

$$\alpha(s) \equiv 1 + \frac{1}{2} s = m \quad ; \quad m \text{ integer} > 0 \tag{1.38}$$

where $\sqrt{s}$ is the center of the mass energy. The integer m is the highest spin of the intermediate particles at this level and one has a family of Regge trajectories spaced by integers, the leading trajectory being given by $J = \alpha(s)$. Conditions (1.34)(1.35) are such that the external particle with mass $-k^2 = -2$ is the lightest particle of the spectrum. Unfortunately this is a tachyon. From (1.38) one sees that the Regge trajectories have a slope $\alpha' = 1/2$. This is due to our particular choice of factors in front of the actions (1.1) and (1.24).

The pole structure is exhibited by changing variables from z_ℓ to N-3 independent $u_{m\ell}$'s. There are many ways to do this which lead to different dual pole configurations. In particular, let us choose $z_N \to \infty$, $z_{N-1} = 1$, $z_1 = 0$. In these limits

$$\begin{aligned} V_{k_1}(z)|0\rangle &\sim |k_1, 0\rangle \\ \langle 0| V_{k_N}(z_N) &\sim z_N^{-2} \langle k_N, 0| \end{aligned} \tag{1.39}$$

where, in general, we introduce the states $|k, 0\rangle$ such that

$$a_m^\mu |k, 0\rangle = 0 \,, \quad m > 0 \quad ; \quad p^\mu |k, 0\rangle = k^\mu |k, 0\rangle \tag{1.40}$$

Equation (1.37) becomes

$$\mathcal{S}_N = \int dz_2 \cdots dz_{N-2} \langle k_N, 0| V_{k_{N-1}}(1) V_{k_{N-2}}(z_{N-2}) \cdots V_{k_2}(z_2) |k_1, 0\rangle \tag{1.41}$$

For L_0, formula (1.13) gives, simply

$$L_0 = \frac{p^2}{2} + \sum_{m>0} a_{-m}^\mu a_m^\mu \tag{1.42}$$

and it is easy to check that (remember condition (1.35))

$$V_k(z) = z^{L_0} V_k(1) z^{-L_0} z^{-1} \tag{1.43}$$

Next, one changes variables by letting

$$z_{N-2}=x_{N-1}\ ;\ z_{N-3}=x_{N-1}x_{N-2}\ ;\ \dots;\ z_2=x_{N-1}\cdots x_3 \tag{1.44}$$

which is such that

$$0\leq x_\ell \leq 1 \quad ; \quad \prod_\ell \frac{dz_\ell}{z_\ell}=\prod_\ell \frac{dx_\ell}{x_\ell} \tag{1.45}$$

One can now integrate out the x variables obtaining

$$\mathcal{J}_N=\langle k_N,0|\,V_{k_{N-1}}(1)\frac{1}{L_0-1}V_{k_{N-2}}(1)\frac{1}{L_0-1}\cdots\frac{1}{L_0-1}V_{k_2}(1)\,|k_1,0\rangle \tag{1.46}$$

In this formula the so-called multiperipheral pole structure is explicit. The associated diagram is drawn on figure (1):

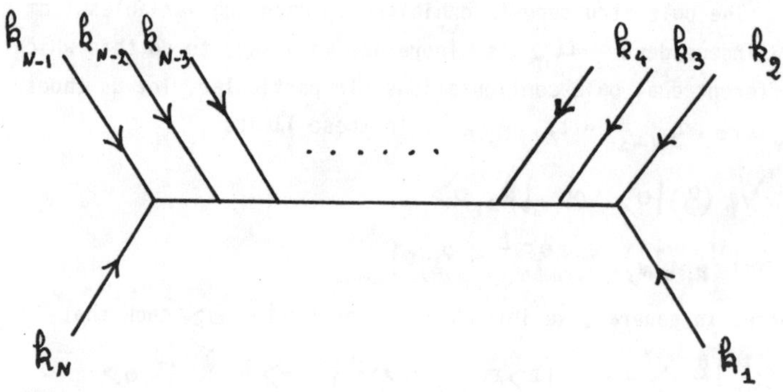

Fig. (1)

Next, a fundamental property of $V_k(z)$ for general k is its conformal covariance. One can check that

$$[L_m, V_k(z)]=z^m\left[z\frac{\partial}{\partial z}+(m+1)\frac{k^2}{2}\right]V_k(z) \tag{1.47}$$

For $k^2=2$ (condition (1.35)) we can write

$$(L_m-L_0-m)V_k(1)=V_k(1)(L_m-L_0) \tag{1.48}$$

Moreover, it is simple to verify that

$$(L_m - L_0 + 1)\frac{1}{L_0 - 1} = \frac{1}{L_0 + m - 1}(L_m - L_0 - m + 1) \tag{1.49}$$

and, therefore

$$(L_m - L_0 + 1)\frac{1}{L_0 - 1}V_k(1) = \frac{1}{L_0 + m - 1}V_k(1)(L_m - L_0 + 1) \tag{1.50}$$

Consider inserting $(L_m - L_0 + 1)$ in the left of a propagator $1/(L_0 - 1)$ in (1.46). One can move it to right step by step using (1.50) until it reaches the state $|k_1, 0\rangle$ on the left hand side. There

$$\begin{aligned} (L_0 - 1)|k_1, 0\rangle &= (\tfrac{1}{2}k_1^2 - 1)|k_1, 0\rangle = 0 \\ L_m|k_1, 0\rangle &= 0, \quad m > 0 \end{aligned} \tag{1.51}$$

One thus sees that for $m > 0$, $L_m - L_0 + 1$ gives zero when it is inserted in the left of any propagator $1/(L_0 - 1)$ in (1.46). The poles of $\mathcal{S}_N$ arise from the so-called on-shell intermediate states which are eigenvectors of L_0 with eigenvalue 1. From the above argument it follows that they have non vanishing residue only if they are in addition such that $(L_m - L_0 + 1)$ applied to them gives zero. Hence the poles only appear from the so-called on-shell physical states which satisfy

$$L_m|phys\rangle = 0 \qquad (L_0 - 1)|phys\rangle = 0 \tag{1.52}$$

As we already noticed, the Fock space of the oscillators a_m^μ is not positive definite. One hopes that (1.52) does define a positive definite physical subspace. Let us check this at the lowest levels. L_0 is of the form

$$L_0 = \frac{p^2}{2} + N \tag{1.53}$$

where N has positive integer or zero eigenvalues. From (1.52) these are related to the mass μ^2 through the relation

$$(\frac{\mu^2}{2} + 1)|phys\rangle = N|phys\rangle \tag{1.54}$$

The ground state $|0, k\rangle$ is such that

$$N|0, k\rangle = 0$$

and (1.54) gives $\mu^2 = -2$. It is the lightest particle which is a tachyon. For N=1, we have the state

$$|k,\varepsilon\rangle \equiv \varepsilon_\mu a^\mu_{-1}\,|k,0\rangle \tag{1.55}$$

Its mass is zero according to (1.54). The rest of conditions (1.52) is trivially satisfied except for n =1 which gives

$$k^\mu \varepsilon_\mu = 0 \tag{1.56}$$

This state is a massless spin one particle and, as usual, condition (1.56) ensures the positivity of the norm

$$\langle k,\varepsilon|k,\varepsilon\rangle = \varepsilon^2 \tag{1.57}$$

The next level is more intriguing. For N=2, the general eigenstate of N is of the form

$$|k,\alpha,\beta\rangle \equiv \left(\tfrac{1}{2}\alpha_{\mu\nu}\, a^\mu_{-1} a^\nu_{-1} + \beta_\lambda\, a^\lambda_{-2}\right)|k,0\rangle \tag{1.58}$$

Condition (1.54) gives k^2 = -2 and k being time like, we go to its rest frame $k=(\sqrt{2},\vec{0})$. $|k,\alpha,\beta\rangle$ is physical if α and β satisfy

$$\sqrt{2}\,\beta_\mu = \alpha_{\mu 0}\ ; \quad 5\,\alpha_{00} = \alpha_{ii} \tag{1.59}$$

This leaves us with a spin 2 state with components

$$\alpha^{(2)}_{ij} = \alpha_{ij} - \delta_{ij}\frac{\alpha_{\ell\ell}}{D-1}\ ; \quad \alpha^{(2)}_{00} = 0 \tag{1.60}$$

and a spin zero state with $\alpha^{(0)}_{ij}$ = 0, $\alpha^{(0)}_{00} \neq 0$. After some computation, one finds

$$\langle k,\alpha,\beta|k,\alpha,\beta\rangle = \frac{1}{2}\left(\alpha^{(2)}_{ij}\right)^2 + \left(\alpha^{(0)}_{00}\right)^2 \frac{26-D}{2(D-1)} \tag{1.61}$$

which is positive definite only if $D \leq$ 26. The free string formalism and the tree amplitude (1.46) are acceptable only for $D \leq 26$. At D = 26 the scalar state has zero norm and one can show that it decouples from the S-matrix. As a matter of fact the bosonic string theory we have just described is completely consistent only at D = 26. In the present covariant formalism, this appears only at the level of loops where unitarity breaks down except at this critical dimension. This can be seen by looking at the so-called one loop Pomeron diagram which has an unphysical cut except at $\mathcal{D} = 26$.

The particular role of $\mathcal{D} = 26$ is already apparent at the level of free strings as the above discussion shows. In general, for $\mathcal{D} \neq 26$, conditions (1.52) remove one string degree of freedom while at $\mathcal{D} = 26$, only $26-2 = 24$ degrees of freedom remain. A general argument which indicates that only $\mathcal{D} = 26$ is consistent will now be given following Kato and Ogawa[8]. This is based on the BRS transformations associated with the Virasoro algebra. Consider in general a Lie algebra G with generators T^a and structure constants C^{ab}_c

$$[T^a, T^b] = C^{ab}_c T^c \tag{1.62}$$

The BRS transformation is constructed from a ghost field η_a which is such that, under BRS,

$$\delta\eta_a = -\frac{\omega}{2} C_a^{bc} \eta_b \eta_c \tag{1.63}$$

η_a is anticommuting, ω is the infinitesimal anticommuting parameter. For a quantity ψ which transforms under G as $\delta_G \psi = \varepsilon_a T^a \psi$ the BRS transformation reads

$$\delta\psi = \omega \eta_a T^a \psi \tag{1.64}$$

It is just the group action with $\omega\eta_a$ as parameter. It is easy to check that (1.63) and (1.64) are such that the BRS transformation is nilpotent ($\delta^2 = 0$). This only follows from the Lie algebra (1.62) together with the Jacobi identity. If we introduce an additional ghost field ρ such that

$$\{\rho^a, \rho^b\} = 0 \qquad \{\rho^a, \eta_b\} = \delta_{ab}. \tag{1.65}$$

we can write the generator of (1.63), (1.64) as

$$Q = \sum_a T^a \eta_a - \frac{1}{2} \sum_{abc} C_a^{bc} \rho^a \eta_b \eta_c \tag{1.66}$$

Going back to the Virasoro algebra we notice two fundamental features: the Lie algebra (1.22) has a central term and there are an infinite number of generators. Let us proceed, anyhow. We now have two ghost fields η_n , ρ_m , n , m integers such that

$$\{\eta_n , \rho_m\} = \delta_{n,-m} \tag{1.67}$$

Following ref.8), we generalize (1.66) as

$$Q = \sum_{m} L_m \eta_{-m} + \sum_{n,m} m : \rho_n \eta_m \eta_{-n-m} : - \alpha_0 \eta_0 \tag{1.68}$$

The last term is the ambiguity arising from normal ordering. In order to compute Q^2 it is useful to rewrite

$$Q = \frac{1}{2\pi i} \oint dz/z \left[\frac{1}{2} : P^2(z) : \eta(z) + : \rho(z) \eta(z) z \eta'(z) : \right] \tag{1.69}$$

where

$$\eta(z) = \sum_{m} \eta_m z^{-m} \quad ; \quad \rho(z) = \sum_{m} \rho_m z^{-m} \tag{1.70}$$

The calculation proceeds very much like the above derivation of $[L_n, L_m]$. One needs the Wick contraction for ρ and η . Specify the vacuum to be such that

$$\rho_n|0\rangle = 0 \quad ; \quad \eta_n|0\rangle = 0 \quad ; \quad n > 0 \tag{1.71}$$

define the normal ordering of zero modes as

$$: \rho_0 \eta_0 : = - : \eta_0 \rho_0 : = \rho_0 \eta_0 - \eta_0 \rho_0 \tag{1.72}$$

Then

$$\begin{aligned} \rho(x)\eta(y) &= : \rho(x)\eta(y) : + \frac{1}{2} \frac{x+y}{x-y} \\ \eta(x)\rho(y) &= : \eta(x)\rho(y) : + \frac{1}{2} \frac{x+y}{x-y} \end{aligned} \tag{1.73}$$

In computing Q^2 from (1.69) by Wick's theorem one must again order the contours and one picks up only the pole at x=y. This gives

$$Q^2 = \frac{D-26}{48\pi i} \oint dz\, (z\eta)''' z\rho + \frac{1-\alpha_0}{2\pi i} \oint dz\, \eta\eta' \tag{1.74}$$

This vanishes only if D = 26; α_0 = 1. It follows from (1.71) that the on shell physical states(1.52) satisfy

$$Q\,|phys\rangle = 0 \tag{1.75}$$

and Q is really the relevant operator for the consistent string quantization. In the next section we discuss string quantization from the gauge theory view point. In this type of approach BRS invariance is known to play a crucial role. Hence it is natural that the present string theory only makes sense at D = 26, α_0 = 1.

2. Functional methods - Gauge approach.

From (1.27) and (1.37), one can rewrite the dual amplitude as a functional integral over surfaces; indeed, standard manipulations show that

$$\langle 0|V_{k_N}(z_N)\cdots V_{k_1}(z_1)|0\rangle \propto \int \mathcal{D}X\, e^{\frac{i}{\pi}\int d\sigma d\tau \left[\frac{1}{2}(\dot{X}^2 - X'^2) + J.X\right]} \tag{2.1}$$

$$J^\mu = \pi \sum_e k_e^\mu\, \delta(\sigma)\, \delta(\tau - \tau_e) \tag{2.2}$$

$$z_e = e^{i\tau_e} \tag{2.3}$$

The right member of (2.1) involves the self energies of the external sources which are infinite. Disregarding them is equivalent to the normal ordering prescription (1.27). The most important point about (2.1) is that it involves all D components of X^μ while we know that at D =26 only 24 degrees of freedom are physical. This decoupling comes in naturally if we functionally integrate over geometric surfaces in D dimensions and not over D functions of σ, τ . Indeed a geometrical surface being reparametrization invariant only depends on $\mathcal{D}-2$ functions. Reparametrization invariance is used as a gauge principle. One starts from an action invariant under reparametrization which is the two dimensional version of general coordinate invariance (2-$\mathcal{D}$ general relativity). Denote by g_{ab} ($a,b=\sigma,\tau$) the 2x2 metric tensor of the surface. The associated 2-D Einstein action $\sqrt{g}\,R$ is a total derivative and can be forgotten. Under reparametrization $\sigma \to \tilde{\sigma}(\sigma,\tau)$, $\tau \to \tilde{\tau}(\sigma,\tau)$; the X^μ transform as

$$\tilde{X}(\tilde{\sigma},\tilde{\tau}) = X(\sigma,\tau) \tag{2.4}$$

They behave as 2-D spin zero matter fields and the associated invariant action reads

$$S_{BDH} = \frac{1}{2\pi}\int d\sigma d\tau \sqrt{g}\; g^{ab}\partial_a X^\mu \partial_b X^\mu \tag{2.5}$$

$$g = -\det(g_{ab}) \qquad g^{ab} = (g^{-1})_{ab} \tag{2.6}$$

This action proposed by Brink, Di Vecchia and Howe[9] is also invariant under Weyl transformation

$$X \to X \qquad ; \qquad g_{ab} \to \Lambda(\sigma,\tau)\, g_{ab} \tag{2.7}$$

where Λ is an arbitrary function. The point of (2.5), as compared to the Nambu Goto action introduced earlier, is that one can regularize it without breaking reparametrization invariance. One can, for instance, introduce Pauli-Villars regulators Y^μ_ℓ, $\ell=1,..,N$, which are 2-D matter fields as X^μ but with masses $M_\ell \neq 0$. One adds to (2.5) the action

$$S_R = \frac{1}{2\pi}\int d\sigma d\tau \sqrt{g}\sum_\ell \left[g^{ab}\partial_a Y^\mu_\ell \partial_b Y^\mu_\ell - M^2_\ell\, Y^\mu_\ell Y^\mu_\ell\right] \tag{2.8}$$

with the understanding that M_ℓ will tend to infinity at the end. The Y^μ_ℓ are alternatively bosons and fermions so as to cancel the infinities. By adjusting the masses M_ℓ we can remove the divergences. S_R is still reparametrization invariant but the mass term breaks the Weyl invariance

(2.7) and an anomalous term appears in general. In the functional integral

$$\int \mathcal{D}g\,\mathcal{D}X\,\delta(F_1)\delta(F_2)\,\Delta_{FP}\,e^{iS_{BDH}} \tag{2.9}$$

one only introduces gauge fixing terms for the reparametrization invariance. The choice of F_1 and F_2 specifies the parametrization of the surface. The basic point here is that all two-dimensional manifolds are conformally flat and one can choose orthonormal coordinates on the surface. However, one considers g_{ab} and X^μ as independent variables. Hence we have two independent objects which behave as metric tensors, i.e. g_{ab} and

$$G_{ab} = \partial_a X^\mu \partial_b X^\mu \tag{2.10}$$

One can follow two paths from this point:

a) Following Polyakov[10], we can choose the gauge where

$$F_1 = g_{\tau\tau} + g_{\sigma\sigma} \quad ; \quad F_2 = g_{\sigma\tau}$$

$$g_{ab} = e^{\varphi}\begin{pmatrix} 1 & 0 \\ 0 & -1 \end{pmatrix} \tag{2.11}$$

With this choice, it is easy to integrate X^μ out, since the action (2.5) is quadratic in X^μ. One can regularize using (2.8). A counter term is needed:

$$S_c = \frac{\mu}{\pi}\int d\sigma\, d\tau \sqrt{g} \tag{2.12}$$

where μ behaves as $\sum_e \varepsilon_e M_e^2 \ln(M_e)$; $\varepsilon_e = \pm 1$, for $M_e \to \infty$. One further determines the Faddeev Popov determinant. The final result reads[10], for the partition function,

$$\int \mathcal{D}X\,\mathcal{D}g\, e^{i(S_{BDH}+S_c)}\,\delta(g_{\tau\tau}+g_{\sigma\sigma})\,\delta(g_{\sigma\tau})\,\Delta_{FP} = \int \mathcal{D}\varphi\, e^{iS_L} \tag{2.13}$$

$$S_L = \frac{26-d}{48\pi}\int d\sigma\, d\tau \left[\frac{1}{2}(\dot{\varphi}^2 - \varphi'^2) - e^{\varphi}\right] \tag{2.14}$$

S_L is the so-called Liouville action. We put it in a nice looking form by choosing the coefficient of e^{φ} equal to 1 inside the bracket. Originally it was a divergent quantity but we can give it any value by shifting φ by a constant. Obviously, at $\mathcal{D} = 26$, S_L disappears and it seems that the anomaly becomes irrelevant, thus supporting the direct operator treatment described in sect. 1. This discussion is still rather formal however. One integrates over X^μ with fixed φ thus ignoring the quantum effects on this latter field. Moreover possible difficulties in choosing the gauge (2.10) are ignored. We shall have more to say on this in section 3.

b) Following A. Neveu and myself[11], one can choose the gauge fixing functions F_1 and F_2 to be only functions of X^μ. Then it is possible to integrate over g_{ab} point by point since the action (2.5) does not involve any of its derivatives. We shall proceed very formally so that $\mathcal{D} \neq 26$ will not be needed explicitly. Since no derivative of g appears in the action, the functional integral over g can be rewritten as

$$\prod_{\sigma,\tau} \int d[g_{ab}]\; e^{\frac{i}{\pi}\left[d\sigma d\tau \sqrt{g}\, g^{cd} G_{cd}(\sigma,\tau) + \mu \sqrt{g}\, d\sigma d\tau\right]}$$

$$\equiv \prod_{\sigma,\tau} F\left(G_{\ell m}(\sigma,\tau)\right) \tag{2.15}$$

Consider one particular factor of this formal infinite product. It is given by the ordinary integral

$$F(X,Y,Z) = \int dx\, dy\, dz\; \exp\left\{\frac{i}{\sqrt{xy-z^2}}\left[Xy + Yx - 2zZ\right]\right\} \tag{2.16}$$

where we have let

$$X = \frac{d\sigma d\tau}{2\pi} G_{11}\;;\quad Y = \frac{d\sigma d\tau}{2\pi} G_{22}\;;\quad Z = \frac{d\sigma d\tau}{2\pi} G_{12} = \frac{d\sigma d\tau}{2\pi} G_{21}$$

$$x = g_{11}\;;\quad y = g_{22}\;;\quad z = g_{12} = g_{21} \tag{2.17}$$

We next separate the integral over the determinant of g by inserting into (2.16) the expression

$$1 = \int d\rho\; \delta(\rho - xy + z^2) \tag{2.18}$$

which can also be cast into the form

$$1 = \int d\rho\, dk\; e^{ik(\rho - xy + z^2)} \tag{2.19}$$

Integrating over x, y, z, for fixed ρ and k, one obtains

$$F(X,Y,Z) = \int d\rho \int \frac{dk}{k^{3/2}}\; e^{i\left(k\rho + \frac{D}{k\rho}\right)} \tag{2.20}$$

where

$$D = Z^2 - XY. \tag{2.21}$$

As is usual for path integral with real time, the remaining integrations are only semi-convergent. The integral over k will be replaced by the convergent integral

$$\int_0^\infty \frac{dk}{k^{3/2}}\; e^{-\left(k\rho + \frac{\lambda}{k\rho}\right)} \qquad \lambda = -D \tag{2.22}$$

which can be obtained by deforming the contour of integration. From the formula[12)]

$$\int_0^\infty \frac{dx}{\sqrt{x}}\; e^{-\left(x + \frac{\lambda}{x}\right)} = \sqrt{\pi}\; e^{-2\sqrt{\lambda}} \tag{2.23}$$

we deduce

$$\int_0^\infty \frac{dk}{k^{3/2}}\; e^{-\left(k\rho - \frac{D}{k\rho}\right)} = i\sqrt{\frac{\pi\rho}{D}}\; e^{2i\sqrt{D}} \tag{2.24}$$

Recalling (2.17), we rewrite this as

$$F(G) \propto \frac{e^{\frac{i}{\pi}\sqrt{G}\, d\sigma\, d\tau}}{\sqrt{G}\; d\sigma\, d\tau}$$

$$G = -\det(G_{ab}) \tag{2.25}$$

and equation (2.9) becomes

$$\int \mathcal{D}X \frac{\delta(F_1)\delta(F_2)}{\prod\limits_{\sigma,\tau} d\sigma d\tau \sqrt{G}} \Delta_{FP}\, e^{\frac{i}{\pi}\int d\sigma d\tau \sqrt{G}} \tag{2.26}$$

We have found back the original Nambu-Goto action. The measure is however non trivial. It precisely coincides with the one proposed by Sakita and myself[13] from the study of the light-cone gauge in functional integral. Indeed the additional factor $\prod\limits_{\sigma,\tau}(d\sigma d\tau \sqrt{G})^{-1}$ is precisely needed to integrate out two components in this gauge in order to remain with only physical degrees of freedom. We now summarize this point briefly. Define the light-cone notation as

$$X^{\pm} = \frac{1}{\sqrt{2}}(X^0 \mp X^1) \; ; \; \underset{\sim}{X} = X^{\mu} \quad \mu = 2,\dots,\mathcal{D}-1 \tag{2.27}$$

Choose the gauge fixing conditions

$$F_1 \equiv X^+ - f(\sigma,\tau) \; ; \; F_2 = (\dot{X} - X')^2 \tag{2.28}$$

f is a function which we shall suitably determine below. Consider the generating functional

$$Z(J) \equiv \int \mathcal{D}X \frac{\delta(X^+ - f)\delta((\dot{X} - X')^2)}{\prod d\sigma d\tau \sqrt{G}}\, e^{\frac{i}{\pi}\int d\sigma d\tau [\sqrt{G} + J\cdot X]} \tag{2.29}$$

The integration over X^+ is of course trivial. In order to also integrate out the X^- component, we choose f in such a way that after substituting $X^+ = f$ into the exponent, all dependence in X^- disappears from the action including the source term. This is possible since, due to the F_2 gauge condition, the Nambu-Goto action is now linearized:

$$\frac{1}{\pi}\int d\sigma d\tau \sqrt{G} = \frac{1}{2\pi}\int d\sigma d\tau \left[(\dot{X})^2 - (X')^2\right] \tag{2.30}$$

Introducing the free field Green function $N(\vec{\xi},\vec{\xi}')$ such that

$$\left(\frac{\partial^2}{\partial\sigma^2}-\frac{\partial^2}{\partial\tau^2}\right)N(\vec{\xi},\vec{\xi}')=-i\delta_2(\vec{\xi}-\vec{\xi}') \tag{2.31}$$

$$\partial_{n_\xi}N(\vec{\xi},\vec{\xi}')\Big|_{\text{boundaries}}=\rho(\vec{\xi}') \quad ; \quad \vec{\xi}\equiv(\sigma,\tau) \tag{2.32}$$

we choose

$$f(\vec{\xi})=-i\int d_2\vec{\xi}'N(\vec{\xi},\vec{\xi}')\,J^+(\vec{\xi}') \tag{2.33}$$

and the integration over X^- reduces to

$$\int\mathcal{D}X^-\,\frac{\delta((\dot{X}-X')^2)\,\Delta_{FP}}{\prod\limits_{\sigma,\tau}d\sigma d\tau\sqrt{G}}=1 \tag{2.34}$$

This is proven in ref.13). One is left with

$$Z(J)=\exp\left\{-\frac{1}{\pi}\int d_2\vec{\xi}\,d_2\vec{\xi}'\,J^-(\xi)J^+(\xi')\,N(\vec{\xi},\vec{\xi}')\right\}$$
$$\int\mathcal{D}\underset{\sim}{X}\,\exp\left\{\frac{i}{\pi}\int d_2\xi\left[(\dot{\underset{\sim}{X}}^2-\underset{\sim}{X}'^2)\frac{1}{2}+\underset{\sim}{J}\cdot\underset{\sim}{X}\right]\right\} \tag{2.35}$$

and the functional integral is only over the $\mathcal{D}$ -2 physical degrees of freedom. The gauge choice $X^+=f$ may seem peculiar since it explicitly depends upon the external sources J^+. Next we briefly show that, on the contrary, this dependence is precisely needed to describe string scattering in the frame where X^+ is taken as time. With this motivation we come back for a moment to the free string formalism of sect. 1 but where one explicitly eliminates the $X^\pm$ components following ref. 14). In the covariant operator formalism one works in a Fock space larger than the physical one which is defined by $(L_0-1)|phys\rangle=L_m|phys\rangle=0$. In the GGRT formalism[14)] one imposes as operator identities

$$L_0=1 \qquad L_m=0 \qquad m\neq 0 \tag{2.36}$$

From (1.26) one sees that this corresponds to

$$\frac{1}{2}:(\dot{X}\pm X')^2:-1=0 \tag{2.37}$$

Moreover one eliminates the a_m^+ modes by letting

$$X^+ = p^+\tau \tag{2.38}$$

These last two conditions determine X^- as

$$X^- = q^- + p^-\tau + i\sum_{m\neq 0}\frac{1}{m}\tilde{L}_m\, e^{-im\tau}\cos m\sigma \tag{2.39}$$

$$\tilde{L}_m = \frac{1}{2}\sum_{n} :\underset{\sim}{a}_n\, \underset{\sim}{a}_{m-n}: \tag{2.40}$$

The only quantum degrees of freedom are the transverse modes $\underset{\sim}{X}$ and q^-, p^+. p^- is determined as

$$p^- = \frac{1}{p^+}\left(\underset{\sim}{L}_0 - 1\right) \tag{2.41}$$

Going back to the interacting string theory we remark, following Mandelstam[15] that (2.38) does not allow to adjust the surface representing different strings since the definition of τ depends on p^+. Hence in the above free string formalism we perform a rescaling of σ and τ by p^+ in such a way that equation (2.38) becomes simply

$$X^+ = \tau \tag{2.42}$$

and the new σ variable now goes from 0 to $\pi p^+ = \alpha$. With this parameters string interactions become very simple. Strings break or combine as shown, typically, on figure (2)

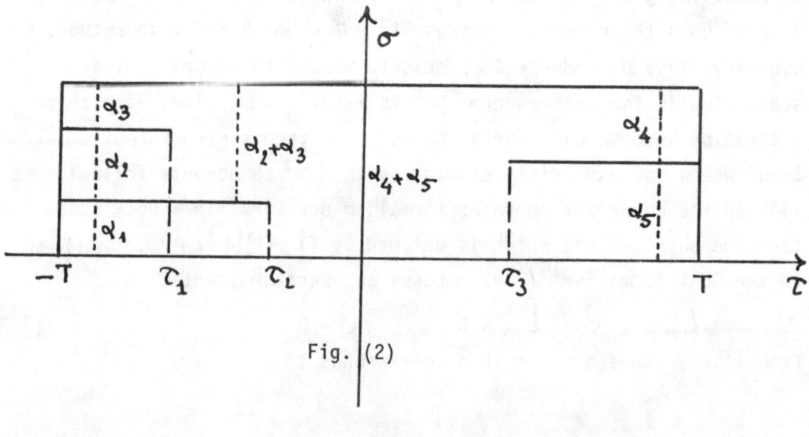

Fig. (2)

Since $\alpha = \pi p^{+}$ is a conserved quantity, the domains of variation of the various σ automatically adjust at each interaction in such a way that the total range of σ variation remains the same. The scattering process associated with fig. (2) is obtained from (2.35) by putting as sources momenta distributions at -T and +T which represent the initial and final string states. For each incoming (time -T) or outgoing (time T) strings we can use the free field formalism if T is so large that all interactions take place between -T and +T. Equation (2.30) shows that the corresponding p^{+} density is

$$p^{+} = \frac{1}{\pi}\frac{\partial X^{+}}{\partial \tau} = \frac{1}{\pi} \tag{2.43}$$

We set, accordingly $J^{+} = 1$ at $\tau = \pm T$ in (2.35). At this point it becomes more convenient to work in the Euclidean formulation where (2.31) becomes

$$\Delta N(\vec{\xi}, \vec{\xi}') = -\delta_2(\vec{\xi} - \vec{\xi}') \tag{2.44}$$

so that for $\vec{\xi} \sim \vec{\xi}'$

$$N(\vec{\xi}, \vec{\xi}') \sim \frac{1}{4\pi} \ln\left[(\vec{\xi} - \vec{\xi}')^2\right] \tag{2.45}$$

With this information, it is easy to show that, according to (2.33),

$$\partial_\tau f = i \qquad \text{at} \quad \tau = \pm T \tag{2.46}$$

Moreover, (2.31) shows that f has a vanishing Laplacian inside the domain $-T \lesssim \tau \lesssim T$ since J^{+} is only at the boundary. This combined with (2.32) shows that $f \equiv i\tau$ everywhere and in the domain of fig (2), the F_1 gauge condition is precisely equivalent to the light-cone condition (2.42) after Wick's rotation. By conformal transformation we can map fig (2) to different σ, τ domains where X^{+} will not be equal to $i\tau$ in general. This leads back to the general form (2.28).

Going back to fig (2) we remark that the factor in front of (2.35) can be written as

$$\exp\left[-\frac{1}{\pi}\int d\xi\, J^{-}(\xi) f(\xi)\right] = \exp\left\{-2TE\right\}$$

$$E = \sum_{\substack{e \\ \text{initial}}} p_e^{-} = \sum_{\substack{e \\ \text{final}}} p_e^{-} \tag{2.47}$$

This equation follows immediately from the fact that $f(\xi)$ is a constant on the boundary $\tau = \pm T$ so that only the total J^- appear. Finally, the functional integral (2.35) computed in the domain of fig (2) represents the transition probability between the initial and the final state with strings joining at times τ_1, τ_2 and later breaking at τ_3. In the light-cone formalism, p^- plays the role of the energy. Hence, the factor (2.47) is precisely such that one gets a finite limit for $T \to \infty$. This is the standard way of deriving the S-matrix from the finite time evolution operator. One has to further integrate over the interaction times τ_1, τ_2, τ_3. In the limit $T \to \infty$, time translation invariance is recovered and only interaction time differences are relevant. This leaves us with 2 = 5-3 variables.

Finally one can map fig (2) onto fig (3) by a transformation which is conformal inside the domain

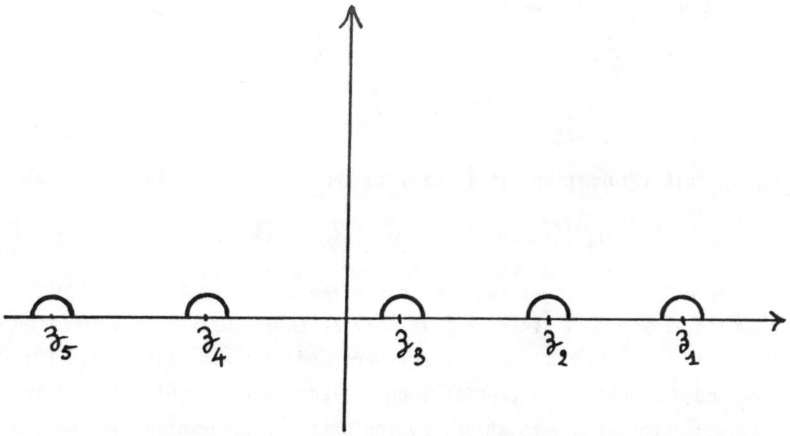

Fig. (3)

The new domain is the upperhalf plane minus the inside of the half circles which shrink to points $z_1, \cdots, z_5$ as $T \to \infty$. We therefore find back the procedure discussed in sect. 1 since the momentum distributions are located on these small circles. Integrating over the time differences becomes equivalent to the integrating over the N-3 Koba-Nielsen variables of sect. 1.

3. Beyond free two-dimensional field theories.

As we just saw, conformal transformations play a crucial role. Using complex variables z and z^* a general conformal transformation reads

$$z' = F(z) \tag{3.1}$$

where F is a function of a single variable. In general a quantity $\mathcal{O}(z,z^*)$ is called conformally covariant[16] if it transforms according to

$$\mathcal{O}'(z',z'^*) = \left(\frac{dF}{dz}\right)^{-\delta}\left(\frac{dF^*}{dz^*}\right)^{-\bar{\delta}} \mathcal{O}(z,z^*) \tag{3.2}$$

where δ and $\bar{\delta}$ are parameters depending on the quantity considered which are called conformal weights. This notion was rediscovered recently[17] and the covariant fields were called primary. If we separate the real and imaginary parts according to

$$z = x_1 + i x_2 \tag{3.3}$$

the differential transforms as

$$dx'_\ell = \mu R_{\ell m}(\theta)\, dx_m \tag{3.4}$$

where $R_{\ell m}$ is the rotation matrix with angle θ and where μ is the dilatation factor. These two quantities are given by

$$\frac{\partial x'_1}{\partial x_1} = \frac{\partial x'_2}{\partial x_2} = \mu \cos\theta \qquad \frac{\partial x'_1}{\partial x_2} = -\frac{\partial x'_2}{\partial x_1} = \mu \sin\theta \tag{3.5}$$

Formula (3.2) becomes

$$\mathcal{O}'(x'_1, x'_2) = \mu^{-d} e^{-iJ\theta} \mathcal{O}(x_1,x_2)$$
$$d = \delta + \bar{\delta} \quad ; \quad J = \delta - \bar{\delta} \tag{3.6}$$

Since μ and θ are the local dilatation and rotation parameters, d is the dimension and J is the spin of the quantity considered.

In a conformally invariant field theory the improved energy momentum tensor is traceless symmetric and conserved. For two-dimensional field theories this has strong consequences. If we have in real σ, τ space

$$\dot{T}^0_0 + T'^1_0 = \dot{T}^0_1 + T'^1_1 = 0$$
$$T^0_0 + T^1_1 = 0 \quad ; \quad T^1_0 = T^0_1 \tag{3.7}$$

we conclude that

$$\left(\frac{\partial}{\partial\tau} \mp \frac{\partial}{\partial\sigma}\right)\left(T_0^0 \pm T_0^1\right) = 0 \tag{3.8}$$

Imposing periodicity in σ with period 2π leads to

$$2\pi\left(T_0^0 + T_0^1\right) = \sum_m L_m z^{-m} \quad ; \quad z = e^{i(\tau+\sigma)}$$

$$2\pi\left(T_0^0 - T_0^1\right) = \sum_m \bar{L}_m \bar{z}^{-m} \quad ; \quad \bar{z} = e^{i(\tau-\sigma)} \tag{3.9}$$

In the same way as in the free case of section 1, the L_m and $\bar{L}_m$ are the Virasoro generators. In quantum theory, a covariant operator transforming according to (3.2) satisfies

$$[L_m, \theta] = z^m\left[z\frac{\partial}{\partial z} + (m+1)\delta\right]\theta$$

$$[\bar{L}_m, \theta] = \bar{z}^m\left[\bar{z}\frac{\partial}{\partial \bar{z}} + (m+1)\bar{\delta}\right]\theta \tag{3.10}$$

In conformally invariant quantum field theories there exist several such covariant operators θ_α with weights δ_α and $\bar{\delta}_\alpha$ which depend upon the model considered. At the classical level the above structure is realized by Poisson brackets. The weights are in general modified by quantum effects. An example is the operator $V_k(z)$ of section 1 (formula (1.27)) which has $\delta = \bar{\delta} = 0$ at the classical level and $\delta = k^2/2$, $\bar{\delta} = 0$ in the quantum theory.

One must consider all covariant operators such that, together with all their derivatives, they form a family $\mathcal{A}^{(M)}$ which is closed by short distance expansion

$$\mathcal{A}^{(M)}(z)\,\mathcal{A}^{(N)}(z') \underset{z\to z'}{\sim} \sum_L C_L^{MN}(z-z')\,\mathcal{A}^{(L)}\left(\frac{z+z'}{2}\right) \tag{3.11}$$

In this expansion, the singularity structure of the c-number coefficients C_L^{MN} at short distance is determined by the difference between the dimension of the operator $\mathcal{A}^{(L)}$ and the sum of the dimensions of $\mathcal{A}^{(M)}$ and $\mathcal{A}^{(N)}$.

As is well known[16)17)] the derivative of a covariant operator is not covariant in general. This is why the expansion (3.11) does not include covariant operators only. An important exception is the case of an operator of vanishing weight. Its derivative with respect to z (or z^*) is a covariant operator with weights $\delta=1$, $\bar{\delta}=0$ (or $\delta=0$, $\bar{\delta}=1$). Conversely, let us assume there exists an operator I(z) with weights $\delta=1$, $\bar{\delta}=0$. It is obvious that

$$\left[L_m, \oint dz\, I(z)\right] = 0 \tag{3.12}$$

At this point it is useful to recall our derivation of the Virasoro algebra for the free case of section 1. Basically we used the short distance expansion of the product $\mathcal{U}(z)\mathcal{U}(z')$. It is clear that the same computation can be repeated, once the short distance expansion is given, quite generally. Deforming the contour will always peak up the pole term which therefore gives the commutator of the operators considered. Classically $\mathcal{U}$ and $\bar{\mathcal{U}}$ have weights $\delta=2$, $\bar{\delta}=0$ and $\delta=0$, $\bar{\delta}=2$ respectively. Let us now show that if this is also true at the quantum level the mode coefficients of $\mathcal{U}$ and $\bar{\mathcal{U}}$ satisfy the Virasoro algebra. Consider $\mathcal{U}$ for instance. Its short distance expansion reads (A and B are constant C-numbers)

$$\mathcal{U}(z)\,\mathcal{U}(z') \underset{z\to z'}{\sim} A\left[\frac{zz'}{(z-z')^2}\right]^2 + B\,\frac{zz'}{(z-z')^2}\,\mathcal{U}\left(\frac{z+z'}{2}\right)+\cdots \tag{3.13}$$

Following the same path as in the free case, one starts from

$$L_n L_m = \frac{1}{(2\pi i)^2}\oint \frac{dx}{x}\frac{dy}{y}\, x^n y^m\, \mathcal{U}(x)\,\mathcal{U}(y)$$
$$L_m L_n = \frac{1}{(2\pi i)^2}\oint \frac{dx}{x}\frac{dy}{y}\, x^n y^m\, \mathcal{U}(y)\,\mathcal{U}(x) \tag{3.14}$$

From the positivity of the energy it follows that the product $\mathcal{U}(x)\,\mathcal{U}(y)$ is analytic in y/x for $|x| > |y|$ and again we have to choose $|x| > |y|$ and $|y| > |x|$ in the first and in the last equations (3.14) respectively. The commutator is again given by the pole at $x = y$ which is solely specified from (3.13). One obtains

$$[L_n, L_m] = \frac{B}{2}(n-m)\, L_{n+m} + \frac{A}{12}\, \delta_{n,-m}\, (n^3-n) \qquad (3.15)$$

If $B \neq 2$, we may redefine L_n so as to obtain the usual form. The crucial point is that (3.13) holds, namely $\mathcal{V}$ has its naive dimension and there are no logarithmic corrections to the short distance expansion.

Conformally invariant field theories in two dimensions have two different physical applications. First we can consider a statistical system without boundary by taking the variables x_1 and x_2 of formula (3.3) as coordinates. This may experimentally describe a thin film. With this choice of coordinates, d and J are the physical dimension and spin. $L_0 + \bar{L}_0$ is the generator of rescaling of x_1 and x_2 and a statistical system at a critical point becomes scale invariant. It has been argued by Polyakov that critical systems are in fact invariant under the full conformal group and hence are associated with conformally invariant field theories. It is easy to see that L_{-1} and $\bar{L}_{-1}$ generate x_1 and x_2 translations. The ground state of the system is thus annihilated by these operators. It must also be scale invariant. This is sufficient to determine the two point function of any covariant operator. One finds, assuming for instance that $\delta = \bar{\delta}$,

$$\langle 0 |\, \mathcal{O}(z, z^*)\, \mathcal{O}(z', z'^*)\, | 0 \rangle \propto \left(|z - z'|^2 \right)^{-2\delta} \qquad (3.16)$$

and it follows that, in general, d and J are critical exponents. In the scaling limit of discrete statistical systems, only operators of non negative dimensions survive. In the continuous theory all physical operators must have non negative dimension. Hence the short distance expansion (3.11) must close only with this type of operators.

If an operator of conformal weight 1 exists, formula (3.12) shows that its integral is conformally invariant. It can be added to the action with an arbitrary coefficient without destroying the conformal invariance. As a result there exists a family of critical models and one has a critical line instead of a critical plot in the space of the couplings. The appearance of these so-called marginal operators is frequent in two dimensional models.

The other physical application is in dual theories. There also conformal invariance is crucial since it allows to remove the ghost states due to the Lorentz metric. We have seen in section 1 an example of the ghost-killing mechanism in the covariant approach where the key point was to choose $k^2=2$ in such a way that V_k has $\delta = 1$. One can summarize the above discussion by saying that if $k^2=2$, $\oint \frac{dz}{z} V_k(z)$ becomes a marginal operator which commutes with $L_m - L_0$ up to boundary terms. In the free Bose case of section 1 this means that we have a tachyon. Suppose that we have other conformally covariant fields beside X^μ. Then in the dual amplitude (1.35) we can replace $V_k(z)$ by

$$\mathcal{V}_k^\alpha(z) = V_k(z)\, \mathcal{O}_\alpha(z) \tag{3.17}$$

where $\mathcal{O}_\alpha(z)$ is one of the covariant operators of the additional field theory. The complete Virasoro generators are, now

$$J_m = \mathcal{L}_m + L_m \quad ; \quad \mathcal{L}_m = \frac{1}{2}\sum_n :a_n^\mu a_{m-n}^\mu: \tag{3.18}$$

where L_m represent the generators of the additional fields. The ghost elimination only requires that $\mathcal{V}$ has dimension 1 with respect to J_m, namely

$$\frac{k^2}{2} = 1 - \delta_\alpha \tag{3.19}$$

δ_α is the weight of $\mathcal{O}_\alpha$. For each δ_α, $\mathcal{O}_\alpha$ we will have a whole family of Regge trajectories, formula (3.19) describing the emission of its lightest members. If we want to avoid tachyon we will like to have

$$\delta_\alpha \geq 1 \tag{3.20}$$

and this condition selects the conformally invariant field theories for which all covariant fields have weights larger than one. From the view point of critical systems, this condition is unusual since it means that the corresponding two point function has a Fourier transform which has at most a logarithmic singularity at zero momentum. We shall come back to this later.

In section 1 we argued, by looking at the BRS transformation, that the model based on free bosonic fields is consistent only at $\mathcal{D} = 26$.

Obviously equation (1.68) is now replaced by

$$\mathcal{Q} = \sum_{m} J_m \eta_{-m} + \sum_{m n} m : \rho_m \eta_n \eta_{-m-n} : - \eta_0 \alpha_0 \tag{3.21}$$

By a calculation exactly similar to section 1, one now finds

$$C + \mathcal{D} = 26 \tag{3.22}$$

$$\alpha_0 = 1 \tag{3.23}$$

where C is the central charge of the additional field theory. If $C > 1$ this allows us to go beyond $\mathcal{D} = 26$. A word of caution is needed here. We assume that the additional fields do not themselves bring in new ghosts to be removed. For instance, adding Neveu-Schwarz fields does precisely this and one must then use a larger algebra (the superconformal one) to remove the ghosts. We stick to bosonic string in order to avoid technical complications as much as possible.

At this point is is useful to recall some general properties of representations of the Virasoro algebra

$$[L_n, L_m] = (n-m) L_{n+m} + \frac{C}{12}(n^3 - n)\delta_{n,-m} \tag{3.24}$$

From the group theory viewpoint, it can be regarded as being in a Weyl-Cartan basis, L_0 being the only operator of the commuting subalgebra, and L_n with $n \neq 0$ being step operators. Hence a highest weight vector will be such that

$$L_0 |\varepsilon, 0\rangle = \varepsilon |\varepsilon, 0\rangle \; ; \quad L_n |\varepsilon, 0\rangle = 0, \; n > 0 \tag{3.25}$$

An irreducible representation is characterized by the values of ε and C. The corresponding vector space which is called a Verma module, is spanned by all vectors of the form

$$|\varepsilon, \{m\}\rangle = \prod_{k>0} (L_{-k})^{m_k} |\varepsilon, 0\rangle \tag{3.26}$$

where m_k are arbitrary positive integers. They are eigenstates of L_0

$$L_0 |\varepsilon, \{m\}\rangle = (\varepsilon + \sum_{k>0} k m_k) |\varepsilon, \{m\}\rangle \tag{3.27}$$

All eigenvectors with the same eigenvalues are said to belong to the same level N. Kac[19] has considered the matrix of all inner products in a given module which is entirely determined from the Virasoro algebra together with the hermiticity condition

$$L_{-m} = L_m^+ \tag{3.28}$$

Hence it is purely algebraic and only depends upon the values of $\mathcal{E}$ and C. It obviously factorizes into products of finite matrices at each level. Kac has obtained a closed formula for each finite determinant. Define the quantity

$$\mathcal{E}(p,q) \equiv \frac{1}{48}\Big[(13-C)(p^2+q^2) - 24\,pq - 2(1-C) + (p^2-q^2)\sqrt{(C-1)(C-25)}\Big] \tag{3.29}$$

where p and q are arbitrary integers. If we consider a highest weight representation with $\mathcal{E} = \mathcal{E}(p,q)$ for some given p,q both larger than zero, the Kac determinant vanishes at the level N= pq. This vanishing shows that the metric of the Verma module need not be positive definite. The unitarity of the representation is thus in question. It is easy to show that the negative values of the highest weights are all excluded. For positive values we note that, for $C > 1$, $\mathcal{E}(p,q)$ is always negative for $p > 1$, $q > 1$, i.e. when it corresponds to a zero of a Kac determinant. Hence, in this region, the Kac determinants never change signs for positive $\mathcal{E}$ and one can show by explicit construction[20] that there exists a unitary representation for all $\mathcal{E} > 0$, $C > 1$. For $C < 1$, on the contrary, there are Kac zeroes for positive $\mathcal{E}$ and the positivity of the metric is not assured. It has been shown that unitary representations only exist for[21]

$$C = 1 - 6/(r+1)r \quad , \quad r \geq 3$$

$$\mathcal{E} = \mathcal{E}(p,q) \qquad 1 \leq p \leq q < r \tag{3.30}$$

where r,p,q are integers. Hence the allowed values of $\mathcal{E}$ precisely coincide with zeroes of Kac determinants.

Going back to conformally invariant field theories we recall that, given a covariant operator $\mathcal{O}$ with weight δ, it is easy to see that the state

$$\lim_{z\to 0} O_\alpha(z)|0\rangle \tag{3.31}$$

is a highest weight vector with $\varepsilon = \delta_\alpha$. The spectrum of highest weights coincides with the set of conformal weights which, in general, involves more than one value. The representation of the Virasoro algebra is thus reducible since the Hilbert space is the sum of the corresponding Verma modules. The covariant operators are intertweening operators between the different irreducible representations. For arbitrary ε and C one can construct an infinite family of covariant operators with weights given by formula (3.29) for all p and q positive or negative[22] as a natural byproduct of the exact quantum solution of quantum Liouville theory. These operators are not all physical since formula (3.29) is not always positive. This shows nevertheless that the set of critical dimensions must coincide with Kac formula in general. From the above discussion, it is clear that one has to distinguish three regions for the possible values of C.

a) The region $C < 1$.

As we already pointed out, the Kac zeroes occur for positive highest weights in this case. A systematic discussion has been given[17] which uses this fact, and shows the existence of special values of C

$$C = 1 - 6\,\frac{(r+s)^2}{rs} \tag{3.32}$$

where r and s are integers of opposite signs. The virtue of this formula is that then

$$(C-1)(C-25) = 36\,\frac{(r^2-s^2)^2}{r^2 s^2} \tag{3.33}$$

is the ratio of squares of integers. In view of formula (3.30), unitarity is satisfied only for r+s = -1. This subseries remarkably reproduces a whole set of standard critical models[21]. In particular for r=3 and 5, one recovers the Ising (or 2-state Potts) model and the 3-state Potts model. A simple calculation shows that if we introduce

$$\sqrt{Q} = 2\cos\left[\frac{\pi}{12}(C-1)\left(1+\sqrt{\frac{C-25}{C-1}}\right)\right] \tag{3.34}$$

we obtain the correct number of spin components, i.e. Q=2,3 for the Ising and the three-state Potts model respectively. The Q-state Potts model can be defined for continuous values of Q if we transform it into the random cluster model. For $C < 1$ there exist various equivalent critical models. In particular the Q-state critical Potts model is equivalent to a Coulomb gas model. For $C < 1$ we have $Q < 4$ and one is in a Coulomb phase. The point Q=4 corresponds to a point of transition of Kosterlitz-Thouless. Above Q=4 one enters into the plasma phase and the transition becomes first order. We shall come back to this below.

b) The region $C > 25$.

This region has some similarities with the region $C < 1$ since in both cases the square root of formula (3.29) is real. A different approach is needed, however, since now the Kac determinants do not vanish for positive ε. The region $C > 1$ is naturally covered by the quantum Liouville field theory since its central charge is given by[23]

$$C = 1 + 3/\hbar \tag{3.35}$$

where $\hbar$ is the Planck constant. The region $C > 25$ corresponds to $\hbar < 1/8$, i.e. to the weak coupling regime of Liouville theory which is connected to the semi-classical limit $\hbar \sim 0$. In the exact quantum solution[23], special values of C were again found

$$C = 1 + 6(N+1)^2/N \tag{3.36}$$

They can be put under the form of equation (3.32) continued to r=N and s=1. The spectrum of weights is again given by formula (3.29) with

$$\varepsilon = \varepsilon(1, 2m - N) \;, \quad 0 \leq m \leq \nu \qquad N = 2\nu + 1$$

$$\varepsilon = \varepsilon(1, 2m - 1 - N) \,, \quad 1 \leq m \leq \nu \qquad N = 2\nu \tag{3.37}$$

It is easily checked that all these values are larger than 1, and condition (3.20) is satisfied. The associated string model has no tachyon. However formula (3.22) shows that $\mathcal{D} < 1$ (!) so that one has not gained much from this viewpoint. The existence of conformally invariant field theories for these values of C does however suggest that there are new critical models. These are models of a new type since all the known critical models have $C < 1$, and a spectrum of conformal weights between 0 and 1. The unusual feature of the new models is, as we already pointed out, that the two-point functions are at most logarithmically divergent. The experimental feature of the transition is thus rather different from the standard ones.

c) The region $1 < C < 25$.

In this case formula (3.29) gives complex values except when $p=\pm q$. The choice $p=q$ is unacceptable except for $p=q=1$, since it leads to negative $\mathcal{E}$. For three special values of C

$$C = 7\ ,\ 13\ ,\ 19 \tag{3.38}$$

local fields have been constructed[24] such that the spectrum of weights is given by formula (3.29) for

$$p = -q = 1,2,\ldots \tag{3.39}$$

Condition (3.20) is again satisfied and the associated string theory has no tachyon. Condition (3.22) leads to

$$\mathcal{D} = 19\ ,\ 13\ ,\ 7 \tag{3.40}$$

and there exist new string models for these values of $\mathcal{D}$. From the viewpoint of statistical models, one therefore predicts isolated points of second order phase transition for the above values of C. This may be a bit surprising since for $C > 1$ one is outside of the Coulomb phase. One can directly see, however, that these points must enjoy special properties. Formula (3.34), when continued for $1 < C < 25$, leads to Q complex in general. For the values (3.38) one obtains

$$\sqrt{Q} = 2\cos\left[\frac{\pi}{2} + i\pi\frac{\sqrt{3}}{2}\right]$$

$$\sqrt{Q} = 2\cos\left[\pi + i\pi\right]$$

$$\sqrt{Q} = 2\cos\left[\frac{3\pi}{2} + i\pi\frac{\sqrt{3}}{2}\right] \tag{3.41}$$

and one can verify that the three special values are the only ones for which Q is real even though the argument of the cosine is complex.

Finally we note that, by combining formulae (3.22) and (3.35), one obtains

$$\frac{1}{\hbar} = \frac{25-\mathcal{D}}{3} \tag{3.42}$$

while formula (2.14) which follows Polyakov's computation[10], leads, instead, to

$$\frac{1}{\hbar_{\rho}} = \frac{26 - \mathcal{D}}{3} \tag{3.43}$$

As we pointed out in section 2, this last expression does not take into account the quantum fluctuation of the Liouville field. This explains the discrepancy.

The conclusion of this paragraph is that the study of conformally invariant field theories from the double viewpoint of string theories and critical systems has unravelled an interesting structure. The string theories discussed here are not based on free field theories in two dimensions and, hence, the dual amplitudes are difficult to determine. The common feature of all the new conformally invariant field theories discussed here is the appearance of operators of dimension 1. For the associated string theories, it corresponds to the existence of a massless string state. This fact should play a key role in the complete understanding of these models. On the other hand, we must say that, at the present time, the relevance of these new models to particle physics is not yet clear. The supersymmetric version of the present discussion has been worked out in all details[25)].

REFERENCES

1 J.-L. Gervais, B. Sakita, Nucl.Phys. B34 (1971) 832.

2 V. Alessandrini, D. Amati, M. Le Bellac, D. Olive, Phys.Rep. 1C (1971) 269.

3 J.H. Schwarz, Phys.Rep. 8C (1973) 269.

4 S. Mandelstam, Phys.Rep. 13C (1974) 259.

5 J. Scherk, Rev.Mod.Phys. 47 (1975) 123.

6 J. Schwarz, Phys.Rep. 89 (1982) 223.

7 The First 15 Years of Superstring Theories, ed. by J. Schwarz, World Scientific.

8 M. Kato and K. Ogawa, Nucl.Phys. B212 (1983) 443.

9 L. Brink, P. di Vecchia, P. Howe, Phys.Lett. 65B (1976) 471.

10 A.M. Polyakov, Phys.Lett. 103B (1981) 207. For a review, see ref. 14).

11 Proceeding of the International Conference on High Energy Physics, Brighton, 1983, review talk by J.-L. Gervais.

12 I. Gradshteyn, I. Ryzhik, Tables of Integrals, Ac. Press.

13 J.-L. Gervais, B. Sakita, Phys.Rev.Lett. 30 (1973) 716.

14 P. Goddard, J. Goldstone, C. Rebbi, C. Thorn, Nucl.Phys. B56 (1973) 109.

15 S. Mandelstam, Nucl.Phys. B64 (1973) 205.

16 J.-L. Gervais, B. Sakita, Nucl.Phys. B34 (1973) 205.
17 A. Belavin, A. Polyakov, A. Zamolodchikov, Nucl.Phys. B241 (1980) 333.
18 D. Friedan, Les Houches, lecture notes 1982.
19 V. Kac, Proceedings of the International Congress of Mathematicians, Helsinsky, 1978; Lecture Notes in Physics, vol.94, p. 441, Springer Verlag.
20 J.-L. Gervais, A. Neveu, Com.Math.Phys. 100 (1985) 15.
21 D. Friedan, Z. Qiu, S. Shenker, in Vertex Operator in Mathematics and Physics, ed.J. Lepowsky et al., Springer; Phys.Rev.Lett. 52 (1984) 1575.
22 J.-L. Gervais, A. Neveu, Nucl.Phys. B257, FS14 (1985) 59.
23 J.-L. Gervais, A. Neveu, Nucl.Phys. B224 (1983) 329; B238 (1984) 125.
24 J.-L. Gervais, A. Neveu, Phys.Lett. 151B (1985) 271.
25 J.-F. Arvis, Nucl.Phys. B212 (1983) 151; B218 (1983) 303; O. Babelon, Nucl.Phys. B258 (1985) 680.

VARIATIONAL STUDY OF LATTICE QCD *

Guo Shuo-hong

Department of Physics, Zhongshan University, Guangzhou, China

Lattice QCD is studied by the method of canonical transformation and variational approximations. We present results on the fermion condensate, the mesonic masses, the decay constants f_π and f_ρ, diquark and Δ-N mass difference and variational calculation of hadron spectrum. The results are in reasonable agreement with experiments.

1. INTRODUCTION AND FORMULATION

In recent years many works have been done on the study of hadron spectrum by lattice gauge theories. However, the status of analytical calculation of hadron spectrum is still unsatisfactory. In this paper we present an analytical approach to lattice QCD in strong coupling and intermediate coupling regions. Many aspects of low energy hadron physics are incorporated in this approach.

Chiral symmetry is an essential feature of low energy strong interacfion physics. In order to preserve this symmetry, we work with the naive fermion theroy on the lattice. The hamiltonian is

$$H = H_G + H_\psi$$

$$H_G = \frac{g^2}{2a}\sum_l E_l^2 - \frac{1}{ag^2}\sum_p T_r\,(U_p + U_p^+)$$

$$H = \frac{1}{2a}\sum_x \bar\psi(x)\,\gamma_k\,V(x+k) + m\sum_x \bar\psi(x)\,\psi(x) \qquad (1)$$

* Work supported in part by the Foundation of Zhongshan University Advanced Research Center.

where the symbols have their usual meanings. We represent the fermion field by

$$\psi(x) = \begin{pmatrix} \xi(x) \\ \eta^{+}(x) \end{pmatrix} \tag{2}$$

where $\xi(x)$ and $\eta^{+}(x)$ are 2-spinor operators. The $|0\rangle$ state is defined by

$$\xi(x)|0\rangle = \eta(x)|0\rangle = E_{\ell}^{2}|0\rangle = 0 \tag{3}$$

Since the light quark mass m is small, configurations inside a hadron will contain a large number of pairs in this representation. We make use of a canonical transformation to suppress these pairs.

$$H \rightarrow H' = e^{-iQS} H e^{iQS} \tag{4}$$

The idea is similar to the Melosh transformation from current quarks to constituent quarks. The lowest approximation to S is

$$S = \frac{1}{\sqrt{6}} \sum_{x} \psi^{+}(x) \gamma_{k} U(x,k) \psi(x+k) \tag{5}$$

Assuming the physical vacuum state in the new representation by $|0\rangle$, then the same state in the Original representation is

$$|\Omega\rangle = e^{i\theta S}|0\rangle \tag{6}$$

Minimizing the vacuum energy $\mathcal{E}_{\Omega} = \langle\Omega|H|\Omega\rangle$, we obtain $\theta = \theta(g^{2})$. Hadronic states are then built from the vacuum state by appling fermion operators connected by gauge strings.

2. RESULTS

2.1. The fermion condensate $\langle\bar{\psi}\psi\rangle$ [1]

Working out $\mathcal{E}_\Omega$ in the tree approximation, we obtain

$$\mathcal{E}_\Omega = \frac{g^2 C}{2a}\sum_{n=o}^{\infty}\frac{(-1)^n (2Q)^{2n+2}}{(2n+2)!}\sum_{i=o}^{n}\binom{2n}{2i} T_{i+1}\, T_{n-i+1}$$

$$-\frac{\sqrt{6}}{2a}\sum_{n=o}^{\infty}\frac{(-1)^n (2Q)^{2n+1}}{(2n+1)!}\, T_{n+1} - m\sum_{n=o}^{\infty}\frac{(-1)^n (2Q)^{2n}}{(2n)!}\, T_{n+1} \tag{7}$$

where C is the Casimir invariant of the gauge group, Ti is the Catalan number. Q is determined by minimizing $\mathcal{E}_\Omega$. The fermion condensate is the expection value of $\bar{\psi}\psi$ in the vacuum state $|\Omega\rangle$. The calculated value of $\langle\bar{\psi}\psi(m)\rangle$ at $1/g^2 = 0.9$ is consistent with the Monte Carlo results. [2]

2.2. The effective hamiltonian and the mesonic masses [3]

Expanding H' to order Q^2, we obtain

$$H_{eff} = H^o + \Delta H \tag{8}$$

$$H^o = \frac{g^2}{2a}\sum_{l} E_l^2 + \left[m(1-2Q^2) + \frac{\sqrt{6}}{a}Q\right]\sum_{x}\bar{\psi}(x)\,\psi(x)$$

$$+ \frac{Q^2}{6N}\frac{g^2 C}{2a}\sum_{x,\pm k}\psi_\alpha^+(x)\,\gamma_k\,\psi_\beta(x+k)\,\psi_\beta^+(x+k)\,\gamma_k\,\psi_\alpha(x) \tag{9}$$

ΔH includes the magnetic term and 2-link terms. H_{eff} can be considered as an approximation to (4). Alternatively, it can also be considered as a lattice hamiltonian with the same continuum limit as H'. Many sensible results can be obtained from H^o in the strong coupling region. The 4-quark term in (9) represents spin dependent quark-quark interactions. Writing meson creation operators as $a_m^+ = \psi^+(1-\gamma_4)\,\Gamma_m\,\psi$, with $\Gamma_\pi = \gamma_5$ and $\Gamma_\rho = \gamma_i$, another canonical transformation (Bogoliubov transformation)

$$a_m' = a_m \,\mathrm{ch}\, u_m + a_m^+ \,\mathrm{sh}\, u_m, \qquad M = \pi, \rho . \tag{10}$$

is required to diagonalize H^o. The meson masses in the strong coupling limit are given by

$$m_\pi = \left(\frac{24m}{ag^2c} + 4m^2 \right)^{\frac{1}{2}}, \qquad m_\rho = \frac{4\sqrt{2}}{ag^2C} \tag{11}$$

2.3. Meson decay constants f_π and f_ρ [4]

The decay constants are defined by

$$\sqrt{2m_\pi}\ \langle\Omega|\, j_0^{5a}(o)|\,\pi^b\rangle = m_\pi f_\pi\, \delta_{ab} \tag{12a}$$

$$\sqrt{2m_\rho}\ \langle\Omega|\, j_i^{a}(o)\,|\,\rho_j^b\rangle = f_\rho \delta_{ij}\ \delta_{ab} \tag{12b}$$

Writing

$$|\Omega\rangle = R\,|0\rangle, \qquad |\pi\rangle = R\ a_\pi^+\,|0\rangle, \qquad |\rho\rangle = R\ a_\rho^+\,|0\rangle \tag{13}$$

where R is the operator for the transformation (10), we obtain

$$f_\pi\, m_\rho / f_\rho = \sqrt{\frac{m_\rho}{m_\pi}}\ e^{u_\rho - u_\pi} = 0.58 \tag{14}$$

The experimental value is 0.64. We have thus succeeded in a unifying description of the mass ratio m_π/m_ρ and the ratio of decay constants.

2.4. Diquark and the Δ-N mass difference [5]

Lattice QCD gives a natural description of diquark configurations inside baryons. Ignoring gauge invariance for a moment, we consider the static diquark with spin 0,

$$D = \sum_x \psi^+_{a_1\alpha_1}(x)\,\gamma_5\,\gamma_2\,\psi^+_{a_2\alpha_2}(x) \tag{15}$$

D commutes with H, it would have effective mass zero. While gauge invariance prevents the existence of the static diquark (15), it is expected that the

S=0 diquark will have an effective mass considerably lower than that of the S=1 diquarks.

Working with an effective hamiltonian of the form (9) appropriate for the baryon spectrum, we confirm that the 4-quark term in H_{eff} leads to a strong attraction for the S=0 diquark and a much weaker repulsion for the S=1 diquark. Therefore, the S=0, I=0 diquark forms a substructure inside a hadron. A simple model calculation gives the correct order of magnitude for the Δ-N mass difference.

2.5. Variational calculation of the hadron spectrum [6]

We use the hamiltonian (9) for a vaiational calculation of the hadron spectrum in the intermediate coupling region. Configurations up to 4-links are included. The calculated masses in Mev at $1/g^2=0.9$ are

	a^{-1}	m	m_π	m_ρ	m_ϵ	m_A	m_B	m_N	f_π
cal.	439	4.1	175	623	1279	1257	1217	input	input
expt.			140	770	1300	1100	1235	940	93

Δ-N mass difference has not been considered in this calculation.

3. CONCLUSION

Results of variational calculations on lattice QCD are in general agreement with experiment. It is remarkable that a simple lattice hamiltonian can account for a large variety of hadron phenomena in the strong coupling region and/or the intermediate coupling region. The results are encouraging for further studies.

REFERENCES

1 Guo S.H. Chen Q.Z., Liu J.M. and Hu L. Commun in Theor. Phys. (Beijing) 3 (1984) 481.
2 Hu L., Guo S.H, to be published in Phys. Energ. Fortis.et Phys. Nucl.
3 Guo S.H., Liu J.M., Chen Q.Z. and Hu L., Commun. in Theor. Phys. (Beijing) 3 (1984) 575.
4 Guo S.H., Chen Q.Z., to be published in Commun. in Theor. Phys. (Beijing).
5 Guo S.H. Chen Q.Z., to be published in Commun. in Theor. Phys. (Beijing).
6 Hu L., Ph.D. Thesis, Zhongshan University.

THE MIXING EFFECTS OF Z° AND TOPONIUM

T.H.Ho, J.Liu, X.Zhang and Z.Y.Zhu
Institute of Theoretical Physics, Academia Sinica,
P.O.Box 2735, Beijing, P.R.China

Recently, some two-jet events were observed in UA1 and were interpreted as the evidence of top-quark with mass around 30-50 Gev. If this is true, the mixing effects between Z° and toponium will be important.

Basing on the Standard Model, the mixing element in the mass-squared matrix of Z° and toponium is, up to the first order of e,

$$f=\frac{e}{2}\left(\frac{5}{3}\tan\theta_w-\cot\theta_w\right)\sqrt{3m_{t\bar{t}}}\,\psi(0)$$

where $m_{t\bar{t}}$ and $\psi(0)$ are the mass and the zero-point wave function of toponium respectively.

Consider toponium $T^\circ(1^{--})$. f is about 12 $(\mathrm{Gev})^2$, so the mass shift is less than 0.065 Gev, which is the maximal shift reached in the case the mass difference between Z° and T° vanishes, and is negligible. The mixing angle is sensitive to the mass difference. Furthermore, the decay properties of physical Z is almost the same as that of unmixed Z°, but the mixing effects are significant to some decay modes of toponium if the mass difference is less than 20 Gev. When the mass difference is in the domain 0–10 Gev, the mixing will give the main contribution to the decay of physical toponium.

For example, when the mass difference is 8 Gev, the mixing angle is about 8×10^{-3}, then

$$\Delta\Gamma_{T\to g\bar{g}}/\Gamma_{T^\circ\to g\bar{g}}=16$$

$$\Delta\Gamma_{T\to\gamma\bar{\gamma}}/\Gamma_{T^\circ\to\gamma\bar{\gamma}}=3.4$$

The detailed results are given

GAUGE GROUP COCYCLES HIRARCHY

Hou Bo-Yu
Institute for Modern Physics
Northwest University at Xian

Group cohomology cocycles characterize the different phase structures of group representations. The 0-cocycles are invariant on each component of group. The 1-cocycles are the phases of induced representations. The 2-cocycles are the phases of projective representations. We show that the 3-cocycles constitute the associative mode dependent phase of "representations".

For the translation group, the group cocycles are simply the de Rham cocycle forms on space. The existence of m-th order cocycle implies the vanishness of (m+1)-th order cocycles.

We show that the gauge group cocycles constitute Cech-de Rham double cohomology. They are not only related by the desent equation, but also have the same topological index. By using stokes theorem we succeed to turn the contribution of m-order cocycles on m+2 patches in n dimensional space to m+1 order cocycle on the common n-1 dimensional boundary of these m+2 patches.

It is wellknown that since topological index are related to the analytical index, this hirarchy of cocycles realizes various anomalies in physics with fermions. For example zero cocycle describes vaccum angle, 1-cocycles describes anomalous divergence of currents, 2-cocycles describes anomalous commutators. The double cohomology indicate that nontrival lower cocycles implies higher cocycles as a boundary effect. Therefore there will be 3-cocycles —violation of Jacobi identity, or nonassociativeness in topological nontrival background field such as Skyrmion and monopole.

An Unified Treatment of Ordinary Hadrons and New Particles as Bound States in Quark Model

Ning Hu
Institute of Theoretical Physics, Peking University
Institute of Theoretical Physics, Academia Sinica.

1. Introduction and general discussions

It has been pointed out that [1] the mass spectra of ground and excited states of ordinary and new hadrons of angular momenta J = L + S are given by the following expression.

$$m^2 = A + w(L + 2n) \qquad (1)$$

where L and S are respectively orbital and spin angular momenta. n is the radial quantum number, and w = 1 (GeV)^2. A takes different values for different kinds od hadrons. Deviations from (1) of other hadrons with J > L + S can be explained by a tensor force of the following form

$$V_t = g_1[\tfrac{1}{2}(\vec{\sigma}_1\cdot\vec{\sigma}_2) - \frac{1}{r}(\vec{\sigma}_1\cdot\vec{r})(\vec{\sigma}_2\cdot\vec{r})] \qquad (2)$$

plus a spin-orbit force of the form

$$V_{LS} = \frac{g_2}{2}[J(J+1) - L(L+1) - S(S+1)] \qquad (3)$$

(1) represents the set of eigen values of the following relativistic wave equation:

$$[A' - \nabla^2 + \frac{w^2}{16}r^2 - m^2]\Psi = 0 \qquad (4)$$

Fig. 1 shows J - m^2 plots for ordinary meson states. For pseudo-scalar mesons J = L. Excited states of each kind of these mesons lie on one straight line represented by (1). Straight lines for different kinds of mesons are nearly parallel. For vector mesons of S = 1, only J = L + S states lie on one straight line. Other excited states of the same L but different J are connected by dotted lines.

Radial excited states of ordinary mesons are rare. ρ' and ϕ' seem to be the only radial excited states of vector mesons. The difference in m^2 values between these states and their ground states is two times larger than that of correspondint successive orbital excited states, in agreement with (1).

Isomeric states of small angular momenta and large energy of excitation are unstable in nuclear physics. The rareness of radial excited states may be explained by the same mechanism.

Similar plots for baryons are given in Fig. 2 - 7. States lying on solid

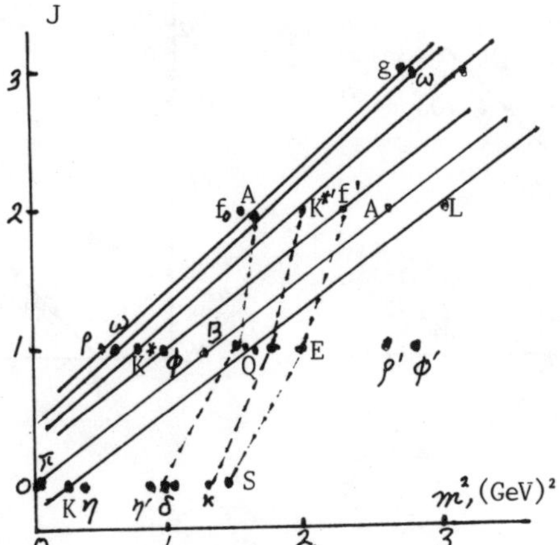

Fig. 1*

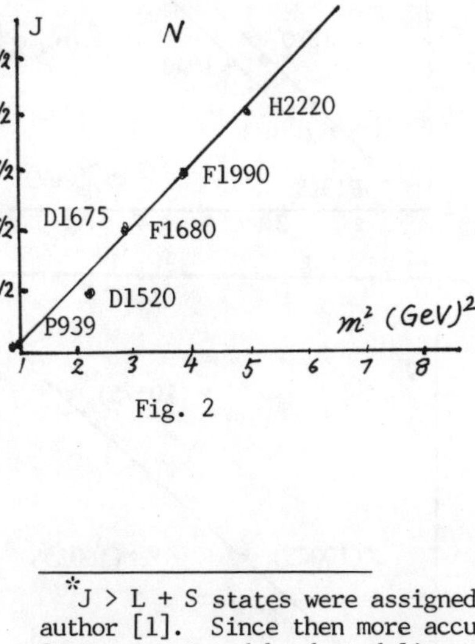

Fig. 2

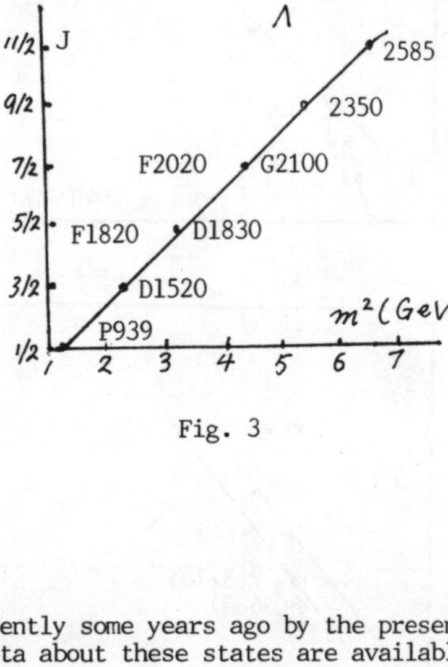

Fig. 3

*J > L + S states were assigned differently some years ago by the present author [1]. Since then more accurate data about these states are available. States connected by dotted lines in Fig. 1 represent J multiplets assigned when new observed results are taken into account.

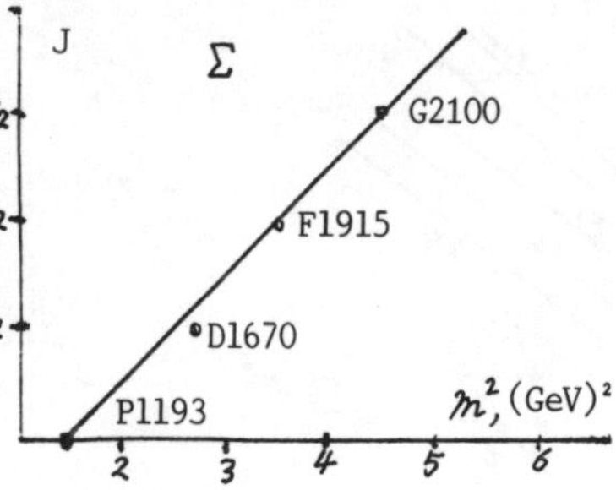

Fig. 4

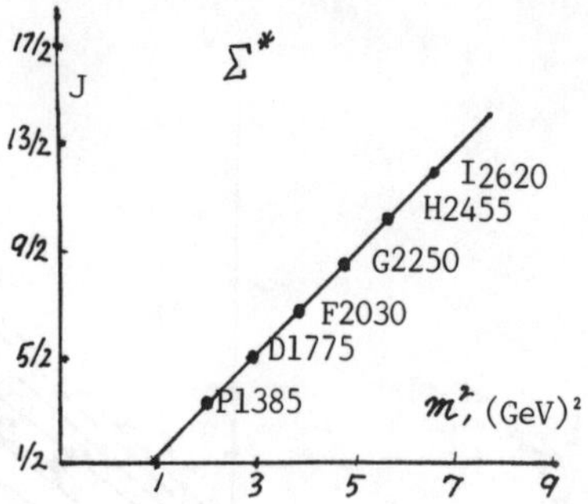

Fig. 5

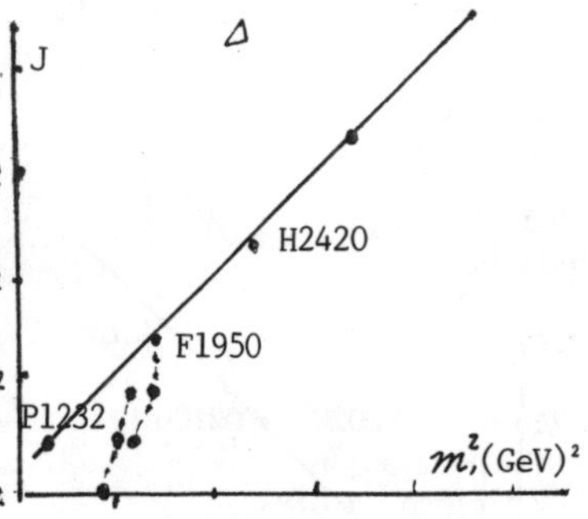

Fig. 6

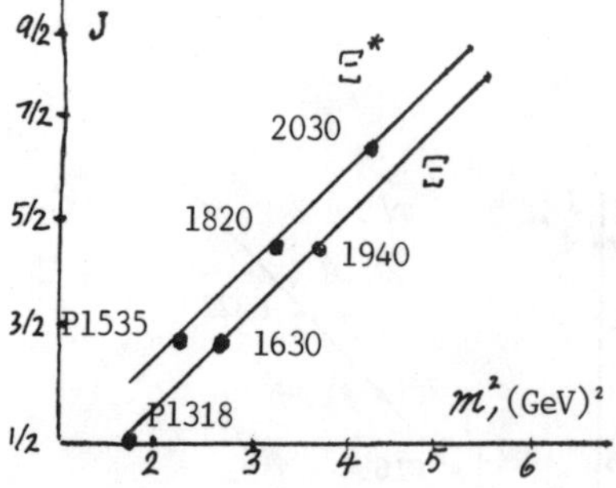

Fig. 7

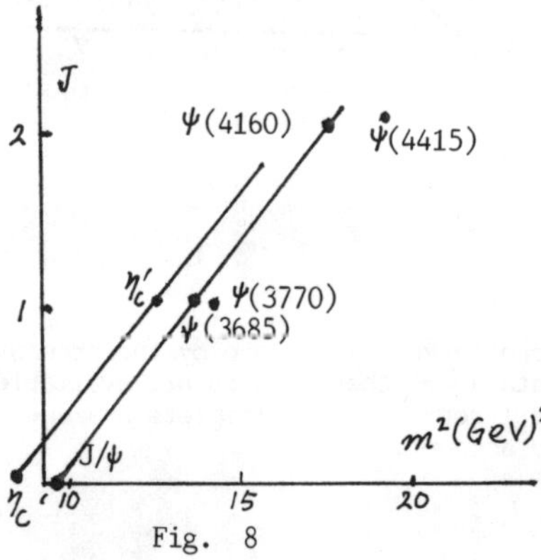

Fig. 8

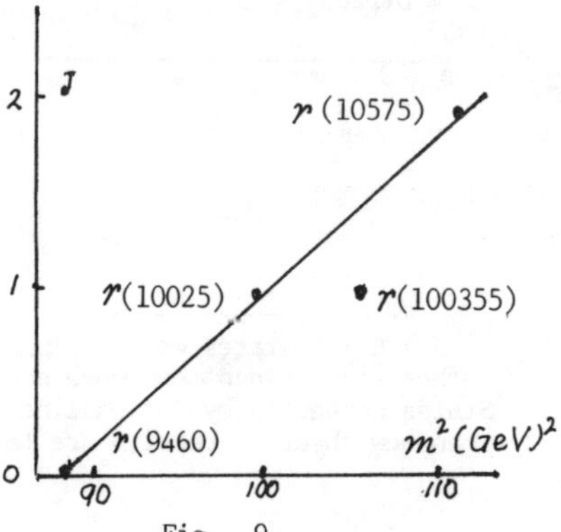

Fig. 9

straight lines are presumably also states with J = L + S. Possible J multiplets are also connected by dotted lines. It is seen that all these solid sttaight lines are nearly parallel. Thus for all ordinary hadrons w has approximately the same value 1 $(GeV)^2$. It will be seen in Section 3 that w for baryons is smaller than that of mesons by a factor 0.92. This is qualitatively in agreement with the experiment.

J - m^2 plots for ground and excited states of J/ψ and Υ particles are shown in Fig. 8 and 9. Unlike ordinary hadrons, radial excited states of new particles are stable under the strong interaction up to n = 2. This is due to the fact that new particles cannot decay into ordinary hadrons through strong interaction. The lightest particles an excited state of J/ψ can decay to are D + $\bar{D}$, the total energy of which is more than 4 GeV, just sufficient to make the excited state n = 3 unstable. It is seen in Fig. 8 that the ground state J/ψ and its excited states ψ(3685) and ψ(4160) lie on one straight line. The value of w determined from this straight line is 2 $(GeV)^2$. In preparing these plots, the state ψ(4030) has not been considered as an excited state of J/ψ, but as a molecule formed from D* and $\bar{D}$*. The reason for this is that ψ(4030) decays predominently into D* and $\bar{D}$*, although the state D + $\bar{D}$ has much larger phase space [3, 4] The states ψ(3685) and ψ(4160) are assigned as n = 1 and n = 2 states respectively. The state ψ(3770) and ψ(4415) on the right side of the straight line may be considered as D wave states admixed with small S wave components which enable these excited states to be produced through one photon intermediate state in e^+e^- collisions. This admixing of S and D waves can be produced by the tensor force as will be explained in next section.

The pseudo-scalar counter part η_c of J/ψ and its radial excited state η'_c lie on a straight line parallel to that of J/ψ. The w value of these states is therefore also given by 2 $(GeV)^2$.

Fig. 9 shows that Υ(9460), Υ(1oo25) and Υ(10578) also lie on one straight line represented by (1) with w = 5.6 $(GeV)^2$. These are states with n = 0, 1 and 2 respectively. The other state Υ(10350) may be considered as a D state admixed with a small component of S wave which enables this state to be produced in e^+e^- collisions.

Calculations in Section 2 will show that the shift of m^2 levels of D wave excited states of J/ψ and Υ particles produced by tensor and spin orbit forces agrees quite well with the observed values.

Equa. (4) with r^2 replaced by $r^2-c^2t^2$ can be derived from string model, Same solutions are obtained when motion in t direction is suppressed.

The bound state problem of new particles has also been treated non-relativistically using a potential different from that of (4). By setting $m = A^{\frac{1}{2}} + E$, (1) and (4) give

$$E + E^2/2A^{\frac{1}{2}} = (E/2A^{\frac{1}{2}})(L + 2n)$$

$$[-\frac{1}{2M}\nabla^2 + \frac{w^2}{16M}r^2 - E(1 + E/2A^{\frac{1}{2}})]\phi = 0, \quad M = A^{\frac{1}{2}}.$$

Above is the non-relativistic wave equation of a simple harmonic oscillator, when $E/2A^{\frac{1}{2}}$ can be neglected in comparison with 1. It is found from solutions of (4) that $E/2A^{\frac{1}{2}}$ is equal to 0.1 and o.2 respectively for n = 1 and n = 2 excited states of ψ particles, and equal to 0.03 and 0.06 respectively for n = 1 and n = 2 excited states of Υ particles. Therefore non-relativistic approximation can be used in the present case. However this approximation cannot be applied to ordinary hadron states since the motion of quarks inside these states is known to be relativistic.

It should be pointed out that since only a few low lying excited states of new particles have been observed, phenomenological determination of interaction potential cannot be unique. This explains why a potential different from that of equation (4) can give rise to the same mass spectra for new particles. In the present unified and relativistic treatment, both ordinary hadrons and new particles are shown to be described by wave equations of the same form (4).

2. Wave equations for new particles

In order to explain the splitting of J multiplets of P wave excited states of J/ψ, both tensor and spin-orbit forces are needed as in the case of ordinary vector mesons. The radial wave function for the state J can be written in the following form:

$$\Psi_{JM} = R_{J-1}F_{J,J-1,M} + R_{J+1}F_{J,J+1,M} \quad (5)$$

where $F_{J,L,M}$ is the irreducible spherical tensor, and R_L is the radial wave function of orbital angular momentum L. (2) gives

$$V_t F_{J,L,M} = -g(r)\sum_{L'} t_{J,L,L'} F_{J,L',M}$$

with $t_{J,J-1,J-1} = (J-1)/(2J+1)$, $t_{J,J,J} = -1$, $t_{J,J+1,J+1} = (J+2)/(2J+1)$

$$t_{J,J-1,J+1} = t_{J,J+1,J-1} = 3J^{\frac{1}{2}}(J+1)^{\frac{1}{2}}/(2J+1)$$

It is seen from (5) that the tensor force will introduce a D wave component into the S wave state, and also an S wave component into the D wave state when J = 1. The spin orbit force effects only the D wave component.

Radial wave equations for J = 1 states may be written

$$\frac{1}{\rho^2}\frac{d}{d\rho}\left(\rho^2\frac{d}{d\rho}R_0\right) - \rho^2 R_0 + \sqrt{2}\, g_1' R_2 \pm \quad R = 0$$
$$\frac{1}{\rho^2}\frac{d}{d\rho}\left(\rho^2\frac{d}{d\rho}R_2\right) - \rho^2 R_2 - \frac{6}{\rho^2}R_2 - g'R + \sqrt{2}g'R - \frac{3}{2}g'R + \quad R = 0 \tag{6}$$

where g' = 2g /w, g' = 2g /w and $\lambda = 2(m^2 - A)/w$,

$$\lambda_{Ln} = 2(m_{Ln}^2 - A)/w = 2(L + 2n + 3/2) \tag{7}$$

is the eigen value of (6) when $g_1 = g_2 = 0$. m_{Ln}^2 is given by (1) for the state L, n. R_0 and R_2 may be expanded into the following series:

$$R_0 = \sum_n a_n R_{0n}, \quad R_2 = \sum_n b_n R_{2n}, \tag{8}$$

(6) then become

$$(\lambda - \lambda_{0s})\, a_s + \sqrt{2}\sum_n b_n \langle R_{0s}\, g_1'\, R_{2n}\rangle = 0, \tag{9}$$
$$(\lambda - \lambda_{2s} - g_1' - 3g_2'/2)\, b_s + \sum_n \sqrt{2}\, a_n \langle R_{2s}\, g'\, R_{0n}\rangle = 0, \tag{10}$$

where

$$\langle R_{Ls}\, g_i'\, R_{L'n}\rangle = \int \rho^2 R_{Ls}^*\, g_i'\, R_{L'n}\, d\rho \tag{11}$$

(7) gives $\lambda_{0,n+1} = \lambda_{2,n}$. For simplicity g_1 and g_2 may be considered as constants. Numerical values of (11) can be obtained by inserting known wave functions R_{Ln}. Results are given in Table 1.

	Os = 00	01	02	03
2n = 20	1.10	-1.89	0	0.19
21	0.59	0.72	-1.07	0
22	0	0	0.163	-0.54

Table 1: numerical values of $\sqrt{2}\langle R_{Ls}\, R_{L'n}\rangle$

On inserting

$$\lambda = \lambda_{01} - \lambda' g_1' \tag{12}$$

(9) and (10) become

$$4(1 - s)/g_1' - \lambda']\, a_s + \sqrt{2}\sum_n b_n \langle R_{0s}\, R_{0n}\rangle = 0$$
$$\sqrt{2}\sum_n a_n \langle R_{2s}\, R_{0n}\rangle + [- 4s/g_1' - 1 - 3g_1/g_2 - \lambda']\, b_s = 0 \tag{13}$$

When only a_0, a_1 and b_0 are considered as different from zero, the condition for existing a solution different from zero is:

$$\begin{vmatrix} 4/g' - \lambda' & 0 & 1.10 \\ 0 & -\lambda' & -0.09 \\ 1,10 & -o.89 & -1 - \lambda' - 3g'/2g' \end{vmatrix} = 0. \tag{14}$$

When a_2 and b_1 are also considered as different from zero, the above condition becomes

$$\begin{vmatrix} 4/g_1' - \lambda' & 0 & 1.10 & 0.59 & 0 \\ 0 & -\lambda' & -1.89 & 0.72 & 0 \\ 1.10 & -1.89 & - \ -3g_1'/2g_2' - \lambda' & 0 & 0 \\ 0.59 & -0.72 & 0 & -4/g-1-3g\ /2g-\lambda' & -1.0 \\ 0 & 0 & 0 & -1.07 & -4/g_1' - \lambda' \end{vmatrix} = 0 \tag{15}$$

When a_3 and b_2 are further included , the above condition becomes

$$\begin{vmatrix} 4/g'-\lambda' & 0 & 1.10 & 0.59 & 0 & 0.191 & 0 \\ 0 & -\lambda' & -0.89 & 0.72 & 0 & 0.234 & 0 \\ 1.10 & -1.89 & -1-3g'/2g'-\lambda' & 0 & 0 & 0 & 0 \\ 0.59 & 0.72 & 0 & -4/g-1-3g'/2g-\lambda' & -1.07 & 0 & 0 \\ 0 & 0 & 0 & -1.10 & -4/g-\lambda' & 0.163 & 0 \\ 0.19 & 0.234 & 0 & 0 & 0.163 & -8/g-3g/2g-\lambda' & -0.54 \\ 0 & 0 & 0 & 0 & 0 & -0.54 & -8/g-\lambda' \end{vmatrix} = 0 \tag{16}$$

Numerical values in above equations are taken from Table 1. In the following only (15) will be considered. Let $\lambda_1 < \lambda_2 < \lambda_3 < \lambda_4 < \lambda_5$ be solutions of (15), λ_1 and λ_2 should represent respectively the states J/ψ and $\psi(3685)$. The difference in m^2 of these two states is $3.981(\text{GeV})^2$, which is related to $\lambda_2 - \lambda_1$ by the following relation:

$$3.981 = K(\lambda_2 - \lambda_1) \tag{17}$$

Mass values of three other states corresponding to λ_3, λ_4 and λ_5 are given by the following relation:

$$m_i^2 = 3.685 + K(\lambda_i - \lambda_2) \quad (i = 3, 4, 5). \tag{18}$$

Equation (15) contains two unknown parameters g_1' and g_2'. These parameters may be determined fron the split of levels of P wave J multiplets $\chi_c(3415)$, χ_c (3510) and $\chi_c(3555)$ produced by tensor and spin-orbit forces (2) and (3). Since the separation of these levels is small, perturbation calculation will be sufficient for the purpose. By using (2) and (3) the following relations are obtained.

$$m^2_{J=1} - m^2_{J=0} = -3g_1 + g_2 \tag{19}$$

$$m^2_{J=2} - m^2_{J=1} = \frac{6}{5} g_1 + 2g_2 \tag{20}$$

Inserting observed masses of P wave excited states, one obtains

$$g_1 = -0.15, \quad g_2 = 0.24 \tag{21}$$

Inserting (21) in (15) and solving the resulting equations, one obtains the following table.

state	n=0,L=0	n=1,L=0	n=0,L=2	n=2,L=0	n=1,L=2
λ	-53.35868	-1.31432	2.72695	52.75764	55.32168
a_0	.99978	-.00064	-.02011	-.00545	-.00005
a_1	-.01189	-.82280	-.56818	.01070	.ooo21
b_0	.01622	-.56818	.82271	.00769	.00016
a_2	.00263	.00647	.00018	.47379	.88061
b_1	.00478	.01147	.00030	.88052	.47384
m,Calc.	imput	imput	3730MeV	4.20GeV	4.23GeV
m,Obs.	3097MeV	3686MeV	3770MeV	4160MeV	4415 MeV

Table 2 (A and w are determined by two imput masses).

The calculated mass values in last two collumns deviate from the observed values by about 1% and 4%. This may be due to the fact that the assumption g_1 and g_2 are constants is inaccurate at large r.

Above calculation can also be applied to obtain mass spectra of Υ particles. As in the case of ψ particles, g_1 and g_2 may be determined from P wave excited states of Υ . Excited P wave n = 0 and n = 1 states have been observed and are listed in the following table:

P wave states	n	J	P wave states	n	J
χ^b(9875)	0	0	χ^b(10235)	1	0
χ^b(9895)	0	1	χ^b(10255)	1	1
χ^b(9915)	0	2	χ^b(10270)	1	2

Table 3

J values of above states have not been determined experimentally, but are given tentatively. It will be noted from above table that intervals between two neighboring J levels are the same and equal to 0.4$(\text{GeV})^2$ using

(19) and (20), one obtains

$$g_1 = -0.0556, \quad g_2 = 0.23.$$

Results obtained from solutions of (15) are given in the following table:

n	0	1	0, D wave	2	1, D wave
λ	-71.96044	-.61834	5.83235	71.73144	77.36754
a_0	.99989	-.00034	-.01426	-.00396	-.00003
a_1	.00468	.95137	.30789	-.00885	-.00013
b_0	-.01348	.30791	-.95132	-.00300	-.00005
a_2	.00079	.00194	-.00004	.19348	.98110
b_1	.00388	.00913	-.00018	.98105	-.19349
m, Calc.	input	input	10.08GeV	10.57GeV	10.61GeV
m, Obs.	9.46GeV	10.025GeV	10.355GeV	10.575GeV	-

Table 4

It is seen from above table that the calculated value of m for n = 0, D wave state is too small. The n = 1, D wave state has not been observed yet.

The facts shown in Figs. 8 and 9 that J/ψ(3100), ψ(3685) and ψ(4160) lie on one straight line, that η_c(2980) and η_c'(3590) lie on a straight line parallel to that of J/ψ, and also the fact that Υ(9460), Υ(10025) and Υ(10350) lie on one straight line show that the internal motions of these new particles, like those of ordinaary hadrons, are simple harmonic oscillations.

Transition probabilities of electrmagnetic and weak decay processes of ordinary hadrons do not depend sensitively on the form of wave functions which fall down smoothly with increase of r. The normalized wave function of n = 1 radial excited state obtained from wave equation (4) is 3/2 times larger than that of the ground state at r = 0, and has a node at ρ = 3/2. The observed e^+e^- width of ψ(3685) is equal to 0.42 times that of J/ψ. It will be of interest to find out whether these decay widths obtained from wave functions agree with those observed. It can be shown that the ratio of e^+e^- decay width of n=1 excited state to that of the ground state calculated from wave functions is $R = (3.097/3.685)^2(3/2) = 1.059$. R can further be reduced when one notices that only S wave component of the wave function can contribute to the e^+e^- decay process, and finds from Table 2 that only $(82.3)^2$ percent of the state is is in S wave. The resulting decay width is then equal to $1.059(823)^2 = 0.50$. which is not far from the observed value 0.42.

3. The wave equation for baryon states

Baryons are considered as bound states of three quarks. The wave equation for these states can be obtained in the following way: Let r_1 and r_2 be coordinates of quark and anti-quark respectively in a meson state. in center of mass coordinate system, the relative coordinate is $r = r_1 - r_2 = 2\, r_1 = 2\, r_2$. Equation (4) may be written

$$[\ \tfrac{A}{2} - \nabla_1^2 - \nabla_2^2 + \tfrac{w^2}{32}(2r_1^2 + 2r_2^2) - m^2\ ]\Psi = 0 \tag{22}$$

which may be broken into two equations

$$[\ \tfrac{A}{4} - \nabla_i^2 + \tfrac{w^2}{16} r_i^2 - \tfrac{m^2}{4}\ ]\psi_i = 0 \quad (\ i = 1,\ 2\ ). \tag{23}$$

(23) may be considered as the wave function for a single quark or a single anti-quark in a meson state. The eigen value of this equation is the square of m/2, representing the fact that each particle contributes one-half of the total mass to the meson state.

The SU(3) symmetry requires that the potential between two quarks should be equal to one-half of the potential between a quark and an anti-quark. The wave equation of the i-th quark in the baryon state may be written

$$[\tfrac{A}{4} - \nabla_i^2 + V_i - \tfrac{m^2}{9}]\psi_i = 0\ (\ i = 1,\ 2,\ 3\ ), \tag{24}$$

where V_i is the potential acted on the i-th quark. Adding above euation to similar equations for other two quarks in the baryon state, one obtains

$$[\tfrac{A}{3} - \nabla_1^2 - \nabla_2^2 - \nabla_3^2 + V_{12} + V_{23} + V_{31} - \tfrac{m^2}{3}]\Psi = 0. \tag{25}$$

Now the potential between quarks i and j is

$$V_{ij} = \tfrac{w^2}{32} r_{ij}^2 \tag{26}$$

where r_{ij} is the distance between quark i and quark j. Since

$$r_{12}^2 + r_{23}^2 + r_{31}^2 = 3\ (\ r_1^2 + r_2^2 + r_3^2\ ) \tag{27}$$

one obtains

$$V_{12} + V_{23} + V_{31} = (3/32)\ w^2\ (\ r_1^2 + r_2^2 + r_3^2\ ). \tag{28}$$

Introduce further the following change of variables

$$\vec{r}_1 = \frac{1}{\sqrt{3}}\vec{\rho} + \vec{R}, \quad \vec{r}_2 = -\frac{1}{2\sqrt{3}}\vec{\rho} + \frac{1}{2}\vec{r} + \vec{R}, \quad \vec{r}_3 = -\frac{1}{2\sqrt{3}}\vec{\rho} + \frac{1}{2}\vec{r} + \vec{R}$$
$$\vec{p}_1 = \frac{2}{\sqrt{3}}\vec{p}_\rho + \frac{1}{3}\vec{P}, \quad \vec{p}_2 = -\frac{1}{\sqrt{3}}\vec{p}_\rho - \vec{p}_r + \frac{1}{3}\vec{P}, \quad p_3 = -\frac{1}{\sqrt{3}}\vec{p}_\rho + \vec{p}_r + \frac{1}{3}\vec{P} \tag{29}$$

The wave function for a baryon state becomes

$$A/3 - 2\nabla_\rho^2 - 2\nabla_r^2 + 3w^2/64\,(\rho^2 + r^2) - m^2/3\,]\,\Psi = 0 \tag{30}$$

This equation can be separated into following two equations:

$$[A/6 - 2\nabla_\rho^2 - (3/64)w^2 r^2 - m^2/6\,]\psi(\rho) = 0 \tag{31}$$
$$[A/6 - 2\nabla_r^2 - (3/64)w^2 r^2 - m^2/6\,]\psi(r) = 0 \tag{32}$$

Accordingly the baryon state is described by two independent simple harmonic oscillators. The mass value is given by

$$m^2 = A + 0.92\, w[(\;L + 2n + 3/2) + (\;L + 2n + 3/2)] \tag{33}$$

Comparing expression (1) for mesons, one obtains

$$w_{baryon} / w_{meson} = 0.92 \tag{34}$$

This shows that the level spacing of excited states of baryons is slightly smaller than that of mesons. This is qualitatively in agreement with the experiment. Values of w determined for Λ and ρ are given by 1.05 $(GeV)^2$ and 1.14 $(GeV)^2$ respectively. The ratio is just 0.92. When the vector meson is replaced by a pseudo-scalar meson, the ratio reduces to 0.77.

Actually different quarks inside baryon states have different properties. These two oscillators will have slightly different mass levels. It is seen from Fig. 3 that the orbital excited states F1815 and D 1830 and also F2020 and G2100 lie very closely on the same position of the straight line. In Fig. 2 similar pair of states D1675 and F1680 also lie very closely on the same position of the straight line. D1830 decays mainly into $\Sigma\pi$, showing that the oscillator not containing the strange quark is excited, so the decay product does not contain a strange meson. The other state F1815 decays mainly into N K, showing that the oscillator containing a strange quark is excited.

As it has been stressed in Section 1, the potential determined phenomenologically cannot be unique when only a few low lying levels of the bound states are known. One important advantage of equation (4) is that it provides a unified description of both ordinary hadrons and new particles. A similar potential in (4) wuth r^2 replaced by $x^2 = r^2 - c^2t^2$ has been obtained from the string theory,but in order to get solutions for bound states, one has to freeze the temporal component of oscillation.

REFERENCES

1. Hu Ning, Scientia Sinica, 22 (1979) 295
2. Rujula, A. D., George, H., Glashow, S. I. Phys. Rev. Lett. 38 (1977) 317
3. Hu, Ning Communications in Theor. Phys. (Beijing) 1 (1982) 59
4. Review of Particle Properties, Rev. Mod. Phys. April (1984)

MULTI-PARTICLE PRODUCTION AND THE THREE FIRE-BALL MODEL

Huang Chao-shang

Institute of Theoretical Physics, Academia Cinica, Beijing, China

We, my collaborators: Cai Xu, Chao Wei-qin, Chou Kuang-chao, Liu Lian-sou, Meng Ta-chung, Sa Ban-hao and myself, made a systematic analysis of the experimental data [1] for the non-jet non-diffractive hadron-hadron collision processes. I shall, in this talk, briefly summarize some of the results we obtained.

1. The Model

The underlying physical picture of our model [2,3,4] is the following: In a nondiffractive collision event the projectile (P) and the target (T) go through each other and appear as leading particles. During the collision process, a considerable amount of kinetic energy of P and T are converted into excitation energies which materialize —— in general into a number of clusters which subsequently decay into hadrons. These energies are randomly distributed in three distinct systems which are characterized by their locations in rapidity space. We call them "fireballs" C^*, P^* and T^*. (Here C^* stands for central rapidity region, P^* and T^* stand for projectile-and target-fragmentation regions, respectively.) Each fireball receives its energy from two independent energy-sources (i.e., the colliding objects P and T). It is assumed that the fireball "forgets its history" after its formation in the sense that the probability of forming such a fireball characterized by its energy (i.e., its mass in the rest frame) does not depend on the specific way of its formation.

The basic assumption of the model implies that the energy E_i of the fireball is distributed in the following way [3] :

$$\langle E_i \rangle P(E_i) = 4 \frac{E_i}{\langle E_i \rangle} \text{sxp} \left(-2 \frac{E_i}{\langle E_i \rangle}\right), \quad i=C^*, P^*, T^*. \qquad (1)$$

(Note that we denote the average value of A by $\langle A \rangle$ throughout this talk.)

2. Multiplicity Distributions

A. Multiplicity Distributions in the Central Rapidity Range

From Eq.(1), one obtains[2]

$$\langle n_i \rangle P(n_i) = \frac{4n_i}{\langle n_i \rangle} \exp(-2 \frac{n_i}{\langle n_i \rangle}), \quad i = C^*, P^*, T^*. \quad (2)$$

In central rapidity region ($|y| < 1.5$ approximately) the products of the C^* fireball dominate, i.e., $n_c \approx n_{c^*}$. Then the multiplicity distribution has the following scaling form

$$\langle n_c \rangle P(n_c) = \psi(z_c), \quad (3)$$

$$\psi(z_c) = 4z_c \exp(-2z_c) \quad (z_c = \frac{n_c}{\langle n_c \rangle}). \quad (4)$$

The comparison between theory and experiment is shown in Fig.1.

B. Multiplicity Distributions in the full Rapidity Range

In the full rapidity region, we have

$$P_{nd}(n_{nd}) = \int 2\delta(\sum_i n_i - n_{nd}) \prod_i P(n_i)\, dn_i , \quad (5)$$

where n_{nd} is the observed multiplicity in nondiffractive collision processes and $P(n_i)$ is given by Eq.(2). The results of computation with Eq.(5) are given in Ref 2,3 and indeed give a good description of the data.

C. Multiplicity Distributions in Different Rapidity Intervals

Experiments [5] show:

The distribution with respect to the scaled multiplicity changes appreciably with the pseudorapidity interval ("rapidity window") W. In particular, the corresponding KNO-plot changes its form in the following way: (a) The position of the maximum value, z_{max}, moves to smaller values for smaller pseudorapidity windows. (b) The curve of $\psi_w(z_w)$ at large z values becomes more and more flat when the size of the windows decreased.

These observed features of rapidity interval-dependence can be reproduced in our model. Let us consider the contributions from the C^* fireball which gives the dominating contribution, especially in the central rapidity range. According to our model, we have a binomial distribution for n_w (n_w is the number of positively, or negatively, charged hadrons in W) [4] :

$$P_w(n_w) = \sum_{N_c} P(N_c) \frac{N_c}{(n_w/2)!(N_c - n_w/2)!} q_{cw}^{n_w/2} (1-q_{cw})^{N_c - n_w/2} , \quad (6)$$

where q_{cw} is the probability for a hadron produced by one of the N_c clusters of the C^* fireball to be in the rapidity interval W and $P(N_c)$ is $P(N_c) = 4 \frac{N_c}{\langle N_c \rangle} \exp(-2 \frac{N_c}{\langle N_c \rangle})$.

In Fig.2 we show the results calculated from Eq.(6) [4,6] . The comparison between theory and experiments shows that Eq.(6) gives a good description of the data for $\eta_w \leq 3$. The discrepancy at $\eta_w = 5$ is due to the neglect of the P^* and T^* fireballs. We note, in particular, the two observed characteristic features of rapidity-interval-dependence are nicely reproduced by this formula. We have also calculated the η_w -dependence of the moments of the multiplicity distributions. The results are in agreement with data. See Fig.4 of the Ref. 4 .

3. Rapidity Distributions

In our model the growth of the average multiplicities of the fireballs, roughly speaking, implies the increase in height of the central rapidity plateau. The quantative computations [7], taking the energy-momentum conservation into account (assuming that the about 50% of the avaliable energy is carried away by the leading particles) and neglecting the transverse momenta of the fireballs, show that the results are in agreement with the data. See Fig.3. The comparison between theory and experiments for the rapidity distribution in events of different multiplicity n_{nd} can be found in Ref. 7 .

4. Long Range Correlations

We now consider the long-range correlation between the forward-and backward-multiplicities n_F, n_B of charged hadrons measured in the following rapidity intervals ($-b \leq y \leq -a$; $a \leq y \leq b$) where a and b are positive real constants. We shall call the rapidity regions $-b \leq y \leq -a$ and $a \leq y \leq b$, B and F respectively; and call the rest of the kinematically allowed rapidity region R. It is assumed that the distribution $Q(n_B, n_F)$ is a trinomial:

$$Q(n_B, n_F) = \sum_{N_c \geq (n_F + n_B)/2} P(N_c) \frac{N_c!}{(n_B/2)!\,(n_F/2)!\,[N_c - (n_F + n_B)/2]!} q_{cB}^{n_B/2}\, q_{cF}^{n_F/2}\, q_{cR}^{[N_c - (n_B + n_F)/2]} \qquad (7)$$

which is the simplest nontrival discrete distribution for three possible outcomes. In Eq.(7), q_{CB} and q_{CF} are the probabilities for a charged hadron produced by C^* to be in B, F, respectively, and $q_{CR} = 1 - q_{CB} - q_{CF}$. The calculation with Eq.(7) and corresponding data are given in Ref. 4 .

From Eq.(7), we obtain immediately

$$d_s^2(n_F) = n_s/2 \quad , \tag{8}$$

$$\langle Z^2 \rangle_s = 2n_s \quad , \tag{9}$$

where $n_s = n_F + n_B$, $Z = n_F - n_B$ and $d_s^2(n_F)$ is the dispersion of the distribution describing the probability of finding events with n_F charged hadrons in F for fixed n_s. It should be pointed out that the Eq.(9) is first obtaind by Chou and Yang [8] by analyzing the UA5-data. The comparison between Eq.(8) and data is shown in Fig.4.

We expect to see that long-range correlations in pion-and kaon-nucleon reaction at comparble energies will be the same as those discussed here. This is because, in our model, the projectile-target asymmetry (viewed from the c.m.s. frame) only effects the leading particles but not the fireballs C^*, P^* and T^*.

References

1 See, for example, K.Alpgard et al.(UA5 Collaboration), Phys. Lett. 123B(1983)361; P.Carlson, in Proceeding of the 4th Topical Workshop on Proton-Antiproton Collider Physics, Bern, March 1984 as well as the reference given therein.

2 Liu Lian-sou and Meng Ta-chung, Phys. Rev.D27(1983)2640.

3 Chou Kuang-chao, Liu Lian-sou and Meng Ta-chung, Phys. Rev. D28(1983)1080.

4 Cai Xu, Chao Wei-qin, Huang Chao-shang and Meng Ta-chung, preprint, FUB/HEP 85-4 (1985).

5 G.Alner et al.(UA5 Collaboration), preprint CERN-EP/85-61 (1985).

6 Cai Xu, Huang Chao-shang, Meng Ta-chung and Sa Ban-hao, Chinese Phys. Lett. 2(1985) 101.

7 Cai Xu, Liu Lian-sou and Meng Ta-chung, Phys. Rev. D29(1984) 869.

8 T.T.Chou and C.N.Yang, Phys. Lett. B135(1984)175.

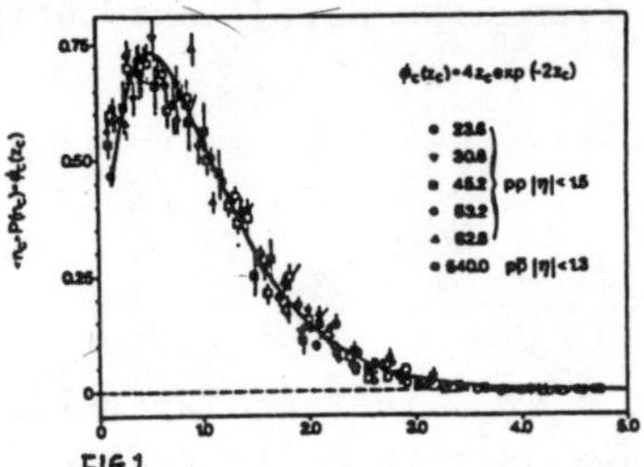

FIG 1

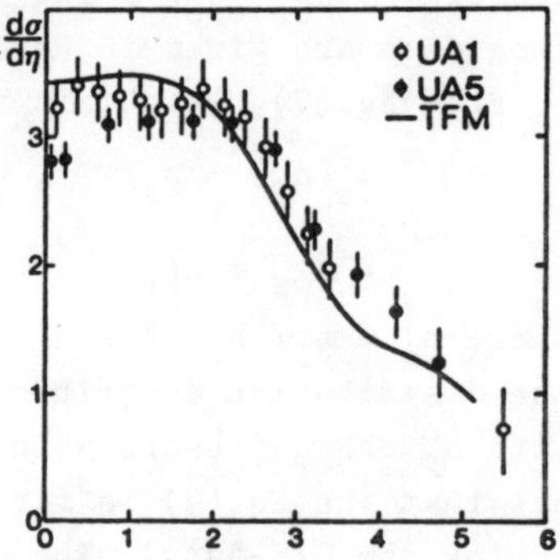

FIG 3

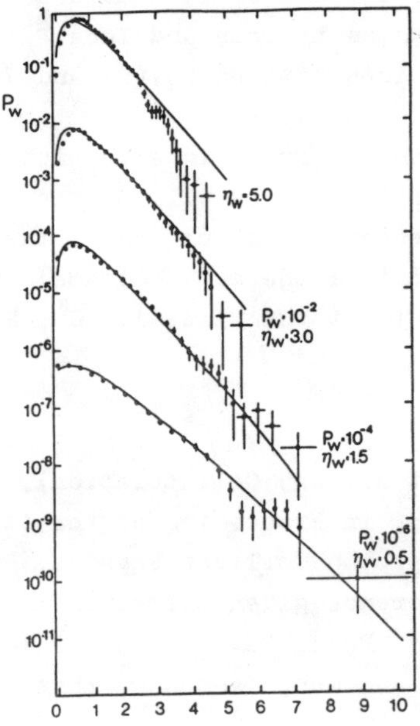

FIG. 2

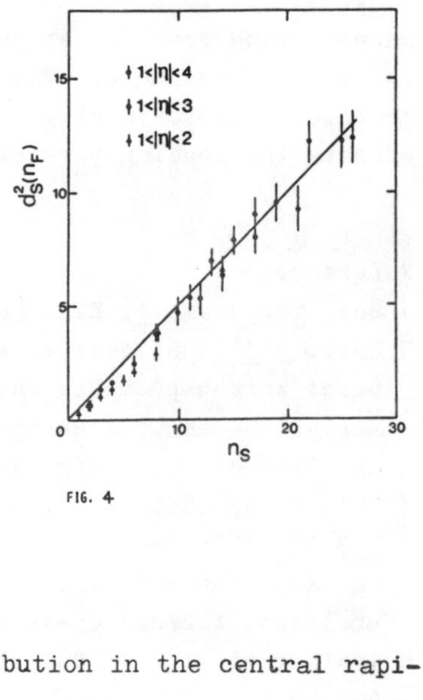

FIG. 4

Figure Captions

Fig.1: The scaled multiplicity distribution in the central rapidity range.

Fig.2: The scled multiplicity distribution in the pseudorapidity intervals $|\eta| < n_w = 5.0,\ 3.0,\ 1.5$ and 0.5.

Fig.3: Comparison between data and the calculated rapidity distribution at $\sqrt{s} = 540$ Gev.

Fig.4: The dispersion d_s squared as a function of n_s for three different forward-backward regions each with a gap $(-1 < \eta < 1)$ between them.

THE HADRONIC STRUCTURE AND SUM RULES IN QCD

Tao Huang

Institute of High Energy Physics, Academia Sinica, Beijing, China

Abstract

Sum rules for the hadronic structure and for moments of the hadronic distribution amplitude $\phi(x_i,Q)$ are discussed by using Fock state basis $\{\psi_n(x_i,\vec{k}_{\perp i},\lambda_i)\}$. These sum rules show that the influence of the non-perturbative effects on the hadronic wavefunction is very important. The initial form $\phi(x_i,Q)$ can differ greatly in its asymptotic form $\phi_{as}(x_i)$ and the x distribution of meson's wavefunction is greatly different for the different mesons. The valence wavefunction of a hadron is more compact than the hadronic radius.

It has been shown that there is a natural and consistent framework for describing bound states in gauge theory analogous to the Fock state.[1,2] The Fock basis $\{\psi_n(x_i,\vec{k}_{\perp i},\lambda_i)\}$, which is defined at equal $\tau = t+z$, is applied to derive the factorization of perturbative and nonperturbative dynamics. In particular, for the exclusive processes. One can use the wave functions to calculate many observable processes, and can obtain some general and some phenomenological constraints upon them. For example, in the case of the pion we have derived two important constraints on the wave function:[3,4]

$$\int_0^1 dx \int \frac{d^2\vec{k}_\perp}{16\pi^3}\,\psi_\pi(x,\vec{k}_\perp) = \frac{f_\pi}{2\sqrt{n_c}} \tag{1}$$

and

$$\int_0^1 dx\,\psi_\pi(x,\vec{k}_\perp=0) = \frac{\sqrt{n_c}}{f_\pi} \tag{2}$$

Then one can predict other exclusive processes by using these sum rules. In the case of the baryon wavefunction one can also obtain some nontrivial constraints on the form of the 3-quark valence wavefunction by computing some observable exclusive processes.[2]

However, these constraints on the hadronic wavefunction only give some information of the normalization constant and parameters, and they do not tell us the shape of the hadronic wavefunction. The shape of the hadronic wavefunction is determined by the large distance non-perturbative interactions. The initial form $\phi(x_i,Q_o)$ can differ greatly in its asymptotic form $\phi_{as}(x_i)$. If the asymptotic form $\phi^{as}(x)$ is used in the physical processes, all of theoretical results are smaller than the data. but for $\chi \to 2\pi$ processes which have more singular T_H (as $x \to 0.1$) the theoretical predictions are smaller than experiments by two orders of magnitude. Therefore this means that the $\phi(x, Q_0^2)$ should be much wider than the asymptotic form.[5]

In order to consider non-perturbative effects we assume that the physical vacuum is greatly different from the perturbative vacuum. This physical vacuum can be considered as a classical average effect and described by means of classical background field interaction. Therefore, the quarks (gluons) propagate, of course, not in an empty space, but through the physical vacuum, filled with the non-perturbative quark and gluon fields.[8] The interaction of quarks (gluons) with the non-perturbative vacuum fluctuations can be described by a convenient method of operator expansion in the external background field.[6] These non-perturbative corrections are expressed through the vacuum expectation values of the quark condensate $\langle 0|\bar{\psi}\psi|0\rangle$ and the gluon condensate $\langle 0|G_{\mu\nu}^2|0\rangle$. Therefore, one can use the classical background fields to account for all those nonvanishing expectation values. The perturbative effects can be described by the quantum fluctuation. More precisely, we make the following substitutions in the QCD Lagrangian and all the Green's function,[6]

$$
\begin{aligned}
A_\mu^a(x) &\longrightarrow A_\mu^a(x) + \phi_\mu^a(x) \\
\psi(x) &\longrightarrow \psi(x) + \eta(x) \quad ,
\end{aligned}
\tag{3}
$$

where $A_\mu^a(x)$, $\psi(x)$ denote gluon and quark background field respectively, and $\phi_\mu^a(x)$, $\eta(x)$ stand for their quantum fluctuations. The quantization of $\phi_\mu^a(x)$ and $\eta(x)$ have been discussed when the background fields are intoduced.[6]

From the effective Lagrangian, one can derive the propagators of quark and gluon in background fields. If fixing the gauge freedom of the background field $A_\mu^a(x)$ by the use of the Schwinger gauge or "fixed point gauge", one can express $A_\mu^a(x)$ in the series of gauge invariant operators at some point.

Thus we can write down the propagators as a perturbative series in a gauge invariant form. A detailed calculation gives the expressions of the quark and gluon propagators in the background fields up to dimension-6 operator in both coordinate space and momentum space.[9]

As an example, we consider a two-point correlation function,

$$
\Pi_{\mu\nu}^{2n,0}(q^2, z\cdot q) = i\int d^4x\, e^{iqx} {}_{Phys}\langle 0|T(j_\mu^{(2n)}(x)\, j_\nu^{(0)}(0))|0\rangle_{phys} \; , \tag{4}
$$

where

$$
j_\mu^{(2n)}(x) = \bar{d}(x)\gamma_\mu\gamma_5 (i z^\rho \overleftrightarrow{D}_\rho)^{2n} u(x) \tag{5}
$$

with $\overleftrightarrow{D}_\mu = \overleftrightarrow{\partial}_\mu - igA^a_\mu T^a$. $|0\rangle_{phys.}$ is the physical vacuum state. One finally gets the expression of the two-point Green's function which is a sum of the function of the background field operators multiplying by coefficients of Wilson expansions, given by the calculations of the quantum field contractions.

Introducing the Borel transformation and neglecting the higher dimensional operators and higher excited states, one can get the sum rules for the moments of pion. Taking account of all those nonvanishing expectation values,[8] one can obtain the values of moments of the distribution amplitude,[9] ($Q^2 = 3GeV^2$, $x = x_1 - x_2$)

$$\langle x^0 \rangle = 0.83 \; , \qquad \langle x^2 \rangle = 0.46 \; , \qquad \langle x^4 \rangle = 0.29 \; . \tag{6}$$

These moment values provide the information of the shape of the hadronic wave function. It is found that the influence of the non-perturbative effects on meson distribution amplitudes is vary important. Consequently, the x distribution of the meson's wavefunction is greatly different for the different meson. The shape is broader than the asymptotic form for the pion; the shape is quite narrower than the pion and it is closed to the asymptotic form for the ρ meson; the shape is much narrower than the asymptotic form for the heavy quark mesons and it will be a δ-like function.

In principle, as soon as we know all values of moments, one can determine the distribution amplitude. However we have only obtained a few of lower moments and it is not enough to get the distribution amplitude. The relatively large values for the second and fourth moments of the pion imply that the pion distribution amplitude is quite broad. Here we try a simple model to satisfy the QCD sum rules. A possible ansatz for the pion's nonperturbative valence wavefunction is

$$\psi_{q\bar{q}}(x, \vec{k}_\perp) = A\,(x_1 - x_2)^2 \exp\left[-b^2 \frac{\vec{k}_\perp^{\,2} + m^2}{x_1 x_2}\right] , \tag{7}$$

where A an b can be determined from Eqs.(1) and (2), $A = \frac{3\sqrt{3}}{f_\pi}$ and $b^2 = \frac{3}{5}\frac{1}{(4f_\pi)^2}$. Then we can use it to analyze experimental data. It has been shown that it fits data well for the pion form factor, as well as for the processes $\chi^0 \to 2\pi$ and for $\chi^2 \to 2\pi$. We get that the probability of finding the valence $q\bar{q}$ state in the pion is 9/28 and a radius of the valence state is 0.52fm. Therere we conclude that the valence wavefunction of a hadron is more compact than the hadronic radius and the probability for finding the valence $q\bar{q}$ state is about 30%. In this way one hopes to develop a deeper understanding of the detailed structure of hadrons.

Acknowledgement

The author would like to thank S. J. Brodsky, GOU Liang, G. P. Lepage, SHI Zhui-jian, WANG Xin-nian, XIANG Xiao-dong, XIE Yi-cheng for helpful discussions and conversations.

References

[1] G. P. Lepage, S. J. Brodsky, T. Huang, O. B. Mackenzie, "Particles and Fields", 2,83,(1983), (ed by A. Z. Cappi and A. N. Kamal).

[2] S. J. Brodsky, T. Huang and G. P. Lepage, Ibid, 143(1983).

[3] T. Huang, Proceedings of the XXth Int. Conf. on High Energy Physics, Madison, Wisc(1980); S. J. Brodsky, T. Huang and G. P. Lepage, SLAC-PUB-2540, (1980).

[4] S. J. Brodsky and G. P. Lepage, Phys. Lett., 87B, 359,(1979); A. V. Efremov and A. V. Radyshkin, Phys. Lett., 94B, 245,(1980); A. Duncan and A. Mueller Phys. Rev. D21, 1636,(1980).

[5] X. N. Wang, X. D. Xiang and T. Huang, BIHEP-TH-25, (1984).

[6] J. Govaerts et. al., Phys. Lett., 128B, 262 (1983).

[7] V. L. Chernyak and A. R. Zhitnitsky, Nucl. Phys. B201, 492,(1982); X. D. Xiang, X. N. Wang and T. Huang, BIHEP-TH-29 (1984).

[8] M. A. Shifman, A. I. Vainshtein and V. I, Zakharov, Nucl. Phys. B142(1979)358, 448.

[9] X. D. Xiang, X. N. Wang and T. Huang, BIHEP-TH-29 (1984); T. Huang, X. D. Xiang and X. N. Wang, Chinese Phys. Lett., 2, 76,(1985); BIHEP-TH-4, (1986).

[10] S. J. Brodsky, C. Carlson, T. Huang, and G. P. Lepage, SLAC-PUB-3559,(1985). (to be published in Phys. Rep.).

POTENTIAL-MODEL FITTING OF QUARKONIUM STATES

C.J. KOH

Department of Physics
National University of Singapore
Kent Ridge, Singapore 0511.

and

W. Hellman

Boston University
Boston, U.S.A.

Energy levels of a quarkonium system are determined using the linear potential modified by a Coulomb term. When applied to ψ and Υ states, the agreement is as good as those of more sophisticated models. The approximate bare-mass ratio of ψ and Υ is also well accounted for.

1. INTRODUCTION

In the years that followed the discovery of the J/ψ resonance, a lot was done in an effort to understand the energy levels and some decay properties of resonances in the J/ψ family using the charmonium model(1). In most cases non-relativistic potential models were used. They are either motivated by QCD, or simply made up to achieve the best fit with experimental data. In general, the fitting of energy levels was reasonably good, though elaborate computer programs were often needed for computation. We shall show here that a good fit for the ψ(and Υ) resonances can be obtained with a simple potential model based on the linear potential, with a Coulomb term added as a perturbation. The system is treated nonrelativistically, assuming the quark masses involved are large.

2. THE LINEAR POTENTIAL

Consider $q\bar{q}$ bound states formed with the linear potential $V(r) = Kr$. The radial part of the Schrödinger equation is

$$\frac{1}{m}[-\frac{1}{r}\frac{d^2}{dr^2}r + \frac{\ell(\ell+1)h^2}{r^2} + Kr - E]R(r) = 0$$

For $\ell = 0$, the equation can be solved exactly and the energy eigenvalues are related to the zeros of the Airy function. The masses of the $q\bar{q}$ resonances can be expressed as

$$M_n = 2m + \left(\frac{k}{m}\right)^{1/3} b_n$$

where $-b_n$ is the nth zero of the Airy function and $k = \frac{2m}{h^2} K$. The result so obtained does not fit the observed ψ system very well for any choice of the parameters m and k[2].

3. ADDITION OF A COULOMB TERM

With an additional Coulomb term the potential has the form

$$V(r) = Kr - \frac{C}{r} \quad (C > 0)$$

It is assumed that the effect of the Coulomb term is sufficiently small so that perturbation technique can be applied. The S-states of the linear model have wave functions of the form

$$U_n(r) = N_n \frac{Ai(X_n)}{r}$$

where $X_n \equiv k^{1/3}r - b_n$ and $N_n^2 = (4\pi)^{-1}k^{1/3}[Ai'(-b_n)]^2$. The corresponding energy eigenvalue is

$$E_n = \left(\frac{k^2}{m}\right)^{1/3} b_n$$

The 1st-order correction of the energy eigenvalue E_n in the stationary-state perturbation theory is

$$\langle H_c \rangle_n = \langle n | -\frac{C}{r} | n \rangle = -4\pi N_n^2 C \int_0^\infty Ai^2(X_n) r^{-1} \, dr$$

The Airy function has no explicit form, so the integral form

$$Ai(x) = \frac{1}{2\pi} \int_{-\infty}^{\infty} \exp\left[i\left(\frac{\omega^3}{3} + \omega x\right)\right] d\omega$$

is used to calculate $\langle H_c \rangle$.

4. THE MASS FORMULA

With the Coulomb perturbation the approximate energy levels are therefore

$$E'_n = E_n + \langle H_c \rangle_n = \alpha\, b_n - \beta\, d_n$$

Here

$$\alpha \equiv k^{-1/3}\, K$$

$$\beta \equiv k^{1/3}\, C$$

$$d_n \equiv \langle n | \frac{1}{X_n + b_n} | n \rangle$$

d_n can be evaluated analytically using integral transforms.

The mass of the nth $q\bar{q}$ resonance is taken to be

$$M_n = M_1 + E'_n - E'_1$$

To compare with experimental data we look at the difference

$$M_n - M_1 = E'_n - E'_1$$

$$= \alpha(b_n - b_1) - \beta(d_n - d_1)$$

For the $c\bar{c}$ system[4] the best fit is obtained by setting α_c = 2000 MeV; β_c = 946 MeV. For the $b\bar{b}$ system (Υ) the best choice for α and β are α_b = 132 MeV; β_b = 1300 MeV. The results are shown in Table I.

	Observed (MeV)	Calculated (MeV)
$\psi(1S)$	3097	3097
$\psi(2S)$	3686	3684
$\psi(3S)$	4030	4075
$\psi(4S)$	4415	4392
$\Upsilon(1S)$	9460	9460
$\Upsilon(2S)$	10019	10018
$\Upsilon(3S)$	10351	10350
$\Upsilon(4S)$	10573	10603

Table I. S-states ψ and Γ.

The simple model used also enable us to determine the ratio of the constituent masses for the $b\bar{b}$ and $c\bar{c}$ systems. From the chosen α values, and assuming that the strength of the linear potential is flavor independent, we find

$$\frac{m_b}{m_c} = 3.4$$

This is roughly what we would have expected.

5. LEPTONIC-DECAY BRANCHING RATIOS

The ratio of the leptonic decay widths as given by the Van Royan-Weisskopf formula is

$$\frac{\Gamma_n}{\Gamma_1} = \frac{M_1^2|U_n(o)|^2}{M_n^2|U_1(o)|^2}$$

For S-states,

$$|U_n(o)|^2 = \frac{m}{2\pi h^2} \langle\frac{dV}{dr}\rangle$$

In this case

$$\frac{dV}{dr} = K + \frac{C}{r^2}$$

In this 1st-order approximation the ratios calculated for the $c\bar{c}$ system is shown in Table II.

Branching ratios	Observed	Calculated
$\Gamma(2S)/\Gamma(1S)$	0.45	0.52
$\Gamma(3S)/\Gamma(1S)$	0.16	0.37
$\Gamma(4S)/\Gamma(1S)$	0.11	0.25

We have thus seen that the very simple model of a linear potential with Coulomb perturbation gives an account of the heavy quarkomium systems almost as good as those of more complicated potential models. The simple form of the potential also lends itself to simpler physical interpretations.

REFERENCES

1. K. Heikkila, N.A. Tornquist and Seiji Ono, Phys. Rev. D 29, 110 (1984), and the references quoted therein.
2. B.J. Harrington, S.Y. Park and A. Yildiz, Phys. Rev. Lett. 34, 706 (1975).

MULTIPLICITY DISTRIBUTIONS IN LIMITED PSEUDORAPIDITY INTERVALS

C.S. Lam and M.S. Zahir

Physics Department, McGill University, Montreal, Quebec

Canada H3A 2T8

Abstract

We calculate the multiplicity distributions in fixed pseudorapidity intervals $|\eta| \leqslant \eta_c$ for $p\bar{p}$ collisions at $\sqrt{s} = 540$ GeV from the total multiplicity distribution. Agreements with experimental data are good where expected. This calculation may be regarded as giving a relation between the number of clusters lying inside $|\eta| \leqslant \eta_c$ and the partition temperature $T_p(N)$ introduced by Chou, Yang and Yen for N produced particles.

Non-singly diffractive multiplicity distribution for $p\bar{p}$ collisions at $\sqrt{s}$ = 540 GeV have been measured by the UA5 collaboration[1]. Data are available for P(N), the probability of producing N particles of any pseudorapidity, and for $P_c(n_c)$, the probability of producing n_c particles all lying within the pseudorapidity interval $|\eta| \leqslant \eta_c$, for a number of η_c's. The purpose of this note is to apply the method of Ref. 2 to calculate $P_c(n_c)$ from P(N). Results of the calculation agree well with the data, except in the region of small η_c where our formula is not expected to valid. Implications of this calculation will be discussed at the end of this note.

Let $\bar{N} = \sum NP(N)$ and $\bar{n}_c = \sum n_c P_c(n_c)$. Although KNO scaling is not assumed, it will be convenient to use the KNO variables $z = N/\bar{N}$ and $z_c = n_c/\bar{n}_c$, as well as the KNO functions $\psi(z) = \bar{N}P(N)$ and $\phi_c(z_c) = \bar{n}_c P_c(n_c)$. The main assumption of Ref. 2 is that in calculating ϕ_c from ψ, we can regard n_c approximately as a one-valued function of N and η_c. We will now discuss the validity of this assumption. This assumption is certainly not true event-by- event, as fluctuations can produce different n_c's for the same N and η_c. But when averaged over many events, this assumption would be expected to be approximately valid for large n_c, unless unexpectedly large correlations are present. This can be seen as follows.

For a fixed N, let $Q_N(n_c)$ be the multiplicity distribution of particles falling within $|\eta| < \eta_c$. The k th moment of this distribution will be denoted by $\langle n^k \rangle_N$ (subscript c and dependence on η_c are understood). The width of this distribution is $\langle\Delta n\rangle_N = [\langle n^2_N\rangle - \langle n\rangle^2_N]^{1/2}$. If it is small compared to $\langle n\rangle_N$, viz. if

$$\frac{\langle\Delta n\rangle^2_N}{\langle n\rangle^2_N} \ll 1 \tag{1}$$

then it would be a good approximation to assume n_c to be a one-valued function of N and η_c, as is done in Ref. 2. Now let

$$\rho^N_1(\eta) = \frac{d\sigma_N}{d\eta}\frac{1}{\sigma_N}$$

$$\rho^N_2(\eta_1,\eta_2) = \frac{d^2\sigma_N}{d\eta_1 d\eta_2}\frac{1}{\sigma_N} \tag{2}$$

be the one- and two-particle semi-inclusive distributions respectively when the total number of particles present is N. From the sum rules

$$\int_{-\eta_c}^{\eta_c} \rho^N_1(\eta)\,d\eta = \langle n\rangle_N$$

$$\int_{-\eta_c}^{\eta_c} \rho^N_2(\eta_1,\eta_2)\,d\eta_1 d\eta_2 = \langle n(n-1)\rangle_N \tag{3}$$

we obtain the following relation between $\langle\Delta n\rangle_N$ and the correlation function $C^N(\eta_1,\eta_2) = \rho^N_2(\eta_1,\eta_2) - \rho^N_1(\eta_1)\rho^N_1(\eta_2)$.

$$\int_{-\eta_c}^{\eta_c} C^N(\eta_1,\eta_2)\,d\eta_1 d\eta_2 = \langle\Delta n\rangle^2_N - \langle n\rangle_N \tag{4}$$

We therefore see that (3) would be approximately valid if the correlation integral on the left hand side of (4) is small compared to $\langle n\rangle_N^2$, and if $\langle n\rangle_N$ is much larger than unity. Barring unexpectedly large correlations, this means that the approximation used in Ref. 2 is expected to be valid when n_c is large.

We return to the calculation of $\phi_c(z_c)$ from $\psi(z)$. Note that by this assumption, the number $n_c = \langle n\rangle_N$ determined from a given N and η_c is not always an integer. Hence it is convenient to use the continuous KNO variable $z_c = n_c/\bar{n}_c$. The assumption just discussed implies that z_c can be regarded as a function of z and η_c ,

$$z_c = f(z,\eta_c) \quad ; \quad z = f^{-1}(z_c,\eta_c) \tag{5}$$

and that the probabilities P(N) and $P_c(n_c)$ can be equated through the formula

$$dP = \psi(z)\,dz = \phi_c(z_c)\,dz_c \tag{6}$$

Eqs. (5) and (6) allow us to calculate ϕ_c from ψ using the equation

$$\phi_c(z_c) = \psi(f^{-1}(z_c,\eta_c))\,\frac{\partial f^{-1}(z_c,\eta_c)}{\partial z_c} \tag{7}$$

once f is known. As shown in Ref. 2, f can be calculated from the semi-inclusive pseudorapidity distribution by making use of eq. (3):

$$f(z,\eta_c) = \frac{1}{\bar{n}_c}\int_{-\eta_c}^{\eta_c} \frac{1}{\sigma_N}\frac{d\sigma_N}{d\eta}\,d\eta \tag{8}$$

Data for $(d\sigma_N/d\eta)\,\sigma_N^{-1}$ at $\sqrt{s}$ = 540 GeV are known[3]. For our calculation we use the parametric form given by Chou, Yang and Yen[4]. In this way we calculate the function f through (8), and the result is displayed for η_c = 0.5 and 5.0 in Fig.1. Similar results for η_c = 1.5 and 3.0 are not shown for the sake of brevity. Also shown are the parametrizations of f in the form

$$z_c = f(z,\eta_c) = \frac{A(\eta_c)\,z^2}{1+B(\eta_c)\,z} \tag{9}$$

The values of A and B for different values of η_c (i.e., η_c =0.5,1.5,3.0 and 5.0) are shown in Fig. 2. Both of them as functions of η_c can very well be fitted by second order polynomials in η_c.

$$A(\eta_c) = 0.24\,\eta_c^2 - 0.22\,\eta_c + 1.91$$

$$B(\eta_c) = 0.24\,\eta_c^2 - 0.26\,\eta_c + 1.11 \tag{10}$$

Using f and eq.(7), $\phi_c(z_c)$ is calculated for a range of η_c. For ψ we use the negative binomial form

$$\psi(z) = \bar{N}P(N) = \binom{N+k-1}{k-1}\left(\frac{\bar{N}/k}{1+\bar{N}/k}\right)^N \frac{1}{(1+\bar{N}/k)^k} \tag{11}$$

$$z = N/\bar{N}$$

where k = 3.69 and N = 28.3 were used by the UA5 group to fit the data at $\sqrt{s}$ = 540 GeV[1]. The result is shown in Fig. 3. We see that the calculation agrees very well with the experimental data except for small η_c and hence small $n_c = \langle n \rangle_N$. As discussed before, this is precisely the region where Eqs.(7) and (8) are not expected to be valid. At this point it is worthwhile to make a comment about the parametrisations of A and B we obtained. If η_c is increased to cover the whole pseudorapidity space, z_c and ϕ_c should be identical with z and ψ respectively. This indeed is the case as we can see from Eqs.(9) and (10) since A/B -> 1.0 for large η_c.

Alternatively we can compare the moments calculated from (7) and those obtained from experimental data (Table 1). To be explicit, the parametric forms of Eq.(10) were used to predict values for different η_c in Table 1. Once again we have good agreements except for small η_c. The fifth moment is also not satisfactorily reproduced presumably because of its sensitivity to the fits. In summery, the approximate formula (7) gives us a satisfactory method of calculating $\phi_c(z_c)$ from $\psi(z)$, except when η_c is small. This then confirms the validity of the method of Ref. 2 even when the calculation is compared with a large amount of high statistics data over various ranges of pseudorapidity intervals. In Ref. 2 the method was used to predict $\phi_c(z_c)$ for one value of η_c only, and comparison was made only with data of lower statistics.

There is another way to look at these calculations, which is perhaps more physical and more interesting. $\phi_c(z_c)$ and $\psi(z)$ may be fitted by a negative binomial distribution[1,5,6] with a η_c-dependent parameter k which under certain circumstances may be regarded as the initial number of clusters produced in the collision. Experimentally k increases with η_c, which is to be expected because some of the produced clusters may lie outside the specific interval $|\eta| \leq \eta_c$; and the larger η_c is the less likely this would occur. Now eq. (7) allows us to calculate $\phi_c(z_c)$ from $\psi(z)$. For a given $\psi(z)$, this is equivalent to a calculation of $k(\eta_c)$ from the single particle inclusive distribution $\rho_1^N(\eta)$. But according to Chou, Yang and Yen[4] the latter can be calculated if the partition temparature T_p (N) is known. At the present moment, both k and T_p must be considered as parameters to be determined by experimental fits. What eq. (7) effectively accomplishes is to relate these two sets of parameters k and T_p. This is interesting since these two sets of parameters describe a priori very different physical objects and concepts.

This research is supported in part by the Natural Sciences and Engineering Research Council of Canada and the Quebec Department of Education.

REFERENCES

1. UA5 Collaboration: Phys. Lett. 121B, 209(1983); 123B, 361(1983); 138B, 304(1984); and the following two preprints dated 30th April, 1985: 'An investigation of Multiplicity Distributions in different Pseaudo- rapidity Intervals in pp Reactions at C.M.S Energy of 540 GeV' and 'A New Empirical Regularity for Multiplicity Distributions in Place of KNO Scaling'.

2. C.S. Lam, Phys. Rev. D28, 1228(1983).

3. J.G. Rushbrook, in Proc. of the 14th International Symposium on Multiparticle Dynamics, Lake Tahoe, 1983, ed. by J.F. Gunion and P.M. Yager (World Scientific, Singapore,1984).

4. T.T. Chou, C.N. Yang and E. Yen, Phys. Rev. Lett 54, 510(1985)

5. P. Carruthers and C.C. Shih, Phys. Lett 127B, 242(1983).

6. D.C. Hinz and C.S. Lam, McGill Preprint, 'A Dynamical Basis for KNO Scaling and its Violations'.

Figure Captions

Fig. 1 The circled points are the experimental values calculated from the right-hand-side of Eq. (8) as explained in the text showing f as a function of z for (a) η_c =0.5 and (b) η_c =5.0. The continuous line is the best fit having the form in Eq. (9), hence determining A and B.

Fig. 2 The values of $A(\eta_c)$ and $B(\eta_c)$ are presented by circles for η_c =0.5, 1.5, 3.0 and 5.0. The solid lines represent Eq. (10) showing goodness of the parametrization.

Fig. 3 Experimental values of charged multiplicity distributions (Ref. 1) in the pseudorapidity intervals $|\eta| \leq 0.5$, 1.0, 3.0 and 5.0 plotted in the variables $\bar{n}_c P_c(n_c)$ versus $z_c = n_c/\bar{n}_c$. The continuous lines are predictions of the model from Eq. (7).

Table Caption

Table 1 The calculated moments of the multiplicity distributions predicted by Eq. (7) in central pseaudorapidity intervals $|\eta| \leq \eta_c$. The values in the parentheses are the ones calculated from data (Ref. 1). $D_2 = \left[\langle n_c - \bar{n}_c\rangle^2\right]^{1/2}$ and the C-moments are defined by $C_q = \langle n_C^q\rangle/\bar{n}_c^q$.

TABLE 1

η_c	$\bar{n}_c$	D_2	$\bar{n}_c/D_2$	C_2	C_3	C_4	C_5
0.2	1.49 (1.16)	0.86 (1.42)	1.35 (.82)	2.12 (2.48)	4.56 (8.50)	12.35 (39.0)	40.41 (230)
0.5	3.17 (3.01)	2.27 (2.90)	1.33 (1.04)	1.68 (1.93)	3.64 (5.20)	9.98 (18.0)	33.03 (80)
1.0	6.22 (6.17)	4.70 (5.22)	1.31 (1.18)	1.60 (1.72)	3.45 (4.00)	9.39 (12.1)	30.79 (44)
1.5	9.44 (9.49)	7.09 (7.46)	1.34 (1.27)	1.55 (1.62)	3.26 (3.55)	8.58 (9.90)	26.80 (33)
2.0	12.64 (12.80)	9.26 (9.50)	1.38 (1.34)	1.50 (1.56)	3.03 (3.24)	7.63 (8.40)	21.28 (26)
2.5	15.62 (15.90)	11.09 (11.30)	1.43 (1.40)	1.45 (1.51)	2.82 (3.00)	6.77 (7.40)	16.08 (21)
3.0	18.50 (18.90)	12.72 (12.90)	1.49 (1.47)	1.41 (1.46)	2.65 (2.79)	6.08 (6.40)	12.01 (17)
3.5	20.90 (21.40)	13.96 (14.10)	1.53 (1.52)	1.38 (1.43)	2.51 (2.62)	5.55 (5.80)	9.31 (14.8)
4.0	23.00 (23.60)	14.98 (14.90)	1.58 (1.58)	1.35 (1.40)	2.40 (2.49)	5.15 (5.30)	7.45 (12.9)
4.5	24.53 (25.20)	15.64 (15.50)	1.61 (1.63)	1.33 (1.38)	2.32 (2.37)	4.86 (4.90)	6.33 (11.4)
5.0	25.67 (26.4)	16.07 (15.70)	1.64 (1.68)	1.32 (1.35)	2.25 (2.28)	4.62 (4.60)	5.60 (10.3)

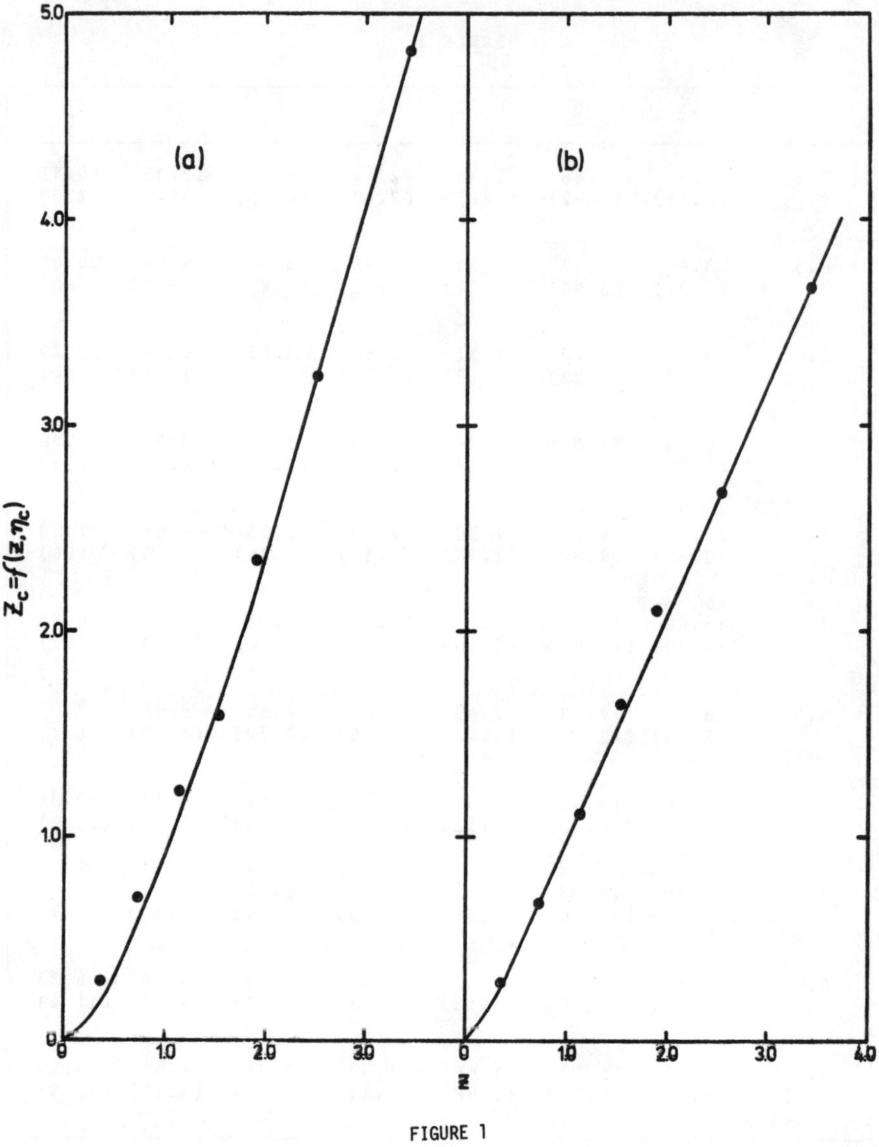

FIGURE 1

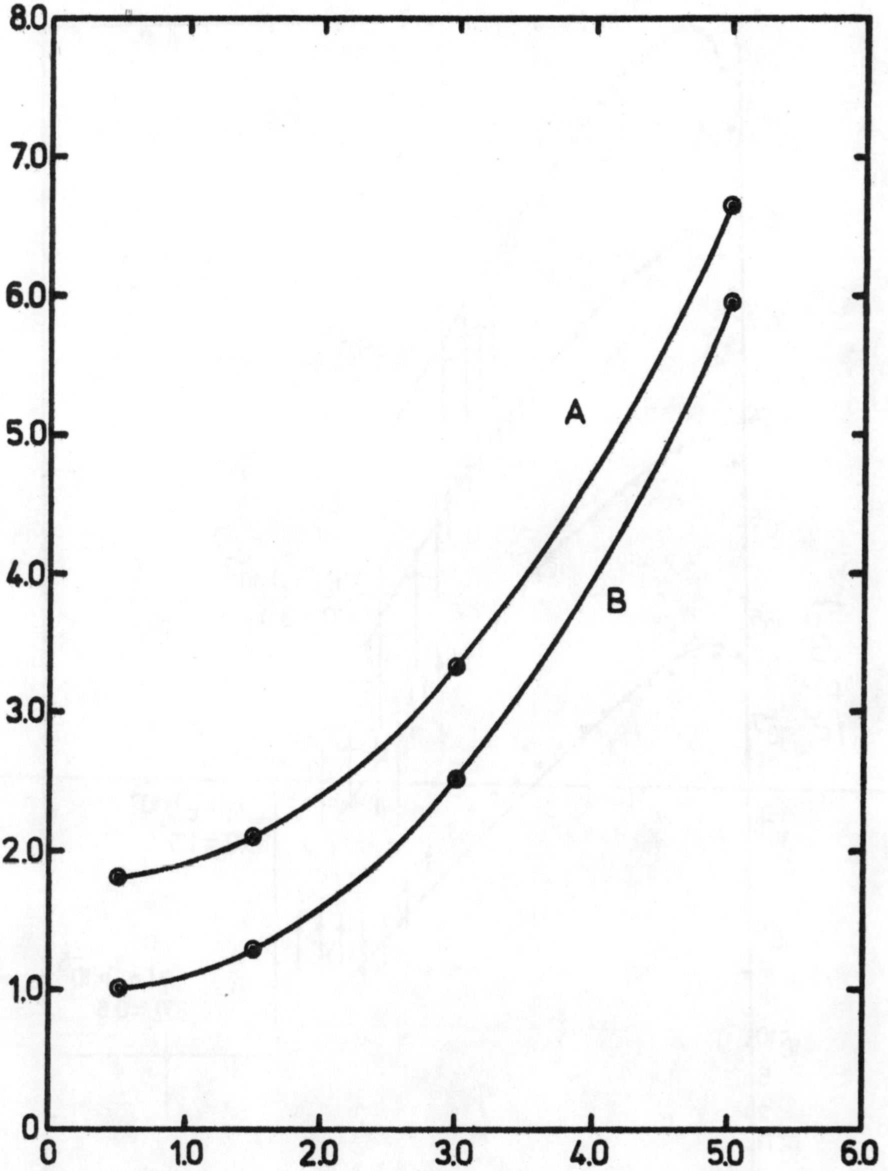

FIGURE 2

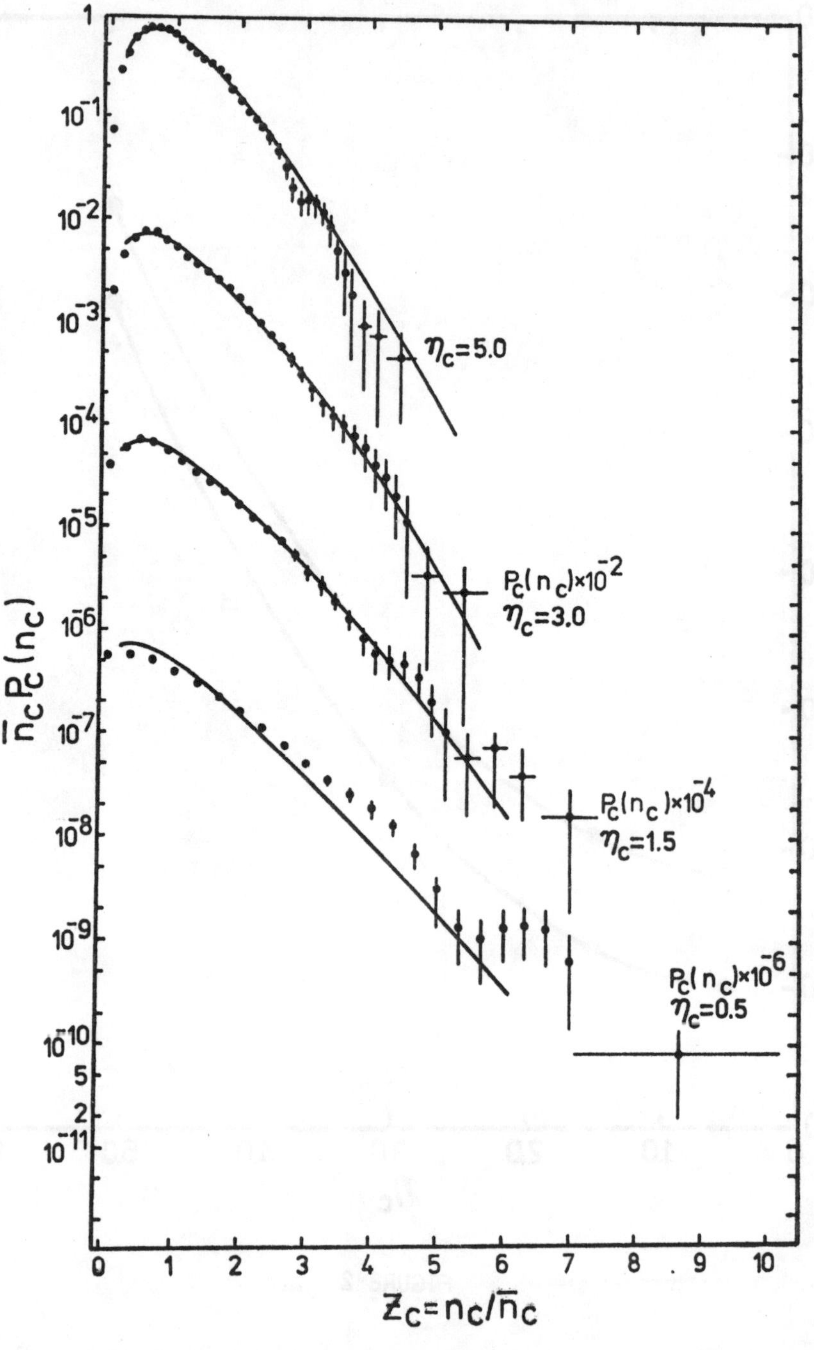
$\bar{n}_c P_c(n_c)$
$Z_c = n_c/\bar{n}_c$
$\eta_c=5.0$
$P_c(n_c)\times10^{-2}$
$\eta_c=3.0$
$P_c(n_c)\times10^{-4}$
$\eta_c=1.5$
$P_c(n_c)\times10^{-6}$
$\eta_c=0.5$

FIGURE 3

Four Quark States and Glueballs

Bing-An Li
Institute of High Energy Physics, Academia Sinica, Beijing, China
Fundamental Physics Center, Univ. of Science and Technology of China, Heifei, Anhui

Abstract

In this talk a possible evidence of $Q^2\bar{Q}^2$ mesons in $\gamma\gamma \to \rho^o\rho^o$ and $\rho^+\rho^-$ reactions is presented. Also the effect of charm quark mass in the process $J/\psi(\psi) \to \gamma +$ glueball is discussed. This effect is used to explain the suppressed radiative decay mode of ψ' (3684).

In hadron physics the spectroscopy of the s-wave $Q^2\bar{Q}^2$ mesons has been studied in both the MIT bag model [1] and the potential model [2]. The salient features of these states are the following: (1) The wavefunctions of the $Q^2\bar{Q}^2$ states consists of two parts. In one part, the $Q\bar{Q}$ pairs are in color singlet representation and in another part the $Q\bar{Q}$ pairs are in the color octet representation. (2) Their decays obey the OZI rule. Most of the $Q^2\bar{Q}^2$ states can "fall apart" into two constituent color-singlet $Q\bar{Q}$ mesons, making them too broad to be observed. However, there are some exceptions which decay to two vector mesons dominantly and whose masses just beyond the threshold of corresponding two vector mesons, hence their widths are not too broad to be detected as ordinary "mass bump".

The data from TASSO [3] and CELLO [4] revealed a large cross section of $\gamma\gamma \to \rho^0\rho^0 \to \pi^+\pi^-\pi^+\pi^-$ (in the range $1.3 < W_{\gamma\gamma} < 2.5$ GeV) even below the threshold of $\rho^0\rho^0$. According to the studies in the MIT bag model in the mass range of 1.3-2.5 GeV there are three 2^{++} and three 0^{++} $Q^2\bar{Q}^2$ mesons which decay to $\rho\rho$ and $\omega\omega$ dominantly. Their calculated masses in the MIT bag model [1] are only 100 MeV above the VV threshold. This small phase space makes them perhaps narrow enough to be detected. We can use VDM to calculate the cross section of the process $\gamma\gamma \to VV \to Q^2\bar{Q}^2 \to VV$. In the mechanism of "fall apart" this is s-wave decay. The cross section can be written as [5]

$$\sigma = \frac{1}{8W} \left\{ \frac{7}{3} \frac{\Gamma_{2^{++} \to \rho\rho}(W)}{a^2} |A_{2^+}|^2 + 2 \frac{\Gamma_{0^{++} \to \rho\rho}(W)}{a^2} |A_{0^+}|^2 \right\} \tag{1}$$

The data [3] shows in the process $\gamma\gamma \to \rho^o\rho^o$ as $W > 1.8$ GeV 2^{++} dominate and as $W < 1.8$ GeV 0^{++} dominate. These are consistent with our scheme. Let's see the contribution of three 2^{++} $Q^2\bar{Q}^2$ states. Taking $M_{2^+} = 1.8$ GeV (theoretical value is 1.65 GeV) at $W = M^i_{2^+}$ in the cross section (1) the parameter a^2 is cancelled. There is no new parameter in σ at $W = M^i_{2^+}$. The theoretical value of σ is

$$\sigma(w = M_{2^+}) = 96 \text{ nb} \tag{2}$$

which is in very good agreement with data. If choosing the masses of 0^{++} and 2^{++} $Q^2\bar{Q}^2$ to be 1.45 GeV and 1.8 GeV respectively we can fit the data very well [5].

Let's discuss the process $\gamma\gamma \to \rho^+\rho^-$. In three 0^{++} or 2^{++} $Q^2\bar{Q}^2$ there are one I = 2 state and two I = 0 states respectively. According to isospin analysis we have

$$|\rho^0\rho^0\rangle = \frac{1}{\sqrt{3}}|I = 0\rangle + \frac{\sqrt{2}}{\sqrt{3}}|T = 2\rangle\ ,\ |\rho^+\rho^-\rangle = \sqrt{\frac{2}{3}}|I = 0\rangle - \frac{1}{\sqrt{3}}|I = 2\rangle \quad (3)$$

In the scheme of four quark states from eq. (2) we can obtain that the relative sign between $|I = 0\rangle$ and $|I = 2\rangle$ is positive. Therefore in $\gamma\gamma \to \rho^0\rho^0$ there is constructively interference hence the cross section is large, but in $\gamma\gamma \to \rho^+\rho^-$ there is destructively interference, hence we predict a small cross section of $\gamma\gamma \to \rho^+\rho^-$. If we choose the masses of three 0^{++} and 2^{++} are the same respectively we get $\sigma(\gamma\gamma \to \rho^+\rho^-) \approx 0$. This prediction was proved by the data of JADE [6]. So far the $Q^2\bar{Q}^2$ scheme provides a unique explanation for the processes of $\gamma\gamma \to \rho^0\rho^0$ and $\rho^+\rho^+$. In order to identify the existence of $Q^2\bar{Q}^2$ states other evidences are needed especially the evidence of the existence of I = 2 state like $\rho^+\rho^+$ or $\rho^-\rho^-$ resonance.

Let's go to another subject. One of important predictions of QCD is the existence of glueball. The state $\iota(1440)$ is a good candidate of 0^{-+} glueball[7]. In order to identify ι as a glueball more evidences are needed. Experimentalists have found a suppressed production of ι in radiative decay of the charmonium 2S state ψ' (3684)

$$R = \frac{B(\psi' \to \gamma\iota)}{B(J/\psi \to \gamma\iota)} < 3\% \quad (4)$$

Usually we expect this ratio is about 12%. Let's see if we can explain this phenomena in glueball picture.

The wavefunction of a 0^{-+} glueball is

$$\langle 0|T\{A^a_\alpha(x_1)A^b_\beta(x_2)\}|G_0{+-}\rangle = (2E_G)^{-1/2}G(x)\varepsilon_{\mu\nu\alpha\beta}x^\mu P_G{}^\nu \exp(-iP_G X)\delta_{ab} \quad (5)$$

where E_G is the energy of the glueball, P_G is momentum of glueball, $x = x_1 - x_2$ $X = 1/2\ (x_1 + x_2)$, and $G(x)$ is internal wavefunction. Neglecting internal motion of glueball we use the wavefunction at origin $G(o)$ instead of $G(x)$ in the calculation of the width of the process $(cc)_V \to \gamma + 0^{-+}$. In this calculation two gluons exchange is dominate. We obtain [8]

$$\Gamma((c\bar{c})_V \to \gamma + 0^{-+}) = (2^{11}/81)\pi^2\alpha\alpha_s{}^2\psi(0)^2G^2(0)\frac{(M^2 - M_G{}^2)^3}{M^2M_c{}^8(M^2 - 2M_G{}^2 + 4M_c{}^2)}$$

$$\times [2M^2 - 3M_G{}^2(1+2M_c/M) - 16\ M_c{}^3/M]^2 \quad (6)$$

where $\psi(0)$ is the wavefunction $(c\bar{c})_V$ at the origin, M is the mass of $(c\bar{c})_V$, and M_c is the charm quark mass which comes from the charm quark propagator. Actually, in perturbative QCD α_s is running coupling constant and M_c is effective charm quark mass (usually it is called as current quark mass). Using eq. (6) we can get the ratio R in which $\psi(0)$ and G(0) are cancelled the only parameter is M_c. It is found out that in the range 1.45 GeV < M_c < 1.7 GeV the inequality (4) is satisfied. This mass range is reasonable. Why can the glueball picture explain the suppressed production of ℓ in ψ' radiative decay? The reason is that the glueball picture provides a dynamical cancellation shown in eq. (6) for different mass of M we obtain a very different result.

Experiments also found that [9]

$$B(\psi' \to \gamma\eta')/B(J/\psi \to \gamma\eta') < 2.6\%, \quad B(\psi' \to \gamma\eta)/B(J/\psi \to \gamma\eta) < 1.8\% \qquad (7)$$

If η and η' have some glueball component there ratios can be understood in the same way as production in a reasonable M_c range, but for η the range is much more narrow.

In ref. [9] we also discussed exotic glueball production, like 1^{-+} and 3^{-+}. The corresponding ratios are much greater than one in a reasonable M_c range. Therefore, ψ' radiative decay might be a good place to search for exotic glueballs.

References

[1] R. L. Jaffe and K. Johnson, Phys. Lett. 60B (1976) 201; R. L. Jaffe, Phys. Rev. D15 (1977) 281.

[2] N. Isgur and G. Karl, Phys. Rev. D18 (1979) 4187; D19 (1979) 2653; D20 (1979) 1191; K. F. Liu and C. W. Wong, Phys. Rev. D21 (1980) 1350.

[3] TASSO Collaboration, R. Brandelik et al., Phys. Lett 97B (1980) 448; M. Althoff et al., Z. Phys. C16 (1982) 13.

[4] CELLO Collaboration, H. J. Behrend et al., DESY Report No. 83-081.

[5] Bing An Li and K. F. Liu, Phys. Rev. Lett. 51 (1983) 1510; Phys. Rev. D30 (1984) 613.

[6] J. Dainton, in Proceedings of European Physics Society Conference, Brighton, England, 1983.

[7] J. Donoghue, K. Johnson, and B.A.A., Phys. Lett. 99B (1981) 416; M. Chanowitz, Phys. Rev. Lett. 46 (1981) 981; K. Ishikawa, Phys. Rev. Lett 46 (1981) 978.

[8] Bing An Li and Qi Xing Shen, Phys. Lett. 157B (1985) 453.

[9] R. A. Lee, SLAC-PUB-3676.

STOCHASTIC QUANTIZATION AT POSITIVE TEMPERATURE

S. C. Lim

Physics Department
Universiti Kebangsaan Malaysia
Bangi, Selangor
Malaysia

Stochastic quantization scheme of Nelson is applied to scalar field at positive temperature. The relatiosnhip between the resulting field and the finite temperature field obtained by conventional methods is discussed.

Recently, it has been shown that there exists an equivalent class of quantum field theories at finite temperature, each of which produces the same statistical averages [1]. The real time generalization of the perturbation rules used by Matsubara [2] for the imaginary time can be obtained by the use of the ordered products along a path in complex time. Quantum field theory can be constructed based on such rules and the perturbation rules are expressed in terms of the time-ordered product. A different choice of the path will result in a different quantum field theory at finite temperature. However, all these theories lead to the same physical results for equilibrium systems. In all these real time formalisms it is necessary to double the degrees of freedom in order to obtain a consistent theory [3].

In this paper, we shall report some preliminary results of quantum field at finite temperature based on the stochastic mechanics of Nelson [4]. The motivation for using Nelson stochastic quantization scheme is that it is a real time formalism and one would like to see whether it is possible to obtain a consistent real time finite temperature field theory without having to double the degrees of freedom.

First we consider the one-dimensional harmonic oscillator with Hamiltonian:

$$H_{cl} = p^2/2 + \omega^2 q^2/2 \tag{1}$$

According to Nelson stochastic quantization scheme, the ground state wave function of the harmonic oscillator

$$\psi_o(x,t) = \left(\frac{\omega}{\pi}\right)^{1/4} \exp\left[\frac{\omega^2 x^2}{2} - \frac{i\omega t}{2}\right] \tag{2}$$

is associated to a Gaussian Markov process $q_o(t)$ with expectations:

$$\langle q_o(t) \rangle = 0$$

$$\langle q_o(t) q_o(t') \rangle = \frac{e^{-\omega|t-t'|}}{2\omega} \tag{3}$$

$q_o(t)$ satisfies the following stochastic differential equation:

$$dq_o(t) = -\omega q_o(t)dt + dW(t), \tag{4}$$

where $W(t)$ is a Wiener process with mean zero and covariance given by

$$\langle dW(t)\, dW(t) \rangle = 2\nu\, dt \tag{5}$$

ν denotes the diffusion parameter, which is usually taken as equal to 1/2.

One can easily check that

$$\xi_\alpha(t) = q_o(t) + q_\alpha(t) \tag{6}$$

also satisfies eq.(4) if $q_\alpha(t)$ is the solution of the deterministic equation

$$\frac{d^2 q_\alpha(t)}{dt^2} = -\omega^2 q_\alpha(t), \tag{7}$$

where $\alpha = \{p_o, q_o\}$ denotes the initial condition for each solution $\{p(t), q(t)\}$ of the classical equation of motion:

$$\frac{dq}{dt} = p, \qquad \frac{dp}{dt} = -\omega^2 q \tag{8}$$

Now $\xi_\alpha(t)$ is the superposition of two independent processes, namely the deterministic process $q_\alpha(t)$ and the random fluctuation of the ground state oscillator process $q_o(t)$. The wave function that corresponds to $\xi_\alpha(t)$ turns out to be the coherent state of the oscillator:

$$\psi_\alpha(x,t) = \left(\frac{\omega}{\pi}\right)^{1/4} \exp\left[\frac{-\omega}{2}(x-q(t))^2 + ixp(t) - \frac{i}{2}p(t)q(t) - \frac{i}{2}\omega t\right] \tag{9}$$

The use of coherent states facilitates the generalization of stochastic quantization to the thermal equilibrium state. The density matrix in the coherent state representation at temperature β^{-1} is given by

$$\rho = \int_\Gamma |\psi_\alpha\rangle\langle\psi_\alpha|\, d\mu_\beta(\alpha) \tag{10}$$

where the measure $d\mu_\beta$ in the classical phase space Γ is

$$d\mu_\beta(\alpha) = \hat{\beta}\, e^{-\hat{\beta} E_{cl}}\, dE_{cl}\, d\theta/2\pi \tag{11}$$

with $\hat{\beta} = (e^{\beta\omega} - 1)/\omega$; E_{cl} is the classical energy and θ is the phase angle variable. Note that $\hat{\beta} \to \beta$ as $\beta \to 0$, then $d\mu_\beta$ is proportional to the Boltzman factor. To the thermal mixture state at temperature β^{-1} corresponding to the density matrix (10), it is possible to associate a stochastic process $\xi_\beta(t)$ defined through the generic average:

$$\langle F(\xi_\beta(t_1),\dots,\xi_\beta(t_n)) \rangle = \int \langle F(\xi_\alpha(t_1),\dots,\xi_\alpha(t_n)) \rangle\, d\mu_\beta(\alpha) \tag{12}$$

This results in a stochastic process given by the sum of two independent Gaussian processes:

$$\xi_\beta(t) = q_\beta(t) + q_o(t), \tag{13}$$

where $q_o(t)$ is again the ground state oscillator process and $q_\beta(t)$ is a Gaussian process determined by

$$\langle q_\beta(t)\rangle = 0$$
$$\langle q_\beta(t)q_\beta(t')\rangle = \frac{1}{\beta\omega^2}\cos\omega(t-t'). \tag{14}$$

Following the same method as for zero temperature stochastic field [5], one can find that the stochastic field at finite temperature associated to the stochastic process $\xi_\beta(t)$ is a Gaussian random field with mean zero and covariance given by

$$\langle\phi_\beta(\vec{x},t)\phi_\beta(\vec{x}',t')\rangle = \frac{1}{(2\pi)^3}\int\left[\frac{e^{-2\nu\omega|t-t'|}}{2\omega^2} + \frac{\cos\omega(t-t')}{\omega(e^{\beta\omega}-1)}\right]e^{i\vec{k}(\vec{x}-\vec{x}')}d\vec{k} \tag{15}$$

with $\omega = \sqrt{(k^2+m^2)}$. In order to compare the above result with that from the conventional quantization schemes, it is convenient to rewrite eq. (15) in the following form:

$$\langle\phi_\beta(\vec{x},t)\phi_\beta(\vec{x}',t')\rangle = \frac{1}{(2\pi)^4}\int\left[\frac{e^{2i\nu k_o(t-t')+i\vec{k}.(\vec{x}-\vec{x}')}}{\omega^2+k_o^2} + \frac{2\pi\delta(k_o^2-\omega^2)}{e^{\beta\omega}-1}e^{-ik.(x-x')}\right]d^4k$$

where $k.x = k_o t_o - \vec{k}.\vec{x}$ (16)

If we allow the diffusion parameter to take its usual value, $\nu = 1/2$, then eq. (16) becomes

$$\langle\phi_\beta(\vec{x},t)\phi_\beta(\vec{x}',t')\rangle_{\nu=1/2} = \frac{1}{(2\pi)^4}\int\left[\frac{e^{ik_o(t-t')+ik.(x.x')}}{\omega^2+k_o^2} + \frac{2\pi\delta(k_o^2-\omega^2)e^{-ik.(x-x')}}{e^{\beta\omega}-1}\right]d^4k \tag{17}$$

The above result shows that stochastic quantization of classical field at finite temperature givesrise to a Gaussian random field with two independent components - the "quantum random fluctuation" part which is temperature independent, and the "classical thermal" part which depends on temperature. It is interesting to note that the term associated with the quantum fluctuation in eq. (17) can be identified with the Euclidean field at zero temperature, whereas the classical thermal term is Minkowskian in form. The result thus rules out the possibility of using Nelson stochastic quantization to provide a physical interpretation to the Euclidean fields (see [6]).

If we allow ν to take imaginary values (refer to Davidson [7] for justification), say $\nu = i/2$, then eq. (16) becomes

$$\left\langle \phi_\beta(\vec{x},t)\,\phi_\beta(\vec{x}',t')\right\rangle_{\nu=i/2} = \frac{1}{(2\pi)^4}\int e^{-ik.(x-x')}\left[\frac{i}{k_o^2-\omega^2+i\varepsilon} + \frac{2\pi\delta(k_o^2-\omega^2)}{e^{\beta\omega}-1}\right] d^4k \tag{18}$$

which is just the real time causal Green's function [8]. For $\nu = -i/2$ we get the antitime causal Green's function. Thus, one can recover the ordinary real time finite temperature field theory (without doubling the degrees of freedom) at the expense of stochastic interpretation.

Finally, we shall consider the connection between ϕ_β and the other real time formalisms which require a doubling of the degrees of freedom. In thermo field dynamics [3a], this doubling is introduced through the tilde field $\tilde{\phi}$. The two-point Green's function is then given by the matrix:

$$\left\langle T \begin{bmatrix}\phi(x)\\ \tilde{\phi}^{\dagger}(x)\end{bmatrix}\left[\phi^{\dagger}(x')\ \tilde{\phi}(x')\right]\right\rangle$$

$$= \frac{1}{(2\pi)^4}\int d^4k\, e^{-ik.(x-x')}\begin{bmatrix} \dfrac{i}{k_o^2-\omega^2+i\varepsilon} + \dfrac{2\pi\delta(k_o^2-\omega^2)}{e^{\beta\omega}-1} & \dfrac{-2\pi e^{\beta\omega/2}}{e^{\beta\omega}-1} \\ \dfrac{-2\pi e^{\beta\omega/2}}{e^{\beta\omega}-1} & \dfrac{-i}{k_o^2-\omega^2-i\varepsilon} + \dfrac{2\pi\delta(k_o^2-\omega^2)}{e^{\beta\omega}-1}\end{bmatrix} \tag{19}$$

The diagonal terms are the same as $\langle\phi_\beta(x)\ \phi_\beta(x')\rangle_{\nu=i/2}$ and $\langle\phi_\beta(x)\phi_\beta(x')\rangle_{\nu=-i/2}$. If we use the same analogy for the off-diagonal terms, one gets

$$\left\langle\phi_\beta(x,\nu=i/2)\phi_\beta(x,\nu=-i/2)\right\rangle = \frac{1}{(2\pi)^4}\int d^4k\, e^{ik.(x-x')}\,\frac{2\pi\delta(k_o^2-\omega^2)}{e^{\beta\omega}-1} \tag{20}$$

which differs from that of eq. (19) by a factor of $-\exp[\beta\omega/2]/2$. A similar relationship exists for other real time formalism that involves a doubling of degrees of freedom [3b].

Our discussians can also be applied to gauge fields by generalizing the corresponding zero temperature stochastic fields as given in [9]. Other properties of ϕ_β such as the KMS condition and Markov property will be discussed elsewhere. Further work needs to be done in order to determine whether the above formalism will serve as a useful alternative to the finite temperature field theory.

<u>References</u>

[1] H. Matsumoto, Y. Nakano, and H. Umezawa, J. Math. Phys. <u>25</u>, 3076 (1984).

[2] T. Matsubara, Prog. Theor. Phys. <u>14</u>, 351 (1955).

[3a] H. Umezawa, H. Matsumoto, and M. Tachiki, <u>Thermo Field Dynamics</u> and Condensed States (North Holland, Amsterdam, 1982).

[3b] Kuang-chao Chou, Zhao-bin Su, Bai-lin Hao and Lu Yu, Phys. Rep. <u>118</u>, 1 (1985).

[4] E. Nelson, <u>Dynamical Theories of Brownian Motion</u> (Princeton Univ. Press, Princeton, 1967).
E. Nelson, <u>Quantum Fluctuations</u> (Princeton Univ. Press, Princeton, 1985).

[5] S. C. Lim, Lett. Math. Phys. <u>7</u>, 469 (1983).

[6] S. C. Lim, to appear in Phys. Rev. D<u>33</u> (1986).

[7] Mark Davidson, Lett. Math. Phys. <u>4</u>, 101 (1980).

[8] L. Dolan and R. Jackiw, Phys. Rev. <u>D9</u>, 3320 (1974).

[9] S. C. Lim, Phys. Lett. <u>149B</u>, 201 (1984); Phys. Rev. <u>D32</u>, 1384 (1985).

EMC EFFECT AND THE DISTORTION OF VACUUM IN NUCLEI

LIU Lian-sou

Institute of Particle Physics
Hua-Zhong Normal University, Wuhan, China

Basing on the arguement that the bag radius of a bounded nucleon is larger than the free nucleon's due to the change of vacuum property in nucleus, a good fit is obtained for the EMC effect.

Lately people are very interested in the EMC effect. A lot of papers with different explanations have been published, among which the scale change model proposed by Jaffe et al./1/ is noticed widely. It is assumed in their model that the effective single nucleon structure function in a nucleus with A nucleons and that in deuteron are related by

$$F_2^A(x,Q^2) = F_2^D(x,\xi_A Q^2) , \tag{1}$$

where ξ_A is related to the bounded nucleon bag radius R_A by

$$\xi_A(Q^2) = \left(\frac{R_A}{R_0}\right)^{2\ln(Q^2/\Lambda^2)/\ln(\mu^2/\Lambda^2)} , \tag{2}$$

where R_0 is the bag radius of a free nucleon.

Choosing proper values for ξ_A($\xi_A > 1$ means $R_A > R_0$), both the EMC data for deep inelastic scattering of muon off Fe target and the SLAC data of electron off He, Be, Al, Fe and Au targets fulfil Eq.(1) quite well/2/. It is natural to ask why $R_A > R_0$?

Recently, we proposed an explanation/3/ starting from the soliton bag model of T.D.Lee. Basing on the color antiscreening effect in QCD, Lee introduced a phenomenological scalar field σ to describe the color dia-electric property of the vacuum and proposed a relation between σ and the color dielectric constant κ as

$$\sigma = \sigma_{vac}(1-\kappa) \quad . \tag{3}$$

In the physical vacuum outside the hadronic bags $\sigma = \sigma_{vac}$ (κ=0), while inside the bags σ=0 (κ=1).

When this model is applied to nucleons bounded in a nucleus, it should be noticed that a nucleus is packed with many nucleon bags. There are A-1 nucleon bags embeded in the vacuum around any one of the nucleon bags in the nucleus. Inside the space occupied by these bags σ=0 (κ=1) instead of $=\sigma_{vac}$, therefore the vacuum outside a bounded nucleon is a complex domain consisting of domains with two different σ-values.

For a nucleus with A nucleons we introduce an effective color dielectric constant κ_A:

$$\kappa_A = \frac{A^{\alpha} - 1}{\dfrac{3A^{\alpha}}{4\pi\rho_A R_A^3} - 1} \quad , \tag{4}$$

where ρ_A is the nucleon number density in the nucleus. Eq.(4) is simply an average of the two kind of regions with κ=0 and κ=1 surrounding the considered nucleon bag. In doing so, a parameter $\alpha<1$ is introduced in Eq.(4) to avoid overestimating the effect of those bags far away from the considered bag.

Since $\kappa_A > 0$, $\sigma_A < \sigma_{vac}$, it can be seen from Fig.1 that the energy difference per unit volume inside and outside the bag is depleted by ΔB_A. Expanding $V(\sigma)$ around σ_{vac} to the first nonvanishing term:

$$V(\sigma) = b(\sigma - \sigma_{vac})^2 \quad . \tag{5}$$

Combining with Eq.(3) we obtain

$$\Delta B_A = \sigma_{vac}^2 b\,\kappa_A^2 \equiv G\kappa_A^2 \quad . \tag{6}$$

The relation between ΔB_A and R_A is determined from Eqs.(4) and (6). On the other hand, ΔB_A and R_A can also be related by the equilibrium condition of the presures inside and outside the bag:

$$B_A = \frac{3MR_0}{16\pi}\left(\frac{1}{R_0^4} - \frac{1}{R_A^4}\right) \quad , \tag{7}$$

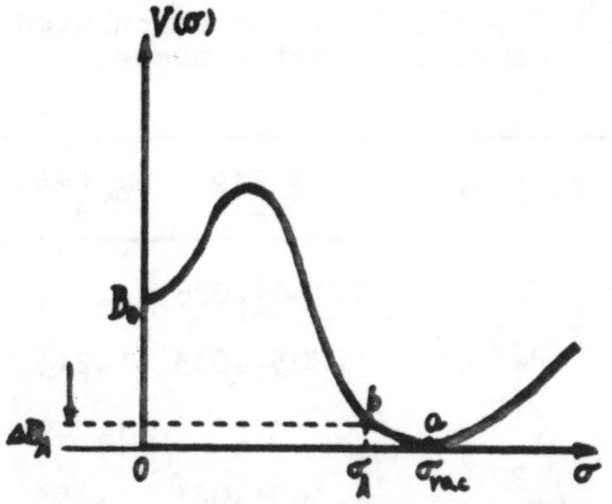

Fig.1 The potential energy of σ-field and the change of bag constant due to the change of σ-field in the vacuum

where M is the mass of the free nucleon. From Eqs.(4),(6) and (7) a self-consistent equation for determining the nucleon bag radius R_A in the nucleus is obtained:

$$G\left(\frac{A^{\alpha}-1}{\frac{3A^{\alpha}}{4\pi\rho_A R_A^3}-1}\right)^2 = \frac{3MR_0}{16\pi}\left(\frac{1}{R_0^4}-\frac{1}{R_A^4}\right). \quad (8)$$

There are 3 parameters in Eq.(8): R_0, $G(\equiv \sigma_{vac}^2 b)$ and α. Fitting experimentally determined ξ_A^{exp} of nuclei He, Be, Al, Fe and Au the 3 parameters obtained are

$$R = 0.742 \text{ fm}, \quad G = 0.285 \text{ GeV}\cdot\text{fm}^{-3}, \quad \alpha = 0.541, \quad (9)$$

with $\chi^2/DF = 1.04$. The calculated ξ_A's for the above mentioned five nucei are listed in Table I, which fit the experimental data very well. The predicted values of ξ_A for C, Ca and Ag are also listed in the table.

Two comments are in order: (1) The obtained R=0.742 fm in our model is smaller than the MIT bag radius, but is comparible with the core radius obtained in chiral bag model. (2) The possible existence of two length scales , i.e. the confinement radius and the chiral radius, has not been considered in our model. It is worth while further studying.

Table I Experimental and calculated values of ξ_A for varius nuclei.

Nucleus	ξ_A^{exp}	ξ_A^{cal}
He^4	1.160±.026	1.114
Be^9	1.205±.033	1.213
C^{12}	——	1.249
Al^{27}	1.360±.021	1.353
Ca^{40}	——	1.404
Fe^{56}	1.430±.024	1.448
Ag^{107}	——	1.539
Au^{197}	1.695±.046	1.641

REFERENCES

1 R.L.Jaffe,F.E.Close,R.G.Roberts and G.G.Ross, Phys.Lett. 134B (1984) 449.

2 Liu Feng, Li Jia-rong and Liu Lian-sou, Chinese Phys. Lett. 1 (1984) 43.

3 Liu Lian-sou, Peng Hung-an and Chao Wei-qing, Scientia Sinica A28 (1985) 1184.

THE FRACTIONAL FERMION NUMBER IN (1+1) DIMENTIONS

Guang-jiong Ni, Rong-tai Wang

Physics Department, Modern Physics Institute,
Fudan University, Shanghai, China

The fractional fermion number in (1+1) dimensions is discussed as a result of topological distortion in vacuum. By using the correspondence principle we present two theorems, the node theorem and the theorem of zero mode. An analytical expression of fractional charge distribution is also obtained

1. INTRODUCTION

The investigations in particle physics as well as in condensed matter physics reveal the existence of quantum states with fractional fermion number Q_f and/or fractional charge induced by topological solitons[1,2]. The purpose of this short note is to emphasize some general features of fractional fermion number in (1+1) dimensions.

2. CORRESPONDENCE PRINCIPLE

It is well known that one can build a dictionary between (1+1) dimensional fermion field $\psi(x)$ and boson field $\phi(x)$:[3,4,5]

$$\begin{aligned} & i\bar{\psi}\not{\partial}\psi = \tfrac{1}{2}\partial_\mu\phi\,\partial^\mu\phi \\ & j_\mu = \tfrac{1}{\sqrt{\pi}}\epsilon_{\mu\nu}\partial^\nu\phi\ , \qquad j_\mu^{(5)} = \tfrac{1}{\sqrt{\pi}}\partial_\mu\phi \\ & \bar{\psi}\psi = 2K^2 : \cos 2\sqrt{\pi}\phi : \qquad \bar{\psi}\gamma_5\psi = 2iK^2 : \sin 2\sqrt{\pi}\phi : \end{aligned} \tag{1}$$

where $K^2 = c\mu/2\pi$, μ is an arbitrary mass parameter and c is a constant, $c = e^{\gamma}/2 \approx 0.8905$. It is worth pointing out that the commutation relations

$$\begin{aligned} & [\psi^+(x)\ ,\ \phi(y)] = \sqrt{\pi}\,\theta(x-y)\,\psi^+(x) \\ & [\psi(x)\ ,\ \phi(y)] = -\sqrt{\pi}\,\theta(x-y)\,\psi(x) \end{aligned} \tag{2}$$

imply that $\psi^+(x)$ ($\psi(x)$) is a soliton creation (annihilation) operator which increases (decreases) the value of ϕ by $\sqrt{\pi}$ in the region $x>y$, i.e, creates a positive (negative) kink in the vicinity $x=y$.

3. A MODEL

Consider the Lagrangian of a coupled system of massless fermion $\psi(x)$ interacting with a scalar field $\sigma(x)$

$$\mathcal{L} = i\bar{\psi}\not{\partial}\psi - g\sigma\bar{\psi}\psi + \frac{1}{2}\partial_\mu\sigma\,\partial^\mu\sigma - V(\sigma) \tag{3}$$

If we choose $V(\sigma) = 2g^2(\sigma^2 - m^2/g^2)^2$ (4)

and assume $m/g \gg 1$ (m is the mass of fermion after symmetry breaking), we can neglect the interaction between ψ and σ to get the static solutions of $\sigma(x)$ as

$$\sigma_c^{\pm} = \pm m/g \tag{5}$$

$$\sigma_K^{\pm} = \pm(m/g)\tanh 2m(x - x_0) \tag{6}$$

Then by means of (1), the effective Hamiltonian of fermion field may be recast in the boson version

$$H_{eff} = \int dx\,[\frac{1}{2}\phi'^2(x) + G\sigma(x)\cos 2\sqrt{\pi}\,\phi(x)] \tag{7}$$

where $G \equiv 2K^2 g$

4. INTEGER FERMION NUMBER

For solution (5), e.g., $\sigma_c = -m/g$, the variational eqation derrived from (7) reads as a sine-Gordon equation:

$$\phi''(x) - 4\sqrt{\pi}K^2 m \sin[2\sqrt{\pi}\,\phi(x] = 0 \tag{8}$$

Its solution

$$\phi(x) = \frac{2}{\sqrt{\pi}}\tan^{-1} e^{(4c\mu m)^{\frac{1}{2}}x} + n\sqrt{\pi} \tag{9}$$

describes a soliton which corresponds to a creation of fermion with fermion number density

$$\rho_f(x) = \frac{1}{\sqrt{\pi}}\phi_s'(x) = \frac{1}{\pi}(4c\mu m)^{1/2}\,\mathrm{sech}\,(4c\mu m)^{1/2}x \tag{10}$$

while

$$Q_f = \int_{-\infty}^{\infty}\rho_f(x)\,dx = 1 \tag{11}$$

5. FRACTIONAL FERMION NUMBER

On the other hand, if we adopt the solution (6), then instead of (8), we have

$$\phi''(x) + 4K^2\sqrt{\pi}\,m\tanh 2mx\,\sin[2\sqrt{\pi}\,\phi(x)] = 0 \tag{12}$$

with rigorous solution

$$\phi(x) = \sqrt{\pi}/4 + (1/2\sqrt{\pi})\sin^{-1}\tanh(2mx) \tag{13}$$

This time $\rho_f(x) = (m/\pi)\,\mathrm{sech}\,(2mx)$ (14)

$$Q_f = \int_{-\infty}^{\infty} \rho_f(x)\,dx = 1/2 \qquad (15)$$

It is interesting to see the mass parameter μ is fixed as

$$\mu = m/\sqrt{2}c \approx 0.794\,m$$

in contrast with the integer fermion number case where μ is unfixed.

6. NODE THEOREM

It can be seen in general that the fermion number is controled by the sign of back ground field at infinity

$$Q_f = \pm\tfrac{1}{4}[\,\mathrm{sign}\,\sigma(+\infty) - \mathrm{sign}\,(-\infty)\,] \qquad (16)$$

So only when the background field $\sigma(x)$ developes a soliton with a node can the fractional fermion number be induced. This conclusion can be checked by the above example as well as many others.

7. THEOREM OF ZERO MODE

Let us go back to the original fermion version. After second quantization, the fermion field can expanded as follows

$$\hat{\psi}(x) = \hat{b}_0\varphi_0(x) + \sum_{k\neq 0}[\,\hat{b}_k\varphi_k(x)e^{-iE_kt} + \hat{d}_k^{\dagger}\sigma_3\varphi_k^{*}(x)e^{iE_kt}\,] \qquad (17)$$

where $\varphi_k(x)$ is the eigen function of Dirac equation in (1+1) dimensions

$$[\alpha\hat{p} + g\beta\sigma(x)]\,\varphi_k(x) = E_k\varphi_k(x) \qquad (18)$$

with $\hat{p} = \frac{1}{i}\frac{d}{dx}$, $\alpha = \sigma^2$, $\beta = \sigma^1$. It is the presence of zero mode solution $\varphi_0(x)$ ($E_0 = 0$) which is responsible for the appearance of fractional fermion number

$$Q_f = \frac{1}{2}\int(\hat{\psi}^{\dagger}\hat{\psi} - \hat{\psi}\hat{\psi}^{\dagger})\,dx = \hat{b}_0^{\dagger}\hat{b}_0 - \frac{1}{2} + \sum_{k\neq 0}(\hat{b}_k^{\dagger}\hat{b}_k - \hat{d}_k^{\dagger}\hat{d}_k) \qquad (19)$$

Formally

$$\varphi_0(x) = \begin{pmatrix} c_1 \exp\left(-g\int_0^x \sigma(\xi)\,d\xi\right) \\ c_2 \exp\left(g\int_0^x \sigma(\xi)\,d\xi\right) \end{pmatrix} \qquad (20)$$

but for a locally bound state $\varphi_0(x)$ does exist, either the following condition (a) or (b) has to be satisfied:

$$\text{(a)} \quad \begin{cases} \sigma(x) > 1/2gx & (x \to \infty) \\ \sigma(x) < 1/2gx & (x \to -\infty) \\ c_2 = 0 \end{cases} \qquad (21)$$

$$\text{(b)} \qquad \begin{cases} \sigma(x) < -1/2gx & (x \to \infty) \\ \sigma(x) > -1/2gx & (x \to -\infty) \\ c_1 = 0 \end{cases} \qquad (22)$$

8. DISCUSSION

The above discussion shows that the emergence of fractional charge is a quite general feature resulting from the distortion of the fermion vacuum induced by background field. Similar expression can also be derived by another method , e.g., the Green's function method implemented by the Levinson theorem[7]. The using of the latter method to (3+1) dimensional case, especially to the background of t'Hooft-Polyakov monopole is discussed in Ref.6. These results are in conformity with that of Ref 8,9.

REFERENCE

1. R.Jackiw and C.Rebbi, Phys. Rev. D31 (1976) 3398
2. W.P.Su, J.R.Schrieffer and A.J.Heeger, Phy. Rev. B22 (1980) 2099
3. S.Coleman, Phys. Rev. D11 (1975) 2088
4. J.Kogut and L.Susskind, Phys. Rev. D10 (1974) 3468
5. S.Mandelstam,,Phys. Rev. D11 (1975) 3026
6. R-t Wang and G-j Ni, Chinese Phys. Lett. 2 (1985) 565
7. R-t Wang and G-j Ni, Kexue Tongbao (Science Bulletin) 22 (1985) 1693
8. A.J.Niemi, M.B.Paranjape and G.W.Semenoff, Phys. Rev. Lett. 53 (1984) 515
9. A.J,Niemi and G.W.Semenoff, Phys. Rev. D30 (1984) 809

GLUON AND MASSLESS GLUINO SCATTERING USING N=2 SUPERSYMMETRY [1]

Stephen J. Parke

Fermi National Accelerator Laboratory[2]

P.O. Box 500, Batavia, IL 60510.

Abstract

The use of N=2 supersymmetry to calculate the matrix elements for the scattering of gluons and massless gluinos, at tree level in perturbative QCD, is discussed. Four, five and six particle processes are used as examples.

1 INTRODUCTION

Calculations in perturbative QCD quickly become extremely complicated as the number of vertices increases in the Feynman diagrams. This occurs not only for loop calculations but also for tree level calculations. The reasons for this are the increase in the number of complicated three gluon vertices of QCD and the unphysical degrees of freedom which must be removed from the external particles in physical processes. Consider the amplitude for scattering of two gluons into three gluons. At the tree level, there are 25 Feynman diagrams contributing to this process, with each diagram giving $\sim 6^3$ terms. Thus, the amplitude involves thousands of terms which must be squared and then summed over the physical polarizations of the gluons. Such a calculation is only tractable using algebraic manipulation programs.

In this paper, I show how N=2 supersymmetry[1,2] can be used to make such calculations considerably simpler. This is achieved by relating the gluon amplitudes to amplitudes involving scalar particles which have fewer three gluon vertices and, of course, scalar particles have no unphysical degrees of freedom. Using this idea, the matrix element is calculated for many four and five particle processes. In particular, the two gluon to three gluon amplitude is given in only six terms compared to the thousands using standard techiques. Finally, the two gluon to four gluon amplitude is discussed.

[1]Invited talk at the International Symposium on Particle and Nuclear Physics, Beijing, Sept. 2-7, 1985.

[2]Fermilab is operated by the Universities Research Association Inc. under contract with the United States Department of Energy.

2 SUPERSYMMMETRY REVISITED

The particle content of N=2 Yang-Mills theories, consists of a gluon, two species of gluino and a complex scalar. On the mass shell, the gluon and each gluino have two helicities (± 1) and the complex scalar also has two degrees of freedom, which for convenience will be refered to as helicity (± 1). To relate the creation or annihilation operators for these particles, two hermitan operators are formed from the usual generators of N=2 supersymmetry, Q^a_α, as follows

$$Q^a(\eta) = \overline{\eta}^\alpha Q^a_\alpha \tag{1}$$

where η is an arbitrary Majorana spinor. The commutation relations for this $Q^a(\eta)$ with the annihilation operators for the gluon ($g_\pm$), the gluinos ($\lambda^1_\pm, \lambda^2_\pm$) and the scalars ($\phi_\pm$) are

$$\begin{aligned} [Q^a, g_\pm] &= \mp \Gamma^\pm \lambda^a_\pm \\ [Q^a, \lambda^b_\pm] &= \mp \Gamma^\mp \delta^{ab} g_\pm \mp i\, \Gamma^\pm \epsilon^{ab} \phi_\pm \\ [Q^a, \phi_\pm] &= \pm\, i\, \Gamma^\mp \epsilon^{ab} \lambda^b_\pm \end{aligned} \tag{2}$$

where

$$\Gamma^+(p,\eta) = (\Gamma^-(p,\eta))^* = \sqrt{2E}(\eta_1 cos\tfrac{1}{2}\theta e^{i\beta/2} + \eta_2 sin\tfrac{1}{2}\theta e^{i\beta/2}) \tag{3}$$

if the representations for the momentum, p, and η are

$$p = E(\,1\,,\, sin\theta cos\beta\,,\, sin\theta sin\beta\,,\, cos\theta) \tag{4}$$

$$\eta = (\eta_1 + \eta_2^*\,,\, -\eta_1^* + \eta_2\,,\, -\eta_1 + \eta_2^*\,,\, -\eta_1^* - \eta_2). \tag{5}$$

Here, η_1 and η_2 are just arbitrary complex numbers.

These commutation relationships can be used to find relationships between different processes in this theory. Consider the scattering of m particles into n particles. If supersymmetry is unbroken, $Q^a(\eta)|vac>= 0$, then the following identity holds for the creation and annihilation operators of these particles,

$$\begin{aligned} 0 &= <vac|\,[Q^a\,,\, a_{m+1}\ldots a_{m+n} a_1^\dagger \ldots a_m^\dagger]\,|vac> \\ &= <vac|\,[Q^a\,,\, a_{m+1}] a_{m+2}\ldots a_{m+n} a_1^\dagger \ldots a_m^\dagger\,|vac> + \ldots \\ &\quad + <vac|\, a_{m+1}\ldots a_{m+n}[Q^a\,,\, a_1^\dagger] a_2^\dagger \ldots a_m^\dagger\,|vac> + \ldots \end{aligned} \tag{6}$$

This equation contains relationships between various processes which are exact for N=2 Yang-Mills theories but only hold to tree level in QCD. One can see this by comparing the Feynman diagrams for both theories at tree level or by appealing to R symmetry.

3 TWO TO TWO PROCESSES

The first process I consider is the scattering of two gluons into two gluons. The commutator of Q^a with $g_4 - g_3 - g^\dagger_{2+}\lambda^\dagger_{1+}$ leads to the following equation

$$\Gamma_1^+ < | \, g_4 - g_3 - g^\dagger_{2+} g^\dagger_{1+} \, | > \; + \Gamma_2^- < | \, g_4 - g_3 - \lambda^\dagger_{2+}\lambda^\dagger_{1+} \, | >$$
$$+ \; \Gamma_3^- < | \, g_4 - \lambda_3 - g^\dagger_{2+}\lambda^\dagger_{1+} \, | > \; + \Gamma_4^- < | \, \lambda_4 - g_3 - g^\dagger_{2+}\lambda^\dagger_{1+} \, | > \; = \; 0 \qquad (7)$$

where $\Gamma_i^\pm = \Gamma^\pm(p_i, \eta)$. Remember, Γ^+ depends on η_1, η_2 and Γ^- on η_1^*, η_2^* and since each of these is arbitrary, eqn(7) is really four equations. One of these is simply

$$< | \, g_4 - g_3 - g^\dagger_{2+} g^\dagger_{1+} \, | > = 0. \qquad (8)$$

Similar manipulations also lead to

$$< | \, g_4 - g_3 + g^\dagger_{2+} g^\dagger_{1+} \, | > = 0. \qquad (9)$$

Finally, all the nonzero helicity amplitudes for this process can be obtained by crossings from

$$\mathcal{M}(g_{1+}, g_{2+}; g_{3+}, g_{4+}) \;=\; e^{i\theta_{1234}} \mathcal{M}(\lambda_{1+}, \lambda_{2+}; \lambda_{3+}, \lambda_{4+})$$
$$=\; e^{i\psi} \mathcal{M}(\phi_{1+}, \phi_{2+}; \phi_{3+}, \phi_{4+}). \qquad (10)$$

The phase factor $e^{i\theta_{1234}}$ accounts for the different spin statistics properties of the gluons and gluinos whereas $e^{i\psi}$ is an overall phase factor and is irrelevant.

This result is now used to calculate the cross section for gluon gluon scattering using the N=2 Yang-Mills Lagrangian,

$$\mathcal{L} \;=\; -\frac{1}{4} F^a_{\mu\nu} F^{a\mu\nu} - (D_\mu \phi)^{a\dagger} (D^\mu \phi)^a$$
$$-\frac{1}{2} g^2 (f^{abc} \phi^b \phi^{c\dagger})(f^{ade} \phi^{d\dagger} \phi^e) + fermion\ terms. \qquad (11)$$

Scalar scalar scattering in this theory is very simple and it is easily shown that

$$\mathcal{M}(\phi_{1+}, \phi_{2+}; \phi_{3+}, \phi_{4+}) \;=\; -2ig^2 \, f_{x13} \, f_{x24} \, \frac{(1 \cdot 2)}{(1 \cdot 3)} \;+\; (3 \Leftrightarrow 4) \qquad (12)$$

where $(i \cdot j) \;=\; (p_i \cdot p_j)$. After squaring, and suitably summing and averaging over helicity and color, the well known result is obtained

$$|\mathcal{M}_{gg \to gg}|^2 \;=\; \frac{4g^4 N^2}{(N^2 - 1)} \left[3 - st/u^2 - tu/s^2 - us/t^2\right] \qquad (13)$$

where s, t and u are the Mandelstam variables.

The next process to consider is gluino gluino production by gluon gluon fusion. Taking the commutator of Q^a with $g_{4+}g_{3+}g^{\dagger}_{2+}\lambda^{\dagger}_{1+}$, gives

$$\Gamma_1^+ <|\, g_{4+}g_{3+}g^{\dagger}_{2+}g^{\dagger}_{1+}\,|> + \Gamma_2^- <|\, g_{4+}g_{3+}\lambda^{\dagger}_{2+}\lambda^{\dagger}_{1+}\,|>$$
$$+\ \Gamma_3^+ <|\, g_{4+}\lambda_{3+}g^{\dagger}_{2+}\lambda^{\dagger}_{1+}\,|> + \Gamma_4^+ <|\, \lambda_{4+}g_{3+}g^{\dagger}_{2+}\lambda^{\dagger}_{1+}\,|> \;=\; 0. \tag{14}$$

The Γ^+ terms give two equations in three unknowns, hence

$$|\mathcal{M}(\lambda_{1+}, g_{2+}; \lambda_{3+}, g_{4+})| \;=\; \sqrt{\frac{(1\cdot 4)}{(3\cdot 4)}}\,|\mathcal{M}(g_{1+}, g_{2+}; g_{3+}, g_{4+})|. \tag{15}$$

Again, after squaring and suitably summing and avervaging over helicity and color, gives the following

$$|\mathcal{M}_{gg\to\lambda\lambda}|^2 \;=\; \frac{g^4N^2}{(N^2-1)}\,[tu(t^2+u^2)(s^2+t^2+u^2)/s^2t^2u^2]. \tag{16}$$

This agrees with DEQ[3] for a massless gluino.

4 TWO TO THREE PROCESSES

For the scattering of two gluons into three gluons, all helicity amplitudes are crossings of $\mathcal{M}(g_{1+}, g_{2+}; g_{3\pm}, g_{4+}, g_{5+})$ or are zero to tree level. Once again supersymmetry can be used to relate this amplitude to an amplitude involving scalars,

$$|\mathcal{M}(g_{1+}, g_{2+}; g_{3\pm}, g_{4+}, g_{5+})| \;=\; (1\cdot 2)^{\pm 1}(4\cdot 5)^{\mp 1}\,|\mathcal{M}(\phi_{1+}, \phi_{2+}; g_{3\pm}, \phi_{4+}, \phi_{5+})|. \tag{17}$$

The evaluation of the scalar amplitude is quite simple if one uses the polarization tensors, $\epsilon_\mu(p_3, p_m)$, of Xu, Zhang and Chang,[4] where

$$\epsilon_\mu p_n^\mu \;=\; \pm\frac{<m:n><3:n>}{\sqrt{2}<m:3>} \tag{18}$$

where p_3 is the gluon momentum , p_m is the reference momentum and $<m:n>$ is the spinor inner product of the two momenta p_m and p_n. The only property of this product which is of interest to us here is

$$<m:n>^*<m:n> \;=\; 2(m\cdot n). \tag{19}$$

The result for the scalar matrix element is

$$
\begin{aligned}
\mathcal{M}(\phi_{1+}, \phi_{2+}; g_{3+}, \phi_{4+}, \phi_{5+}) \sim \\
& f_{x1y}\, f_{3y4}\, f_{x25} \frac{<1:2><4:5>}{<1:3><3:4><2:5>} \\
& + f_{x2y}\, f_{3y5}\, f_{y14} \frac{<1:2><4:5>}{<2:3><3:5><1:4>} \\
& + f_{x14}\, f_{y25}\, f_{3yx} \frac{<1:2>^2<4:5>}{<1:3><2:3><1:4><2:5>} \\
& + (4 \Leftrightarrow 5)
\end{aligned}
\tag{20}
$$

and eqn(17) can be used to get the purely gluonic matrix element. Note the simplicity of this expression! There are six independent color factors for this process and six terms in the final result. The linear independence of the color factors implies that each term is gauge independent and therefore each term can be evaluated with a different (convenient) reference momentum.

The third and sixth terms in this expression are the only complications, a part from the color factors, caused by the non-abelian nature of the gluon. The rest is just scalar QED. Also, all denominators are, up to a phase factors, square roots of poles. Therefore the required cancellation of all double poles[5] for the purely gluonic matrix element squared has been achieved at the matrix element level!

The square of this expression is easy obtained and after summing over color indices gives

$$
|\mathcal{M}(g_{1+}, g_{2+}; g_{3+}, g_{4+}, g_{5+})|^2 \sim (1\cdot 2)^4 \sum_P \frac{1}{(1\cdot 2)(2\cdot 3)(3\cdot 4)(4\cdot 5)(5\cdot 1)} \tag{21}
$$

where $\sum_P$ is the sum over all permutations of 1 ... 5. After the appropriate sum and average of helicity and color, the final expression is

$$
|\mathcal{M}_{gg\to ggg}| = \frac{-g^6 N^3}{240(N^2-1)} [\sum_P (1\cdot 2)^4] \sum_P \frac{1}{(1\cdot 2)(2\cdot 3)(3\cdot 4)(4\cdot 5)(5\cdot 1)}. \tag{22}
$$

Again the power of this method can be demonstrated by relating this result to that of the process of two gluons scattering into two massless gluinos and a gluon. The appropriate relationships can be obtained as before,

$$
|\mathcal{M}(g_{1+}, g_{2+}; \lambda_{3-}, g_{4+}, \lambda_{5+})| = \sqrt{\frac{(3\cdot 4)}{(4\cdot 5)}}\, |\mathcal{M}(g_{1+}, g_{2+}; g_{3-}, g_{4+}, g_{5+})| \tag{23}
$$

$$
|\mathcal{M}(g_{1+}, g_{2-}; \lambda_{3+}, \lambda_{4-}, g_{5+})| = \sqrt{\frac{(2\cdot 3)}{(2\cdot 4)}}\, |\mathcal{M}(g_{1+}, g_{2-}; g_{3+}, g_{4-}, g_{5+})|. \tag{24}
$$

The final result for this process is

$$|\mathcal{M}_{gg\to g\lambda\lambda}|^2 = \frac{-g^6N^3}{20(N^2-1)}\left[\,\left[\,(1\cdot 4)^3(1\cdot 5) + (1\Leftrightarrow 2) + (1\Leftrightarrow 3)\,\right] + (4\Leftrightarrow 5)\,\right] \sum_P \frac{1}{(1\cdot 2)(2\cdot 3)(3\cdot 4)(4\cdot 5)(5\cdot 1)}. \tag{25}$$

An interested reader can now enjoy calculating $|\mathcal{M}_{\lambda\lambda\to g\lambda\lambda}|^2$ in a similar manner.

5 TWO TO FOUR PROCESSES

In this section I consider the two gluon to four gluon process. Here I define all helicities as if all particles are incoming. Once again supersymmetry can be used to show that to tree level

$$|\mathcal{M}(g_{1-},g_{2-},g_{3-},g_{4-},g_{5-},g_{6-})| = 0, \tag{26}$$

$$|\mathcal{M}(g_{1-},g_{2-},g_{3-},g_{4-},g_{5-},g_{6+})| = 0, \tag{27}$$

$$|\mathcal{M}(g_{1-},g_{2-},g_{3-},g_{4-},g_{5+},g_{6+})| = \frac{s_{56}}{s_{23}}\,|\mathcal{M}(g_{1-},\phi_{2-},\phi_{3-},g_{4-},\phi_{5+},\phi_{6+})|, \tag{28}$$

and

$$\begin{aligned}|\mathcal{M}(g_{1-},g_{2-},g_{3-},g_{4+},g_{5+},g_{6+})| = \frac{1}{s_{23}s_{56}}\Big|&(s_{12}+s_{23}+s_{13})^2\mathcal{M}(g_{1-},\phi_{2-},\phi_{3-},g_{4+},\phi_{5+},\phi_{6+})\\ &-2i\sqrt{-s_{14}}(s_{12}+s_{23}+s_{13})(p_5^x+p_6^x-ip_5^y-ip_6^y)\mathcal{M}(\lambda_{1-},\phi_{2-},\phi_{3-},\lambda_{4+},\phi_{5+},\phi_{6+})\\ &-s_{14}(p_5^x+p_6^x-ip_5^y-ip_6^y)^2\mathcal{M}(\phi_{1-},\phi_{2-},\phi_{3-},\phi_{4+},\phi_{5+},\phi_{6+})\Big|\end{aligned} \tag{29}$$

where this last equation is given in the center of mass frame of particles 1 and 4. Unfortunately, a compact analytical result for this process has not yet been obtained but fast numerical programs[6] have been written which allow both experimentialists and theorists to study this process. It is worth noting that various checks can be applied to these programs especially permutation symmetry and the absence of double poles when two momenta are made parallel.

Methods similar to that used in the two to three gluonic process might lead to a simple expression for the amplitude which can then be squared without producing an enormous number of terms.

6 DISCUSSIONS

The use of supersymmetry is a powerful tool in reducing the complicated calculations required when dealing with gluons in perturbative QCD. There are many areas one could extent the basic ideas of this paper. Can one include quarks into this scheme? Yes.[2,7] Are higher order calculations simplified by using the supersymmetry relations and subtracting out the extra graphs not found in ordinary QCD? What happens if you include masses which do not break supersymmetry? If they do break supersymmetry? How about extenting these ideas to the complete standard model.

My collaborator, Tom Taylor, deserves special thanks for many discusions on the subject of this talk. I would also like to take this opportunity to thank my hosts, colleagues and friends in China for a very fruitful and stimulating visit to their magnificent country.

References

[1] M. T. Grisaru, H. N. Pendleton and P. van Nieuwenhuizen, Phys. Rev. **D15**, 997 (1977);
M. T. Grisaru and H. N. Pendleton, Nucl. Phys. **B124**, 81 (1977).

[2] S. J. Parke and T. R. Taylor Phys. Lett. **157B**, 81 (1985).

[3] S. Dawson, E. Eichten and C. Quigg, Phys. Rev. **D31**, 1581 (1985).

[4] Z. Xu, D.-H. Zhang and L. Chang, Tsinghua University Preprints TUTP-84/3, 84/4, 84/5.

[5] G. Altarelli and G. Parisi, Nucl. Phys. **B126**, 298 (1977).

[6] S. J. Parke and T. R. Taylor, Fermilab Preprint Pub-85/118-T;
Z. Kunszt, CERN Preprint TH-4319 (1985).

[7] S. J. Parke and T. R. Taylor, Fermilab Preprint Pub-85/162-T.

Signatures of Exotic Fermions and Other New "Low-Energy"* Phenomena in Superstring E_6

Jonathan L. Rosner
Enrico Fermi Institute and Department of Physics
University of Chicago
5640 S. Ellis Ave., Chicago, IL 60637

ABSTRACT

Superstring theories have drawn renewed attention to the exceptional group E_6 as a grand unified theory. We discuss the extra $U(1)$ symmetries (beyond that in the standard $SU(3)\times SU(2)\times U(1)$ model) that arise when E_6 is broken in such theories. These extra $U(1)$'s must lead to at least one extra neutral gauge boson, called Z', for which we give mass limits and branching ratios. Neutral current effects of this Z' and its direct production are discussed briefly. Finally, some signatures are noted of exotic fermions to which Z' can decay.

I. INTRODUCTION

The familiar generations of quarks and leptons,

$$\begin{pmatrix} u \\ d \end{pmatrix} \quad \begin{pmatrix} c \\ s \end{pmatrix} \quad \begin{pmatrix} t \\ b \end{pmatrix}$$
$$; \qquad ; \qquad ; \ldots \tag{1}$$
$$\begin{pmatrix} \nu_e \\ e \end{pmatrix} \quad \begin{pmatrix} \nu_\mu \\ \mu \end{pmatrix} \quad \begin{pmatrix} \nu_\tau \\ \tau \end{pmatrix}$$

are remarkably repetitive. Can there be "orphan" particles, ones that don't belong to a conventional family? Could there exist different numbers of charge $\frac{2}{3}$ and charge $-1/3$ quarks, or different numbers of quarks and leptons? Could unpaired leptons (for example, extra neutral ones) exist? A recent class of theories, based on superstrings,[1] suggests that the answer could be Yes.

* Presented at International Symposium on Particle and Nuclear Physics, Beijing, Sept. 2-7, 1985.

Variations in a pattern are nothing new to physics; they have already provided important clues to a new level of structure. Consider, for example, the repetitive pattern depicted in Fig. 1a:

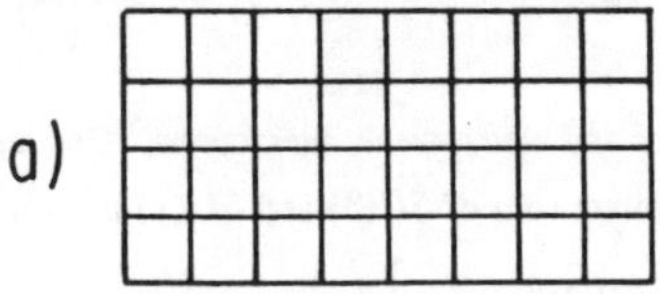

Is this familiar? Perhaps it is more easily recognizable when we split it up (Fig. 1b):

If the pattern is still a mystery, it is no longer so when we see how it varies (Fig. 1c):

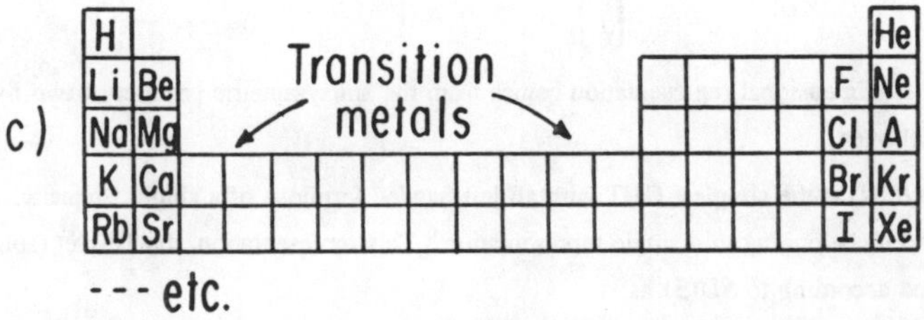

What we are looking for in the present case are the quark-lepton analogues to the transition metals!

In the past two years, superstring theories have drawn renewed attention to the grand unified group[2] E_6 and its subgroups. The fermions in E_6 include leptons and quarks not belonging to traditional "families". In this report I will describe some suggestions[3,4] for observing such particles.

Sec. II is devoted to a brief description of E_6 as a grand unified theory (GUT). The breaking of E_6 in superstring theories as originally proposed leads to at least one extra neutral gauge boson, denoted Z'. Its properties are described in Sec. III, and the neutral current effects to which it can lead are treated in Sec. IV. Prospects for direct production of Z' are noted in Sec. V. The Z' is one gateway to producing exotic fermions whose signatures are noted in Sec. VI. Some problems associated with fermion masses are mentioned in Sec. VII. Sec. VIII concludes.

Since this talk was given, many articles have appeared going into the subjects covered here in greater detail or even modifying the conclusions somewhat. I shall try to be faithful to the content of the original talk while pointing out new developments of which I am aware.

II. E_6 AS A GUT

The low-energy group $SU(3)_c \times SU(2)_L \times U(1)_{Y_w}$ can be embedded into several larger ones, including $SU(5)$, $SO(10)$, and E_6.

A. $SU(5)$ is the minimal GUT unifying color and electroweak interactions.[5] One simply writes its transformations in a 5×5 representation as the direct sum of $SU(3)$ and $SU(2)$ transformations:

$$\left[SU(5)\right] = \left[\begin{array}{c|c} SU(3) & \\ \hline & SU(2) \end{array}\right] \tag{2.1}$$

Quarks and leptons fall into *two* representations of $SU(5)$,

$$5^* = \begin{bmatrix} \bar{d}_1 \\ \bar{d}_2 \\ \bar{d}_3 \\ e^- \\ \nu \end{bmatrix}_L \quad \text{and } 10 = \begin{bmatrix} 0 & \bar{u} & -\bar{u} & u & d \\ & 0 & \bar{u} & u & d \\ & & 0 & u & d \\ & & & 0 & e^+ \\ & & & & 0 \end{bmatrix}_L , \tag{2.2}$$

where the 10-dimensional representation comes from the antisymmetric product of two five-dimensional ones.

B. $SO(10)$ is the simplest GUT with all left-handed fermions of a single "generation" (e.g., e, ν, u, and d) belonging to a single representation.[6] This representation, the 16-plet (spinor), may be decomposed according to $SU(5)$ as

$$16 = 5^* + 10 + 1 \tag{2.3}$$

The $SU(5)$ singlet is new, and as yet unobserved. We shall call it $\bar{N}$.

C. E_6 contains $SO(10)$ and has recently emerged from superstring theory as a plausible subgroup of $E_8 \times E_8{}'$.[7] The latter is one of the few allowed possibilities for an internal symmetry group of the superstring.[8]

The fundamental representation of E_6 is a 27-plet, whose decomposition under $SO(10)$ is

$$27 = 16 + 10 + 1. \tag{2.4}$$

The 16-plet is the usual "generation" (2.3). The 10-plet contains a $5 + 5^*$ of $SU(5)$, while the singlet corresponds to a neutral lepton which we shall label n. The new 5^*-plet of $SU(5)$ consists of

$$5^* = (\bar{h}, E^-, \nu_E), \tag{2.5}$$

where h is a new isosinglet $Q = -1/3$ quark. There will also be a 5-plet

$$5 = (h, E^+, \bar{N}_E). \tag{2.6}$$

D. *GUT-level* E_6 breaking must proceed via several steps through the chain

$$E_6 \to \ldots \to SU(3)_c \times SU(2)_L \times U(1)_{Y_W} . \tag{2.7}$$

In present versions of superstring theories,[9] the breaking of E_6 behaves as if induced by an *adjoint* (78-dimensional) representation of Higgs bosons.[10] This scheme must lead to at least one extra $U(1)$. [11] The proof is very simple. Consider the subgroup

$$E_6 \supset SU(3)_c \times SU(3)_L \times (SU(3)_R , \tag{2.8}$$

where "c" stands for color, "L" stands for "left", and "R" stands for "right". [The $SU(2)_L$ group must be a subgroup of $SU(3)_L$.] The 78-plet of Higgs decomposes under this subgroup as

$$\begin{aligned} 78 = &(8, 1, 1) + (1, 8, 1) + (1, 1, 8) \\ &+ (3^*, 3, 3) + (3, 3^*, 3^*) . \end{aligned} \tag{2.9}$$

The Higgs bosons, preserving $SU(3)_c$, must of course belong to $(1, 8, 1) + (1, 1, 8)$. Now, an octet of $SU(3)$ breaks it down to $SU(2)$ only if an additional $U(1)$ is also preserved.[12] Thus, if $SU(2)_L$ is preserved, so is a $U(1)$ quantum number which we shall call Y_L.

The electromagnetic charge is

$$Q_{em} = I_{3L} + Y_W/2 , \tag{2.10}$$

where Y_W is the weak hypercharge. Can Y_L be the same as Y_W? No, since then all antiquarks would have no charge, and there would be no way to have integrally charged leptons. Under $[SU(3)]^3$, the fundamental 27-plet decomposes into

$$27 = (3^*, 1, 3^*) + (3, 3, 1) + (1, 3^*, 3) , \tag{2.11}$$

where the first set of parentheses refer to antiquarks, the second to quarks, and the third to leptons. Note that antiquarks are singlets under $SU(3)_L$. Thus we need

$$Y_W = Y_L + [SU(3)_R \text{ contribution}] , \tag{2.12}$$

and the two terms on the right-hand side of Eq. (2.12) have to be separately conserved at the GUT scale.

E. *The electric charge in* E_6 must acquire contributions both from $SU(3)_L$ and from $SU(3)_R$ of Eq. (2.8), according to the previous discussion:

$$Q_{em} = Q_L + Q_R . \tag{2.13}$$

The embedding of quarks in the $(3, 3, 1)$ representation of $SU(3)_c \times SU(3)_L \times SU(3)_R$ [Eq. (2.11)] implies that we must take

$$Q_L = I_{3L} + \frac{Y_L}{2} . \tag{2.14}$$

However, three distinct choices are possible for Q_R (essentially corresponding to $U(1)$'s orthogonal to U-spin, V-spin, and I-spin subgroups of $SU(3)_R$).[3] Thus, we may take

$$Q_R = I_{3R} + \frac{Y_R}{2} \text{ } (U\text{–spin singlet}), \tag{2.15a}$$

$$Q_R = -I_{3R} + \frac{Y_R}{2} \text{ } (V\text{–spin singlet}), \tag{2.15b}$$

or

$$Q_R = -Y_R \text{ } (I\text{–spin singlet}). \tag{2.15c}$$

The choices are equivalent unless we specify how the breaking of E_6 to $SO(10)\times U(1)_\psi$ and the subsequent breaking of $SO(10)$ to $SU(5)\times U(1)_\chi$ affect $SU(3)_R$. [With this notation we hereby define $U(1)_\chi$ and $U(1)_\psi$, as in Ref. 13.] With the conventional choice that $SU(2)_R$ is a generator of $SO(10)$, we find that a) Eq. (2.15a) leads to an electromagnetic charge Q_{em} which is a generator of $SU(5)$, and which receives no contribution from $U(1)_\psi$ or $U(1)_\chi$, b) Eq. (2.15b) gives Q_{em} in $SO(10)$ but not wholly in $SU(5)$ (it receives a contribution from $U(1)_\chi$), and c) Eq. (2.15c) gives Q_{em} in E_6 but not wholly in $SO(10)$ or $SU(5)$ (it receives contributions from both $U(1)_\psi$ and $U(1)_\chi$). In what follows, we shall choose (2.15a), so that

$$Q_{em} = I_{3L} + \frac{Y_L}{2} + I_{3R} + \frac{Y_R}{2} \quad . \tag{2.16}$$

III. THE SECOND Z

A. *A new* $U(1)$ charge.

Comparing Eqs. (2.10) and (2.16), we find that

$$Y_W = 2I_{3L} + Y_L + Y_R \quad . \tag{3.1}$$

In the standard electroweak theory, Y_W is conserved down to modest ($\sim M_W$) energy scales. In the most straightforward version of superstring theories, we have seen that Y_L must be *separately* conserved at the E_6-breaking scale. Now, the photon and (standard) Z^0 each couple to a different linear combination of I_{3L} and Y_W. The separate conservation of Y_L means that there will be a third neutral boson (a second Z; call it Z') coupling to a *different linear combination* of $I_{3R} + Y_R/2$ and Y_L than that in Eq. (3.1). If the $U(1)$ couplings for $Y_L/2$ and $I_{3R} + Y_R/2$ are equal, as they frequently are in the simplest symmetry-breaking chains, one finds that this new linear combination is

$$Q' = \frac{1}{2\sqrt{15}}\{3(I_{3R} + \frac{Y_R}{2}) - 6Y_L\} \ . \tag{3.2}$$

Here we have normalized Q' in such a way that

$$\sum_{27} Q'^2 = \sum_{27} I_{3L}{}^2 = 3. \tag{3.3}$$

The combination (3.2) may also be expressed in terms of the charges $Q(\psi)$ and $Q(\chi)$ associated with the $U(1)$'s in $E_6 \to SO(10)\times U(1)_\psi \to SU(5)\times U(1)_\chi\times U(1)_\psi$.

B. *Fermions in the 27-plet* must be identified in order to calculate Z' branching ratios. They are depicted in Fig. 2. The corresponding values of $2\sqrt{15}\,Q'$ for left-handed fermions are listed in Table I. Notice that ordinary matter has small values of $2\sqrt{15}\,Q' = (1,-2)$, whereas some forms of exotic matter have values up to -5 of this quantity.

C. ***The branching ratios of a Z' within a single generation*** **can be calculated with the charges of Table I. The results are shown in Table II.**

Table I. Values of $2\sqrt{15}\,Q'$ describing coupling of second Z in superstring theories

Quarks	Antiquarks	Leptons
u, d -2	$\bar{h}, \bar{d}$ 1	e^+ -2
h 4	$\bar{u}$ -2	$E^+, \bar{N}_E$ 4
		$\bar{N}_e$ -5
		ν_E, E^- 1
		n -5
		ν_e, e^- 1

Table II. Branching ratios within a single generation for Z' decays

$u\bar{u}$	$3(2^2+2^2) = 24$	13%
$d\bar{d}$	$3(2^2+1) = 15$	8%
e^-e^+	$1+2^2 = 5$	3%
$\nu_e\bar{\nu}_e$	$1 = 1$	0.6%
$\bar{N}_eN_e$	$5^2 = 25$	14%
$h\bar{h}$	$3(4^2+1) = 51$	28%
E^+E^-	$4^2+1 = 17$	9%
$\nu_E\bar{\nu}_E$	$1 = 1$	0.6%
$\bar{N}_EN_E$	$4^2 = 16$	9%
$n_e\bar{n}_e$	$5^2 = 25$	14%
TOTAL	180	100%

In Table II the second column depicts values of $60\,Q'^2$. Because of the large value of $|Q'|$ for left-handed h quarks, the decay $Z' \to h\bar{h}$ is particularly prominent. The weak isosinglet neutral leptons N_e and n_e together make up more than 1/4 of the Z' 's decays to the first generation. There is the possibility of $e_L^+ \leftrightarrow E_L^+$ mixing which can affect the relative e^+e^- and E^+E^- branching ratios, but not their sum.[14]

Fig. 2. Members of the 27-plet of E_6

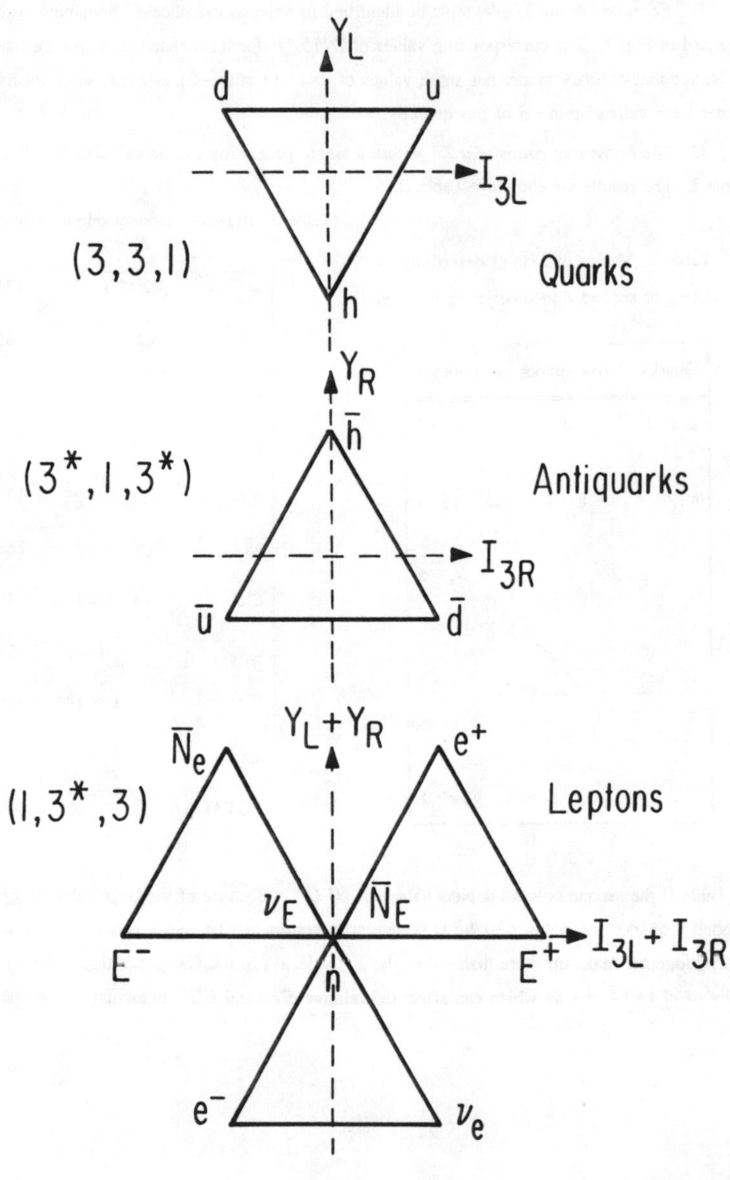

IV. THE Z' IN NEUTRAL CURRENTS

The Z' can be exchanged in all the neutral-current processes leading to tests of the standard electroweak theory. For example, in $\nu + (u,d) \to (u,d)$, the parameters $\varepsilon_{L,R}(u,d)$ receive corrections from Z' exchange. In the limit of a heavy Z' which does not mix appreciably[15] with the Z^0 one finds[16]

$$\begin{aligned}
\varepsilon_L(u) &= \frac{1}{2} - \frac{2}{3}x - \frac{1}{15}z = 0.340 \pm 0.033 \\
\varepsilon_L(d) &= -\frac{1}{2} + \frac{1}{3}x - \frac{1}{15}z = -0.424 \pm 0.026 \\
\varepsilon_R(u) &= -\frac{2}{3}x + \frac{1}{15}z = -0.179 \pm 0.019 \\
\varepsilon_R(d) &= \frac{1}{3}x - \frac{1}{30}z = -0.017 \pm 0.058
\end{aligned} \tag{4.1}$$

Here we have defined

$$x \equiv \sin^2\theta, \quad z \equiv \frac{5}{3}(\sin^2\theta)\,\frac{M_Z^2}{M_Z{}'^2} \,. \tag{4.2}$$

For simplicity we have taken all $U(1)$ couplings equal. The experimental numbers on the right-hand side of (4.1) are taken from the review of Ref. 17; more precise values are expected presently.[18]

A typical error of ± 0.01 in the fit to the quantities (4.1) implies a constraint of order

$$|z| < 0.15 \,, \tag{4.3}$$

corresponding to

$$M_{Z'} > 150 \; GeV \,. \tag{4.4}$$

This agrees with more careful estimates.[16]

V. Z' PRODUCTION

From the couplings in Table I, we find

$$\frac{\Gamma_{u\bar{u}}(Z')/M_{Z'}}{\Gamma_{u\bar{u}}(Z)/M_Z} = \frac{8}{5}\sin^2\theta \simeq 0.4 \tag{5.1}$$

$$\frac{\Gamma_{d\bar{d}}(Z')/M_{Z'}}{\Gamma_{d\bar{d}}(Z)/M_Z} = \frac{10}{15}\sin^2\theta \simeq 0.2 \,, \tag{5.2}$$

again assuming all $U(1)$ couplings are equal. Thus the Z' is harder to produce in pp or $\bar{p}p$ collisions than a "standard" Z of the same mass. A rough estimate is that one should be able to see a Z' of mass up to 200 or 300 GeV/c^2 at the Fermilab Tevatron collider, depending on what decay channels are open, by looking in e^+e^- or $\mu^+\mu^-$ final states.

VI. EXOTIC FERMION SIGNALS

There are a wide variety of characteristic features of the exotic fermions predicted in superstring theories. [We warn the reader right at the outset that recent versions of these theories have been constructed[19] in which such fermions have been pushed to very high energies, so that the effective theory remaining to test at accelerators could be rather conventional.]

A. *Large Z' branching ratios to exotic fermions* are expected. This is a general features of "extra" Z's. They can act as gateways to new forms of matter, coupling more copiously to them than to the conventional quarks and leptons.[20] For example, within a 27-plet of E_6, the branching ratios

$$Z' \to \begin{Bmatrix} h\bar{h} : 28\% \\ d\bar{d} : \ 8\% \end{Bmatrix} \tag{6.1}$$

are to be compared.

B. *Weak isosinglet $Q = -1/3$ quarks*, belonging to the 10-plet of $SO(10)$, are expected as in any E_6 model,[2] unless they have been banished to high energies.[21] Such quarks can mix with isodoublets, and thereby decay, e.g. via

$$(h \leftrightarrow d) \to u + l + \nu \ . \tag{6.2}$$

There are constraints on such mixings in several-generation models.[9,14,16] Moreover, we know that the b quark cannot be a candidate for h, or it would have dilepton decay modes which are too prominent.[22]

As a result of its isosinglet nature, the h will have a purely vectorial coupling to Z^0. Thus, the process $e^+e^- \to h\bar{h}$ will exhibit no forward-backward asymmetry for any energy, in contrast to the processes $e^+e^- \to c\bar{c}$, $b\bar{b}$, for which such asymmetries are well-documented.[23]

C. *New isodoublet leptons* occur in the 10-dimensional representation of $SO(10)$. They consist of the "ordinary" multiplet

$$\begin{pmatrix} \nu_E \\ E^- \end{pmatrix}_L \tag{6.3}$$

and the "mirror" multiplet

$$\begin{pmatrix} E^+ \\ \overline{N}_E \end{pmatrix}_L \ . \tag{6.4}$$

The E^+E^- pair couples to Z^0 purely vectorially, so $e^+e^- \to E^+E^-$ will exhibit no forward-backward asymmetry. Because ν_E and N_E are weak isodoublets, the decays

$$Z^0 \to (\nu_E\bar{\nu}_E,\ \overline{N}_E N_E) \tag{6.5}$$

should occur with the usual rate, corresponding to a partial width of about 170 MeV. The mirror lepton $\overline{N}_E$ decays by mixing with ordinary weak isodoublet neutrinos; its neutral-current decays are enhanced, and the all-neutral decay modes probably exceed 20% of the total.[24]

D. *Neutral isosinglet leptons* occur not only in E_6 models, but also in $SO(10)$, where a "right-handed neutrino" is expected. [For recent discussions of the phenomenology of such neutrinos, see Ref. 25]. In the 27-plet of E_6, two such objects are expected: $\overline{N}_e$ and n. They decay by mixing with standard neutrinos, and their all-neutral decays constitute about 10% of the total.

VII. FERMION MASSES IN E_6

A. *u* quarks are standard in E_6, since there can be no mixing except betwen members of different generations.

B. *d* and *h* quarks are constrained by weak universality[3] to have small mixings with one another, at least for the left-handed components. The right-handed components are both isosinglets, so no such constraint is present.[14]

C. *Charged leptons* can undergo substantial mixing of the form $e_L^- \leftrightarrow E_L^-$. (Both states have the same weak isospin.) In order that weak universality be preserved, this mixing must be the same as for neutrinos. The mixing of $e_R^- \leftrightarrow E_R^-$ must be small, since polarized electron-nucleon scattering experiments at SLAC imply $I_{3L}(e_R^-) = 0$.[26] [The weak isospin of E_R^- is $I_{3L}(E_R^-) = -\, I_{3L}(E_L^+) = -1/2$.]

D. *Neutral leptons* pose an interesting challenge to superstring theories. The mixing of ν_{eL} with ν_{EL} must be the same as for e_L^- with E_L^-, in order to preserve weak universality. But the light (or vanishing) masses of the neutrinos remain a puzzle. Several solutions have been proposed to this puzzle since the conference.

1. *An extra E_6 singlet* could pair with $\overline{N}_e$ to form a massive Dirac fermion,[19] leaving ν_e as a massless Majorana particle. Detailed investigation of this possibility is in progress.[27]

2. *A gauge fermion* belonging to the adjoint representation of E_6 could pair with $\overline{N}_e$.[28]

3. *Other higher-dimension couplings*, induced at high mass scales, could contribute to the neutrino mass matrix.[29]

4. *Selected Yukawa couplings could vanish* or be related to one another.[3,7] The symmetry leading to such relations has not yet been determined.

VIII. SUMMARY

The addition of some modest assumptions based on superstrings[1] to conventional E_6 phenomenology[2] has led one to expect at least an extra $U(1)$ beyond that in standard $SU(3)\times SU(2)\times U(1)$. The Z' coupled to this $U(1)$ can provide a window to a whole new set of exotic fermions, if the mass scale associated with the breaking of this $U(1)$ is low enough.

Typical E_6 signatures we have suggested include:

— Isosinglet quarks

— Mirror leptons

— Neutral leptons.

If there are only two mass scales, that of GUT breaking and that of electroweak breaking, the physics of the new $U(1)$ is more likely to occur at the lower scale and thus to be accessible in foreseeable accelerator experiments. It is important to note that many detailed studies [30] point to an intermediate mass scale of $\sim 10^{11}\ GeV$ as a distinct possibility. If that is the case, one will have to be more ingenious in devising accelerator tests of superstring theories.

ACKNOWLEDGEMENTS

I wish to thank the Theory Group of Fermilab for its hospitality during the preparation of part of these remarks. This work was supported in part by the United States Department of Energy under Contract No. DE-AC02-82ER-40073.

REFERENCES

1. For a review, see J. H. Schwarz, Physics Reports **89**, 223 (1982), and Comments on Nuclear and Particle Physics **15**, 1 (1985).

2. F. Gürsey, P. Ramond, and P. Sikivie, Phys. Lett. **60B**, 177 (1976); Y. Achiman and B. Stech, Phys. Lett. **77B**, 389 (1978); Q. Shafi, *Ibid.*, **79B**, 301 (1979); F. Gürsey and M. Serdaroglu, Lett. Nuov. Cimento **21**, 28 (1978); Nuovo Cimento **65A**, 337 (1981); R. Barbieri, D. V. Nanopoulos, and A. Masiero, Phys. Lett. **104B**, 194 (1981).

3. J. Rosner, Enrico Fermi Institute report EFI 85-34, May, 1985, Comments on Nuclear and Particle Physics, to be published.

4. David London and J. Rosner, Enrico Fermi Institute report, 1986, in preparation.

5. H. Georgi and S. L. Glashow, Phys. Rev. Lett. **32**, 438 (1974).

6. H. Georgi, in *Particles and Fields*, proceedings of the Williamsburg Meeting of the Division of Particles and Fields, American Physical Society, edited by C. E. Carlson (American Institute of Physics, New York, 1975), p. 575; H. Fritzsch and P. Minkowski, Ann. Phys. (N.Y.) **93**, 193 (1975).

7. P. Candelas, G. T. Horowitz, A. Strominger, and E. Witten, Nucl. Phys. **B258**, 46 (1985).

8. D. J. Gross, J. A. Harvey, E. Martinec, and R. Rohm, Phys. Rev. Lett. **54**, 502 (1985); Nucl. Phys. **B256**, 253 (1985); Princeton preprint, June, 1985.

9. E. Witten, Nucl. Phys. **B258**, 75 (1985); M. Dine *et al.*, Nucl. Phys. **B259**, 549 (1985); S. Ceccoti *et al.*, Phys. Lett. **156B**, 318 (1985); J. D. Breit, B. A. Ovrut, and G. Segré, Phys. Lett. **158B**, 33 (1985); S. M. Barr, Phys. Rev. Lett. **55**, 2778 (1985); E. Cohen, J. Ellis, K. Enqvist and D. V. Nanopoulos, Phys. Lett. **161B**, 85 (1985), and **165B**, 76 (1985); P. Binétruy et al.,

Lawrence Berkeley Laboratory and U. of Calif. (Santa Cruz) reports LBL-20317 and UCSC-TH-85/49, October 1985; P. K. Mohapatra, R. N. Mohapatra, and P. B. Pal, Univ. of Maryland report 85-148, 1985; R. Holman and D. Reiss, Phys. Lett. **166B**, 305 (1986); J. L. Hewett, T. G. Rizzo, and J. A. Robinson, Phys. Rev. **D33**, 1476 (1986).

10. The use of Wilson lines for E_6 breaking actually can lead to more general patterns for breaking, as shown by Dine *et al.*, Ref. 9. I am indebted to D. London for a recent discussion on this point.
11. A recent class of exceptions to the need for an extra $U(1)$ has been noted by E. Witten, Princeton preprint, November, 1985. If $E_8 \times E_8$ does not lead to an E_6 at all but to a lower symmetry, the extra $U(1)$ may not be present.
12. L. F. Li, Phys. Rev. **D9**, 1723 (1974).
13. P. Langacker, R. W. Robinett, and J. Rosner, Phys. Rev. **D30**, 1470 (1984).
14. R. W. Robinett, Univ. of Massachusetts report UMHEP-239, October, 1985.
15. P. Langacker, Phys. Rev. **D30**, 2008 (1984).
16. See also Cohen *et al.*, Ref. 9; V. Barger, *et al.*, Univ. of Wisconsin preprint MAD/PH/268, 1985, and Phys. Rev. Lett. **56**, 30 (1985); L. S. Durkin and P. Langacker, Univ. of Pennsylvania report UPR-0287-T, 1986; D. London and J. Rosner, Ref. 4.
17. J. E. Kim *et al.*, Rev. Mod. Phys. **53**, 211 (1981).
18. U. Amaldi *et al.*, in progress.
19. E. Witten, Ref. 11.
20. Examples of this behavior are given by R. W. Robinett and J. L. Rosner, Phys. Rev. **D25**, 3036 (1982); R. W. Robinett, Phys. Rev. **D26**, 2388 (1982); and in Ref. 13.
21. One reason for breaking E_6 to $SO(10)$ and giving 10-plet fermions large masses is to suppress baryon decay rates below present limits. Exchanges of 10-plet fermions can induce proton decay (see Refs. 9 and 11) unless certain Yukawa couplings are set arbitrarily to zero.
22. G. Kane and M. Peskin, Nucl. Phys. **B195**, 29 (1982); P. Avery *et al.*, Phys. Rev. Lett. **53**, 1309 (1984).
23. See, e.g., H. Yamamoto, Univ. of Tokyo report UT-HE-85/15, presented at 1985 International Symposium on Lepton and Photon Interactions at High Energies, Kyoto, Aug. 19-24, 1985.
24. L. J. Hall, A.E. Nelson, and J. E. Kim, Phys. Rev. Lett. **54**, 2285 (1985).
25. M. Gronau, C. N. Leung, and J. L. Rosner, Phys. Rev. **D29**, 2539 (1984); J. L. Rosner, Comments on Nuclear and Particle Physics **14**, 229 (1985); M. Duncan and P. Langacker, Univ. of Penna. report UPR-0293a-T, 1986.

26. C. Y. Prescott *et al.*, Phys. Lett. **77B**, 347 (1978) and **84B**, 524 (1979).

27. C. Albright and J. Rosner, in progress.

28. R. N. Mohapatra, Phys. Rev. Lett. **56**, 561 (1986).

29. S. Nandi and U. Sarkar, Phys. Rev. Lett. **56**, 564 (1986).

30. See, e.g., Dine *et al.*, and Binétruy *et al.*, Ref. 9.

QCD ANALYSIS OF THE NEUTRINO CHARGED CURRENT STRUCTURE FUNCTION xF_3 IN DEEP INELASTIC SCATTERING

Mohammad Saleem
Centre for High Energy Physics
University of the Punjab
Lahore-20, Pakistan

and

The Institute for Basic Research
96 Prescott Street, Cambridge
Massachusetts 02138
U.S.A.

AND

Fazal-e-Aleem
Centre for High Energy Physics
University of the Punjab
Lahore-20, Pakistan.

ABSTRACT

An analytic expression for the neutrino charge current structure function $xF_3(x,Q^2)$ in deep inelastic scattering, consistent with quantum chromodynamics, is proposed. The calculated results are in good agreement with experiment.

Recently, the CCFRR group[1] has measured the neutrino charged current structure functions. The data were obtained using the Fermilab narrow band beam and the Laboratory E neutrino detector[2-3]. The sturcture function $xF_3(x,Q^2)$ is extracted from the neutrino and anti-neutrino event samples by investigating the distributions in the variables x, y and Q^2 where:

$$Q^2=4EE_\mu \quad \sin^2\theta/2, \quad x=Q^2/2ME_H, \quad y=E_H/E = E_H/(E_\mu+E_H)$$

Here θ and E_μ are the outgoing muon angle and energy, E is the incident neutrino energy, E_H is the energy of the final state hadrons and M is the nucleon mass. The data for structure function xF_3[2-4] are plotted against x in Fig.1 and against Q^2 in Fig.2. They do not include systematic errors. They were analysed assuming R=0.1 and making no Fermi motion corrections. Apart from experimental corrections the data are corrected for smearing, are bin-centred, and include isoscalar corrections, charm mass corrections (m_c=1.5 GeV) , and radiative corrections.

A number of attempts have been made to fit the data for lepton-nucleon deep inelastic scattering by using QCD. The analyses are based on fitting of the Altarelli-Parisi(A-P) equations. For a fixed $Q^2=Q_o^2$, the structure function xF_3 is assumed to be of a particular form, so as to give x dependence consistent

---

The financial assistance from the Pakistan Science Foundation under contract No.P-PU/Phys(11/2) is gratefully acknowledged.

with experimental data at Q_o^2. The A-P equations are then solved numerically to yield results which are consistent with experiment. The arbitrariness of Q_o^2 is, however, restricted to sufficiently large values of Q_o^2 for which perturbative calculations can be trusted. In this paper, an analytic expression for $xF_3(x,Q^2)$, consistent with QCD, is proposed.

According to QCD, the non-singlet structure function $xF_3(x,Q^2)$ and its momenta are related to each other[5] by the equation

$$\int_0^1 x^{n-1} F_3^{NS}(x,Q^2)dx = \delta_{NS}^3 A_n^{NS}[\ln\frac{Q^2}{\Lambda^2}]^{-d_{NS}^n} \qquad (1)$$

For n=2, this equation yields [5]:

$$\int_0^1 xF_3(x,Q^2)dx = A_2[\ln\frac{Q^2}{\Lambda^2}]^{-0.427} \qquad (2)$$

The function $F_3(x,Q^2)$ can not be expressed as a product of two functions, one depending upon x only and the other depending upon Q^2 alone, because then the function of Q^2 alone as evaluated by using left hand side of equation(1) would be independent of n. This is not valid because right hand side of equation(1) depends upon n. To incorporate this n dependence, we assume that

$$F_3(x,Q^2) = A(1-x)e^{-3.7ax}$$

$$\text{where} \quad a = (\ln\frac{Q^2}{\Lambda^2})^{0.213}$$

For large value of Q^2, this expression when substituted in equation(1) gives results which are consistent with QCD.

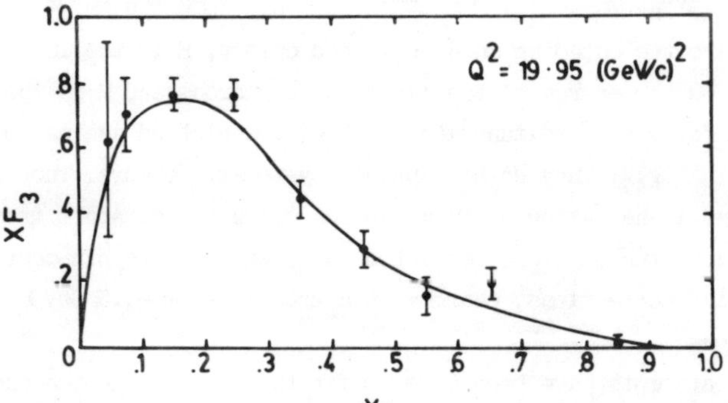

Fig.1. The structure function xF_3 extracted from the CCFRR data versus x for fixed value of Q^2. The solid curve represents the predictions of the model described in the text.

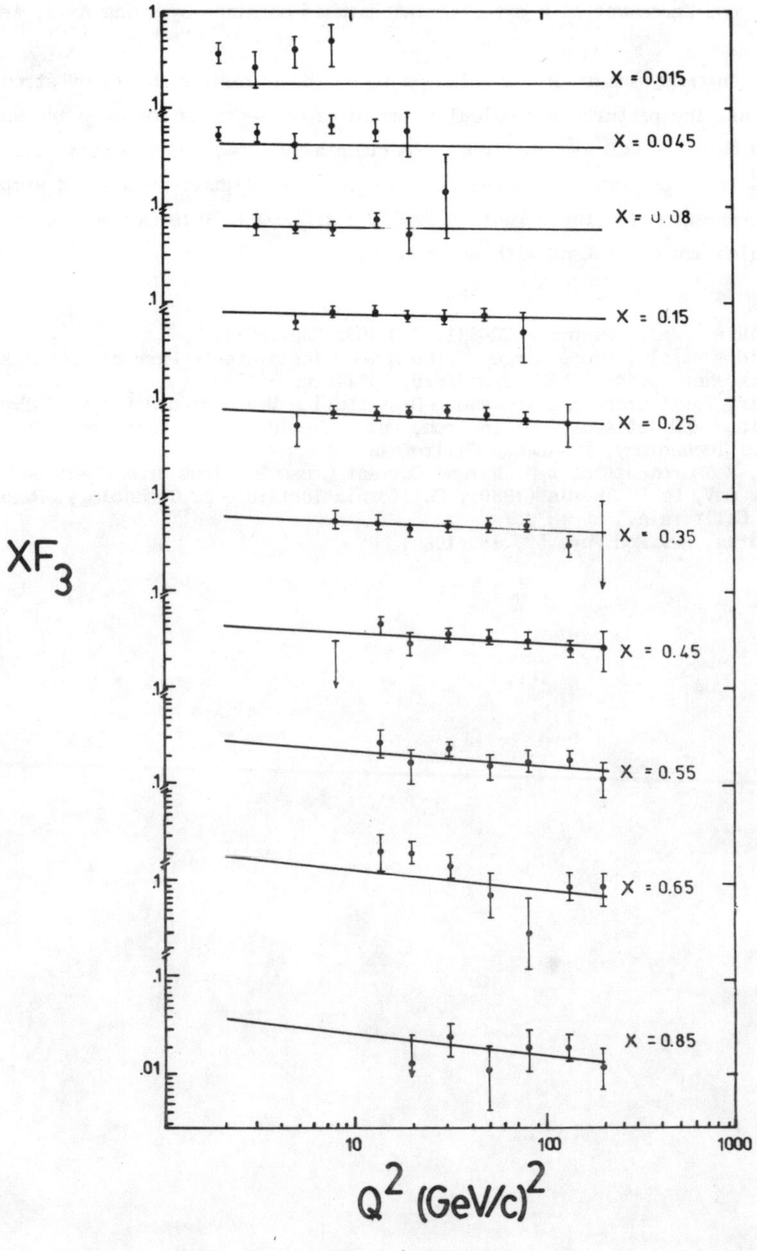

Fig.2. The structure function xF_3 extracted from the CCTRR data versus Q^2 for fixed values of x. The solid curves represent the predictions of the model described in the text.

A very good agreement with experimental data is obtained by using A=12, Λ=0.5 GeV/c.

Quantum chromodynamics is now believed to be the promising theory of strong interactions. The perturbative calculations of cross sections for deep inelastic scattering based on QCD are confirmed experimentally. However, numerical integration has to be performed to obtain various results. We have proposed a simple analytic expression for the structure function $xF_3(x,Q^2)$. This expression yields results which are consistent with experiment.

References

1. R.E.Blair et al., Preprint UR-831, COO-3065-339(1983).
2. P.Rapidis et al., Proceedings of the Summer Institute on Particle Physics, ed. A.Mosher, p.641, SLAC, Standford, California (1982).
3. R.Blair, Total Cross Section and υ-Distribution Measurements for Muon-Type Neutrinos and Antineutrinos in Iron, Ph.D. Thesis (1982), California Institute of Technology, Pasadena, California.
4. J.Lee, Measurements of $\upsilon_\mu N$ Charged Current Cross Sections from E_υ=25 GeV to E_υ=360 GeV, Ph.D. Thesis (1980), California Institute of Technology, Pasadena, California.
5. A.J.Buras, Rev.Mod.Phys.52, 199(1980).

ANALYSIS OF THE COMPTON SCATTERING OF PROTONS

Mohammad Saleem and Fazal-e-Aleem
Centre for High Energy Physics
University of the Punjab
Lahore-20, Pakistan.

ABSTRACT

The experimental data for Compton scattering of protons, including the recent measurements, have been fitted by using a simple Regge pole model.

Compton scattering is one of the fundamental processes involving photons and protons. Till 1970, the data for proton Compton scattering, $\gamma p \rightarrow \gamma p$, were available only up to incident photon energies of 1.5 GeV[1]. In the same year, Anderson et al[2] reported measurements of the differential cross sections for incident photon energies between 5 GeV and 17 GeV and for -t between 0.06 $(GeV/c)^2$ and $1.1(GeV/c)^2$. Results of their measurement are plotted in Fig.1.

The measurements of $d\sigma/dt$ were extended to large -t in 1978 by Shupe et al [3]. Their results for incident energies from 2 to 6 GeV and in the range -t=0.7 to 4.3 $(GeV/c)^2$ are shown in Fig.2.

In 1983, Duda et al[4] reported extension of their earlier measurements of angular distribution[5] to the energy region between 1.2 and 1.7 GeV. Two previous experiments had contributed data at these energies[6,7]. The measured differential cross sections of Ref.4 together with those of Refs.6 and 7 are shown in Fig.3.

It has been traditional to compare the Compton cross section to vector meson photoproduction through the vector meson dominance (VMD) model. This model for Compton scattering relates its amplitude to the scattering amplitudes for vector meson ρ^o, ω and ϕ. The results obtained by using VMD model and using the vector-meson photon coupling constants from the colliding-beam experiments[8] are systematically lower and the slope is steeper than the experimental data. Fuchs[9] has proposed a model for high-energy vector-meson photoproduction and has extended it to Compton scattering. The agreement of the model with high-energy experimental data is good so for as the differential cross sections have the correct slope; but the energy dependence is in disagreement with experiment. It has also been shown[10-12] that experimental data for the proton Compton scattering between 2.2 and 17 GeV can be fitted by using O(3,1) symmetry model.

In this paper we shall show that all the existing differential cross-

---

We are indebted to Dr.Inam ur Rehman and Dr.I.H.Qureshi for the facilities provided at CNS and PINSTECH. The financial assistance from Pakistan Science Foundation under contract No.P-PU/Phys(11/2) is gratefully acknowledged.

section data on Compton scattering can be fitted by using a simple Regge pole model in which the spins of the interacting particles are ignored and the Pomeron and π trajectories are taken as dominant exchanges. The scattering amplitude for this process then can be written as

$$T = \gamma_P \xi_P(t) s^{\alpha_P(t)} + \gamma_\pi \xi_\pi(t) s^{\alpha_\pi(t)} \quad \sqrt{nb}\ \text{GeV}$$

A very good fit with experimental data is obtained by the following choice of parameters and trajectories:

$$\gamma_P = 13.5 e^{2.5t} \sqrt{nb}\ \text{GeV}; \quad \gamma_\pi = 3.96 e^{-1.36t} \sqrt{nb}\ \text{GeV}$$

$$\alpha_P(t) = 1 + 0.175t \text{ and } \alpha_\pi(t) = 0.75t$$

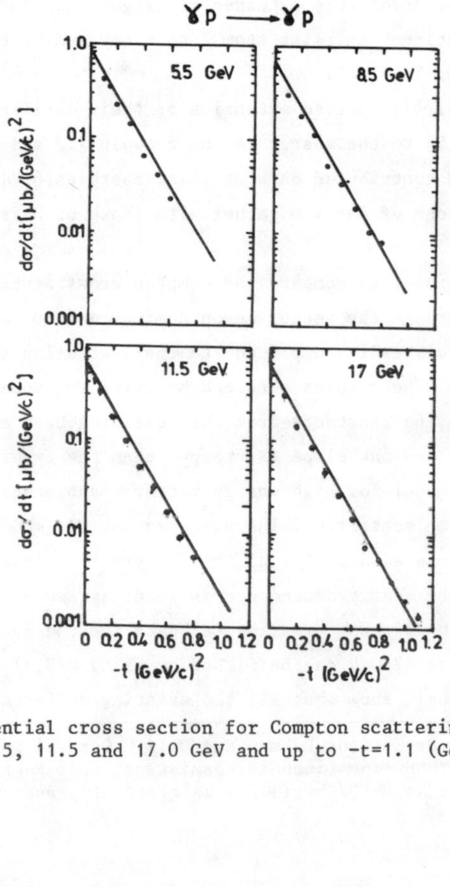

Fig. 1. Differential cross section for Compton scattering of protons at 5.5, 8.5, 11.5 and 17.0 GeV and up to $-t=1.1$ $(GeV/c)^2$.

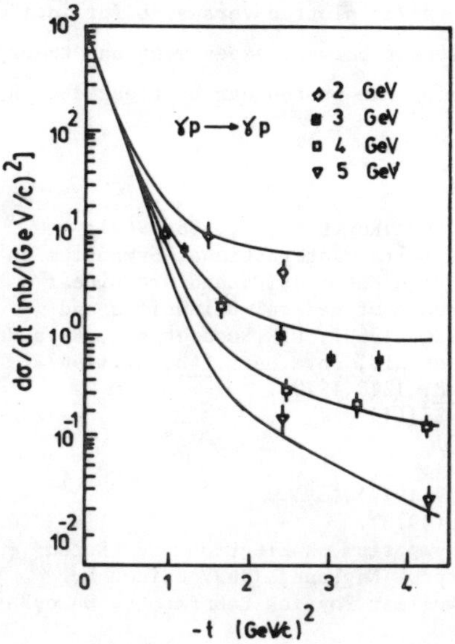

Fig.2. Differential cross section for Compton scattering of protons at 2, 3, 4 and 5 GeV and up to -t=4.3 (GeV/c)².

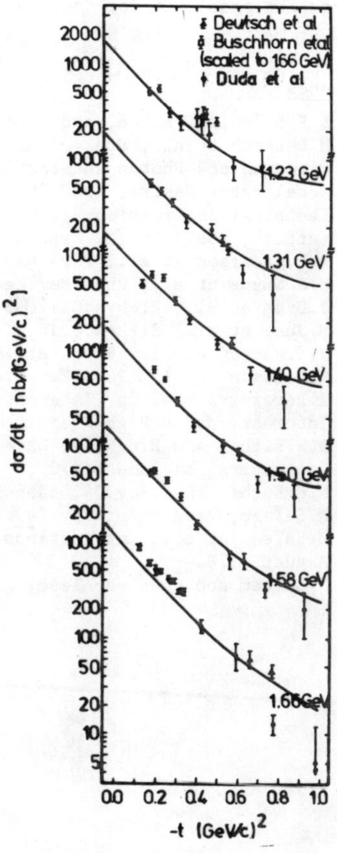

Fig.3. Differential cross section for Compton scattering of protons between 1.2 and 1.7 GeV and up to -t=1 (GeV/c)².

Figs.1 to 3 show differential cross sections dσ/dt plotted versus -t for incident photon energies of 1.2 to 17 GeV. The agreement between experiment and theory reflects the fact that for Compton scattering the photon can be treated as a hadron.

References

1. R.F.Stiening, E.Loh, and M.Deutsch, Phys.Rev.Lett.10, 536(1963); M.Deutsch et al., Proceedings of the Third International Symposium on Electron and Photon Interactions at High Energies, Standford Linear Accelerator Centre, 1967 (Clearing House of Federal Scientific and Technical Information, Washington, D.C.(1968); D.R.Rust et al., Phys.Rev. Lett.15, 938(1965); E.Eisenhandler, et al., Phys.Lett.24B, 347(1967).
2. R.L.Anderson et al., Phys.Rev.Lett.28, 1218(1970).
3. M.A.Shupe et al., Phys.Rev.Lett.10, 271(1978).
4. J.Duda et al., Z.Phys.C.17, 319(1983).
5. M.Jung et al., Z.Phys.C.10, 197(1981).
6. J.S.Barton et al., Phys.Lett.B42, 297(1972).
7. M.Deutsch et al., Phys.Rev.D.8, 3828(1973).
8. J.Perez-y-Jorba, in "International Symposium on Electron and Photon Interactions at High Energies" Liverpool England, 1969. Edited by D.W.Barben and R.E.Rand, Daresbury Nuclear Physics Laboratory, Daresbury, Lancashire, England (1970).
9. N.H.Fuchs, Phys.Rev.D4, 1566(1971).
10. M.Saleem, Prog.Theor.Phys.49, 2071(1971).
11. M.Saleem et al., International Meeting on Frontier of Physics, Singapore, August 1978.
12. M.Saleem and Fazal-e-Aleem, Hadronic J.1, 935(1978).

A UNIFYING METHOD OF OBTAINING THE FEYNMAN, KAC, CAMERON AND ALBEVARIO-HOEGH-KROHN SOLUTION TO THE HEAT OR SCHROEDINGER EQUATION

SHAHARIR

Director of the
Centre of Quantitative Studies
Universiti Kebangsaan Malaysia

A solution to the heat equation is obtained by the method of Fourier transformation. Using suitable transformations from IR^n onto C(a,b) for some a and b, the solution is shown to be of the form similar to that of Feynman (1948), Kac (1949, 1959), Cameron (1954, 1960) and Albevario-Hoegh-Krohn (1976) respectively. Consequently, various respective measures can be identified naturally, and the work of Feynman, Kac, Cameron and Albevario-Hoegh-Krohn are unified in a most natural way.

1. INTRODUCTION

Since Feynman (1948, 1965) had postulated the existence of a "path integral" in Quantum mechanics, many authors had formulated various ways of establishing the integral rigorous study was done by Kac (1949, 1959) where he derived an integral form of the solution of a class of heat equation with a rather long and complicated calculation based on probability theory, which suggests that the Feynman integral is a form of a Wiener integral. Later Ito (1951, 1961, 1967) became interested in the Feynman path integral and it led him to introduce a functional integral now as the Ito Integral and for a limited class of dynamical systems the Feynman integral can indeed be interpreted as a kind of Ito Integral. Cameron (1954, 1960) developed a different approach altogether in deriving the Feynman path integral and similarly Nelson (1964). More recently Albevario-Hoegh-Krohn (1976) formulated the Feynman integral in another different way. There are many others who proposed yet different interoretations of the Feynman integral (see Hafsah (1984)) for a complete up to date literature survey on this).

In this paper we expose the relationships between the Feynman Solution, the Kac Solution, the Cameron Solution and the Albevario-Hoegh-Krohn Solution. This is done by obtaining the solution of the heat equation via the Fourier integral and using "a change of variable" in IR^n a set of n-tuple of real numbers into a variable in $C[a, b]$ a set of continuous on the interval $[a, b]$.

2. THE SOLUTION FOR THE HEAT EQUATION IN THE FORM OF AN INTEGRAL

Consider the heat equation

$$\frac{\partial u}{\partial t} = k \frac{\partial^2 u}{\partial x^2} - Vu \qquad (2.1)$$

$$u(x, o) = \psi (x) \qquad (2.2)$$

where k is a positive number and

$$V : \longmapsto V (x)$$

We assume that $V, \psi, u, \frac{\partial u}{\partial t}, \frac{\partial^2 u}{\partial x^2}$

and continuous and square integrable with respect to the spatial coordinate x (a more general result could be obtained by assuming that they are tempered distribution, but here we will make use of the solution in L_2 only).

By the method of Fourier transformation and the property of complex normal density function (which is rigorously discussed recently by Shaharir (1983)), we obtain for $t > o$, $\gamma(t) = x$,

$$u (x,t) = \frac{1}{\sqrt{4 \pi kt}} \int \psi(y) \exp^{(-(y-x)^2/4kt - \int_o^t V(\gamma (s)) ds)} dy \qquad (2.3)$$

An extension of the above result for the heat equation in an n-dimesional space is possible, but in this paper only equation (2.3) will be used.

3. THE SOLUTION OF THE HEAT EQUATION IN THE FORM OF THE KAC SOLUTION AND THE FEYNMAN-KAC SOLUTION

By using

$$y = x + v\sqrt{2k} \qquad (3.1)$$

equation (2.3) becomes

$$u (x,t) = \int_{-\infty}^{\infty} \psi(x + v\sqrt{2k}) \exp \left(- \int_o^t V(\gamma (s) ds \right) d\mathbf{v} (v) \qquad (3.2)$$

where

$$d\mathbf{v} (v) = h (v, 0; 0, t) dv = \frac{e^{-v^2/2t}}{\sqrt{2\pi t}} dv \qquad (3.3)$$

Note that $d\nu$ is an element of measure in IR where

$$d\nu \ (v \in (a,b)) = \int_b^a h\,(v, o; o, t)\,dv \tag{3.6}$$

which may be interpreted as the prob (particle in (a,b) after some time t if at time t it is at the origin).

Now define

$$\rho_s : C_{[o]}[o,t] \longrightarrow IR \tag{3.7}$$

$$\rho_s(\beta) = \beta(s) \ , \ \beta(o) = y-x \tag{3.8}$$

where $C_{[o]}[o,t]$ is a set of continuous path in $[o,t]$ with value zero at time t ; and define also

$$g = \psi_0(\underline{x} + \mathfrak{J}) \ , \ \underline{x}(\mathfrak{z}) = x \ , \ \mathfrak{J}(\mathfrak{z}) = \mathfrak{z} \tag{3.9}$$

Then,

$$u(x,t) = \int_{C_{[o]}[o,t]} \psi(x + \beta(o)) \exp\left(-\int_0^t V(x + \beta(s))\,ds\right) d\mu(\beta) \tag{3.10}$$

where

$$\mu(\{\beta : \beta(t) \in B \subset IR\}) = \int_B \frac{e^{-v^2/2t}}{\sqrt{2\pi t}}\,dv \tag{3.11}$$

But this is precisely the Kac Solution (1959) if $k = \frac{1}{2}$ and $\lim_{t \to o^+} u(x,t) = \psi(x) = \delta(x)$. Equation (3.10) is also of the same form as the well known Feynman-Kac Solution.

In order to indentify the original Kac measure implicity used in his work (1949, 1959) we have, from (2.3),

$$u(x,t) = \frac{e^{-x^2/2t}}{\sqrt{2\pi t}} \int_{\infty}^{\infty} \exp\left(-\int_0^t V(\gamma(s))\,ds\right) dm(y) \tag{3.12}$$

where

$$dm(y) = \psi(y)\, e^{-y^2/2t \ + \ y\,x/t}\,dy \tag{3.13}$$

By a change of variable as before, we have

$$u(x,t) = \frac{e^{-x^2/_{2t}}}{\sqrt{2\pi t}} \int_{C\,[o,t]} \exp\left(-\int_o^t V(\gamma(s))ds\right) \quad dw \qquad (3.14)$$

where $C\,[o,t]$ is a set of continuous path in the interval $[o,t]$ with γ as its representative

$$w(A) = \text{prob}\ (\gamma : c < \gamma(s) < d \mid \gamma(t) = x)$$

$$= \int_c^d \psi(y)\ e^{-y^2/2t\ +\ x\,y/t} \quad dy \qquad (3.15)$$

for any Borelean set $A \subset C\ [o,t]$. Equation (3.14) for a positive continuous and bounded V was derived by Kac (1949, 1959) in a very lengthy manner. However equation (3.15) is a new result.

4. THE CAMERON SOLUTION

In a lengthy discussion Cameron (1954) had established that

$$u(x,t) = \int_{C_o\,[o,1]} \exp\left\{{}^t/a \int_o^1 U(t(1-s),\ 2\sqrt{{}^t/a}\ \gamma(s) + x)\ ds\right\}$$

$$\psi(2\sqrt{{}^t/a}\ \gamma(1) + x)\ d_W\gamma \qquad (4.1)$$

where $C_o\,[o,1]$ is the set of continuous fuction on $[o,1]$ with value zero at zero, is the solution for some class of the heat equation

$$\frac{\partial^2 u}{\partial x^2} + Uu = a\,\frac{\partial u}{\partial t}\ ,\quad a > o \qquad (4.2)$$

$$\underset{t\ \to\ o^+}{Lt}\ u(x,t) = \psi(x) \qquad (4.3)$$

We will show that the same result can be obtained easily from (2.3). In fact by the change of variable

$$\frac{y - x}{2\sqrt{kt}} = z \qquad (4.4)$$

equation (2.3) becomes

$$u(x,t) = \frac{1}{\pi} \int \psi(x + 2\sqrt{kt}\ \beta(0)) \exp(-z^2 \int_0^t V(\bar{\gamma}(\bar{s}))d\bar{s})dz \quad (4.5)$$

where

$$\bar{\gamma}(\bar{s}) = (t\bar{s}) \quad , \quad \gamma(o) = y \ , \quad \gamma(t) = x \quad (4.6)$$

$$\bar{\gamma}(\bar{s}) = x + 2\sqrt{kt}\ \beta(s) \quad (4.7)$$

$$\beta(1) = o \quad (4.8)$$

$$2\sqrt{kt}\ \beta(o) = y - x \quad (4.9)$$

Meanwhile

$$\int_0^1 V(\bar{\gamma}(\bar{s}))\ d\bar{s} = -\int_0^1 V(\bar{\gamma}(1-\bar{\bar{s}}))\ d\bar{\bar{s}} \quad (4.10)$$

$$= \int_0^1 V(x + \xi(s)\ 2\sqrt{kt})\ ds \quad (4.11)$$

where

$$\xi(s) = \beta(1-s) \quad (4.12)$$

Clearly $\xi \in C_o[o,1]$.

Thus using (4.11), (4.12), equation (4.5) becomes

$$u(x,t) = \frac{1}{\sqrt{\pi}} \int \psi(x + 2\sqrt{kt}\ \xi(1)) \exp(-z^2 - t \int_0^1 V(x + 2\sqrt{kt}\ \xi(s))\ ds\ dz \quad (4.13)$$

$$= \int_{C_o[o,1]} \psi(x + 2\sqrt{kt}\ \xi(1)) \exp(-\int_0^t V(x + 2\sqrt{kt}\ \xi(s))\ ds)\ d\mu(\xi) \quad (4.14)$$

where

$$\mu(A) = \frac{1}{\sqrt{\pi}} \int_A e^{-v^2}\ dv$$

for any Borelean set $A \subset C_o[o,1]$

Equation (4.14) is the Cameron Solution by indentifying

$$a = 1/k, \quad V = U/k$$

5. THE FEYNMAN - ALBEVARIO & HOEGH-KROHN SOLUTION

Using substitution

$$z = y - x \tag{5.1}$$

and assuming that ψ is a Fourier transform of a complex bounded measure m, equation (2.3) becomes

$$u(x,t) = \frac{1}{\sqrt{4\pi kt}} \exp\left(-\int_o^t V(\gamma(s))\,ds\right)\int_{-\infty}^{\infty}\int_{-\infty}^{\infty} \exp(-z^2/4kt + ixw + izw)\, dm(w)\, dz \tag{5.2}$$

$$= \int_{\infty}^{\infty} e^{-w^2kt}\, dp(w) \tag{5.3}$$

where

$$dp(w) = e^{ixw - \int_o^t V(\gamma(s))\,ds}\, dm(w) \tag{5.4}$$

and we have used the Fubini's Theorem and the complex normal distribution (Shaharir (1983)).

Furthermore using $< \, , \, >$ the usual inner product in the Hilbert space

$$\mathcal{H} = (C[o,t] \quad , < \, , \, >) \tag{5.5}$$

where

$$< \alpha , \beta > = Q \int_o^t \dot{\alpha}(s)\, \dot{\beta}^*(s)\, ds \quad , \quad Q > o \tag{5.6}$$

and by the change of variables formula, we have, with

$$\lambda(s) = ws \text{ and } bQ = k \tag{5.7}$$

$$u(x,t) = \int_{\infty}^{\infty} e^{-\langle\lambda,\lambda\rangle b} \, dp(w) \quad (5.8)$$

$$= B \int_{\mathcal{H}} \int_{\mathbb{R}} e^{b\langle\gamma+\lambda,\gamma+\lambda\rangle - b\langle\lambda,\lambda\rangle} \, dp(w)\, d\mu(\gamma) \quad (5.9)$$

for some constant B, since

$$\int_{\mathcal{H}} e^{b} \|\xi_1\|^2 \, d\nu(\xi_1) \quad (5.10)$$

exists for any complex number

$$b = {}^{1}/2\sigma^2, \ \arg(\sigma) \ \varepsilon \left[-\tfrac{\pi}{4}, \ \tfrac{\pi}{4}\right] \bigcup \left[\tfrac{3}{4}\pi, \pi\right] \bigcup \left[-\pi, \tfrac{3}{4}\pi\right] \quad (5.11)$$

and a complex bounded measure ν .

Equation (5.9) may be simplified to become

$$u(x,t) = W \int_{\mathcal{H}} e^{b\langle\gamma,\gamma\rangle - \int_0^t V(\gamma(s))\,ds} \psi(x - 2ik \,\mathrm{Re}\,(\gamma(t) - \gamma(o))\, d\mu(\gamma) \quad (5.11)$$

by assuming that

$$\psi(z) = \int_{-\infty}^{\infty} e^{i z y} \, dm(y)$$

for a bounded complex measure m ; or

$$u(x,t) = W \int \exp\left(\int_0^t \{k|\gamma(s)|^2 - V(\gamma(s))\} \, ds\right) \psi(x - 2ki \,\mathrm{Re}\,(\gamma(t) - \gamma(o)\, d\mu(\gamma) \quad (5.12)$$

This is precisely the same form as the Feynman Integral (1948, 1965) by indentifying

$$q = q(s) = \left(i\hbar/_{m}\right) \gamma(s), \ k = -i\hbar/_{2m}, \ V = {}^{iU}/\hbar \quad (5.13)$$

and the classical Feynman measure

$$dq = W \psi(x - 2ik \,\mathrm{Re})\, (\gamma(t) - \gamma(0))\, d\mu(\gamma) \quad (5.14)$$

We call this measure as the Feynman-Albevario & Hoegh-Krohn measure.

Feference

S.A. Albevario, R.J. Hoegh-Krohn (1976), Mathematical Theory of Feynman Path Integral, Springer Verlag.

R.H. Cameron (1954), The Generalised Heat Flow Equation and a Corresponding Poisson Formula, Annals of Math. 59, 434 - 461.

R.H. Cameron (1960), A Family of Integral Serving to Connect the Wiener and Feynman Integrals, J. Math. and Phys., 39, 126 - 140.

C. DeWitt-Morette, K.D. Elworthy (1979), Stochastic Differential Equations Lecture Notes and problems, preprint University of Texas at Austin.

R. Feynman (1948), Space-time approach to non relativistic Quantum Mechanics. Rev. Mohd. Phys. 20, 367 - 387.

R. Feynman, A.R. Gibbs (1965), Path Integral and Quantum Mechanics, McGraw Hill.

Hafsah bt Abdul Majid, Realisasi Kamiran Feynman di dalam Penyelesaian Persamaan Haba, tesis sarjana, Jabatan Matematik, Pusat Pengajian Kuantitatif, Universiti Kebangsaan Malaysia, Bangi, Selangor, 1984. (unpublished masters thesis).

K. Ito (1951), On Stochastic Differential Eqn. Mem. Amer. Math. Soc. No. 4.

K. Ito (1961), Wiener Integral and Feynman Integral, in Proc. Fourth Berkeley Symp. Math. Stat. and Prob., Uni. California Press, Berkeley, 227 - 238.

K. Ito (1967), Generalised Uniform Complex Measure in the Hilbertian Metric Space with their Application to the Feynman Path Integral, in Proc. Fifth Berkeley Simp. Math. Stat. and prob., Uni. California Press. Berkeley, Vol. II, part I, 145 - 161.

M. Kac (1949), On Distribution of Certain Wiener Functionals, Trans Amer. Math. Soc. 65, 1 - 13.

M. Kac (1959), Probability and Related Topics in Physical Sciences, Interscience Publishers.

J. Nelson (1964), Feynman Integrals and the Schroedinger Equation, J. Math. Phys., 5, 332 - 343.

Shaharir bin Mohamad Zain (1983), On Complex Normal Distribution, pracetak (1983), Jabatan Matematik Universiti Kebangsaan Malaysia, will be Published in Jurnal Sains Malaysiana, Siri Pengajian Kuantitatif (1985).

PRESENT STATUS OF COMPOSITE MODELS IN PARTICLE PHYSICS
-Minimal Composite Model of "Elementary" Particles and Fields-

Hidezumi TERAZAWA

Institute for Nuclear Study, University of Tokyo
Midori-cho, Tanashi, Tokyo 188, Japan

The present status of composite models in particle physics is reviewed with emphasis on the minimal composite model. The subjects to discuss include 1) minimal composite model, 2) generations, 3) mass spectrum and 4) Miyazawa's SUSY and Nambu's SUSY.

1. INTRODUCTION

As we all know, there exist at least thirty-six "fundamental" particles. More specifically, there are at least twenty-four "fundamental" fermions including six flavors of leptons, three generations of neutrinos (ν_e, ν_μ, ν_τ) and three generations of charged leptons (e, μ, τ), and six flavors and three colors of quarks, three generations and three colors of up-like quarks (u_i, c_i, t_i) and three generations and three colors of down-like ones (d_i, s_i, b_i) (where i=1,2,3) although the existence of the top quark is yet to be established. Now we also know that there are at least twelve gauge bosons including the photon (γ), the neutral weak boson (Z), the charged ones ($W^\pm$), and the eight gluons (G^a) (where a=1-8). Moreover, the existence of Higgs scalars (ϕ) is assumed in the standard model of Glashow-Salam-Weinberg for electroweak interactions. These "fundamental" particles are listed in Table 1. I hope you all agree with me at the point that there are too many "fundamental" particles and that there must exist subquarks, the more fundamental particles.[1)]

We should also notice that there are at least nineteen "fundamental" (dynamical) parameters in the standard model. They include "fundamental" fermion masses, at least three charged lepton masses (m_e, m_μ, m_τ), even if all the neutrino masses (m_{ν_e}, m_{ν_μ}, m_{ν_τ}) are assumed to vanish, and at least six quark

Table 1. At least thirty-six "fundamental" particles

ν_e	ν_μ	ν_τ	u_i	c_i	t_i	(i=1,2,3)
e	μ	τ	d_i	s_i	b_i	
γ	Z	$W^\pm$	G^a (a=1-8)	(ϕ)		

masses (m_u, m_c, m_t; m_d, m_s, m_b), and quark mixing matrix elements (U_{mn} where m = u, c, t, ··· and n = d, s, b, ···) whose number of independent degrees of freedom is at least four (in general $(N-1)^2$ for N generations of quarks). They also include at least one gauge boson mass, either one of $m_{W^\pm}$ and m_Z since $m_{W^\pm} = m_Z \cos\theta_W$, even if all the other gauge boson masses (m_γ, m_{G^a}, ···) are assumed to vanish, and at least one Higgs scalar mass (m_η). In addition, there are at least three gauge coupling constants, f for quantum chromodynamics (QCD) of $SU(3)_c$ and (g,g') for quantum flavor dynamics (QFD) of $SU(2)_W \times U(1)$ where the weak mixing angle is defiend by $\tan\theta_W = g'/g$. Furthermore, there is at least one independent Higgs scalar coupling constant (λ). These "fundamental" (dynamical) parameters are listed in Table 2. I hope you all agree again at the point that there are too many "fundamental" (dynamical) parameters and that there must exist the more fundamental dynamics.

Table 2. At least nineteen "fundamental" (dynamical) parameters

$(m_{\nu_e} = m_{\nu_\mu} = m_{\nu_\tau} = 0)$	m_u m_c m_t
m_e m_μ m_τ	m_d m_s m_b
U_{mn} (m=u,c,t,···, n=d,s,b,···)	
$m_{W^\pm} = m_Z \cos\theta_W$	$(m_\gamma = m_{G^a} = 0)$
m_η	
f g g'	$(\tan\theta_W = g'/g)$
λ	

This situation was present already in 1976![2)]

In this talk, I shall review the present status of composite models in particle physics with emphasis on the minimal composite model and show how these too many "fundamental" particles are made of the more fundamental particles and how these too many "fundamental" (dynamical) parameters are determined by the more fundamental dynamics.

2. MINIMAL COMPOSITE MODEL

Let us introduce the minimal composite model of quarks and leptons as a standard of reference. It consists of an iso-doublet of spinor subquarks, w_a (a=1,2) called "wakems" (meaning weak and electromagnetic), and a color-quartet of scalar subquarks, C_α (α = 0,1,2,3) called "chroms" (meaning color). The quantum numbers of these subquarks listed in Table 3 satisfy not only the Nishijima-Gell-Mann rule of

$$Q = I_3 + \frac{B-L}{2}$$

but also the "neutrality relations" of

$$\sum_w Q = \sum_C Q = 0 .$$

The simplest composite states of subquarks contain the "fundamental" bosons,

Table 3. Subquarks and their quantum numbers

	I_3	B	L	Q
w_1	+1/2			+1/2
w_2	-1/2			-1/2
C_0			+1	-1/2
C_i(i=1,2,3)		+1/3		+1/6

$$W^a_b = \bar{w}^a w_b \quad \text{and} \quad V^\alpha_\beta = C^{\dagger\alpha} C_\beta$$

and the "fundamental" fermions in the first generation,

$$f_{a\alpha} = w_a C_\alpha = \begin{pmatrix} \nu_e & u_i \\ e & d_i \end{pmatrix} .$$

Among the fundamental bosons, the familiar gauge bosons are more specifically given by

$$W^+_\mu = \bar{w}_{2L}\gamma_\mu w_{1L} , \quad W^-_\mu = \bar{w}_{1L}\gamma_\mu w_{2L} ,$$

$$A_\mu = \frac{\sqrt{3}}{2} \left(\frac{1}{2} \bar{w}_1\gamma_\mu w_1 - \frac{1}{2} \bar{w}_2\gamma_\mu w_2 - \frac{1}{2} iC_0^\dagger \overleftrightarrow{\partial}_\mu C_0 + \frac{1}{6} iC_i^\dagger \overleftrightarrow{\partial}_\mu C_i\right)$$

$$(= \sin\theta_W A^3_\mu + \cos\theta_W B_\mu)$$

$$Z_\mu = \frac{\sqrt{5}}{2} \left(\frac{1}{2} \bar{w}_{1L}\gamma_\mu w_{1L} - \frac{1}{2} \bar{w}_{2L}\gamma_\mu w_{2L}\right)$$

$$- \frac{3\sqrt{5}}{10} \left(\frac{1}{2} \bar{w}_{1R}\gamma_\mu w_{1R} - \frac{1}{2} \bar{w}_{2R}\gamma_\mu w_{2R} - \frac{1}{2} iC_0^\dagger \overleftrightarrow{\partial}_\mu C_0 + \frac{1}{6} iC_i^\dagger \overleftrightarrow{\partial}_\mu C_i\right)$$

$$(= \cos\theta_W A^3_\mu - \sin\theta_W B_\mu) \quad \text{and}$$

$$G^A_\mu = \sqrt{2}\, C_i^\dagger \overleftrightarrow{\partial}_\mu \left(\frac{\lambda^A}{2}\right)_{ij} C_j \qquad (A=1\text{-}8).$$

These "field-current identities" necessarily reproduce the following Georgi-Glashow relations for the weak mixing angle and the gauge coupling constants without depending on their grand unification assumption:

$$\sin^2\theta_W = \Sigma(I_3)^2/\Sigma Q^2 = (1/2)/(4/3) = 3/8$$

and $$f^2/g^2 = \Sigma(I_3)^2/\mathrm{tr}(\lambda^A/2)^2 = (1/2)/(1/2) = 1.$$

It seems more than a mere coincidence that these relations give the same values both in the standard quark-lepton model and in the minimal composite model. This particular "duality" is caused by the same degrees of freedom due to the four wakems, w_{aL} and w_{aR} (a=1,2), and the four chroms, C_α (α=0,1,2,3). The composite Higgs scalars are also made of a subquark-antisubquark pair as

$$\phi_a = \bar{w}_{2R} w_{aL} \quad \text{etc.}$$

3. GENERATIONS

The existence of quark-lepton generations remains as one of the most misterious misteries in particle physics. As emphasized many times, there are at least three possible ways to explain it.

3.1. "Hakems"

As in our original spinor-subquark model,[2)] let us assume that there exist an additional type of N spinor subquarks, h_n (n=1,2,3,··,N) called "hakams" (meaning horizontal). In this case, the scalar chroms ($C_{n\alpha}$ for n=1-N and α=0, 1,2,3) in the minimal composite model can be taken as di-subquark states ($h_n C_\alpha$) provided that the latter chroms (C_α) are also spinor. The composite quarks and leptons are then made of three spinor subquarks as

$$f_{na\alpha} = w_a h_n C_\alpha \ .$$

Since a hakam is shared by quarks and leptons, lepton-mixing is similar to quark-mixing. This can be tested by neutrino-oscillation experiments although no definite evidence for the oscillation has been reported.

3.2. Excited states

The second possibility for the origin of generations is that the higher generations of quarks and leptons can be taken as excited states whose ground state belongs to the first generation as

$$f_1 = (wC),\ f_2 = (wC)',\ \ f_3 = (wC)'', \cdots \ .$$

If this is the case, the origin of quark-mixing is very transparent. The hadronic weak charged current which had been written in terms of hadrons and which has been written in terms of quarks can be written most fundamentally in terms of wakems as

$$J^{\mu} = \frac{G^{\beta}}{G^{\mu}} \bar{p}\gamma_{\mu}(1 - \frac{g_A^{\beta}}{g_V^{\beta}} \gamma_5)n + \frac{G^{\Lambda}}{G^{\mu}} \bar{p}\gamma_{\mu}(1 - \frac{g_A^{\Lambda}}{g_V^{\Lambda}} \gamma_5)\Lambda + \cdots$$

$$= U_{ud}\bar{u}\gamma_{\mu}(1-\gamma_5)d + U_{us}\bar{u}\gamma_{\mu}(1-\gamma_5)s + \cdots$$

$$= \bar{w}_1\gamma_{\mu}(1-\gamma_5)w_2 \ .$$

The quark-mixing matrix can be taken as the expectation value of this subquark current sandwiched between the composite quark states as

$$<u_m|\bar{w}_1\gamma_{\mu}w_2|d_n> = U_{mn}\bar{u}_m\gamma_{\mu}d_n \ .$$

By using the algebra of subquark current,[3] the unitarity of the quark-mixing matrix can be demonstrated very easily. Furthermore, if the isospin breaking is perturbative, the matrix elements can be calculated to be

$$U_{mn} = \frac{<u_m|H_I|u_n>}{m_{u_m}-m_{u_n}} + \frac{<d_m|H_I|d_n>}{m_{d_n}-m_{d_m}} \qquad \text{for} \quad m \neq n \ ,$$

which indicates the anti-hermiticity relation of

$$U_{mn} = -U^{*}_{nm} \ .$$

This agrees with the experimental data which show

$$U_{us} \cong 0.225 \quad \text{and} \quad U_{cd} \cong -0.23 \quad \text{etc.}$$

It also indicates

$$|U_{cb}| \cong (m_s/m_b)|U_{us}| \cong 0.007\sim 0.02,$$

which is to be compared with the data of $|U_{cb}| \cong 0.059$. Furthermore, if the matrix elements of the perturbative Hamiltonian (H_I) between quark states whose generation difference is larger than one vanish, which may likely happen due to some quantum number conservation, the quark mixing matrix elements U_{ub} and U_{td} can appear as the second order perturbative effects and can be related to the other elements as

$$|U_{ub}| \cong (m_s/m_c)|U_{us}|\cdot|U_{cb}| \cong 0.0017\sim 0.0050$$

and $$|U_{td}| \cong |U_{us}|\cdot|U_{cb}| \cong 0.013.$$

The first relation is consistent with the experimental upper bound of $|U_{ub}| \lesssim 0.0082$ and the second can be checked by future experiments.

3.3. Combinations

The third possibility is that there are more than one composite states of subquarks which have the same quantum numbers as a quark or a lepton. In the minimal composite model, an infinite series of

$$f^{(1)} = wC, \quad f^{(2)} = \overline{www}C, \quad f^{(3)} = wC^\dagger C^\dagger C^\dagger,$$

$$f^{(4)} = \overline{www}C^\dagger C^\dagger C^\dagger,$$

are all candidates for real quarks and leptons. This picture of generations seems closely related to the nobel idea of Miyazawa's "hypersymmetry"[4) in which an infinite number of composite states carrying different particle numbers form a generation multiplet.

In this picture, quark mixing occurs due to the "condensation" of multi-subquark states, (wwww) and (CCCC). If this is the case and if the first, second and third generation of quarks and leptons, f_1, f_2 and f_3, are identified with $f^{(1)}$, $f^{(2)}$ or $f^{(3)}$ and $f^{(4)}$, respectively, the quark mixing matrix should have the structure of

$$U_{mn} = \begin{pmatrix} 1- & \varepsilon & \varepsilon\eta/2 & \cdots \\ -\varepsilon & 1- & \eta & \cdots \\ \varepsilon\eta/2 & -\eta & 1- & \cdots \\ \vdots & \vdots & \vdots & \ddots \end{pmatrix},$$

which indicates the relations of

$$U_{us} = -U_{cd}, \quad U_{cb} = -U_{ts} \quad \text{and}$$

$$U_{ub} = U_{td} = U_{us}U_{cb}/2 \cong 0.0066.$$

The first relation agrees with the experimental data while the last one is consistent with the experimental upper bound of $|U_{ub}| \lesssim 0.0082$.

It should be also noticed that in this picture of the third generation a lepton isodoublet of (ν_τ, τ) are made of the same contents of subquarks as a baryon isodoublet of ($\bar{n}$, $\bar{p}$)! They differ from each other only in dynamical configurations. From this viewpoint, it may not be a mere accident that the average masses of these isodoublets almost coincide, i.e.

$$1781\ \text{MeV} \leqq m_{\nu_\tau} + m_\tau \lesssim 1860\ \text{MeV} \cong m_{\bar{n}} + m_{\bar{p}} = 1878\ \text{MeV}\ ,$$

although this sum rule has not yet been derived from this picture.

4. MASS SPECTRUM

How to explain the existing mass spectrum of quarks and leptons is no doubt the most difficult and challenging problem not only in composite models but also in particle physics in general. By glancing at the face values of the observed quark and lepton masses,

$$m\begin{pmatrix} \nu_e & \nu_\mu & \nu_\tau \\ e & \mu & \tau \\ u & c & t \\ d & s & b \end{pmatrix} \cong \begin{pmatrix} <33\text{eV} & <0.5\text{MeV} & <70\text{MeV} \\ 0.511\text{MeV} & 106\text{MeV} & 1.78\text{GeV} \\ 4\text{-}5\ \text{MeV} & 1.2\text{GeV} & >25\text{GeV} \\ 8\text{-}9\ \text{MeV} & 0.15\text{GeV} & 5\text{GeV} \end{pmatrix},$$

I cannot invent no other simple empirical mass formula than

$$m \cong m_\mu \frac{Q^2(n-1)^4}{(B-L)^3} = m_\mu \begin{pmatrix} 0 & 0 & 0 \\ 0 & 1 & 16 \\ 0 & 12 & 192 \\ 0 & 3 & 48 \end{pmatrix} \cong \begin{pmatrix} 0 & 0 & 0 \\ 0 & 106\text{MeV} & 1.70\text{GeV} \\ 0 & 1.27\text{GeV} & 20.3\text{GeV} \\ 0 & 0.32\text{Gev} & 5.1\text{GeV} \end{pmatrix} .$$

This formula, whose agreement with the experimental data is unsatisfactory especially for m_s and m_t, is yet very suggestive, illustrating the following characteristic properties of the mass spectrum: 1) the electrically neutral fermions, neutrinos, have vanishing or extremely small masses, 2) the up quark is lighter than the down one but the up-like quark in the second or third generation (c or t) is heavier than the corresponding down-like one (s or b) and 3) the mass ratio of the second generation to the first one is about 200 while that of the third to the second is about 20. Concerning the property 1) arises a naive question, "Which comes first, the neutrality or the masslessness?". It is trivial that a neutral particle is not necessarily massless. However, a massless particle seems to be necessarily neutral. Try to prove this theorem!

Concerning the overall property of the mass spectrum, a more serious question is often asked. Why are the masses of composite quarks and leptons much smaller than their size inverses? More precisely, why are the masses of the existing quarks and leptons much smaller than those of the weak bosons, $W^{\pm}$ and Z? To answer this question, there have been proposed the two pictures in which quarks and leptons are taken as 1) almost chiral fermions[5)] and 2) almost[6)] or quasi[7)]-Nambu-Goldstone fermions. However, there have been proposed very few pictures for explaining the mass hierarchy lying in the generations.[8)] In the picture 3.2 for generation, the mass hierarchy is totally due to the unknown subquark dynamics while in the picture 3.3, it may be partially due to different numbers of subquarks. Also worth noticing is that the latter picture may explain the previously mentioned characteristic property of the mass spectrum 2) since quarks in the first generation contain a wakem but those in the second and third generations contain three antiwakems. In any case, much more investigation should be made before this picture becomes subject for evaluation.

Table 4. Composite "fundamental" particles

$$\begin{pmatrix} \nu_e & u_i \\ e & d_i \end{pmatrix} \quad \begin{pmatrix} \nu_\mu & c_i \\ \mu & s_i \end{pmatrix} \quad \begin{pmatrix} \nu_\tau & t_i \\ \tau & b_i \end{pmatrix} \quad \cdots$$

$$= wh_1C \quad = wh_2C \quad = wh_3C \quad \cdots$$

$$\text{or} \quad \text{or} \quad \text{or}$$

$$wC \quad (wC)' \quad (wC)''$$

$$\text{or} \quad \text{or}$$

$$\bar{w}\bar{w}\bar{w}C/wC^{\dagger}C^{\dagger}C^{\dagger} \quad \bar{w}\bar{w}\bar{w}C^{\dagger}C^{\dagger}C^{\dagger}$$

$$W^{\pm} = \bar{w}w \qquad \gamma, Z = \bar{w}w/C^{\dagger}C$$

$$G^a = C^{\dagger}C \quad (\phi = \bar{w}w)$$

5. MIYAZAWA'S SUSY AND NAMBU'S SUSY

The original form of supersymmetry à la Miyazawa[9)] may be respected approximately in the minimal composite model where composite fermions are made of a spinor wakem and a scalar chrom and composite bosons are made of a subquark-antisubquark pair. The relations between the masses of composite fermions and those of composite bosons such as

$$m_\eta = 2(\Sigma m_f^4/\Sigma m_f^2)^{1/2} [\gtrsim 2(\Sigma m_f^2/N_f)^{1/2} = (2/\sqrt{3}) m_{W^{\pm}} \cong 92\mathrm{GeV}] ,$$

where m_η is the Higgs scalar mass and N_f is the total number of quarks and leptons, may be taken as a manifestation of the approximate supersymmetry although they have been derived originally from a dynamical model of the

Nambu-Jona-Lasinio type.[2] Recently, a new type of supersymmetry lying between constituent fermions and composite bosons has been emphasized and extensively discussed by Nambu.[10] The relation between the masses of wakems and those of composite Higgs scalars such as

$$m_\xi : m_w : m_\eta = 0 : 1 : 2 \quad \text{or} \quad m_\xi^2 + m_\eta^2 = 4m_w^2$$

or more precisely[2]

$$m_\eta = 2(\Sigma m_w^4/\Sigma m_w^2)^{1/2}[\gtrsim 2(\Sigma m_w^2/2)^{1/2} = (2/\sqrt{3})m_{W^\pm} \cong 92\text{GeV}]$$

may be taken as a consequence of the supersymmetry à la Nambu.

6. CONCLUSION

I have tried not to assume any particular model for the unknown subquark dynamics although a viable candidate for it is quantum subchromodynamics, described by the Yang-Mills gauge theory of subcolor $SU(4)_{sc}$. Yet, I have tried to demonstrate how too many "fundamental" particles are made of the more fundamental particles and how too many "fundamental" (dynamical) parameters are related with each other. These results are summarized in Tables 4 and 5. In conclusion, I would like to emphasize that much more effort should be devoted to the simple, beautiful and working hypotheses of composite "fundamental" particles, perhaps only a right track in particle physics.

Table 5. Relations between "fundamental" (dynamical) parameters

$$m_\eta = 2\left(\frac{\Sigma m_f^4}{\Sigma m_f^2}\right)^{1/2} \left[\gtrsim 2\left(\frac{\Sigma m_f^2}{N_f}\right)^{1/2} = \frac{2}{\sqrt{3}} m_{W^\pm} \cong 92\text{GeV}\right]$$

$$= 2\left(\frac{\Sigma m_w^4}{\Sigma m_w^2}\right)^{1/2} \left[\gtrsim 2\left(\frac{\Sigma m_w^2}{2}\right) = \frac{2}{\sqrt{3}} m_{W^\pm} \cong 92\text{GeV}\right]$$

$$U_{us} = -U_{cd}, \quad |U_{cb}| \cong \frac{m_s}{m_b}|U_{us}| ,$$

$$|U_{ub}| \cong \frac{m_s}{m_c} |U_{us}|\cdot|U_{cb}| \quad \text{or} \quad \frac{1}{2}|U_{us}U_{cb}|$$

$$U_{cb} = -U_{ts}, \quad |U_{td}| \cong |U_{us}|\cdot|U_{cb}| \quad \text{etc.}$$

$$\sin^2\theta_w = \frac{3}{8}, \quad f^2/g^2 = 1$$

REFERENCES

1) For a recent professional review, see for example H. Terazawa, in Proc. XXII Int. Conf. on High Energy Physics, Leipzig, July 19-25, 1984, edited by A. Meyer and E. Wieczorek (Akademie der Wissenschaften der DDR, Zeuthen, 1984), Vol.I, p.63. For a recent pedagogical review, see for example H. Terazawa, to be published in the AIP Conf. Proc. Mexican School on Particles and Fields, Oaxtepec, Morelos, Dec. 3-14, 1984, edited by J.L. Lusio (AIP, New York, 1985).

2) H. Terazawa, Y. Chikashige and K. Akama, Phys. Rev. D15, 480 (1977); H. Terazawa, Phys. Rev. D22, 184 (1980).

3) H. Terazawa, Prog. Theor. Phys. 64, 1763 (1980).

4) H. Miyazawa, in Proc. Third INS Winter Seminar on the Sub-Structure of Quarks and Leptons, INS, Univ. of Tokyo, Dec. 25-26, 1981, edited by M. Yasuè, INS-J-164 (INS, Univ. of Tokyo, 1982), p.71.

5) G. 't Hooft, in Recent Developments in Gauge Theories, edited by G. 't Hooft et al. (Plenum, New York, 1980), p.135.

6) H. Terazawa, Prog. Theor. Phys. 64, 1763 (1980); W. Bardeen and V. Visnjić, Nucl. Phys. B194, 422 (1982); W.A. Bardeen, T.R. Taylor and C.K. Zachos, Nucl. Phys. B231, 235 (1984); H. Terazawa, INS-Report-485 (INS, Univ. of Tokyo) Dec., 1983.

7) W. Buchmüller, R.D. Peccei and T. Yanagida, Phys. Lett. 124B, 67 (1983); R. Barbieri, A. Masiero and G. Veneziano, Phys. Lett. 128B, 179 (1983); O.W. Greenberg, R.N. Mahapatra and M. Yasuè, Phys. Lett. 128B, 65 (1983).

8) See, however, R.N. Mohapatra, J.C. Pati and M. Yasuè, Phys. Lett. 151B, 251 (1985); S. Takeshita and M. Yasuè, Prog. Theor. Phys. 74, 349 (1985).

9) H. Miyazawa, Prog. Theor. Phys. 36, 1266 (1966).

10) Y. Nambu, Physica 15D, 147 (1985) and a preprint EFI85-71, Nov., 1985.

THE ANALYTICAL APPROACH FOR CALCULATING THE EFFECT OF THE VIRTUAL QUARK LOOPS TO SU_2 AVERAGE PLAQUETTE

Chi Min Wu, Pei Ying Zhao

Institute of High Energy Physics, Academia Sinica, Beijing, China

Using hopping parameter exapansion and cumulant expansion, we calculate the effect of fermion determinant to SU_2 average plaquette $\langle E_p \rangle$ up to k^6 order analytically. The corrections of fermion determinant to $\langle E_p \rangle$ are controllable small in our approximation.

In the lattice gauge theory,[1] as a first step, an important siplifacation of numerical computation is the quenched approximation [2] (neglecting the virtual quark loops). The virtual quark loops are contained in the "fermion determinant" which is very hard to take into account in MC procedure. There are several numerical methods to discuss this effect in hadronic calculation, such as pseudo-fermion method,[3] stochastic method[4] and method of hopping parameter expansion.[5,6] But, we still hope to develop an analytic method to discuss this effect, even if it is more difficult.

In this paper, we discuss the effect of fermion determinant to average plaquette $\langle E_p \rangle$ in SU_2 case. We use Wilson standard action and make the hopping parameter expansion, following the cumulant expansion for completing the group integral for link variables in the analytic way

$$S = S_G + S_F \quad (1)$$

$$S_G = \frac{\beta}{2N} \sum_p \mathrm{Tr}\left[U_p + U_p^+ - 2 \right] \quad (2)$$

$$S_F = k \sum_{nn'} \bar{\psi}_n \Delta_{nn'} \psi_{n'} - \sum_n \bar{\psi}_n \psi_n$$

$$\Delta_{nn'} = \sum_\mu (r + \gamma_\mu)\, U(n',n)\, \delta_{n+\mu,n'} \quad (3)$$

$U(n',n)$ is the gauge field variable sitting on the link $n \to n' = n + \mu$, ($\mu = \pm 1, \pm 2, \dots \pm d$) and $\gamma_\mu \equiv -\gamma_{-\mu}$.

Writing the fermion matrix as $Q = 1 - K\Delta$, the hopping parameter of the fermion determinant is

$$\det Q = \det(1 - K\Delta)$$
$$= \exp\left[-n_f \sum_{j=2}^{\infty} \frac{1}{j} K^j \mathrm{Tr}(\Delta^j)\right] \tag{4}$$

Setting r = 1, we have

$$S = S_G - n_f S_{eff}(K, U) \tag{5}$$

$$S_{eff}(K,U) = -8K^4 \sum_p \mathrm{Tr}(U_p + U_p^+)$$
$$+ \frac{1}{6} K^6 \left\{ -32 \sum_{\square\square} \mathrm{Tr}\, U_{\square\square} - 16 \sum_{\triangle} \mathrm{Tr}\, U_{\triangle} - 16 \sum_{\sqcup} \mathrm{Tr}\, U_{\sqcup} \right\} + \cdots \tag{6}$$

The average plaquette $\langle E_p \rangle$ is

$$\langle E_p \rangle = \frac{1}{N_p} \left\langle \sum_p \left[1 - \frac{1}{2N} \mathrm{Tr}(U_p + U_p^+) \right] \right\rangle = 1 - \frac{\partial W}{\partial \beta} \tag{7}$$

where

$$W = \frac{1}{N_p} \ln Z, \quad Z = \int [dU]\, e^{S_G - n_f S_{eff}(K,U)} \tag{8}$$

Next, we introduce a veriational action S_0 for cumulant expansion. Therefore, partition function Z can be rewritten as

$$Z = Z_0\, e^{\langle S_G - n_f S_{eff} - S_0 \rangle_0} \left\langle e^{S_G - n_f S_{eff} - S_0 - \langle S_G - n_f S_{eff} - S_0 \rangle_0} \right\rangle_0 \tag{9}$$

where

$$Z_0 = \int [dU]\, e^{S_0} \tag{10}$$

$$\langle \cdots \rangle_0 \equiv \frac{1}{Z_0} \int [dU]\, e^{S_0} (\cdots) \tag{11}$$

We choose S_0 as

$$S_0 = \beta \sum_l \mathrm{Tr}(U_l J_l^+ + h.c.) \tag{12}$$

where J is a 2 x 2 variational matrix associated with link l. The single link integral in SU_2 case is known[7]

$$f(J) = \int_{SU_2} dU_l\, e^{\beta \mathrm{Tr}(U_l J_l^+ + h.c.)} = \frac{2 I_1(X)}{X}$$
$$X = 2\beta\sqrt{K_0}, \quad K_0 = \mathrm{Tr}(JJ^+) + \det J + \det J^+ \tag{13}$$

Thus,
$$Z_0 = [f(J)]^{N_\ell}$$

$$\langle S_0 \rangle_0 = N_\ell \frac{x I_2(x)}{I_1(x)}, \quad \langle S \rangle_0 = 4N_p\left(\frac{\beta}{2N} + 8n_f K^4\right)\left[\frac{I_2(x)}{I_1(x)}\right]^4 \tag{14}$$

The expectation value of the exponent in Eq.(9) can be rewritten by cumulant expension:

$$\left\langle e^{S_G - n_f S_{eff} - S_0 - \langle S_G - n_f S_{eff} - S_0 \rangle_0} \right\rangle_0$$
$$= 1 + \frac{1}{2!}\left\langle \left(S_G - n_f S_{eff} - S_0 - \langle S_G - n_f S_{eff} - S_0 \rangle_0\right)^2 \right\rangle_0 + \frac{1}{3!}\left\langle \left(S_G - n_f S_{eff} - S_0 - \langle \cdots \rangle\right)^3 \right\rangle_0 + \cdots \tag{15}$$

Substituting (5) (12) into (15), only the connective diagrams give non-vanish contributions. We calculate the contributions of the first two terms of cumulant expansion in last equation and the hopping parameter expansion up to k^6 order. Thus, we have

$$\begin{aligned} W = {} & \frac{2}{d-1} \ln\left[\frac{2I_1(x)}{x}\right] + 4\left(\frac{\beta}{2N} + 8n_f K^4\right)\left[\frac{I_2(x)}{I_1(x)}\right]^4 \\ & + \frac{256}{3}(4d-5) n_f K^6 \left[\frac{I_2(x)}{I_1(x)}\right]^6 - \frac{2}{d-1}\frac{x I_2(x)}{I_1(x)} \\ & + 2\left(\frac{\beta}{2N} + 8n_f K^4\right)^2 \left\{ 1 + 3\left[\frac{I_3(x)}{I_1(x)}\right]^4 - 4(8d-11)\left[\frac{I_2(x)}{I_1(x)}\right]^8 \right. \\ & \left. + 4(2d-3)\left[1 + 3\frac{I_3(x)}{I_1(x)}\right]\left[\frac{I_2(x)}{I_1(x)}\right]^6 \right\} + \cdots \end{aligned} \tag{16}$$

where x is variational parameter which is determinated by the extreme condition of principal part of Eq.(9). One can find the detail in Ref.[8].

Then, substituting Eq.(16) into (7), we get the average plaquette $\langle E_p \rangle$ of pure SU_2 gauge theory and the corrections of the fermion determinant. (Fig.1) It shows the corrections of fermion is to decrease $\langle E_p \rangle$ about 0.8% ($\beta < 2.8$), to 12%($\beta > 2.8$). ($k = 1/8$).

As we know, neither strong coupling expansion nor weak coupling expansion can give a satisfactory fitting to MC data in both small and large β region. Our results are consistant with MC data.[9] It is expected the discontinuity at $\beta \simeq 2.8$ is due to the fact that these are only the contributions of the first few terms of the expansion.[10]

It is trivial to discuss it in SU_3 case. Also, we will discuss the convergence of the cumulant expansion in the near future.

REFERENCES

[1] K. Wilson Phys. Rev. D10(1974)2445.

[2] For example
H. Hamber, G. Parisi Phys. Rev. Lett. 47(1981)1795,
Phys. Rev. D27(1983)208;
E. Marinari, G. Parisi, C. Rebbi Phys. Rev. Lett. 47(1981)1798;
H. Hamber, E. Marinari, G. Parisi, C. Rebbi Phys. Lett. 108B(1982)314;
D. H. Weingarten Phys. Lett. 109B(1982)57; Nucl. Phys. B215[FS7](1983)1.
[3] H. Hamber, E. Marinari, G. Parisi, C. Rebbi Phys. Lett. 124B(1983)99;
F. Fucito, E. Marinari, G. Parisi, C. Rebbi Nucl. Phys. B180[FS3](1981) 369.
[4] B. Berg, D. Foerster Phys. Lett. 106B(1981)323;
J. Kuti Phys. Rev. Lett. 49(1982)183.
[5] A. Hasenfratz and P. Hasenfratz Phys. Lett. 104B(1981)489
A. Hasenfratz, Z. Kunszt, P. Hasenfratz and C. B. Lang
Phys. Lett. 110B(1982)289, 117B(1982)81.
[6] H. Joos and I. Montvay Nucl. Phys. B225[FS9](1983)565;
I. Montvay Phys. Lett. 132B(1983)393, 139B(1984)70.
[7] X. T. Zheng, C. I. Tan, T. L. Chen Phys. Rev. D26(1982)2843.
[8] Chi Min Wu and Pei Ying Zhao
Commun. in Theore. Phys. in Press (1985).
[9] D. H. Weingarten, D. N. Petcher Phys. Lett. 99B(1981)333.
[10] Chi Min Wu and Pei Ying Zhao
Commun. in Theore. Phys. in Press (1985)

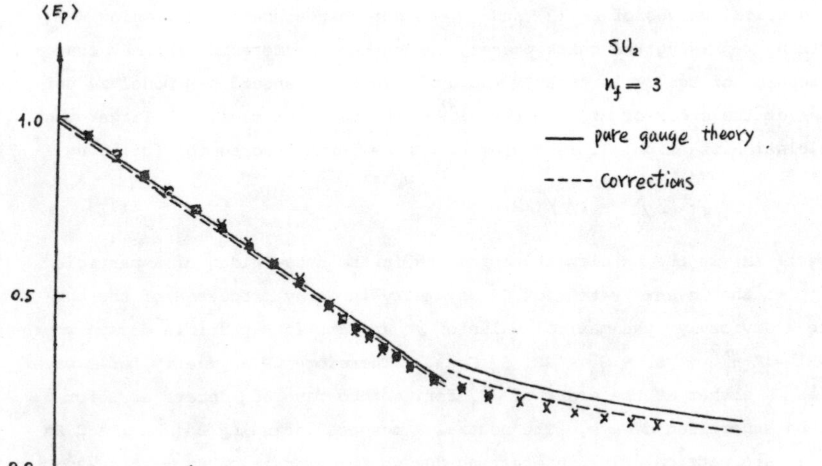

Fig.1

RAPHASING INVARIANTS AND CP IN NEUTRAL B MESONS

Dan-di Wu
Institute of High Energy Physics, Academia Sinica, Beijing, China

It is well known that arbitrary phases of quark field operators do not affect physical quantities. A immediate consequence of this principle is that only rephasing invariants of the KM matrix are of interest. The trivial invariants are the absolute values of the KM elements

$$|K_{i\alpha}|^2 \tag{1}$$

and the CP-nontrivial invariants are made by 4 elements in any 2x2 submatrix in the following specific way[1,2,3]

$$T_{ij;\alpha\beta} = K_{i\alpha} K_{j\beta} K^*_{i\beta} K^*_{j\alpha} \qquad \begin{array}{l}(i,j = 1,2,\ldots n)\\ (\alpha,\beta = 1,2,\ldots n)\end{array} \tag{2}$$

quantity (2) becomes real if the submatrix is unitary or real. In the case of three generation quarks, the K-M matrix is a 3x3 matrix and one can define $\Delta_{k\gamma} \equiv T_{ij;\alpha\beta}$ (ijk and $\alpha\beta\gamma$ cocyclic). $\Delta_{k\gamma}$ is real if the three up quarks (or down quarks) enjoy an SU(2) symmetry, i.e. have a mass degeneracy. The remarkable thing is that[4,5]

$$\mathrm{Im}\,\Delta_{i\alpha} = C_1 C_2 C_3 S_1^2 S_2 S_3 \mathrm{Sin}\,\delta = t. \tag{3}$$

Here $S_i = \mathrm{Sin}\,\theta_i$, $C_i = \mathrm{Cos}\,\theta_i$ and δ are parameters in the standard Kobayashi-Maskawa parameterization of 3x3 unitary matrix[6]. It should be noted that the numerical values of Eq.(2) and (3) do not change when redifining K-M matrix by rephasing the quark phases, though the parameterization may change[7]. The meaning of Eq.(3) is that CP violation in the standard K-M model is universal at the order of t, so the slower the decay process, the larger the CP violation fraction. This statement can be formulated as the following[7]

$$\mathcal{A} = t/\gamma(\mathrm{KM}) \tag{4}$$

where γ(KM) is the KM elements appearing in the decay width of a particle and $\mathcal{A}$ is the event rate of CP asymmetry in decay processes of the particle. Obviously the maximal value of $\mathcal{A}$ appears in B-particle decays where $\gamma(\mathrm{KM}) = |K_{23}|^2 \approx S_2^2 + S_3^2 + 2S_2 S_3 \mathrm{Cos}\,\delta$. Therefore CP asymmetry for B particles is either at the order of S_1^2 for Cabibbo favored decays or 1 for Cabibbo suppressed decays. The neutral B mesons, including Bd($\bar{b}d$) and Bs ($\bar{b}s$), are particularly interesting due to the transition $B^\circ \longleftrightarrow \bar{B}^\circ$, so dynamical suppressions may only play the least role. The rates of some processes for neutral meson systems, such as the rate of $B^\circ \leftrightarrow \bar{B}^\circ$, are proportional to the trasition mass Δm, which for Bd is Cabibbo suppressed comparing to γ(KM). The possibility of sucessfully measuring a specific CP asymmetry parameter[8,9] is not only decided by the amount of the asymmetry but also by the amount of the event rate. An order 1 CP asymmetry in a Cabibbo suppressed decay channel is not much easier to be measured than an order S_1^2 asymmetry in a Cabibbo favored decay channel .

Now, let us turn to some specific asymmetry parameters for exclusive nonleptonic decays. In the following we shall denote the pure B meson by its contents, e.g. ($\bar{b}d$) or ($\bar{b}s$), while leave Bd and B_s for physicsl particles which at time t=0 are ($\bar{b}d$) and ($\bar{b}s$) respectively and subject to evelusion afterwards. Also we shall use ξ for $\Delta m/\gamma$ where γ is the avarage width of B_{short} and B_{Long} and Δm the mass difference between the two.

The process $B \to F^{\pm}\pi^{\mp}$ is the simplest and CP violation comes only from the $(\bar{b}d) \leftrightarrow (\bar{d}b)$ transition

$$A = \frac{B_r(B \to F^+\pi^-) + B_r(\bar{B} \to F^+\pi^-) - B_r(B \to F^-\pi^+) - Br(\bar{B} \to F^-\pi^+)}{B_r(B \to F^+\pi^-) + B_r(\bar{B} \to F^+\pi^-) + B_r(B \to F^-\pi^+) + B_r(\bar{B} \to F^-\pi^+)}$$

$$= \frac{\xi^2}{1+\xi^2} \cdot \frac{Im(M_{12}\Gamma_{12}^*)}{4(|M_{12}|^2 + \frac{1}{4}|\Gamma_{12}|^2)} = \begin{matrix} O(S_1^2) \text{ for } B_s \\ O(S_1^4) \text{ for } B_d \end{matrix} , \tag{5}$$

$$\text{event rate} = A \cdot \begin{matrix} B_r((\bar{b}s) \to F^-\pi^+) & \text{for } B_s \\ B_r((\bar{b}d) \to F^-\pi^+) & \text{for } B_d \end{matrix} . \tag{6}$$

This effect is suppressed by a dynamical factor $\frac{m_c^2}{m_b^2}\left|\frac{\Gamma_{12}}{M_{12}}\right|$.

Processes $B \to F^{\pm}K^{\mp}$ and $B \to D^{\pm}\pi^{\mp}$ are typical in complicated processes due to $(\bar{b}d) \to f$ and $(b\bar{d}) \to f$ interference[11], etc.. As an example, we have, for the process[5,12] $\gamma^* \to B\bar{B} + x \to F^{\pm}K^{\mp}l^{\mp} + x'$

$$A = \frac{\sigma(F^+K^-l^-x') - \sigma(F^-K^+l^+x')}{\sigma(F^+K^-l^-x') + \sigma(F^-K^+l^+x')}$$

$$\approx -\frac{[1+(-1)^l]\, 8\xi Im(M_{12}x^*)/2(1+\xi^2)\Delta m}{\{2(2+\xi^2)\xi^2 + [\xi^4+(2+\xi^2)]\,|x|^2\}/2(1+\xi^2) + (-1)^l\xi^2(1-|x|^2)/(1+\xi^2)^2}$$

$$= O(1) , \tag{7}$$

where $x = T((b\bar{s}) \to F^-K^+)/T((\bar{b}s) \to F^-K^+)$ is the ratio of the pure particle decay amplitudes, and l is the angular momentum of the $B\bar{B}$ system.

$$\text{event rate} = B_r((b\bar{s}) \longrightarrow F^+K^-)\cdot B_r((b\bar{s}) \to l^-x)\cdot[\text{denominator of (7)}]$$
$$= \begin{cases} O(S_1^2) & \text{for } B_s \to FK \\ O(S_1^4) & \text{for } B_d \to D\pi \end{cases} \qquad (8)$$

There are almost no dynamical suppression factors in these processes. Processes $Bd \longrightarrow K^+K^-$, $D^\circ\bar{D}^\circ$ or $B_s \longrightarrow D^+D^-$, $\psi\pi^\circ$, 2π have similar property except that f is self-conjugated here.

Processes $Bd \longrightarrow F^\pm D^\mp$, $\pi^\mp K^\pm$ and $B_s \longrightarrow F^\mp D^\pm$, $\pi^\pm K^\mp$ are characterized by having nontrivial CP phases in decay amplitudes such as $(\bar{b}d) \to \pi^-K^+$. It is similar to the partial width differences[13] for charged B particle decays, if we omit the CP asymmetry due to the $B^\circ \longleftrightarrow \bar{B}^\circ$ trasition and set $\text{Im}(M_{12}\Gamma_{12}^*)/(|M_{12}|^2 + \frac{1}{4}|\Gamma_{12}|^2) = 0$.

For processes involving cascade decay to final states including K_S[14], e.g. the process

$$(\bar{b}d) \xrightarrow{T_{B\bar{D}}} \bar{D} + \pi\text{'s} \xrightarrow{T_{\bar{D}K}} K^O\pi\text{'s} \xrightarrow{1/2p_K} K_S\pi\text{'s},$$

we have[10]

$$A = \frac{\sigma(K_S\pi\text{'s } l^-) - \sigma(K_S\pi\text{'s } l^+)}{\sigma(K_S\pi\text{'s } l^-) + \sigma(K_S\pi\text{'s } l^+)}$$

$$\approx -\frac{2\xi}{(1+\xi^2)^2}\cdot\left[1+(-1)^l\right]\frac{\text{Im}(M_{12}M_{12}^*(K)T_{B\bar{D}}T^*_{B\bar{D}}T_{\bar{D}K}T^*_{\bar{D}K})}{|T_{B\bar{D}}T_{\bar{D}K}|^2\,\Delta m\,\Delta m_K}$$

$$= O(S_1^2), \qquad (9)$$

where p_K, $M_{12}(k)$ and Δm_K are mixing parameters for the $K^\circ - \bar{K}^\circ$ system.

Note all formulas here are explicitly particle rephasing invariant, i. e. invariant when $B^\circ \longrightarrow e^{i\eta}B^\circ$ etc. This invariance gaurantees the formulas to be exclusively expressed by quantities in Eqs.(1,2). Incidently, these properties can also be found in the busiest discussed $K^\circ - \bar{K}^\circ$ system. We have[2,9,15]

$$\epsilon \simeq \frac{1+i}{2}\ \frac{\mathrm{Im}(M_{12}a_o^2)}{\Delta m\,|a_o|^2} \approx \frac{1}{\sqrt{2}}\ e^{i\pi/4}\ \mathrm{Im}\,\Delta_{33}/\ \mathrm{Re}\,\Delta_{33}\ , \tag{10}$$

$$\epsilon' \approx \frac{1}{\sqrt{2}}\ \exp\ i(\pi/2+\delta_2-\delta_o)\ \frac{\mathrm{Im}(a_2a_o^*)}{|a_o|^2}$$

$$\approx \frac{1}{\sqrt{2}}\ \exp\ i(\frac{\pi}{2}+\delta_2-\delta_o)\ \frac{\mathrm{Im}\,\Delta_{33}}{|S_1|^2}\ \frac{a_2\bar{a}^p\ \ln(m_t^2/m_c^2)/\ln(m_c^2/\Lambda_c^2)}{|a_o^s+a^e+a^p+\bar{a}^p|^2}\ , \tag{11}$$

where a_o^s, a^e and a^p (or $\bar{a}^p$) are I=0 amplitudes for spectator, exchange and penguin-like contributions respectively; and a^p and $\bar{a}^p$ are long distance (mainly from a u quark penguin) and short distance penguins respectively, $\bar{a}^p \approx k_{22}{*}k_{21}\ln(m_c^2/\Lambda_c^2)$. Note in Eqs.(10) and (11) there are not any arbitrary phases, i.e. both ϵ and ϵ' should be measurable as complex numbers. Exact formulas for ϵ and ϵ' are given in Ref.2.

References

1. O. W. Greenberg, Phys. Rev. D32(1985)1891.
2. D.-D. Wu, EFI-35-85, May, 1985,to be published.
3. C. Jarlskog, Phys. Rev. Lett. 55(1985)1039
4. The formula in the small angle limit has been noticed by L. L. Chau and W. Y. Keung, Phys. Rev. Lett. 53(1984)1802.
5. Real parts of $\Delta_K\gamma$, see D. S. Du. I. Dunietz and D. D. Wu, BIHEP-TH-24, Sep. 1985.
6. M. Kobayashi and T. Maskawa, T. Prog. Th. Phys. 49(1973)652.
7. About reparameterization independent definition of maximal CP violation, see I. Dunietz, O. W. Greenberg and D. D. Wu, Phys. Rev. Lett. 55(1985)2935
8. L. Wolfenstein, Nucl. Phys. B246(1984)45.
9. D. D. Wu. LBL-19982.
10. D. D. Wu, BIHEP-TH-27.
11. R. Sachs, EFI preprint, April 1985.
12. D. S. Du and I. Duniets, in preparation.
13. N. Bander, D. Silverman and A. Soni, Phys. Rev. Lett. 43(1979)242.
14. A. B. Carter and A. I. Sanda, Phys. Rev. D23(1981)1567; I. Bigi and A. I. Sanda, Nucl. Phys. B193(1981)25.
15. We have refered many previous discussions in Ref. 2 and 9.

THE GENERAL CHERN-SIMONS CHARACTERISTIC CLASSES AND THEIR PHYSICAL APPLICATIONS

Wu Yue-liang, Xie Yan-bo, Zhou Guang-zhao

Institute of Theoretical Physics, Academia Sinica
P.O. Box 2735, Beijing, China

Simple derivation of the general Chern-Simons characteristic classes is presented. Application to construct current conservation anomalies and effective action is disscussed in detail. The effective action of nonlinear σ model where the fermions are interacting with external gauge fields is obtained. The central term of Virasoro algebra is also derived by topological method. Possible existence of a gauge potential in the group manifold and its relation to θ-vacuum is indicated.

GENERAL SOLUTIONS OF WESS-ZUMINO CONSISTENCY CONDITIONS FOR AXIAL-VECTOR CURRENT DIVERGENCE ANOMALY

Xiong Chuan-Sheng, Zhu Zhong-Yuan

Institute of Theoretical Physics, Academia Sinica
P.O. Box 2735, Beijing, China

The general solutions of the Wess-Zumino Consistency conditions for axial-vector current divergence anomaly are derived. The terms different from Bardeen anomaly in the solutions can be eliminated by use of suitable counterterms which don't violate the vector-current conservation. Thus, Wess-Zumino conditions can be considered to be necessary and sufficient for determining the structure of the minimal consistency axial anomaly up to a multiplicative constant factor.

ON VACUUM SOLUTIONS OF CONFORMALLY SYMMETRIC THEORIES

Xu Bo-Wei

Department of physics, Shanghai Jiaotong University,
Shanghai, China

We propose conformally covariant energy momentum tensor in conformally symmetric theories and show that this energy momentum tensor is useful in looking for vacuum solutions. The relations between vacuum solutions and quantum vacuum states are also discussed.

Consider a Lagrangian which is Poincaré invariant as well as dilatation covariant, then under the special conformal transformations

$$X'_\mu = \frac{X_\mu + C_\mu X^2}{\Omega(X)} \qquad \Omega(X) = 1 + 2CX + C^2X^2, \tag{1}$$

it transforms as[1]

$$\mathcal{L}'(\phi'_\alpha, \partial'_\nu \phi'_\alpha) = \Omega^4 \mathcal{L}(\phi_\alpha, \partial_\nu \phi_\alpha) + 2C^\lambda R_\lambda\,, \tag{2}$$

where

$$R_\lambda = -\pi^{\sigma\alpha}(I_{\sigma\lambda\alpha}{}^{\beta}\,\phi_\beta + l\, g_{\lambda\sigma}\phi_\alpha) \tag{3}$$

$$\pi^{\sigma\alpha} = \frac{\partial \mathcal{L}}{\partial \partial_\sigma \phi_\alpha}\,, \tag{4}$$

$I_{\sigma\lambda\alpha}{}^{\beta}$ denotes the spin matrix, and l is the conformal weight of the field. We restrict ourselves to the cases where $R_\lambda = 0$ and $R_\lambda = \partial^\sigma R_{\sigma\lambda}$, the field equations will thus be conformally covariant.

By applying Noether's theorem to conformally symmetric theories, we can obtain canonical energy momentum tensor $T_{\mu\nu}$, but which is not conformally covariant. The natural way is to propose conformally covariant energy momentum tensor[2]-[5]

$$\begin{aligned}\Theta_{\mu\nu} = {} & T_{\mu\nu} - \tfrac{1}{2}\partial^\lambda[(\pi^\alpha_\lambda I_{\mu\nu\alpha}{}^{\beta} + \pi^\alpha_\mu I_{\nu\lambda\alpha}{}^{\beta} + \pi^\alpha_\nu I_{\mu\lambda\alpha}{}^{\beta})\phi_\beta] \\ & - \tfrac{1}{2}(\partial^2 R_{\mu\nu} + g_{\mu\nu}\partial^\lambda\partial^\rho R_{\lambda\rho}) + \tfrac{1}{2}(\partial_\mu\partial^\rho R_{\rho\nu} + \partial_\nu\partial^\rho R_{\rho\mu}) \\ & - \tfrac{1}{6}(\partial_\mu\partial_\nu - g_{\mu\nu}\partial^2)R^\lambda_\lambda\,,\end{aligned} \tag{5}$$

which has properties

$$\partial^{\mu}\theta_{\mu\nu}=0 \qquad \theta_{\mu\nu}=\theta_{\nu\mu} \qquad \theta^{\mu}_{\mu}=0. \tag{6}$$

The conformally covariant energy momentum tensor plays a fundamental role in discussing vacuum solutions of the conformally symmetric theories. From the properties of eq.(6), the general form of $\theta_{\mu\nu}$ can be expressed as[6]-[8]

$$\theta_{\mu\nu}=(4\tau_{\mu}\tau_{\nu}-g_{\mu\nu}\tau_{\lambda}\tau^{\lambda})\,\theta(\tau). \tag{7}$$

Here we take $\tau=\frac{1}{2}cx^2+dx+e$ as a general parameter, and $\tau_{\lambda}=\frac{\partial\tau}{\partial x^{\lambda}}$. Conservation of $\theta_{\mu\nu}$ implies that

$$\tau_{\lambda}\tau^{\lambda}\dot{\theta}+6c\theta=0, \tag{8}$$

where dot refers to differentiation with respect to the parameter τ . It is easy to see that there are only two solutions of eq.(8)

$$\theta(\tau)=0 \qquad \text{instantonlike} \tag{9}$$

$$\theta(\tau)=\mathrm{const}(\tau_{\lambda}\tau^{\lambda})^{-3} \qquad \text{meronlike.} \tag{10}$$

In both cases, we have $E=\int d^3x\,\theta_{00}=0$ (for eq.(10), $\tau_0\neq 0$), i.e. they correspond to vacuum solutions. Many conformally symmetric theories are known to have these types of solutions[8]-[11].

The fact that $E=0$ implies that the classical vacuum solutions can be the candidate for the vacuum in the quantum world. Instead of the naive vacuum state $|0\rangle$, a new vacuum state $|\tilde{0}\rangle$ can be introduced, in which the vacuum expectation value of quantum field is[12][13]

$$\langle\tilde{0}|\hat{\phi}(x)|\tilde{0}\rangle=\phi_{cl}(x). \tag{11}$$

$\hat{\phi}(x)$ is composed of two parts

$$\hat{\phi}(x)=\phi_{cl}(x)+\hat{\phi}'(x), \tag{12}$$

where $\phi_{cl}(x)$ is the classical vacuum solutions of conformally symmetric theories, while $\hat{\phi}'(x)$ is the quantum fluctuation around it. Both $\hat{\phi}(x)$ and $\hat{\phi}'(x)$ can be quantized in the Heisenberg picture by

$$\hat{\phi}(x)=\frac{1}{(2\pi)^{3/2}}\int\frac{d^3k}{\sqrt{2|\vec{k}|}}\left[\hat{a}_{\vec{k}}(t)\,e^{i\vec{k}\cdot\vec{x}}+\hat{a}^{+}_{\vec{k}}(t)\,e^{-i\vec{k}\cdot\vec{x}}\right] \tag{13}$$

$$\hat{\phi}'(x)=\frac{1}{(2\pi)^{3/2}}\int\frac{d^3k}{\sqrt{2|\vec{k}|}}\left[\hat{c}_{\vec{k}}(t)\,e^{i\vec{k}\cdot\vec{x}}+\hat{c}^{+}_{\vec{k}}(t)\,e^{-i\vec{k}\cdot\vec{x}}\right], \tag{14}$$

with

$$[\hat{a}_{\vec{k}}(t), \hat{a}^{+}_{\vec{k}'}(t)] = \delta(\vec{k} - \vec{k}') \tag{15}$$

$$[\hat{c}_{\vec{k}}(t), \hat{c}^{+}_{\vec{k}'}(t)] = \delta(\vec{k} - \vec{k}'), \tag{16}$$

and

$$\hat{a}_{\vec{k}}(t) = f_{\vec{k}}(t) + \hat{c}_{\vec{k}}(t), \tag{17}$$

where $f_{\vec{k}}(t)$ is the Fourier components of $\phi_{cl}(x)$

$$\phi_{cl}(x) = \frac{1}{(2\pi)^{3/2}} \int \frac{d^3k}{\sqrt{2|\vec{k}|}} [f_{\vec{k}}(t) e^{i\vec{k}\cdot\vec{x}} + f^{*}_{\vec{k}}(t) e^{-i\vec{k}\cdot\vec{x}}]. \tag{18}$$

From the definition of $|\tilde{0}\rangle$ of eq.(11), it follows that

$$\hat{c}_{\vec{k}}(t)|\tilde{0}\rangle = 0 \tag{19}$$

$$\hat{a}_{\vec{k}}(t)|\tilde{0}\rangle = f_{\vec{k}}(t)|\tilde{0}\rangle. \tag{20}$$

The new vacuum $|\tilde{0}\rangle$, being an eigenstate of the annihilation operator $\hat{a}_{\vec{k}}(t)$ with eigenvalue $f_{\vec{k}}(t)$, is a coherent state which can be expressed as[14][15]

$$|\tilde{0}\rangle = e^{-\frac{1}{2}\|f\|} e^{(\hat{a}^{+} f)} |0\rangle \tag{21}$$

with

$$\langle\tilde{0}|\tilde{0}\rangle = 1, \tag{22}$$

where

$$(\hat{a}^{+} f) = \int d^3k \, f_{\vec{k}}(t) \, \hat{a}^{+}_{\vec{k}}(t) \tag{23}$$

$$\|f\| = \int d^3k \, f^{*}_{\vec{k}}(t) \, f_{\vec{k}}(t). \tag{24}$$

Eqs.(18) and (21) give the relations between classical vacuum solutions on the one hand and quantum vacuum states on the other hand. If one knew the vacuum solutions, then in principle the vacuum states could be obtained. Partial information about classical fields might yield some insight into the quantized theory. As these solutions are nonperturbative, it is hoped that they may reveal new physical configurations which can not be reached from standard perturbation theory in quantum world.

References

1. M. Flato, J. Simon, D. Sternheimer, Ann. Phys. 61(1970) 78.
2. C.G. Callan, S. Coleman, R. Jackiw, Ann. Phys. 59(1970) 42.
3. B.W. Xu, Jour. Phys. A14(1981) L97.
4. B.W. Xu, Jour. Phys. A14(1981) L125.
5. B.W. Xu, Jour. Phys. A15(1982) L329.
6. A. Actor, Z. Phys. C3(1980) 353.
7. A. Actor, Ann. Phys. 131(1981) 269.
8. A.O. Barut, B.W. Xu, Physica 6D(1982) 137.
9. A.O. Barut, B.W. Xu, Phys. Lett. 102B(1981) 37.
10. A.O. Barut, B.W. Xu, Phys. Rev. D23(1981) 3076.
11. B.W. Xu, Phys. Energ. Fort. Phys. Nucl. 9(1985) 241.
12. S. Fubini, Nuovo Cimento 34A(1976) 521.
13. V. De Alfaro, G. Furlan, Nuovo Cimento 34A(1976) 555.
14. S. Skagerstam, Phys. Rev. D19(1979) 2471.
15. G.J. Ni, Y.P. Wang, Phys. Rev. D27(1983) 969.

DYNAMICAL EQUATIONS AND A NEW ANALYTICAL APPROACH IN LATTICE GAUGE THEORIES

She-sheng Xue
Institute of High Energy Physics, Academia Sinica
P.O. Box 918, Beijing, China

In the scheme of Lattice Gauge Theories (LGT) with stochastic quantization method, we establish evolution equations which dominate stochastic process from non-equilibrium to equilibrium. It is demonstrated that these dynamical equations in stochastic process are equivalent to Schwinger-Dyson equations in LGT with path integral quantization method. Base on these equations, we develop a new analytic approach which allows step by step approximation. The plaquette averages of U(1) and SU(2) lattice system are chosen as examples to show how the approach works. The results are in satisfactary agreement with Monte Carlo ones.

1. Introduction

It is well known that the LGT[1] is the most promising approach for the quantum field theory to go beyond the perturbative region, A widely adoped and fruitful approach in LGT is the Monte Carlo simulation.[2] It is worthwhile and necessary to study stochastic quantization[3] which much coincides with Monte Carlo simulation. However, the Monte Carlo simulation suffers from the exaction of formidable computer capacity as the lattice size increase, as well as its purely numerical nature which hinders the extraction of physical insight intuitively. This is why analytic approaches are still much wanted for exploring various aspects of LGT.

2. Stochastic quantization in LGT

The dynamical variable $U_{n,\mu}(\tau)$ in LGT evolutes in fictious time τ according to Langevin equation proposed by

$$\frac{\partial}{\partial \tau} U_{n,\mu}(\tau) - \frac{1}{2}\hat{E}_{n,\mu}(S)\cdot U_{n,\mu} + i\eta_{n,\mu}\cdot U_{n,\mu} = 0 \qquad (1)$$

where, S is the action of system, and differential operator $\hat{E}_{n,\mu}$ is given by

$$\hat{E}_{n,\mu} = U_{n,\mu}\cdot\frac{\partial}{\partial U_{n,\mu}} - \frac{\partial}{\partial U^{+}_{n,\mu}}\cdot U^{+}_{n,\mu} \qquad (2)$$

The random matrices $\eta_{n,\mu}$ in eq.(1) are in Gaussian distribution

$$\langle \cdots \rangle_\eta = \frac{\int [d\eta](\cdots) \exp\{-\frac{1}{2}\sum_{n,\mu}\int d\tau \, tr(\eta^+_{n,\mu}\eta_{n,\mu})\}}{\int [d\eta] \exp\{-\frac{1}{2}\sum_{n,\mu}\int d\tau \, tr(\eta^+_{n,\mu}\eta_{n,\mu})\}} \tag{3}$$

We verify that the stochastic quantized scheme in LGT is equivalent to usual one,

$$\lim_{\tau\to\infty} \langle \cdots \rangle_\eta = \frac{\int [dU] \; (\cdots) \; e^{-S[U]}}{\int [dU] \; e^{-S[U]}} = \langle \cdots \rangle \tag{4}$$

Corresponding to eq.(1), we derive the Fokker-Planck equation, as well as the evolution equation for average of an arbitrary physical quantity G in fictious time τ direction

$$\frac{\partial}{\partial\tau}\mathbb{P}[U(\tau)] = -\hat{H}_{F.P.}\mathbb{P}[U(\tau)] \tag{5}$$

$$\frac{\partial}{\partial\tau}\langle G\rangle_P = \langle e^{-\frac{1}{2}S}\hat{H}_{F.P.}(G\,e^{-\frac{1}{2}S})\rangle_P \tag{6}$$

$$P[U(\tau)] = e^{-\frac{1}{2}S}\mathbb{P}[U(\tau)] \tag{7}$$

Where, $P[U(\tau)]$ is a distribution function of dynamical variable U, and the Fokker-Planck Hamiltonia $\hat{H}_{F.P.}$ determined by lattice system, for instance, SU(N) theory,

$$\hat{H}_{F.P.} = \frac{1}{2}\sum_{n,\mu} tr(\hat{Q}^+_{n,\mu}\hat{Q}_{n,\mu}) - \frac{1}{N}\sum_{n,\mu}(\hat{q}^+_{n,\mu}\hat{q}_{n,\mu}) \tag{8}$$

with

$$\hat{Q}_{n,\mu} = \hat{E}_{n,\mu} - \frac{1}{2}\hat{E}_{n,\mu}(S)$$
$$\hat{q}_n = tr\,\hat{E}_{n,\mu} - \frac{1}{2}tr\hat{E}_{n,\mu}(S) \tag{9}$$

It is shown eq.(6) turns out to be Schwinger-Dyson equation in LGT when the fictious time τ goes to infinite.

3. A new analytic approach

Consider the problem of evaluating the average value of a physical quantity G:

$$G = \int [dU]\, G\, e^{S(U)}/Z\,, \qquad Z = e^{-W} = \int [dU] e^{S(U)} \tag{10}$$

and W is the free energy of the system. Instead of using the exact action S to evaluate eq. (10) (which, in general as we know, is extremely difficult to be carried out analytically), we somehow introduce a trial action $S_o(U,J)$, depending of $\{U\}$ as well as on a set of parameters $\{J\}$, which should be determined to optimize the approximation. Rewriting eq.(10) in the following form

$$\langle G \rangle = \int [dU]\, G\, e^{S-S_o} e^{S_o} / \int [dU] e^{S-S_o} e^{S_o}$$

$$= \langle G\, e^{S-S_o} \rangle_o \; / \; \langle e^{S-S_o} \rangle_o \qquad (11)$$

where $\langle \ldots \rangle_o$ denotes average with respect to the trial partition function $Z_o(J)$:

$$Z_o(J) = \int [dU] e^{S_o(U,J)} = e^{-W_o} \qquad (12)$$

and by expanding e^{S-S_o} into a Taylor series, we have

$$G = \sum_{i=0}^{\infty} g_i \quad , \qquad (13)$$

where

$$g_o = \langle G \rangle_o \quad ,$$

$$g_1 = \langle G(S-S_o) \rangle_o - \langle G \rangle_o \cdot \langle S-S_o \rangle_o \quad ,$$

$$g_2 = \frac{1}{2}\Big\{ \langle G(S-S_o)^2 \rangle_o - 2 \langle G(S-S_o \rangle_o \cdot \langle S-S_o \rangle_o$$

$$+ \langle G \rangle_o \langle S-S_o \rangle_o^2 \Big\} \quad , \qquad (14)$$

......

and likewise, we can expand the free energy W into the following form:

$$W = \sum_{i=0}^{\infty} w_i \quad , \qquad (15)$$

with

$$w_o = - \ln Z_o \quad ,$$

$$w_1 = - \langle S-S_o \rangle_o \quad , \qquad (16)$$

$$w_2 = - \frac{1}{2}\Big\{ \langle (S-S_o)^2 \rangle_o - \langle S-S_o \rangle_o^2 \Big\}$$

......

There is one important property of the physical variable G, which has been unaware of in earlier studies, namely, in many cases, G statisfies a differential equation (6)

$$\langle D[G] \rangle = 0 \quad , \qquad (17)$$

where D is a differential operator in eq.(8). Analogous to eq.(13), we may expand

$$\langle D[G] \rangle = \sum_{i=0}^{\infty} d_i = 0 \qquad (18)$$

where

$$d_0 = \langle D[G] \rangle_0$$

$$d_1 = \langle D[G]\ (S-S_0) \rangle_0 - \langle D[G] \rangle_0 \langle S-S_0 \rangle_0$$

$$d_2 = \frac{1}{2}\Big\{ \langle D[G]\cdot(S-S_0)^2 \rangle_0 - 2\langle D[G](S-S_0) \rangle_0 \cdot \langle S-S_0 \rangle_0 + \langle D[G] \rangle_0 \cdot \langle S-S_0 \rangle_0^2 \Big\} \qquad (19)$$

......

For convenience, we define

$$D_i = \sum_{j=0} d_j \quad ,$$

$$G_i = \sum_{j=0} g_j \quad , \qquad i = 0,1,2..... \qquad (20)$$

$$W_i = \sum_{j=0} w_j \quad ,$$

and we have

$$\lim_{i\to\infty} G_i = G \quad ,$$

$$\lim_{i\to\infty} W_i = W \quad , \qquad (21)$$

$$\lim_{i\to\infty} D_i = 0 \quad .$$

the last line of eq.(21) is due to eq.(17). It is easy to see that from eq. (21) a step-by-step approximation scheme can be developed in the following way:

(I) Approximate $\langle G \rangle$ by G_i, with i finite:

(II) Determine the parameters $\{ J_i^* \}$ by solving the approximate equation

$$D_i(J_i^*) = 0 \quad ; \qquad (22)$$

(III) In case of several sets of solution $\{J_i^*\}, \{J_i'^*\}, \ldots$, satisfying the same eq.(22), choose the one which gives the lowest "free energy" W_i, i.e.,

$$W_i(J_i^*) < W_i(J_i'^*) < \ \ldots \qquad (23)$$

(IV) Substitue this set of parameters $\{ J_i^* \}$ into the expression of S_0 to calculate G_i .

Of course, as $i \to \infty$, the above approach is exact.

We evaluate the plaquette averages of U(1) and SU(2) system, and results [shown in Fig.1 and Fig.2 respectively] are in good agreement with Monte Carlo ones.[4][5]

References

[1] K. G. Wilson, Phys. Rev. D14(1974)2245.

[2] M. Creutz, L. Jacobs and C. Rebbi, Phys. Rep. 95(1983)201.

[3] G. Parisi and Y. S. Wu Scientia Sinica 24(1981)483.

[4] G. Bhanot and M. Creutz, Phys. Rev. D21(1980)2892.

[5] M. Creutz Phys. Rev. Lett. 43(1979)553.

Figure captions

Fig.1 Comparison of the zeroth and first approximation results (dash line and solid line respectively) to Monte Carlo one[4] in U(1) theory.

Fig.2 Comparison of the zeroth and first approximation results (dash line and solid line respectively) to Monte Carlo simulation[5] (circles and crosses) in SU(2) theory.

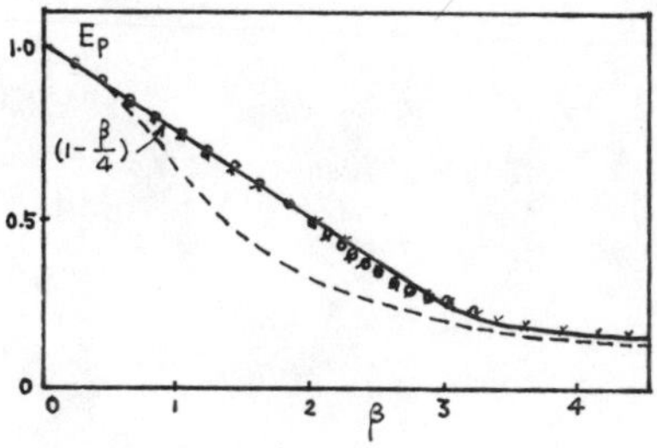

(Fig.2)

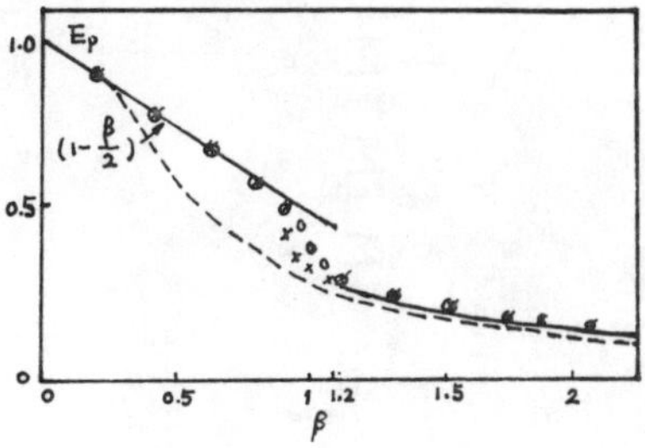

(Fig.1)

IMPLICATIONS OF SPONTANEOUSLY BROKEN CONFORMAL SUPERGRAVITY

ZHA Chao-Zheng

Department of Physics, Xinjiang University
Wulumuqi, Xinjiang, PRC

In N=1 Poincare+conformal supergravity a nonvanishing vev of the lowest component of the compensating chiral supermultiplet signals a spontaneous breakdown of conformal invariance. Meanwhile, the Newton's constant expressed in terms of the vev of the compensating field emerges from the Lagrangian, and the chiral gauge field acquires a mass of Planck scale. The induced cosmological constant is necessarily vanishing when the conformal invariance is spontaneously broken.

THE STATUS OF THE BEIJING ELECTRON-POSITRON COLLIDER (BEPC) AND THE HEAVY ION RESEARCH FACILITY OF LANZHOU (HERFL) UNDER CONSTRUCTION

H. Y. Zhu

Institute of High Energy Physics, Academia Sinica

Before giving a report on these two projects now under construction for experimental high energy physics research and nuclear physics rcsearch, I would like first of all to give a brief account of the historical background.

In 1956, a 12 year plan for developing science in China was formulated. It was decided then to build experimental nuclear physics research facilities in China. As a result, cyclotrons were built in Beijing, Lanzhou, Shanghai and Chengdu. On the other hand, it was decided then not to build experimental high energy physics research facilities in China during this period, but to join the Joint Institute for Nuclear Research at Dubna (JINR) 20% of the budget of JINR was contributed by China for 10 years. But China has to withdraw from that Institute in 1965. It was then decided to build an experimental high energy physics research center in our own country. However, the Cultural Revolution soon swept over our whole country. fundamental scientific research pratically stopped for a long period.

In 1975, it was secided to build a 50 GeV proton synchrotron in Beijing. In 1976, a decision was made to build an heavy ion separated sector cyclotron in Lanzhou. Our national economical readjustment started in 1980. During the course of which it was decided, that the pace for the construction of Heavy Ion Research Facility of Lanzhou (HERFL) should slow down and the project of building the 50 Gev proton synchrotron should be postponed. What remains now of this project is a 35 MeV proton linear accelerator, which was designed as a part of the injector, but is now going to be used for purposes other than high energy physics research.

After extensive discussions, it was decided in 1982 to build an electron-positron collider of 2.2 — 2.8 Gev per beam, the BEPC. The research on charm particle and τ particle physics to be carried out on BEPC is to be the first step to develop experimental high energy physics research in China. Though the budget of BEPC is smaller than that of a 50 GeV proton synchrotron, the technological requirements are more exacting. It needs more accurate magnetic field, much higher vacuum, more complicated R. F. system, more sophisticated control system, etc. It has to operate reliably for long periods. As this research area has been explored for more than one decade, the problems left to be studied are harder to tackle. It needs not only a higher luminosity, but also very good detectors. The data handling work is going to be heavy.

Systematic experimental research on high energy physics started with proton synchrocyclotrons in the late fourties. Experimental high energy physics research with colliders began to dominate the scene only in the seventies. We try to start our experimental high energy physics research with a collider of high luminosity. As a late beginner to take such a first step, we have to be cautions. Of course, we have to use advanced technologies. But to be on the safe side, we have decided to use only those advanced technologies which are well developed.

The layout of BEPC is shown in Fig.1. The injector of BEPC is an electron linear accelerator of about 200 m in length, which is going to supply electrons and positrons of 1.1 — 1.4 GeV. The storage ring has a circumstance of about 240 m, which is going to store electrons and positrons of 2.2 GeV at the beginning. The energy per beam will be raised later to 2.8 GeV. The designed luminosity at 2.8 GeV per beam is $1{,}7 \times 10^{31}$ cm^{-2} s^{-1}. The ground was broken for BEPC in October 1984.

A prototype of a section of the electron linear accelerator of 90 MeV has been built and operates satisfactorily (Fig.2). The maximum pulse current obtained up to now is 500 mA. At 200 mA pulse current, the energy spread is leass than 1%. The accelerator can be turned on quickly and runs stably. A prototype of an energy donbler has been built with $Q = 10^5$ and with a multiplication factor of 1.5 (Fig.3). It is now used successfully at the 90 MeV prototype electron linear accelerator. The pulse current of the electron gun now used is only 1 A. To shorten the positron injection time, an electron gun of 2 A pulse current has been built and is now undergoing test.

The 56 accelerating tubes each of 3 m in length, the 16 energy doublers and many other components are now being manufactured by the machine shop of the Institute of High Energy Physics in Beijing (BIHP). A klystron of higher power and a new modulator were designed. Their prototypes are being tested. The power ontput has now reached 30 MW. Other parts of the electron linear accelerator are being built and delivered by the industries.

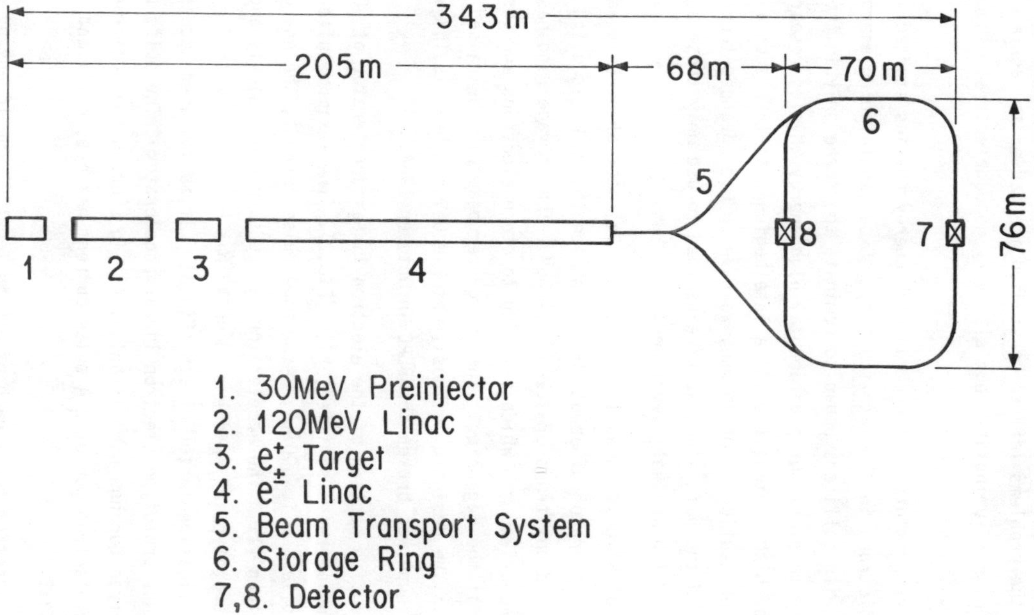

Fig.1. Schematic diagram of BEPC

Fig 2. 90 MeV electron linac

Fig 3. Energy doubler

The prototypes of bending magnets and quadrupole magnets of both the storage ring and the beam transport system have been built. Measurements and tests show that they satisfy the design requirement. The quadrupole magnets of the storage ring are now being manufactured by the machine shop of BIHP. The bending magnets of the storage ring, the bending magnets and quadrupole magnets of the beam transport system are being manufactured and delivered by the industries. The pototypes of kicker magnet and the electrostatic separator as well as a model of the Lambertson magnet have also been built and tested satisfactorily. The power supplies for part of the magnets are now being built by the industries, while the prototypes of the remaining parts are now undergoing test.

One unit of the R. F. transmitter of 30 KW for the storage ring has been built and has run continuously for one month smoothly. A prototype of the R. F. cavity made of copper clad steel has been built and is now undergoing test. A prototype of one section of the designed vacuum chamber of the storage ring has been built and tested without beam but with the magnetic field turned on. The vacuum has reached $2x10^{-10}$ torr.

The civil engineering work is proceeding rapidly. The tunnels housing the electron linear accelerator, the klystron gallery and the tunnels for the beam transport system are ready for the installation of the electron linac system. The experimental halls for the two colliding points, the experimental hall for the medium energy nuclear physics research, the tunnel for the storage ring, the control room, the building housing the R. F. power supplies as well as the power supplies for the magnets are all under construction. According to the plan BEPC will start operating in 1988.

The layout of the Heavy Ion Research Facility of Lanzhou is shown in Fig.4. The injector is a sector focusing cyclotron (SFC) with an energy constant k = 69. It has three sectors and its pole diameter is 1.7 m. It is converted from an old 1.5 m classical cyclotron. The main accelerator is a four 52° separated sector cyclotron with an energy constant k = 540. The sector width, height and length are 330 cm, 508 cm and 540 cm respectively. The maximum radius of the pole face is 358.4 cm. Each sector magnet weights 500 tons. It is expected that carbon ion can be acceleraled to an energy of 100 MeV per nucleon while xenon ion can be accelerated to an energy of 4.8 MeV per nucleon. The beam intensity ranges from 10^{12} to 10^{10} p.p.s. correspondingly. The beam emittance is designed to be 10-15 m m x mrad. The beam transport line has a length of 60 m .

Major components of SSC, such as the sector magnets, the coils, its

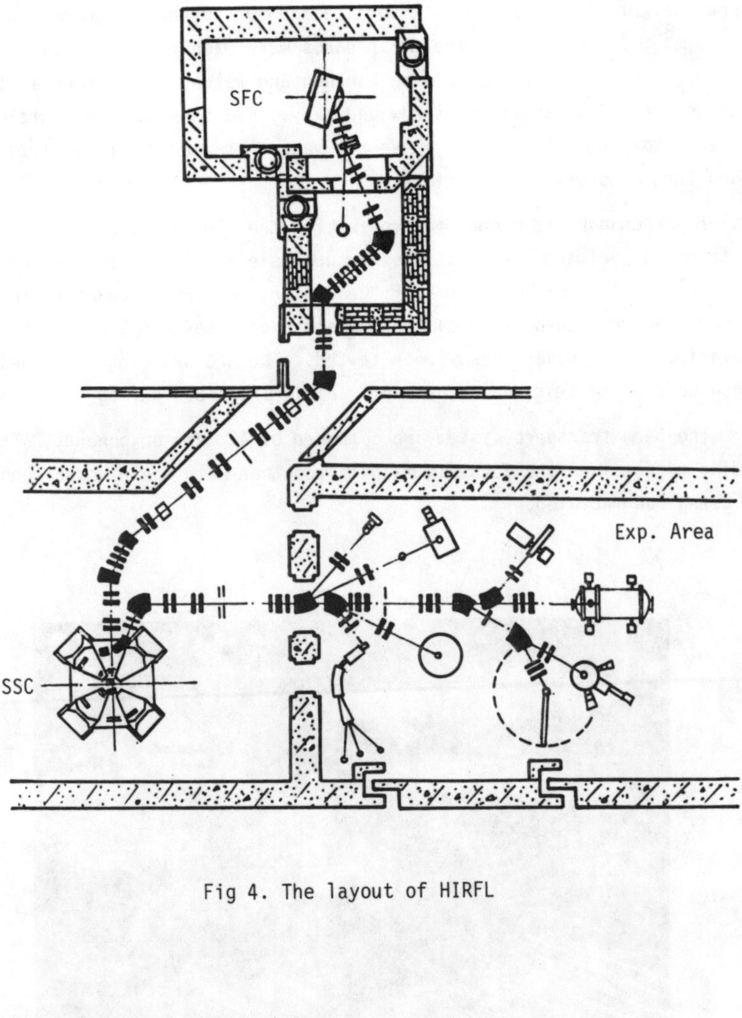

Fig 4. The layout of HIRFL

D. C. power supplies, the R. F. cavity, the R. F. power supply, the vacuum chamber and vacuum pumps were ordered from the industries. Almost all of them have been delivered already. The vacuum chamber of SSC has a monolithic structure with a total volume of 90 m^3 and a vacuum voluum of 60 m^3. Its net weight is 65 tons and its surface exposed to the vacuum has an area of 1100 m^3. Therefore, its four main parts were machined separately by the industry. They were transported to Lanzhou and welded into a monolithic piece on site. Installation is now underway. Fig.5 shows the installation of the sector magnets. Fig.6 shows the vacuum chamber which is already welded into a monolithic structure.

Major components needed for converting the old cyclotron into a sector focusing cyclotron, such as the new pole pieces, the spiral sectors, the new R. F. cavity, the improved R. F. generator, the vacuum chamber etc are all made in the machine shop of the Institute of Modern Physics in Lanzhou. The varions parts have already been tested. The SFC has been assembled and is now undergoing test and adjustment. Fig.7 shows SFC during installation.

The beam transport system are composed of 42 main components. The bending magnets and the focusing magnets together with their power supplies are being manufactured.

Fig.5 The sector magnets during intallation

Fig.6 The vacuum chamber after welding

Fig.7 The installation of SFC

The civil engineering works have already been completed. It includes the accelerator building, the experimental area, the control room, the computer room and various laboratories.

If no unexpected delay happens, the first beam from HIRFL will be obtained arround the end of 1987

During the design and construction of BEPC and HIRFL, we have received helps from many laboratories all over the wored. In particular, physicists and engineers from SLAC and GANIL have offered helps on many aspects of our work. For all these helps we would like to offer our thanks

RESCALING FOR KAON STRUCTURE FUNCTION

Zhu Wei

Department of Physics, East China Normal University,

Shen Jian-guo and Qiu Xi-jun

Institute of Nuclear Research, Academia Sinica,
P.O.Box 8204, 201849 China

We use the valon model to analysi the kynamical rescaling of x in the inelastic structure functions and point out it to be the delta approximation of the valon distribution functions. We make use of the contribution of valence quarks to the mass of a hadron in a bound state to determine a few parameters for the x-rescaling and thus obtain the kaon structure functions:

$$F^{u}_{K}(x,Q'^{2})=F^{u}_{N}(0.75x,\xi_{NK}Q'^{2}=Q^{2})$$
$$F^{s}_{K}(x,Q'^{2})=F^{u}_{N}(0.6x,\xi_{NK}Q'^{2}=Q^{2})$$

with $\xi_{NK}=0.229$, which is consistent with the data well.

In order to explain the EMC effect of the structure functions, Jaffe[1], Close, Roberts and Ross[2] suggested the rescaling for Q^2 in the nucleon structure function ($Q^2\to\xi_{AN}Q'^2$). The scale change originates from the scale change of quark confinement. In the last year it is pointed out in ref.[3] that if the rescaling of x is further assumed, i.e. $x\to b$, the following relation connecting the nucleon structure function and the pion one is obtained:

$$F^{u}_{\pi}(x,Q^2)=F^{u}_{N}(\tfrac{2}{3}x,\xi_{N\pi}Q^2) \tag{1}$$

where b=2/3 is determined by the valence quark number in a hadron. In this letter we shall use the valon model[4] to analyse the x-rescaling mentioned above and suggest another definition of b and generalize it to the kaon structure function.

According to the valon model[4], the distribution function of valence quark in a hadron is given by

$$F^{v}_{h}(x,Q^2)=\int_x^1 dy\,G^{v}_{h}(y)F_{v}^{NS}(x/y,Q^2) \tag{2}$$

in which $F_v^{NS}(z,Q^2)$ is the universal valon structure function and $G^v_h(y)$ is the distribution of valence quark (valon) at the reference point $Q^2=\mu_h^2$ of the QCD evolution. The form for $G^v_h(y)$ can be determined by the experiments in the valon model. In this work we take the delta approximation of $G^v_h(y)$:

$$G^{v}_{h}(y)=\delta(y-a^{v}_{h}), \tag{3}$$

that is, $G^v_h(y)$ has an extreme value at $y=a^v_h$ in which the average momenta of valons v is a^v_h and it is a constant. We can see in the following that the x-rescaling may be gotten by (3). Evidently it is important to select a^v_h correctly. In this work we assume that

$$a^{v}_{h}=E_{v}/E_{tot}. \tag{4}$$

in which $E_v(E_{tot})$ is the contribution of a valon (all the valons) to the mass of a hadron in a bound state. In order to reduce some arbitrariness on the parameter selection, we use the related data provided by the MIT bag model[5]. Neglecting the interaction between two quarks we get

$$a_h^v=\omega(m_v,R_h)/\sum_i(m_i,R_h)=(\chi_v-(m_vR)^2)^{1/2}/\sum_i(\chi_i-(m_iR)^2)^{1/2} \qquad (5)$$

with χ_v satisfying the following equation:

$$\tan\chi_v=\chi_v/(1+m_vR_h-(\chi_v^2+(m_vR)^2)^{1/2}) \qquad (6)$$

in which the meaning of each quantity is referred to those of ref.[5] Using the related data for the mass spectrum in the bag model, it may be obtained that $a_N^u=1/3$, $a_\pi^u=0.5$ $a_K^u=0.445$ and $a_K^s=0.554$. Utilizing each formula mentioned above and noticing the universality of $F_v^{NS}(z,Q^2)$, it is not difficult to get the x-rescaling not only in the formula:

$$F_K^u(x,Q^2)=F_N^u(0.75x,Q^2) \qquad (7)$$

but also in the formula:

$$F_K^s(x,Q^2)=F_K^u(0.6x,Q^2) \qquad (8)$$

Here it should be meticulous that

i) Because of the restriction of the integral limits in the formula (2), the above equalities (1),(7) and (8) are reliable only in the region $x< a_N^u=1/3$. Considering the approximate symmetry on the two sides at the peak of the structure function, however, we may extend the effective region of x to $\lessapprox 2/3$. Because of omittig the contribution of sea quarks, we don't expect the above qualities to suit for the region $x<0.2$.

ii) The deviation between G_h^v and the delta approximation can break up the QCD evolution of the distribution function of the x-rescaling. But Buras and Gaemers[6] had confirmed that as for the QCD evolution of the distribution function of a valence quark, it is enough to use the moments of the 2-nd and 3-rd order to determine its form correctly. Therefore the delta approximation for a true G_h^v doesn't cause too large deviation. For example, as concerns nucleon here, comparing the assumed formula (3) and $G_h^v=\frac{105}{16}y^{1/2}(1-y)^2$ given by the experiments, their 2-nd order moments are equal and the the 3-rd order moemnts differ by twenty seven percent.

Finally the differences among the radii of hadrons are taken into account. As shown in ref.[1], the differences of radii of hadrons can cause the change of the reference momentum of the QCD evolution:

$$R_N/R_K=\mu_K/\mu_N \qquad (9)$$

Hence the Q^2's on the two hand sides of the equalities (7) and (8) are different. The scale differed by them are determined by the following equation in the leading logarithm approximation(LLA):

$$\log(Q^2/\Lambda^2)/\log(\mu_N^2/\Lambda^2)=\log(Q'^2/\Lambda^2)/\log(\mu_K^2/\Lambda^2) \qquad (10)$$

The R in the above formula is still offerred by the MIT bag model[5]. Taking $\mu_N^2=0.75$ $(\mathrm{GeV})^2$ and $\Lambda=250$ MeV, we get

$$F_K^u(x,Q'^2)=F_N^u(0.75x,\xi_{NK}Q'^2=Q^2) \qquad (11)$$
$$F_K^s(x,Q'^2)=F(0.6x,\xi_{NK}Q'^2=Q^2)$$

The experiment[7] for the Drell-Yan process of kaon-nucleus provides us the possibility to check (11). Taking $Q'^2=20$ $(\mathrm{GeV/c})^2$, so $\xi_{NK}=0.229$. The distribution function of the valence quark in a nucleon fitting ref.[6] is inserted into the right hand side of the formula and so $F_K^u(x,Q'^2=20)$ is obtained. The comparision of it with the experiment is shown if Fig.1, in which the data of the pion structure function are taken from ref.[8]. The curves corresponding to each rescaling parameter are simutaneously drawn in Fig.1. Obviously the comparision of it with the experiment in the expected region is satisfaroty well.

The distribution of the strange valence quark is given in Fig.2. The similar approach could be applied to all other hyperons.

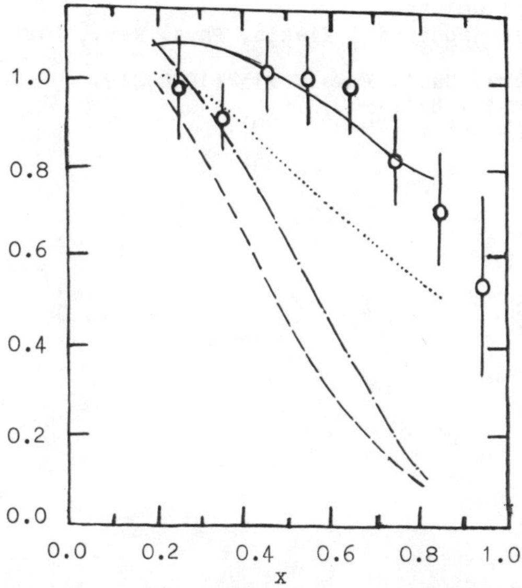

Fig.1(left): the ratio $F^{\mu}_{A}/F^{\mu}_{A'}(x,Q^2=20)$ as function of x for lepton pair production. The data points $F^{\mu}_{K}(x,Q^2=20)/F^{\mu}_{\pi^-}(x,Q^2=20)$ are quoted from ref.[7]. The four curves corresponding to each rescaling parameter are our predictin. The solid curve represents $F^{\mu}_{N}(0.75x,Q^2=4.57)/F^{\mu}_{\pi^-}(x,Q^2=20)$ the dotted one $F^{\mu}_{N}(0.75x,Q^2=20)/F^{\mu}_{\pi}(x,Q^2=20)$, the dotted-dashed one $F^{\mu}_{N}(x,Q^2=4.57)$ $F^{\mu}_{\pi}(x,Q^2=20)$ and the dashed one $F^{\mu}_{N}(x,Q^2=20)/F^{\mu}_{\pi^-}(x,Q^2=20)$.

Fig.2(right): the x dependence of the distribution for a strange valence quark.

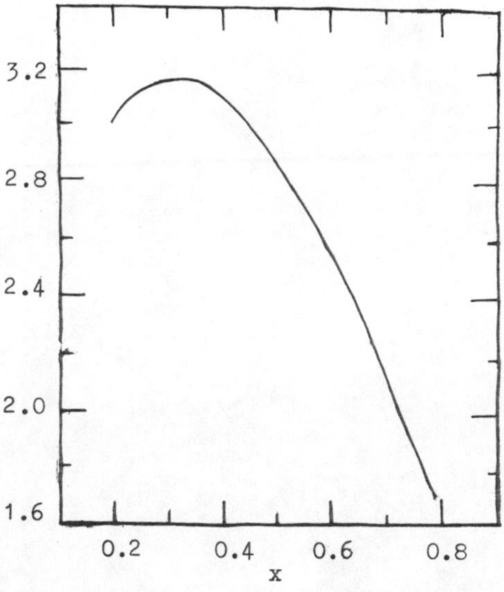

Conclusion: What the valon model is used to analyse the x-rescaling is the delta approximation of the valon distribution function. The contribution of a valence quark to the mass of a hadron in a bound state is used to determine the parameters fjor the x-rescaling and thus the kaon structure functions which agree with the experiment can be obtained.

[1] R.L.Jaffe, Phys. Rev. Lett.,50(1983)228.
[2] F.E.Close, R.G.Roberts and C.G.Ross, Phys. Lett. 129B(1983)346. RL.Jaffe,F.E.Close, R.G.Roberts and C.G.Ross, Phys. Lett.,134B (1984)449.
[3] F.E.Close, R.G.Roberts and C.G.Ross, Phys. Lett., 142B(1984)202

[4] R.C.Hwa, Phys. Rev., D22(1980)759.
[5] T.DeGrand, R.L.Jaffe, K.Johnson and I.Kiskis, Phys. Rev., D12 (1975)2060.
[6] A.J.Buras and R.J.F.Gaemers, Nucl. Phys., B132(1978)249.
[7] J.Badier et al., Phys. Lett., B93(1980)354.
[8] J.Badier et al., CERN preprint EP/79-67(1979).

PART B: NUCLEAR PHYSICS

SDG BOSON MODEL AND ITS APPLICATION

Y. Akiyama

Physics Department, College of Humanities and Sciences,
Nihon University, Tokyo 156, Japan

Abstract: The interacting boson model with both s-, d- and g-bosons is shown to describe collective nuclear states much better than that with only s- and d-bosons. The model is applied to ^{168}Er, in particular, to demonstrate its expanded capability for describing deformed nuclei. A preliminary study on the dependence of moments of inertia on the rotational bands is also reported.

This work is done in collaboration with N. Yoshinaga and A. Arima.

1. Need for g-boson

The interacting boson model (IBM) with s- and d-bosons has been very successful in accounting for various aspects of collective motion in atomic nuclei. Nevertheless, we now have reasons to believe, from both microscopic and phenomenological points of view, that the g-boson, in addition, is important. I would not discuss microscopic studies here, but concentrate upon phenomenological approaches. If the s-, d- and g-bosons are treated on the same footing (sdg IBM), new features, not present in the sd IBM, emerge, and the purpose of this paper is to show that capability of the IBM can be very much expanded with this extension.

Inclusion of the g-boson has been considered to be necessary for deformed nuclei. However, Dukelsky et al.[1)] gave indication that the g-boson is needed even in the spherical region. A series of positive parity states of ^{218}Ra can be considered to form a vibrational band. They assumed that these states were described by the IBM, although there is an alternating interpretation[2)] in terms of the α-cluster model. This nucleus is a 5-boson system in the IBM description. The highest possible angular momentum is 10 according to the sd IBM, whereas the ground vibrational band continues up to I=16 experimentally. It is clear that inclusion of the g-boson helps to solve the situation. Along this line, we parametrize this nucleus by the Hamiltonian, $H=0.4n_d+0.88n_g-0.003C-0.03P(O(15))$, where n_d and n_g stand for the number operators of d- and g-bosons. The C is the Casimir operator of SU(3). The P(O(15)) is the pairing interaction among the s-, d- and g-bosons, which should give rise to the O(15) limit. The results of diagonalization are shown in fig.1. Energy levels are satisfactorily reproduced. The main terms in this Hamiltonian are the single particle energies, which means that this nucleus is an example of the weak coupling limit. It will be interesting to study ^{218}Ra in much more detail, both experimentally and theoretically, to confirm this interpretation. Also, search for other examples would be important to establish the role of the g-boson in higher excitations of spherical nuclei.

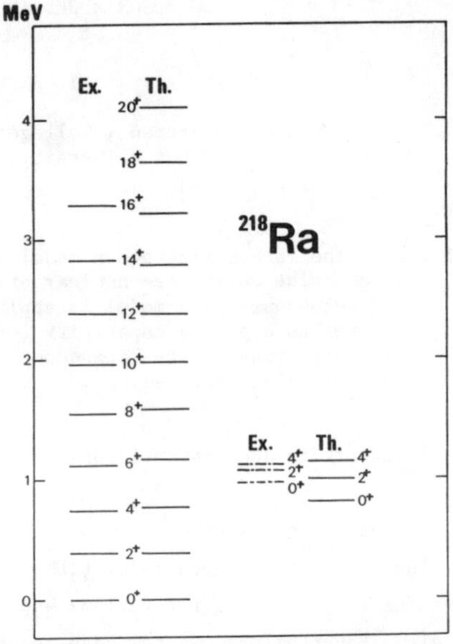

Fig.1. Energy levels of ^{218}Ra in terms of the sdg IBM.

2. SU(3) tensors

Now I turn to deformed nuclei. They are most conveniently described by using the group chain U(15) ⊃ SU(3). Following descriptions of our treatment will clarify some of special features appearing in the sdg IBM.

An interaction energy among bosons is assumed to consist of one- and two-body interactions. They are decomposed into 3 one-body and 32 two-body components, which are classified according to their SU(3) tensor characters. With such a large number of parameters, however, parametrization of a nucleus is impossible. Tensor-decomposition provides us with the best way for reducing their number. SU(3) tensors are defined below.

2.1. Definition

Creation and annihilation operators for bosons are denoted as $b^{\dagger}_{lm}$ and $\tilde{b}_{lm}=(-1)^{l+m}b_{l-m}$ $(l=0,2,4)$. One-body tensors of rank L_0 are defined by

$$T_1(\lambda_0\mu_0K_0L_0M_0)=(b^{\dagger}\tilde{b})^{(\lambda_0\mu_0)K_0L_0M_0}, \quad (1)$$

where SU(3) coupling to $(\lambda_0\mu_0)$ is implied. With two-body operators

$$B^{\dagger}(\lambda\mu)^{KLM}=(b^{\dagger}b^{\dagger})^{(\lambda\mu)KLM} \text{ and } B(\mu\lambda)^{KLM}=(\tilde{b}\tilde{b})^{(\mu\lambda)KLM}, \quad (2)$$

where $(\lambda\mu)=(8,0)$, $(4,2)$ and $(0,4)$, two-body tensors are similarly defined as

$$V(\lambda\mu,\mu'\lambda',\lambda_0\mu_0,\rho K_0L_0M_0)=(B^{\dagger}(\lambda\mu)B(\mu'\lambda'))^{(\lambda_0\mu_0)\rho K_0L_0M_0}. \quad (3)$$

The ρ distinguishes independent modes of coupling, and is needed only when $(\lambda\mu)=(4,2)$ and $(\mu'\lambda')=(2,4)$. For energy operators $K_0=L_0=M_0=0$, in which case these quantum numbers are omitted hereafter.

2,2 SU(3) seniority

The concept of the SU(3) seniority was introduced in ref.3 to distinguish multiple occurrence of a given SU(3) representation $(\lambda\mu)$ in an N-boson system. This scheme works for the case $\lambda+2\mu=4N$, in which the operator

$$S=V(04,40,00) \tag{4}$$

is diagonal, and its eigenvalue is

$$S=(375)^{-\frac{1}{2}}(2N-2w+3), \tag{5}$$

where the w is the number of (0,4)-coupled pairs. In other cases, the S is still used to classify the SU(3) multiplicity; the S is diagonalized within the states of the given U(15) symmetry. However, the quantum number w can no longer be defined, and an expression like eq.(5) cannot be available.

In analogy with eq.(4), two more operators

$$U=V(04,40,22) \tag{6}$$

and

$$Z=V(04,40,44) \tag{7}$$

are introduced. These operators have an interesting and important property that they do not have any effect on the states with $w=0$. By adding them to an SU(3)-preserving Hamiltonian, bands with $w\neq0$ can be adjusted in some way while those with $w=0$ are kept unchanged at their SU(3)-limit positions. The effect of the U is demonstrated in fig.2 for $N=16$ system. Diagonalization is made under the truncated bases described in the next section. The calculated band-structures appear drastically different from those anticipated from the SU(3) limit, and cannot be obtained by the sd IBM. In fact, there is no counterpart of the $w\neq0$ band in the sd description.

3. Basis truncation and restriction of interaction

Since the U(15) group is large, the Hilbert space for a many-boson system is huge. Also, there are 32 parameters for two-body interactions to specify the interaction energy among bosons. To make calculations feasible, it is necessary to restrict both the size of the Hilbert space and the number of interaction parameters.

The SU(3) representations $(\lambda\mu)$ in the U(15) theory belong to a class k, which is defined through $\lambda+2\mu=4N-3k$, where N is the number of bosons and $k=0,1,2,\cdots$. Most of the low-lying bands belong to the $k=0$ class in the SU(3) limit. In my earlier band-mixing calculations[3], only $k=0$ representations were taken into account. However, this restriction has turned out to be too narrow for a detailed description, such as the so-called anharmonicity observed in ^{168}Er. In analyzing this nucleus in the next section, classes down to $k=2$ are taken. Among them, further restrictions are made; in the

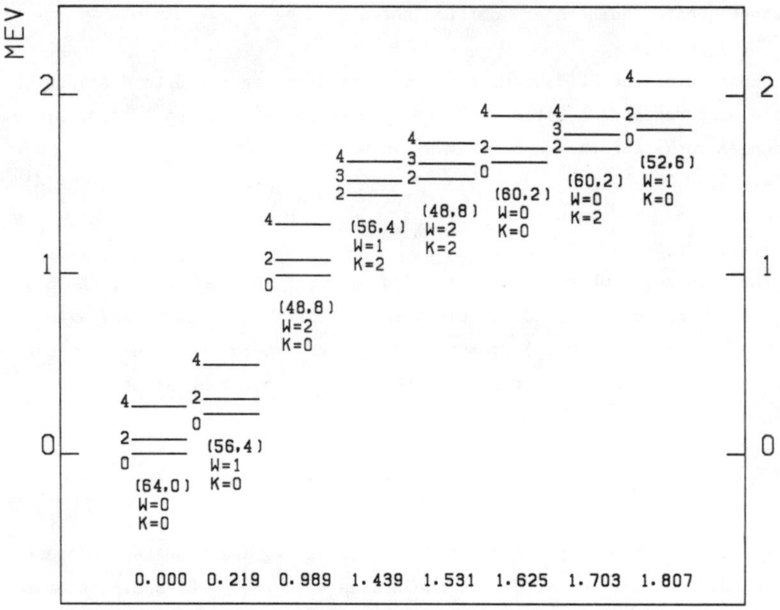

Fig.2, Effect of the U interaction. The Hamiltonian is taken as $H=-0.0043C+0.013L^2-1.2U$. The bands with $w=0$ stay at their SU(3) limit positions. In the $w=1$ member of the (56,4) representations, the $K=0$ band is more strongly pushed down than the $K=2$ one, while the $K=4$ one is so much pushed up that it cannot be shown in the figure. The effect of the U is much stronger for the $w=2$ members. Such an example can be seen in the (48,8) representation.

cases of $k=0$, 1 and 2, SU(3) representations with $\mu \leq 8$, 7 and 6 are taken, respectively.

In a deformed nucleus, components of higher SU(3) tensors in the boson interaction should be small, otherwise rotational level structures would not appear. For simplicity, we assume that the boson interaction can be expressed in terms of the SU(3) tensors with $(\lambda_0\mu_0)=(0,0)$, (2,2) and (6,0)+(0,6). There are 2, 7 and 4 tensors, respectively, for parametrizing energy-bands. Among them, the Casimir operator of SU(3), C, the L^2 force and the S interaction of eq.(4), preserve the SU(3) symmetry. There still remain 10 independent ones, which can be used to describe deviations from the SU(3) limit. A further procedure for reducing the number of parameters is explained in the next section.

The problem of basis truncation is connected with a choice for the interaction energy. By restricting to the above-mentioned tensors, SU(3) bands of lower classes, connected directly by the energy-matrix with the low-lying bands with k=0, are almost exhausted, if they are taken down to those with k=2. This is strictly true for the bands with ($\lambda\mu$)=($4N$,0), ($4N$-4,2), ($4N$-6,3) and ($4N$-8,4) of the k=0 class.

4. The ^{168}Er nucleus

We take ^{168}Er as a typical example of the sdg IBM description, not because it is considered to be strongly deformed but because it exhibits some features which are not subject to standard descriptions of well deformed nuclei. The most prominent is the one which is the so-called anharmonicity; the ratio of the band-head energy of the first K=4$^+$ band to that of the first K=2$^+$ band is experimentally 2.5, whereas ordinary theories predict 2 for this ratio. Wu

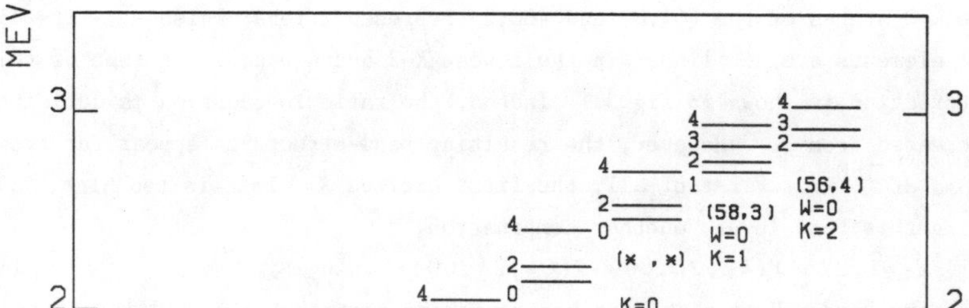

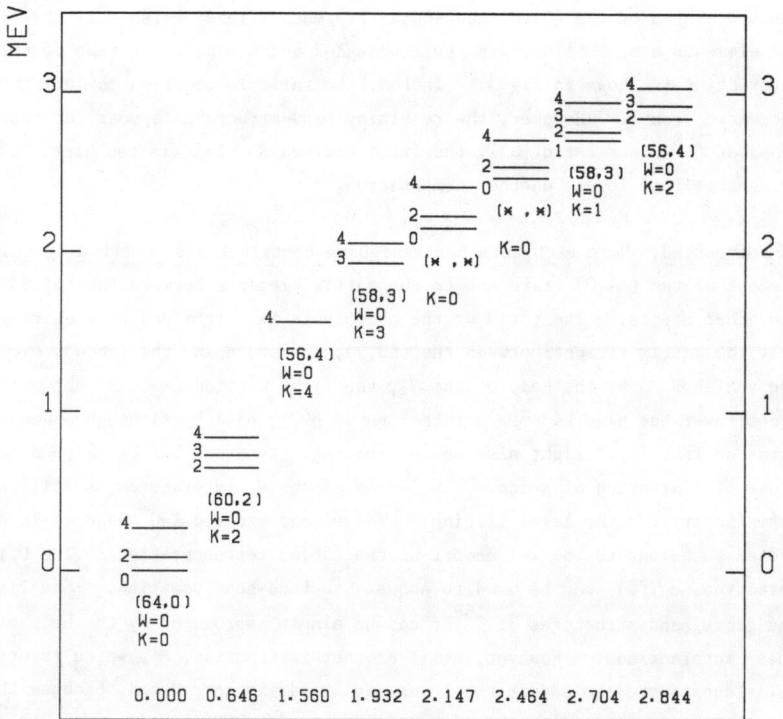

Fig.3. Anharmonicity produced by the H_1 interaction. The Hamiltonian is H=-0.0043C+0.013L^2-0.76H_1. Although the anharmonicity is produced, the lowest K=2 and 4 bands are very much pushed down. Notations (*,*) mean that definite SU(3) labels cannot be assigned to these bands because of strong SU(3)-mixings. It is interesting to note that the K=1 and 3 bands of the (58,3) representation are separated by this interaction.

and Zhou[4] analyzed this nucleus by means of the sdg IBM, but they were unable to reproduce this ratio. Earlier I reported[3] that the anharmonicity could be explained by band-mixing between the w=0 and w=1 members of SU(3) states. This argument was based on a calculation using only k=0 class representations. However, a further calculation, with the expanded basis as described in the preceding section, has brought the ratio back to 2 under a Hamiltonian, which was previously parametrized by using only the k=0 bases. The failure is attributed to mixings between k=0 and lower class states. Thus, a further study is needed to find out an interaction which produces the anharmonicity.

The interaction, which we are searching for, should have as less mixing effects as possible. We make the combination

$$H_1=V(80,24,22)-1.0167V(80,24,60)+\text{h.c.} \tag{8}$$

The ratio of its components is so determined that the matrix element between the K=0 states of the (64,0) and (60,2) representations vanishes. Its diagonal elements are not linear in the lowest K=2 and 4 bands. A test of this interaction is shown in fig.3. Indeed, the ratio in question is definitely increased from 2. However, the resulting band-structures appear far from those of ^{168}Er. First of all, the first excited K=0 band is too high. To bring this band lower, another combination

$$H_2=1.2251V(42,24,60)-V(04,24,60)+\text{h.c.} \tag{9}$$

is introduced. Here each term has vanishing contributions to the diagonal element of the (64,0) state and to the matrix elements between the (64,0) and the other states. The ratio of the components is determined by requiring that the matrix element between the (60,2), K=0 state and the (56,4), w=0, K=0 one vanishes. By the help of the H_2, the first excited K=0 band not only comes lower but also becomes a rather pure (60,2) band. Although a demonstration like fig.3 might also be interesting, it cannot but be skipped because of limitation of space. Inclusion of the H_2 interaction is still not satisfactory for the level fitting. The second excited K=0$^+$ band is interpreted to belong to the w=1 member of the (56,4) representation. The U interaction, eq.(6), can be used to adjust its band-head position. See fig.2. The gross band-structures of ^{168}Er can be almost reproduced by the help of these interactions. However, still another interaction, P_1, which is originally constructed to adjust the lowest K=1 band, is introduced, because the parametrization including this one seemed to reduce SU(3) mixings. The final parametrization is shown in fig.4. Considering the successful reproduction of the anharmonicity, the overall fit is rather good. Subsequent analyses of the electromagnetic transition data in ^{168}Er by using the wave functions obtained here will be reported in a later publication.

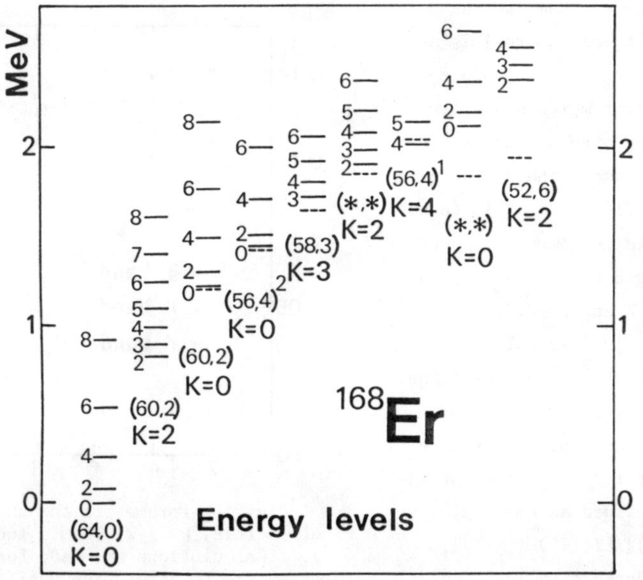

Fig.4. Positive parity levels of ^{168}Er calculated by the sdg IBM. The Hamiltonian, H=$-0.0043C+0.013L^2-0.76H_1+0.41H_2-1.2U+1.0P_1$, is diagonalized. Solid lines show theoretical levels and dashed lines show the experimental band-head energies. The superscripts 1 and 2 on the SU(3) label (56,4) show that the bands belong to w=0 and 1 members. States with asterisks have the same meaning as in fig.3.

5. Moment of inertia

Table 1. Experimental parameters for the moment of inertia, $\kappa=1/(2\mathcal{J})$.

	gr.	γ	β
^{168}Er	0.0133	0.0124	0.0098 (MeV)
^{156}Gd	0.0148	0.0162	0.0135

In the IBM description, all the rotational bands of a deformed nucleus have a common value for their moments of inertia, $\mathcal{J}$, in the SU(3) limit. Experimentally, the moment of inertia changes from band to band within a single nucleus. Some data are shown in table 1. The first excited K=0^+ band (so-called β band) tends to have a large moment of inertia. Since this is a much more detailed problem of broken SU(3) symmetry, as compared to the band-head calculation described so far, a calculation with the truncated bases would be of no use. Here we take a 5-boson system, for which a complete set of bases is available. Following calculations are not intended to reproduce any of the experimental data. Rather, they should be taken as a test whether such a description is at all possible.

Let us take a Hamiltonian $H=-0.1C+0.1L^2+cX$, where X is an operator which should produce band-dependent moments of inertia. Among others, the pairing interaction among the d- and g-bosons, $P(O(14))$, for X is found to give a large moment of inertia for the β band. With this choice for X and $c=-0.2$, the Hamiltonian is diagonalized. Strictly speaking, the resulting level structures deviate from the $I(I+1)$ pattern. Therefore, a parameter for the moment of inertia is defined as $\kappa=1/(2\mathcal{I})=[E(I)-E(K)]/[I(I+1)-K(K+1)]$, which now depends on I. The κ's deduced from the diagonalization are shown in fig.5. The κ's should be all equal to 0.1 without the last term in the Hamiltonian. The calculated (I-dependent) moment of inertia for the β band is definitely larger as compared to the ground or γ band. This increase comes from both diagonal contribution and mixing effect of the $P(O(14))$. It should be noted, however, that a large c value leads to the O(14) limit, which does not correspond to deformed nuclei. Clearly there should be some limitation for adjusting band-dependent moments of inertia. This problem is still open.

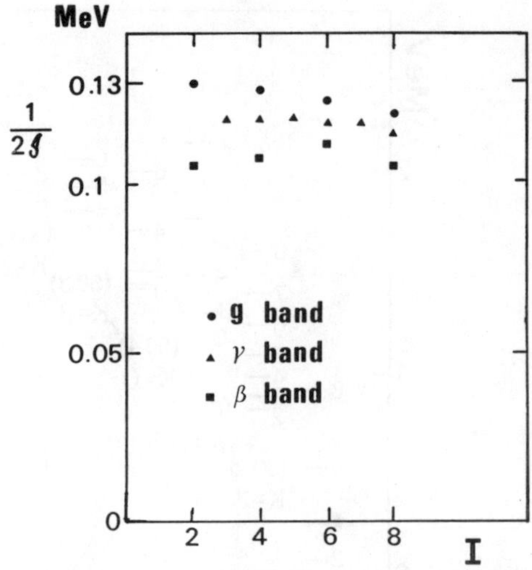

Fig.5. Parameters for the moment of inertia, $\kappa=1/(2\mathcal{I})$, as a function of I. Calculations are made for a 5-boson system, where the Hamiltonian is $H=-0.1C+0.1L^2-0.2P(O(14))$.

Concluding remark

Excellent features of the sdg IBM, for the case of broken SU(3) symmetry in particular, are exemplified. They seem to encourage us in explicitly introducing the g-boson.

References

1) J. Dukelsky, J. Fernandez Niello, H.M. Sofia and R.P.J. Perazzo, Phys. Rev., C28(1983)2183

2) M. Gai, J.F. Ennis, M. Ruscev, E.C. Schloemer, B. Shivakumar, S.M. Sterbenz, N. Tsoupas and D.A. Bromley, Phys. Rev. Lett., 51(1983)646

3) Yoshimi Akiyama, Nucl. Phys., A433(1985)369

4) H.C. Wu and X.Q. Zhou, Nucl.Phys., A417(1984)67

Shape Phase Transition in the Region of Z=40

A. Arima and M. Sugita

Department of Physics, Faculty of Science
University of Tokyo, Bunkyo-ku, Tokyo, Japan

The isotopes of Sr and Zr show a very sharp phase transition from spherical to deformed shape at N=60, where N is the neutron number of an isotope. On the other hand the isotopes of Mo, Ru, and Pd show very slow transitions. Furthermore the phase transition disappears in the Cd isotopes.
In this work, the neutron-proton interaction is found to be responsible for causing this difference. The interaction is assumed to be of quadrupole-quadrupole type. A main origin of the difference is that the $0g_{9/2}$-shell of protons is almost empty in the Sr and Zr isotopes while the shell is halfly filled up in the other isotopes.

1. INTRODUCTION

Since Arseniev et al. [1] first predicted theoretically a new region of deformed neutron-rich nuclei with A $\sim$ 100 and Cheifetz et al. [2] found experimentally large deformations in this region, many theoretical and experimental works have been done (see the review of Hamilton et al. [3]).

In this region, the behavior of Z=40 protons is very interesting as a chameleon as pointed out by Sheline, Ragnarsson and Nilsson [4]. These protons form a submagic shell and give a spherical shape to nuclei when the neutron number N is between 50 and 58, while they are suddenly deformed when N is 60. The similar situation is found in the Sr isotopes. On the other hand this phase transition becomes smooth as Z becomes larger. Especially when Z=48, the transition disappears. These phenomena are clearly seen from fig. 1, where the excitation energies of the first 2^+ states are shown.

In this paper, we try to relate this behavior with the occupancy of the $0g_{9/2}$ proton shell. The deformation due to neutrons splits the $0g_{9/2}$ shell into its subshells. Some of them come down to compete with subshells from the 0f-1p shell. Thus one has level crossings. When $Z \leqq 40$, the $0g_{9/2}$ shell is almost empty if the shape of nuclei is spherical. Being deformed, a few protons jump into the subshells of $0g_{9/2}$. Thus one can expect a sudden change of the shape of nuclei from spherical to deformed. As Z increases, the $0g_{9/2}$ shell is going to be occupied. The change becomes very smooth. The main purpose of this paper is to explain this situation more precisely and mathematically.

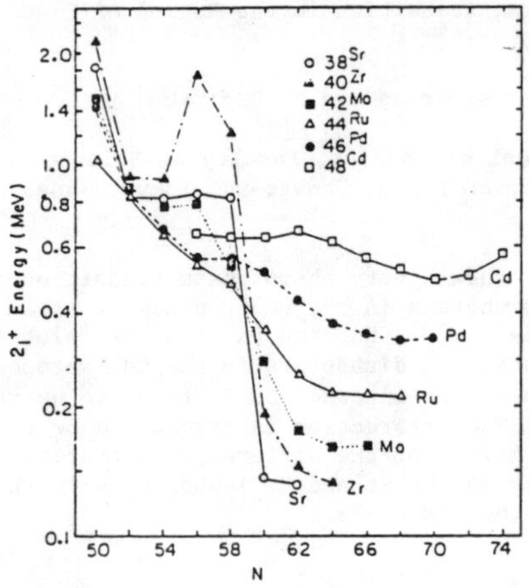

Fig. 1. Systematics of the energies of the first 2^+ states in the region of A = 100 (taken from Whon et al. [5])

2. PAIRING PLUS QUADRUPOLE-QUADRUPOLE INTERACTION MODEL

We take the pairing plus quadrupole-quadrupole force as a residual interaction. The following Hamiltonian is assumed;

$$H = H_S^\pi + H_S^\nu + V_Q, \tag{1}$$

where

$$H_S^\tau = \sum \varepsilon_j^\tau \hat{n}_j^\tau - G_\tau P_\tau^+ P_\tau ,$$

$$V_Q = -\chi_{\pi\nu} Q_\pi \cdot Q_\nu ,$$

$$P_\tau^+ = \sum C_{\alpha\tau}^+ C_{\bar{\alpha}\tau}^+ ,$$

and

$$Q_{\tau,\kappa} = \sum \langle i | r^2 Y_\kappa^{(2)} / b^2 | j \rangle C_{i\tau}^+ C_{j\tau} .$$

Here τ is π or ν, ε_j^τ are the spherical single particle energies, G_τ is the strength of the pairing interaction, $\hat{n}_j^\tau = \sum C_{jm}^{\tau +} C_{jm}^\tau$, where $C_\alpha^+ (C_\alpha)$ are creation (annihilation) operators and $b^2 = \hbar / m\omega_0$, $\hbar\omega_0 = 41.2\ A^{-1/3}$ MeV.

In order to find the lowest state and its energy of the following Schrödinger equation

$$H\psi = E\psi,$$

we use the variation method.

First we introduce a one-body Hamiltonian

$$\hat{H}(D, \Delta, \lambda) = \sum \varepsilon_j \hat{n}_j - \Delta(P + P^+) - \lambda\hat{n} - D\hat{Q}_0, \tag{2}$$

where P is the pair moment, Δ and λ are later determined by subsidary conditions, while $D(= \hbar\omega_0\delta)$ is a deformation parameter. Using the Bogoliubov-Valatin transformation, one finds the lowest eigenstate of (2) as the BCS state

$$|D, \Delta, \lambda\rangle = \prod_{\alpha>0} (u_\alpha + v_\alpha c_\alpha^+ c_{\bar{\alpha}}^+|0\rangle, \tag{3}$$

where C_α^+ is a creation operator of an eigenstate of $\sum \varepsilon_j \hat{n}_j - D\hat{Q}_0$. Here the following subsidary conditions are introduced to determine Δ and λ;

$$N = \langle D, \Delta, \lambda|\hat{n}|D, \Delta, \lambda\rangle$$

$$\Delta = \langle D, \Delta, \lambda|P|D, \Delta, \lambda\rangle = \langle D, \Delta, \lambda|P^+|D, \Delta, \lambda\rangle$$

for protons and neutrons. Here N is the number of valence protons on neutrons. These two equations give definite values to $\Delta(D)$ and $\lambda(D)$ as functions of D.

A trial function for the whole system is given as

$$\begin{aligned} |D_\pi, D_\nu\rangle = &\prod_{\alpha>0} (u_\alpha^\pi + v_\alpha^\pi c_{\alpha\pi}^+ c_{\bar{\alpha}\pi}^+)|0\rangle_\pi \\ &\times \prod_{\alpha>0} (u_\alpha^\nu + v_\alpha^\nu c_{\alpha\nu}^+ c_{\bar{\alpha}\nu}^+)|0\rangle_\nu . \end{aligned} \tag{4}$$

We further define a potential energy surface $E(D_\pi, D_\nu)$ by the expectation value of the pairing plus $Q_\pi . Q_\nu$ Hamiltonian in (1);

$$\begin{aligned} E(D_\pi, D_\nu) = &\langle D_\pi, D_\nu|H_S^\pi|D_\pi, D_\nu\rangle \\ &+ \langle D_\pi, D_\nu|H_S^\nu|D_\pi, D_\nu\rangle \\ &- \chi_{\pi\nu} Q_\pi(D_\pi)\cdot Q_\nu(D_\nu) \end{aligned}$$

$$= E_\pi(D_\pi) + E_\nu(D_\nu) - \chi_{\pi\nu} Q_\pi(D_\pi) \cdot Q_\nu(D_\nu) \tag{5}$$

$$E_\tau(D_\tau) = \langle D_\tau | \sum \varepsilon_j \hat{n}_j^\tau | D_\tau \rangle - \Delta_\tau^2 / G_\tau$$

$$Q_\tau(D_\tau) = \langle D_\tau | \hat{Q}_\tau | D_\tau \rangle$$

$$\tau = \pi, \nu.$$

Stationary points of $E(Q_\pi, Q_\nu)$ and their stabilities are determined by the first and the second derivations of $E(Q_\pi, Q_\nu)$;

$$\partial E / \partial Q_\pi = D_\pi(Q_\pi) - \chi_{\pi\nu} Q_\nu$$

$$\partial E / \partial Q_\nu = D_\nu(Q_\nu) - \chi_{\pi\nu} Q_\pi$$

$$\partial^2 E / \partial Q_\pi^2 = dD_\pi / dQ_\pi$$

$$\partial^2 E / \partial Q_\nu^2 = dD_\nu / dQ_\nu$$

$$\partial^2 E / \partial Q_\nu \partial Q_\pi = -\chi_{\pi\nu}.$$

The condition for stationary points leads to

$$D_\pi(Q_\pi) = \chi_{\pi\nu} Q_\nu$$
$$D_\nu(Q_\nu) = \chi_{\pi\nu} Q_\pi. \tag{6}$$

The stability at this point depends on the sign of Det which is defined as

$$\text{Det} = \partial^2 E / \partial Q_\pi{}^2 \cdot \partial^2 E / \partial Q_\nu^2 - (\partial^2 E / \partial Q_\nu \partial Q_\pi)^2 \tag{7}$$

$$= dD_\pi / dQ_\pi \cdot dD_\nu / dQ_\nu - \chi_{\pi\nu}^2.$$

If Det is positive (negative), this stationary point is a local minimum (saddle) point.

3. CLASSIFICATION OF SHAPE PHASE TRANSITIONS

It is convenient to draw a figure which shows two curves given by the condition for stationary points (6) (See figs. 2, 3).

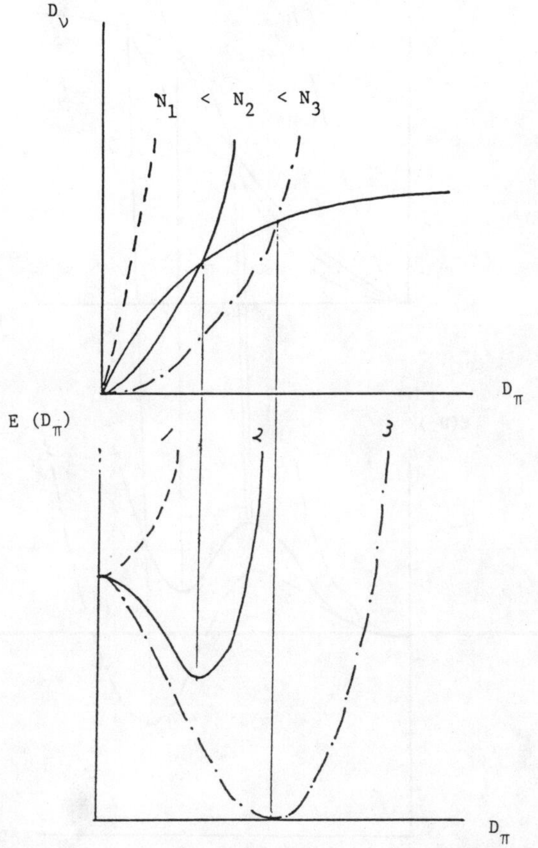

Fig. 2. Shape phase transition case 1. Both protons and neutrons have an upward convex function for Q(D). In the lower part of this figure, a corresponding potential energy curve is plotted.

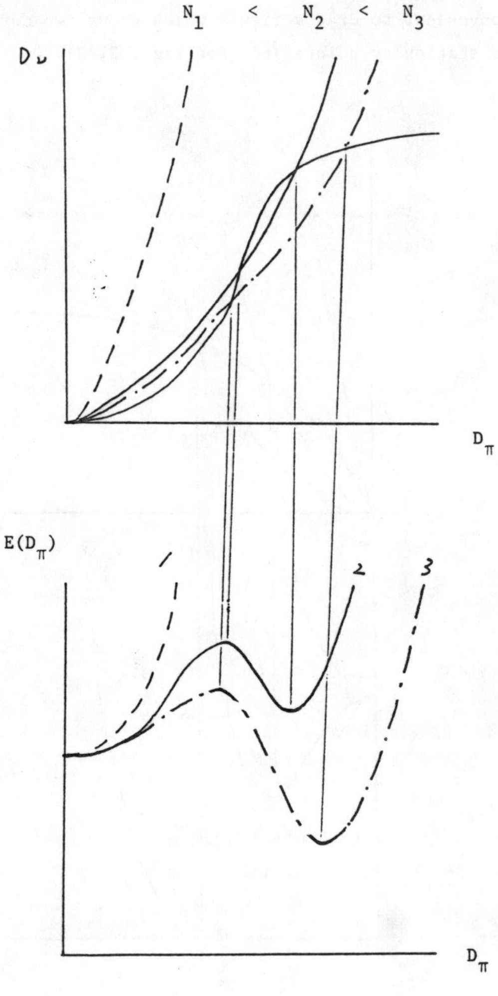

Fig. 3. Shape phase transition Case 2. The quadrupole moment Q(D) for protons is an S-shaped function. In the lower part of this figure, a corresponding potential energy curve is plotted.

We call this figure a self-consistency plot. In this figure $\chi_{\pi\nu}Q_\pi(\chi_{\pi\nu}Q_\nu)$ is plotted from the abscissa (ordinate) as a function of $D_\pi D_\nu$, respectively. The crossing points of these two curves satisfy (6) and are stationary in the potential energy surface $E(D_\pi, D_\nu)$.

In this region of Z, there are typically two kinds of curves $Q_\pi(D_\nu)$; 1) a simple increasing one and 2) an S shape one. On the other hand the function $\chi_{\pi\nu}Q_\nu(D_\pi)$ increases monotonically.

Thus one finds generally either one or two crossing points in the case 1). One crossing point is always found at the origin which corresponds to a spherical shape and the other, if it exists, corresponds to a deformed shape. If two crossing points occur, the origin becomes a saddle point. Thus the other crossing point gives a stable deformation provided that zero-point oscillation is ignored. In this situation the degree of deformation increases gradually as N increases; namely the shape phase transition developes very smoothly with increasing N.

There are three crossing points in the case 2), where the origin and the third crossing points are local minima and the middle crossing point is a saddle point. In this case there occurs a competition between a spherical shape at the origin and a deformed shape at the third point depending on the energy surface $E(D_\pi, D_\nu)$ as shown in figs. 2 and 3. Thus the phase transition from spherical to deformed shape appears very suddenly.

The S shape of $\chi_{\pi\nu}Q_\pi(D_\nu)$ is due to the jump of a proton from the 0f-1p shell to the $0g_{9/2}$-shell because a proton in lower subshells of the $0g_{9/2}$-shell ($K = \pm\frac{1}{2}, \pm\frac{2}{2} \cdots$) has a large positive quadrupole moment, while a proton in higher subshells of the 0f-1p shell has a negative quadrupole moment. If the lower subshells of $0g_{9/2}$ are occupied, the S shape disappears and the normal monotonically increasing shape transition appears.

Combining these findings, one can expect a sudden shape phase transition at $Z \simeq 40$, while a slow smooth one at $Z \geq 42$.

Using the model discribed in §2, we performed a precise calculation. Some results are shown in fig. 4, from which one confirms the explanation described above and sees the sudden phase transition at Z = 38 and 40.

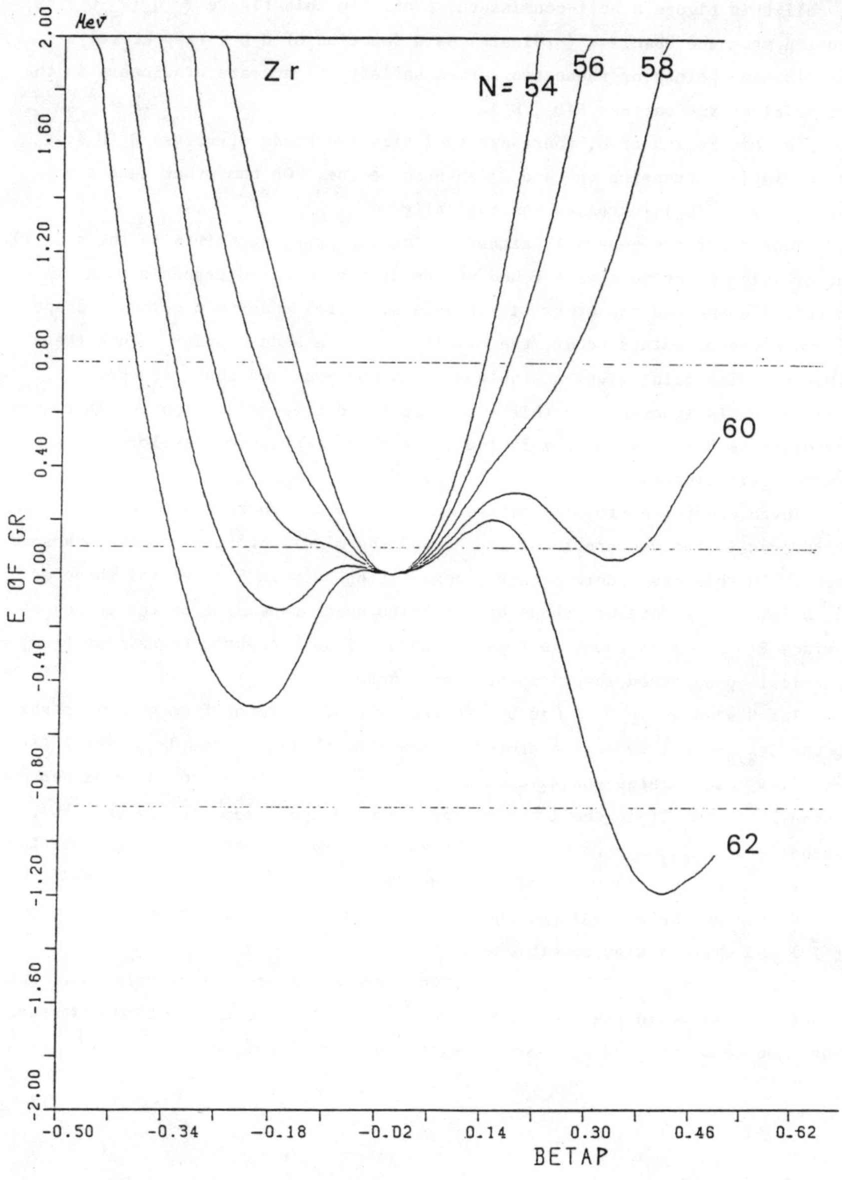

Fig. 4. The potential energy curve for the Zr-isotopes with $54 \leqq N \leqq 62$. The value of $\chi_{\pi\nu}$ is assumed to be 0.218 MeV.

4. CONCLUSION

The origin of the difference between sharp and smooth shape phase transitions is explained. We then can predict where the sharp phase transition occurs. Indeed, the present model can be applied to other regions of the periodic table, for example the mercury isotopes.

There are a few open problems which are as follows;

(1) The present model is based on the BCS approximation. It is therefore the number projection which is very much wanted.

(2) We calculated only the energy surface and no level structures. One has to project out angular momentum eigenstates.

REFERENCES

1) D. A. Arseniev, A. Sobiczewski and V. G. Soloviev, Nucl. Phys. A230 (1974) 302.

2) E. Cheifetz, R. C. Jared, S. G. Thompson, and J. B. Wilhelmy, Phys. Rev. Lett. 25 (1970) 38.

3) J. H. Hamilton, P. G. Hansen and E. F. Zganjar, Reports on Progress in Physics 48 (1985) 631.

4) R. K. Sheline, I. Ragnarsson and S. G. Nilsson, Phys. Lett. 41B (1972) 115.

5) F. K. Wohn et al., Phys. Rev. Lett. 51 (1983) 873.

STRUCTURE OF THE TRANSITIONAL NUCLEI ^{149}Pm, ^{151}Eu AND ^{153}Tb

S. BHATTACHARYA, R. K. GUCHHAIT* AND S. SEN

Saha Institute of Nuclear Physics, Calcutta 700 009, India

The structure of the N = 88 transitional nuclei has been calculated in the framework of (1) quasiparticle anharmonic vibration coupling model and (2) Coriolis coupling model incorporating the variable moment of inertia approach. Results of the calculation are compared with recent experimental findings as well as the results obtained in other thoretical investigations.

The nuclei ^{149}Pm, ^{151}Eu and ^{153}Tb with 88 neutrons were believed to be predominantly spherical. However, recent experimental investigations have revealed the existence of 'normal' ($\Delta I = 1$) positive parity bands based on $d_{5/2}$ and $g_{7/2}$ orbitals and a 'decoupled' ($\Delta I = 2$) negative parity band based on the $h_{11/2}$ orbital in these nuclei. Since these nuclei belong to a transitional region and there is experimental evidence of coexistence of both vibration and rotation like properties, we have studied their level stucture within the framework of i) quasiparticle anharmonic vibration coupling model[1] as well as in the light of Coriolis coupling model incorporating the VMI approach[2]. Some of these isotones have also been studied by other workers in term of simple RPC[3] and IBFM[4]. Some of the results are shown in tables I, II, and fig. I. In the present work a detailed comparative study of these various approaches has been made and this leads to following general observations. The simple RPC is partially successful in explaining the observed level properties where it has been applied. The agreement between the calculated and the observed level energies in IBFM is very good but no results on the electromagnetic properties have been reported. The quasiparticle - anharmonic vibration model is found to reproduce more or less correctly the level energies, the fragmentation of the single particle strengths below 1.5 MeV and almost all the observed B(E2) ↑ in ^{151}Eu. However, it fails to predict the observed high spin positive and negative parity band members as well as the antialigned low-spin negative parity states based on $h_{11/2}$ orbital. The VMI approach gives a very good account of the $\Delta I = 1$ 'normal' and $\Delta I = 2$ 'decoupled' band like structure and also explains a large variety of the static electromagnetic properties and transition probabilities. The simultaneous success of the vibration model and the VMI approach in the present case emphasies the need for studying other transitional nuclei within the framework of such models which may lead to

* Permanent address : Department of Physics, Gurudas College, Calcutta 700 054, India

understanding of the relationship between the Coriolis force and the particle vibration (or fermion - boson) coupling and ultimately may help to determine the nature of transition from spherical to deformed shape.

Table I. Calculated and experimental spectroscopic factors.

A	J^{π}	E(keV)	(2J + 1) S Particle-vibration[1]	RPC[2]	Expt.[5]
^{149}Pm	7/2$^+$	0	3.10	2.90	2.20
	7/2$^+$	789		0.21	0.74
	5/2$^+$	114	2.20	1.60	2.40
	5/2$^+$	211		0.74	0.06
	3/2$^+$	188		0	0.04
	3/2$^+$	414	0.03	0.82	0.78
	3/2$^+$	750	1.04	0.10	0.52
	3/2$^+$	871	0.82	0.06	0.44
	11/2$^-$	240	4.60	4.74	4.30
	11/2$^-$	550	0.47		0.30
	11/2$^-$	1043	0.17	2.18	0.69
^{151}Eu	7/2$^+$	22	2.61	2.41	2.70
	5/2$^+$	0	1.93	1.80	2.40
	5/2$^+$	262		0.09	0.03
	5/2$^+$	3072	0.38	0.21	0.47
	3/2$^+$	335		0.71	2.20
	11/2$^-$	197	5.6	4.39	6.40
	11/2$^-$	886			
^{153}Tb	7/2$^+$	80	1.94	1.68	1.40
	5/2$^+$	0	1.52	1.54	1.90
	5/2$^+$	543	0.56	0.02	0.63
	11/2$^-$	163	6.20	5.42	6.20
	11/2$^-$	883	1.10	1.66	1.60

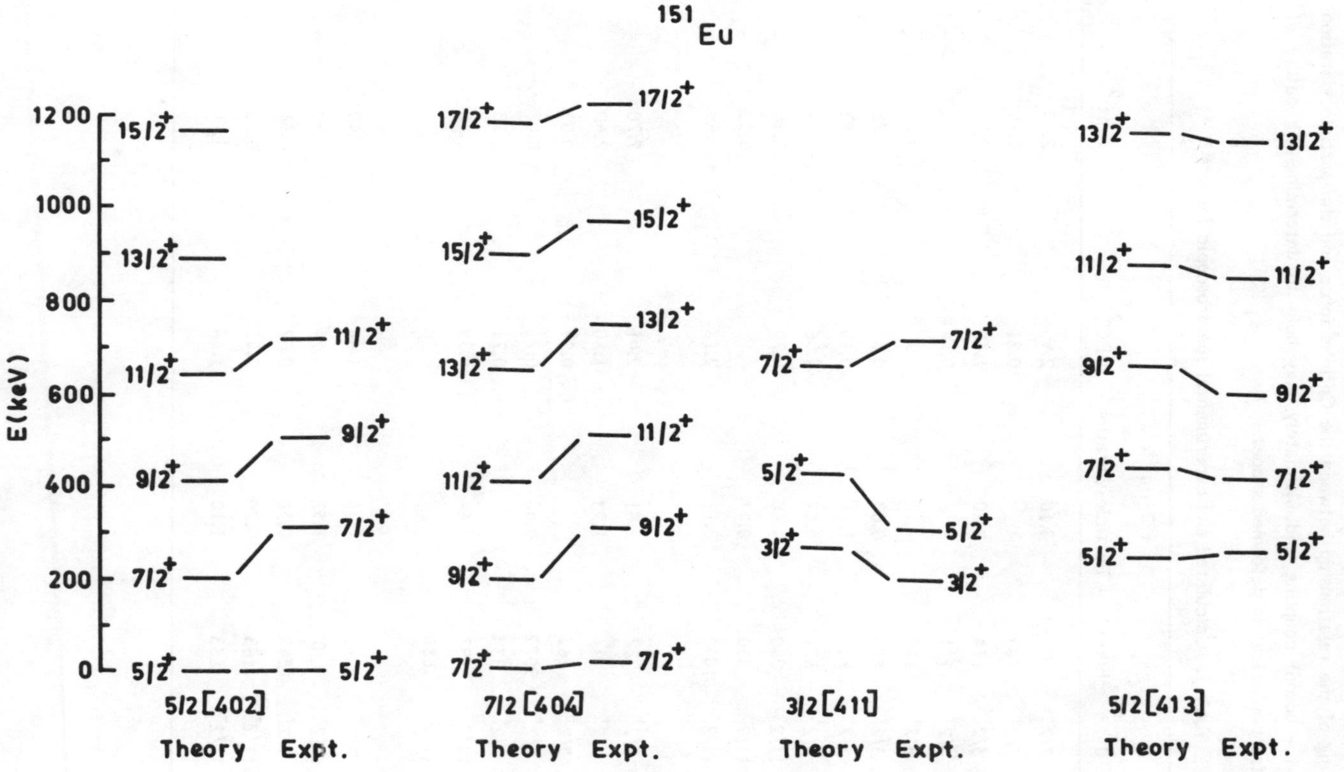

Fig. 1. Calculated and experimental results for the positive parity bands in ^{151}Eu.

Table II. Calculated and experimental B(E 2)↑ values for ^{151}Eu.

				B(E 2) ↑ (e^2b^2)	
E (keV)	J^π	Expt.[3]	RPC[2]	Particle vibration coupling model[1]	Rotational model[3]
22	$7/2^+$	0.045	0.069	0.05	0.03
196	$3/2^+$	0.095	0.053	0.11	0.057
307.2	$5/2^+$	0.072	0.047	0.12	0.131
307.5	$7/2^+$	0.39	0.34	0.40	0.375
307.8	$9/2^+$	0.039	0.04	0.056	0.009
499	$7/2^+$	0.06	0.003	0.10	Very small
503	$9/2^+$	0.22	0.15	0.50	0.42
580	$5/2^+$	0.015	0.018	0.021	
600	$9/2^+$	0.021	0.001		0.002
719	$7/2^+$	0.028	0.015	0.09	

1 R. K. Guchhait, S. Bhattacharya and S. Sen, J. Phys. G **9**, 631 (1983).

2 S. Bhattacharya, S. Sen and R. K. Guchhait, Phys. Rev. C **32**, 1026 (1985).

3 J. R. Leigh et al., J. Phys. G **3**, 519 (1977); G. D. Dracoulis et al., J. Phys. G **3**, 533 (1977).

4 G Lo Bianco et al., J. Phys. G **7**, 219 (1981); O. Scholten and N. Blasi, Nucl. Phys. A **380**, 509 (1982).

5 O. Straume et al., Nucl. Phys. A **266**, 390 (1976).

ENERGIES AND ELECTRIC QUADRUPOLE TRANSITION STRENGTHS OF GROUND STATE BANDS OF EVEN NUCLEI

S. BHATTACHARYYA and S. SEN

Saha Institute of Nuclear Physics, Calcutta 700 009, India

The ground state bands of about 150 even nuclei have been analysed in the framework of an anharmonic vibration model, recently developed by us. Agreement between calculated and experimental energy levels in most of the cases is found to be very good. One of the model parameters is found to bear a definite relationship to the measured B(E 2) value for the $2_1^+ \rightarrow 0_1^+$ transition.

It has been shown recently by us[1] that a good description of the ground state bands of the even nuclei in the spherical to deformed regions can be obtained in the framework of a phenomenological model based on the anharmonic vibrational description of the nuclear collective motion. As a first step, we tried to find an expression which would closely reproduce the energy levels of an anharmonic oscillator. Since we are concerned with the ground state bands only (the sequence J^π = 0^+, 2^+, 4^+, 6^+ ... etc.), in a vibrational picture, the total angular momentum J can be written as $J = 2N$, where N is the number of vibrational quanta. Marshalek[2] showed earlier that the energy spectra of an anharmonic oscillator can be written as,

$$E(J) = aJ + bJ^2 + cJ^3 + \dots \qquad \dots (1)$$

Satpathy and Satpathy[3] derived a similar expression in their shape-fluctuation model. If the above expression is truncated up to any finite number of terms, say upto the terms involving J^2, the coefficients a and b would, in general, have a J dependence. For simplicity, we start with an expression,

$$E(J) = aJ + \frac{J^2}{b(J)} + 1/2\, C\, [b(J) - b_o]^2 \qquad \dots (2)$$

where only the coefficients b is assumed to have a J dependence and the 'potential energy' term has been added to ensure a minimization of E(J) with respect to the parameter b(J), so that

$$\frac{\partial E(J)}{\partial b(J)} = 0 \qquad \text{for each value of J,} \qquad \dots (3)$$

Applying condition (3) to Eq. (2), one gets a cubic equation involving b as in the case of the moment of inertia parameter in the VMI models[4]. For finite positive values of a, b_o, and c, there is only one real root of the cubic equation which gives the equilibrium value of the variable b(J) for each value of J. The ground state bands of about 150 even-even nuclei (covering the entire region of spherical, transitional and rotational excitations) have been analysed in the framework of this model. It is found that agreement between calculated and experimental energy levels in most of the cases is some what better than that achieved in other three parameter models[4]. For meaningful comparison, a quantity D.E. is defined in the following manner

$$D.E. = \quad | E_{exp.}(J) - E_{calc.}(J)| \; / \; (\eta - 3)$$

where | | denotes absolute value and η , number of levels considered. This quantity calculated in different models for some of the nuclei is presented in table I. The qantity $(abo)^{-1}$ can approximately be related to the extent of anharmonicity in the vibration and it is seen that this quantiy varies smoothly with R_4 ($R_4 = E_4 / E_2$) value (fig. I). Another interesting observation is that for each isotopic series, the product $B(E_2, 2_1^+ \rightarrow 0_1^+)_{exp.} \times \sqrt{a}$ comes out to be more or less constant (table II). Detailed analaysis will be published elsewhere.

Table I. Average deviation in energy (D.E.) calculated in different models.

A	R_4	D.E. (keV)			η
		GVMI[4]	VAVM[4]	Present	
^{118}Xe	2.4	22	7	5	5
^{128}Ba	2.69	6	11	0.1	5
^{152}Gd	2.19	43	56	32	8
^{154}Gd	3.02	24	31	4	8
^{154}Dy	2.23	69	26	19	8
^{162}Dy	3.29	11	26	3	9
^{164}Er	3.28	25	20	13	9
^{178}Os	3.02	66	18	10	9
^{184}Pt	2.671	66	38	50	7
^{236}U	3.30	12	8	2	9
^{248}Cm	3.32	24	44	1.5	9

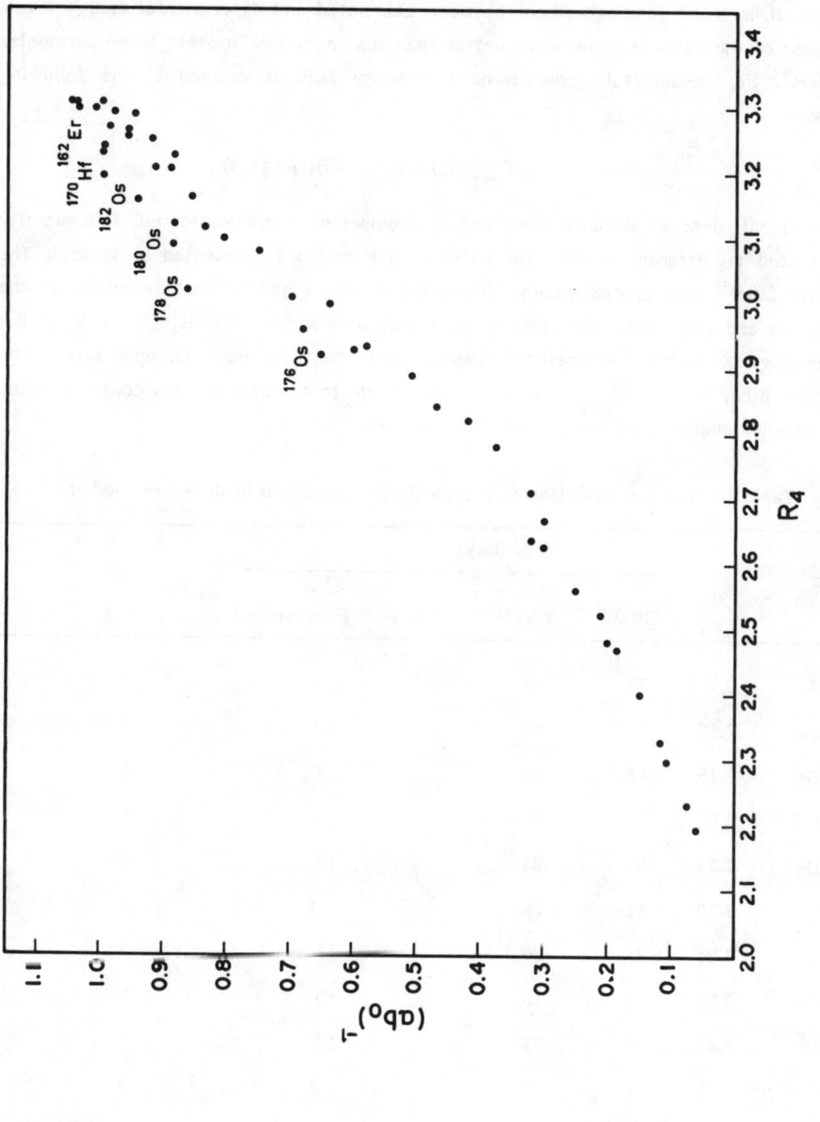

Fig. 1. Correlation between R_4 and $(ab_0)^{-1}$

Table II. Relation between the parameter 'a' and the B(E 2) values for the $2_1^+ \rightarrow 0_1^+$ transition

Isotope	R_4	E_2 (keV)	a (keV)	B(E 2) ($e^2.b^2$)	B(E 2) x $\sqrt{a}$
^{152}Gd	2.194	344.3	152.1	0.31	3.82
^{154}Gd	3.014	123.1	26.0	0.75	3.82
^{156}Gd	3.242	88.9	14.61	0.93	3.55
^{158}Gd	3.288	79.5	13.45	0.99	3.63
^{160}Gd	3.298	75.26	12.69	0.96	3.42
^{160}Yb	2.627	243.0	76.57	0.46	4.03
^{162}Yb	2.926	166.3	38.08	0.732	4.52
^{164}Yb	3.117	123.8	23.34	0.918	4.44
^{166}Yb	3.228	102.4	18.60	1.05	4.53
^{168}Yb	3.266	87.7	15.14	1.10	4.28
^{170}Yb	3.293	84.3	14.67	1.05	4.02
^{172}Yb	3.305	78.8	12.94	1.25	4.50
^{174}Yb	3.309	76.48	12.24	1.20	4.20

1 S. Bhattacharya and S. Sen, Phys. Rev. C **30**, 1014 (1984).

2 E. R. Marshalek, Phys. Rev. C **4**, 1006 (1971).

3 M. Satpathy and L. Satpathy, Phys. Lett. **34B**, 377 (1971).

4 D. Bonatsos and A. Klein, Atom. Data and Nucl. Data Tables **30**, 27 (1984).

Pion Scattering and Charge Exchange from Deformed Nuclei

H. C. Chiang
Institute of High Energy Physics, Academia Sinica, Beijing, China

We extend the theory of scattering to study pion scattering and charge exchange from deformed nuclei. We evaluate the angular distributions of pion scatterings from both unpolarized target and oriented nuclei. We find that the orientation asymmetry $A_S(\theta)$ for charge exchange reaction has a strong, selective sensitivity to the difference between the neutron and proton distribution deformations, β_2^N and β_2^C. We show that measurement of the asymmetry for pion charge exchange reaction can lead to a determination of the deformation, β_2^N, of the excess neutron distribution. We also calculate the energy dependence of $A_S(0°)$ and conclude that single charge exchange from ^{165}Ho at $T_\pi = 160$ MeV would be an interesting case to study experimentally. This study has been extended to the low lying excitation of the vibrational nucleus.

How well do we understand the neutron distribution in nuclei? We know that the determination of the charge distribution in nuclei has been successful with electromagnetic probes. The measurement made on nuclei having intrisic ground-state deformation has provided a way to extract the quadrupole deformation of the charge distribution. On the contrary, the situation for the determination of the neutron distribution in nuclei is much worse. The difficulty comes from the limitation that the forces are less well understood than in the electromagnetic case. The uncertainties have led to especially large uncertainties in the determination of the deformation of the neutron distribution.[1,2]

The selective sensitivity of the (π^+, π^0), (π^+, π^-) pion charge-exchange reactions to the neutrons has led us to reexamine the problem of determining the deformation of the neutron distribution. To learn about β_2^N we must focus on an observabe that has some chance of being sensitive to the deformation but insensitive to the less well-known aspects of the reaction dynamics.

The important quantity for determining β_2^N is the orientation asymmetry variable $A_s(\theta)$

$$A_S(q) = \frac{d\sigma^{\perp}/d\Omega - d\sigma^{\parallel}/d\Omega}{d\sigma^{\perp}/d\Omega + d\sigma^{\parallel}/d\Omega} , \tag{1}$$

where $d\sigma^{\perp}/d\Omega$ and $d\sigma^{\parallel}/d\Omega$ are the cross sections for scattering from deformed nuclei oriented perpendicular and parallel, respectively, to the direction of the incident pions.

For highly deformed nuclei there exist a large number of low-lying levels which could be easily excited by the incident pion. To take the virture excitation into account we use the closure approximation, which implies that the orientation of the nucleus is not changed during the scattering. The scattering amplitude from state $|I\ M\ K\rangle$ to state $|I'M'K'\rangle$ is

$$\langle I'M'K'|\ F(q,\Omega)\ |IMK\rangle = \int d\Omega \psi^*_{I'M'K'}(\Omega)\ F(q,\Omega)\psi_{IMK}(q,\Omega) , \tag{2}$$

where Ω are the Euler angles, and $|I\ M\ K\rangle$ is the eigenstate wave function of the collective motion. Our procedure will be to calculate first the amplitude for elastic and inelastic scattering from the initial state to a given final state. We generate the charge exchange amplitude to the analogs of the final state from this result by applying isospin invariance. We define

$$\vec{\tau} = \vec{\phi} + \vec{T} . \tag{3}$$

Here $\vec{\phi}$ is the pion and $\vec{T}$ the nuclear isotopic spin operator. $\vec{\tau}$ is total isospin.

We calculate F_τ in the eikonal theory by using pion optical potential U_τ, which is the projection of U onto the state of total isospin τ,

$$U_\tau(b,z,\Omega) = \langle \tau |\ u_0 + u_1 \vec{\phi}\cdot\vec{T} + u_2(\vec{\phi}\cdot\vec{T})^2\ | \tau \rangle , \tag{4}$$

and where u_0, u_1 and u_2 describe the isoscalar, isovector, and isotensor components of U.

For the nuclear wave functions we make the usual assumption, that is, ψ_{IMK} is an eigenstate of the rigid rotator Hamiltonian. These wave functions are just the rotation matrices $D^I_{MK}(\alpha,\beta,\gamma)$. If the nucleus is polarized along the Z axis, one has

$$\psi^{(\parallel)}_{IIK}(\Omega) = [(2I+1)/8\pi^2]^{1/2}\ D^I_{IK}(\phi,\theta,0) , \tag{5}$$

and in the case that the nucleus is aligned along the X axis, one finds,

$$\psi^{(\perp)}_{IIK}(\Omega)=[(2I+1)/8\pi^2]^{1/2}\sum_M D^{I}_{IM}(0,\pi/2,0)\,D^{I}_{MK}(\phi,\theta,0)\,. \tag{6}$$

We have the following expressions for the amplitudes

$$F^{\parallel}_{\tau}(q)=\tfrac{1}{2}(2I+1)\int_0^{\pi}\sin\theta\, d\theta\,\left|d^{I}_{IK}(\theta)\right|^2 F_{\tau}(q,\theta)\,, \tag{7}$$

$$F^{\perp}_{\tau}(q)=\tfrac{1}{2}(2I+1)\sum_M \left|d^{I}_{IM}(\tfrac{\pi}{2})\right|^2\int_0^{\pi}\sin\theta\, d\theta\,\left|d^{I}_{MK}(\theta)\right|^2 F_{\tau}(q,\theta) \tag{8}$$

where

$$F_{\tau}(q,\theta)=iK\int_0^{\infty} b\,db\,J_0(qb)\,[1-G_{\tau}(b,\theta)]\,, \tag{9}$$

$$G_{\tau}(b,\theta)=\frac{1}{2\pi}\int_0^{2\pi} d\phi'\, e^{i\chi_{\tau}(b,\theta,\phi')}\,, \tag{10}$$

$$\chi_{\tau}(b,\theta,\phi')=\frac{1}{2K}\int_{-\infty}^{\infty} dz\, U_{\tau}(b,z,\theta,\phi')\,. \tag{11}$$

In our calculation a local form of the first order optical potential is used. We evaluate the elastic and inelastic scattering from ^{152}Sm and compare with the data. Comparison of the measurements with theoretical model has been in general quite favorable.[3)]

We are particularly interested in the specific case of aligned ^{165}Ho. This nucleus is known to be highly deformed. Studies with μ^--atom techniques have determined the parameters of the charge density in the ground state of ^{165}Ho. A woods-Saxon shape for the charge distribution was used in the analysis,

$$\rho_C(r,R) = \frac{\rho_0}{1 + e^{(r-R)/a}}\,, \tag{12}$$

with the half-density radius R mapped onto an elipsoidal surface

$$R = R_0[1 + \beta_2^C\, Y_{20}(\theta)]\,. \tag{13}$$

The results of the analysis give $\beta_2^C = 0.32$ with

$$R_0 = 6.5\ \text{fm}\,, \qquad a = 0.49\ \text{fm}\,. \tag{14}$$

The spin quantum number of the ground state of ^{165}Ho is $I = K = 7/2$.

We have calculated the differential cross sections and the asymmetries for pion incident energies ranging from 120 to 250 MeV and for different values of R_0^N and β_2^N to demonstrate the sensitivity of the cross sections and the asymmetry to these quantities

Fig.1 shows the asymmetry plotted as a function of β_2^N/β_2^c at T_π = 180 MeV for elastic, SCX and DCX scatterings. Charge exchange displays a striking sensitivity to the neutron deformation.

We conclude from our results that charge exchange scattering from aligned deformed nuclei is a sensitive measure of the excess neutron deformation in rare-earth rigion.

This work was performed in a collaboration with M.B. Johnson from Los Alamos National Laboratory. The author would like to thank the hospitality of LAMPF.

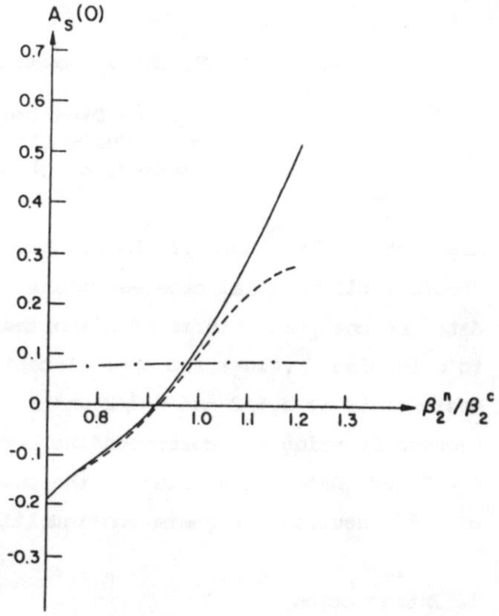

FIG.1. The asymmetries at 0° from ^{165}Ho plotted as a function of β_2^n/β_2^c at T_π=180 MeV. Solid curve, single charge exchange (π^+, π^0); dashed curve, double charge exchange (π^+, π^-); dash-dotted curve, result for π^+ elastic scattering. We have taken β_2^c to be fixed at the value given in Eq.(14).

References

1). H. C. Chiang and M. B. Johnson, Phys. Rev. Lett. 53(1984)1996.
2). H. C. Chiang and M. B. Johnson, Phys. Rev. 31(1985)2140.
3). Z. Huang and H. C. Chiang, to be published.

CONFIGURATION MIXING IN THE GROUND STATE OF ^{96}MO

M. Shafi Chowdhury

Physics Department
Dhaka University
Dhaka-2, Bangladesh

Abstract: The ^{96}Mo(d,p)^{97}Mo reaction has been studied with 12 MeV deuterons. Absolute differential cross-sections have been measured. Angular distribution data are analysed interms of distorted wave Born approximation calculations to determine ℓ values and spectroscopic factors. Ambiguity in the spin assignments of d 5/2 and d 3/2 which is allowed in ℓ= 2 (d,p) transition is removed by using the corresponding L-value of the ^{95}Mo(t,p)^{97}Mo reaction at E_t=12 MeV. Determined value of the sum of spectroscopic factors for transfers of d 5/2 neutrons suggests configuration mixing in the ground state of ^{96}Mo.

1. Introduction

Molybdenum nuclei have been the subject of several theoretical and experimental studies in the recent past as they have very interesting spectroscopic features. This mass region at one end includes nuclei that can be described in terms of simple shell model configurations like ^{92}Mo[1] and at the other end it also includes classic rotors like the neutron-rich ^{104}Mo and ^{106}Mo. The intermediate nuclei ^{98}Mo, ^{100}Mo and ^{102}Mo exhibit characteristic features which are typical of shape transition from spherical to deformed. Experimental spectrum of ^{96}Zr[2] indicates that N=56 is a fairly good closed subshell for neutrons. But ^{98}Mo spectrum[2] does not show the closure of 2d 5/2 orbital at N=56. It would therefore be extremely interesting to study the ^{96}Mo(d,p)^{97}Mo reaction in order to obtain the level structure of ^{97}Mo and also to derive spectroscopic strengths.

$^{97}_{42}$Mo is an even-odd nucleus. The ground state of ^{97}Mo is known[3] to have $J^{\pi}=5/2^{+}$. The ^{96}Mo(d,p)^{97}Mo reaction[4] was studied at E_d= 16 MeV by Medsker and Yntema. They reported some levels in ^{97}Mo to have values of spin and parity as $(3/2,5/2)^{+}$ arising out of ℓ_n=2 transitions. Such assignments are allowed for ℓ_n=2 transition in a single nucleon transfer reaction leading to an even-odd nucleus. In the present study of the (d,p) stripping reaction on ^{96}Mo an attempt is made to remove the ambiguity in these assignments by correborating the results of the ^{95}Mo(t,p)^{97}Mo reaction regarding the levels d 5/2 and d 3/2.

Absolute differential cross-sections and the spectroscopic factors S(d,p) have also been deduced.

2. Experimental procedure

The $^{96}Mo(d,p)^{97}Mo$ and the $^{95}Mo(t,p)^{97}$ experiments were carried out at AWRE Aldermaston using the tandem Van de Graaff accelerator and the multichannel magnetic spectrograph[5]. The ^{96}Mo target was enriched to 96.77% and was approximately 100 $\mu g\ cm^{-2}$ thick and the isotopic composition is shown in Table 1.

Table-1: Isotopic composition in Target ^{96}Mo

^{92}Mo	^{94}Mo	^{95}Mo	^{96}Mo	^{97}Mo	^{98}Mo	^{100}Mo
0.58	0.19	1.06	96.77	0.64	0.67	0.09

The reaction products were detected by means Ilford K2 emulsions 50μm thick mounted in the focal plane of each channel of the spectrograph. The incident beam energies for deuteron and triton were 12 MeV. The energy resolution for the (d,p) reaction was 12 keV (FWHM) and for the (t,p) reaction 15 keV (FWHM). Polythene absorbers 0.75 mm thick were placed in front of the emulsions to eliminate deuteron, triton and α-particle tracks and to improve the quality of the proton tracks. The exposed zones of the plates were scanned in strips 0.24 mm wide at intervals of 0.25 mm. The absolute differential cross-sections were measured by using the technique described by Brown et al.[6]

3. DWBA analysis

The experimental results were analysed in terms of distorted wave Born approximation calculations. The DWBA calculations were performed at AWRE Aldermaston using a computer programme based on code Julie[7]. The optical model parameters for deuterons in the DWBA calculations were obtained from a separate experiment $^{94}Mo(d,d)$ and are shown in Table 2. The proton parameters shown in Table 2 are obtained from ^{90}Zr (p,p) experiment[8]. Theoretical calculation for the $^{95}Mo(t,p)$ ^{97}Mo reaction is in progress. The experimental angular distributions of the (d,p) reaction for the energy levels at 0.0(g.s.), 0.723, 1.263, 1.285, 1.564, 2.151, 2.310 and 2.369 MeV are shown in Figure 1 and the solid curves represent the predictions of the DWBA calculations. The experimental data points for the (t,p) angular distributions are shown in Figures 2,3 and 4 and the solid lines are mere guides to the eye. The level at 1.263 MeV is populated as a doublet in the (t,p) reaction and the angular distributions of the components are shown in Figure 3. The dominant component of the unresolved doublet exhibits characteristic feature of L_n=2.

Table-2: Optical model parameters used in DWBA calculations.

Particle	U (MeV)	W (MeV)	a_u (fm)	a_w (fm)	r_u (fm)	r_w (fm)
Deuteron	98.8794	9.380	0.767	0.6836	1.15	1.6282
Proton	51.8	8.6	0.48	0.48	1.3	1.3

4. Results and Discussions

In this paper angular distributions, absolute differential cross-sections and the spectroscopic factors for the eight energy levels upto an excitation energy range 2.369 MeV which proceed via ℓ_n=2 transition in the (d,p) reaction are presented. A summary of the results of the present work and of the previous (d,p) work of Medsker and Yntema is shown in Table 3.

As observed in Table 3 seven energy levels except the ground state are each assigned $J^{\pi}=(3/2,5/2)^+$ in the previous (d,p) work. Ambiguity regarding the levels d 5/2 and d 3/2 is removed in the present work by comparing the results of the $^{96}Mo(d,p)^{97}Mo$ reaction with those of the $^{95}Mo(t,p)^{97}Mo$ reaction leading to the same final state. A comparison of the two reactions determines the final state spin and parity resulting from an ℓ_n=2 transition in the (d,p) reaction uniquely. Such (d,p) transitions are equivalent to L_n=0+2+4 (with 0 predominant) in the $^{95}Mo(t,p)^{97}Mo$ reaction if the final state is $5/2^+$ whereas L_n=2+4 (with 2 predominant) if the level is $3/2^+$. The levels at 0.723, 1.285 and 1.564 MeV are assigned the spin and parity values $5/2^+$ and the levels at 1.263, 2.151, 2.310 and 2.369 MeV are assigned $3/2^+$ on the basis of the above argument.

Table-3: Summary of results.

	Present work					Previous	(d,p) results	
		(d,p)			(t,p)			
Ex(MeV)	ℓ_n	J^{π}	$(d\sigma/d\Omega)_{max}$ mb/st	S(d,p)	L_n	Ex(MeV)	ℓ_n	J^{π}
0.0	2	$5/2^+$	3.19	0.339	0	0.0	2	$5/2^+$
0.723	2	$5/2^+$	1.395	0.127	0	0.720	2	$(3/2,5/2)^+$
1.263	2	$3/2^+$	1.772	0.211	2	1.265	2	$(3/2,5/2)^+$
1.285	2	$5/2^+$	0.247	0.020	0	1.286	2	$(3/2,5/2)^+$
1.564	2	$5/2^+$	0.434	0.035	0	1.564	2	$(3/2,5/2)^+$
2.151	2	$3/2^+$	1.453	0.173	2	2.152	2	$(3/2,5/2)^+$
2.310	2	$3/2^+$	0.144	0.017	2	2.215	2	$(3/2,5/2)^+$
2.369	2	$3/2^+$	0.300	0.036	2	2.378	2	$(3/2,5/2)^+$

The ground state of ^{97}Mo has been measured[3] to have spin and parity $5/2^+$ in agreement with shell model predictions. In the present study the ground state transition in the (d,p) reaction is an $\ell_n=2$ and in the (t,p) reaction is an $L_n=0$ which confirms the spin and parity value for the ground state as $5/2^+$ and this is also in support of the argument put forward for the assignment of the spin and parity of the reported levels.

Peak value of the absolute differential cross-sections and the spectroscopic factors S(d,p) for each level reported in this work are shown in Table 3. Of great significance is the remarkably large value of $2j+1\sum S(d\ 5/2)=3.13$ which is about 56% greater than the theoretical upper limit predicted by the sum-rule if ^{96}Mo is assumed to be a good shell model nucleus with four d 5/2 neutrons in its ground state. It would appear therefore that the ground state of ^{96}Mo contains an appreciable amount of other configurations mixed into its ground state, probably $(d\ 3/2)^2$ if one believes that the $S(s\frac{1}{2})$ strength is exhausted.

5. Conclusion

Definite values of spin and parity have been assigned for the d 5/2 and d 3/2 levels upto an excitation energy range 2.369 MeV from a comparison of results of the (d,p) reaction on ^{96}Mo with those of the (t,p) reaction on ^{95}Mo. It appears extremely likely from the sum of spectroscopic factors of d 5/2 states that mixing of configurations like $\nu d\ 3/2$ exists in the ground state of ^{96}Mo.

The author is grateful to Professor G. Brown for his help with the plates and for hospitality at the University of Bradford, to the computer personnel at Aldermaston and to Dr. W. Booth for helpful discussions.

R References

(1)D.H. Gloeckner, Nucl. Phys. A253(1975)301

(2)E. Cheifetz, R.C. Jared, S.G. Thompson and J.B. Wilhelmy, Phys. Rev. Lett. 25(1970)38

(3)L.R. Medsker, Nucl. Data B10(1973)1

(4)L.R. Medsker and J.L. Yntemma, Phys. Rev. C Vol 9(1974)664

(5)R. Middleton and S. Hinds, Nucl. Phys. 77(1962)365

(6)G. Brown, A. Denning and A.E. Macgregor, Nucl. Phys. A153(1970)145

(7)R.H. Bassel, R.M. Drisco and G.R. Satchler, ORNL3240(1962)

(8)R.N. Glover, AWRE Aldermaston, Private communication.

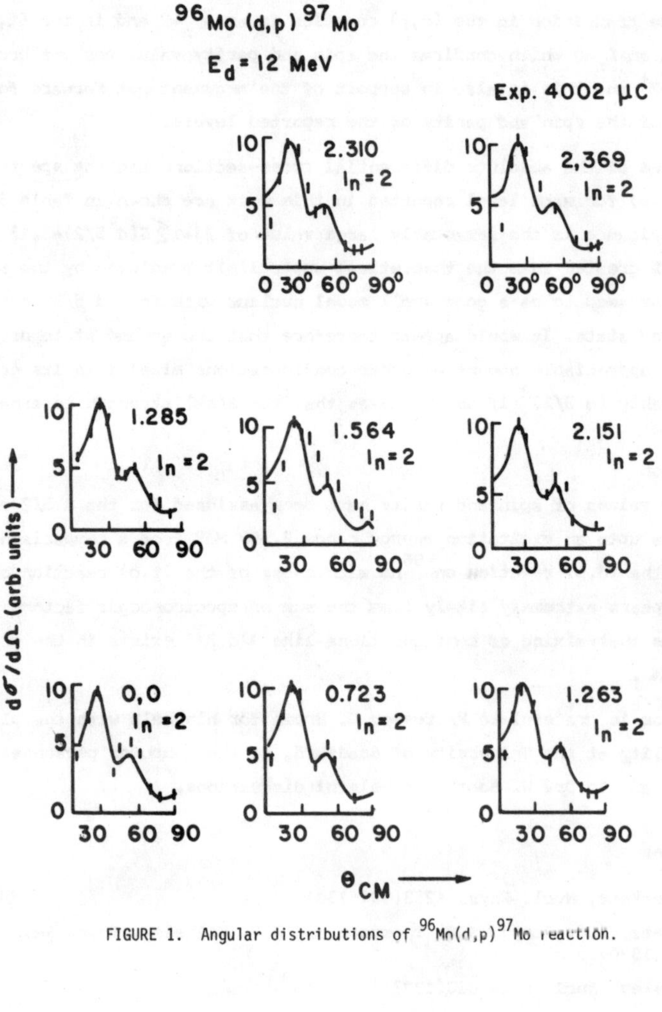

FIGURE 1. Angular distributions of ^{96}Mo(d,p)^{97}Mo reaction.

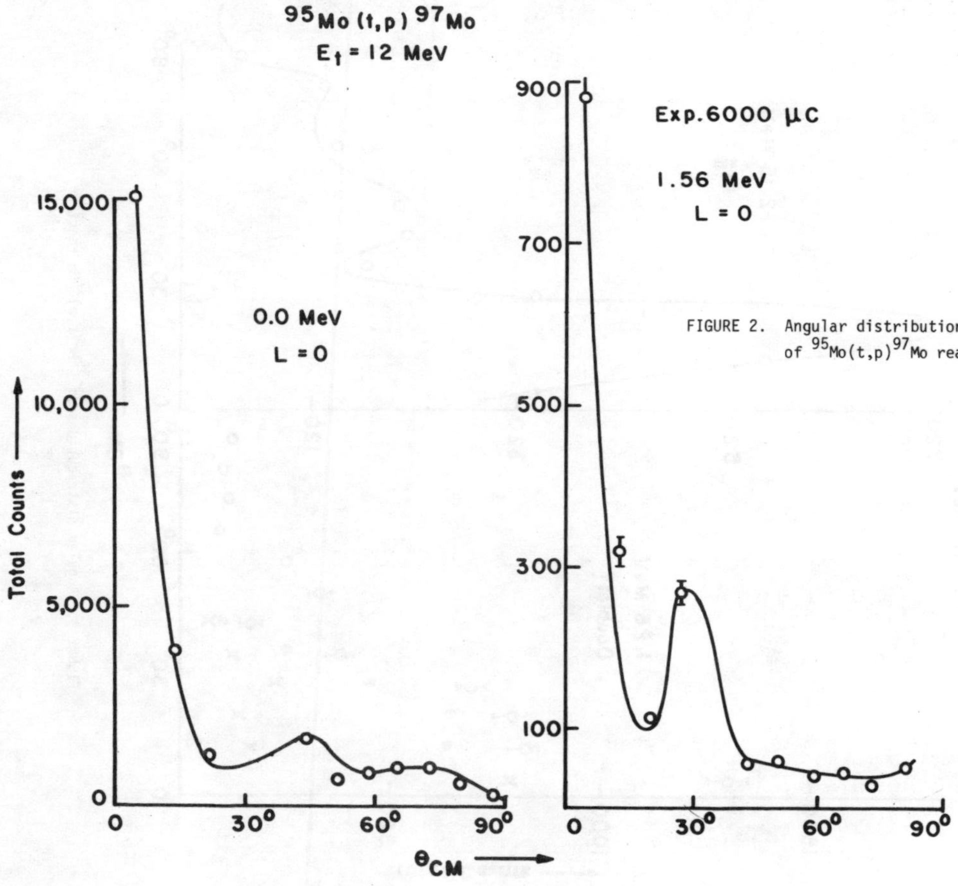

FIGURE 2. Angular distributions of $^{95}Mo(t,p)^{97}Mo$ reaction.

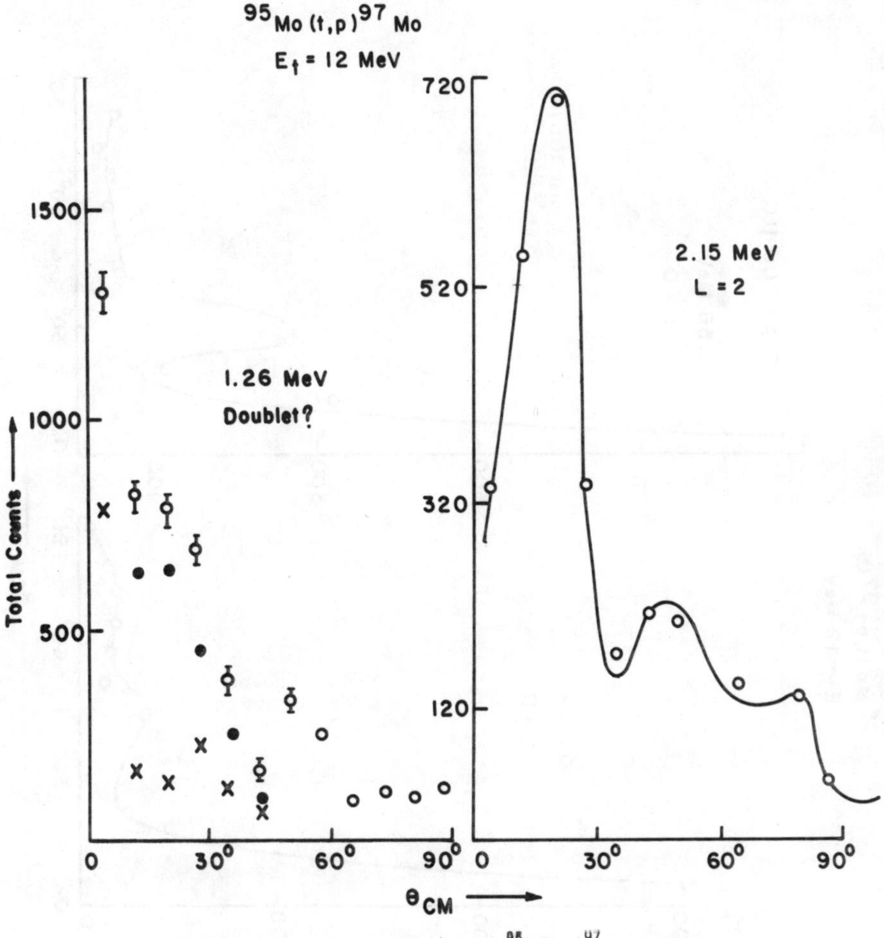

FIGURE 3. Angular distributions of $^{95}Mo(t,p)^{97}Mo$ reaction.

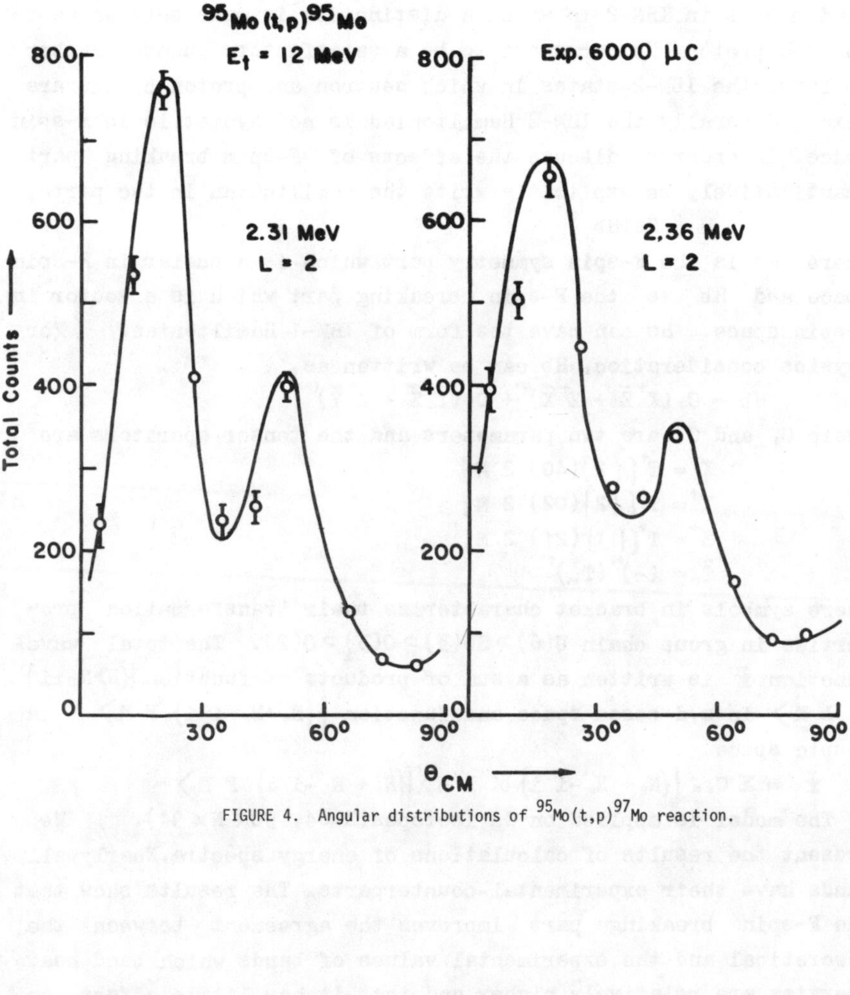

FIGURE 4. Angular distributions of ^{95}Mo(t,p)^{97}Mo reaction.

F-SPIN BREAKING AND Gd ISOTOPES ENERGY SPECTRA

Ding Xiao-Nan

Department of Physics, Xu Zhou Teachers College,
Xuzhou, Jiangsu, People's Republic of China

The F-spin was introduced by Arima et al to distinguish two kind bosons in IBM-2 in which a distinction is made between neutron and proton. It turns out to be a satisfactory quantum number to label the IBM-2 states in which neutron and proton bosons are mixed. Generally the IBM-2 Hamiltonian is not symmetric in F-spin space. In order to discuss the effects of F-spin breaking part quantitatively we explicitly write the Hamiltonian in two parts,

$$H = Hs + Hb$$

where Hs is the F-spin symmetry part which is a scalar in F-spin space and Hb is the F-spin breaking part which is a vector in F-spin space. Hs can have the form of IBM-1 Hamiltonian. For physics consideration, Hb can be written as,

$$Hb = C_1 (X^+ \tilde{Z} + Z^+ \tilde{X})^{(0)} + C_2 (Y^+ \tilde{Z} + Z^+ \tilde{Y})^{(0)}$$

where C_1 and C_2 are two parameters and the tensor operators are

$$X^+ = T^+[\{2\}(40)\ 2\ M]$$
$$Y^+ = T^+[\{2\}(02)\ 2\ M]$$
$$Z^+ = T^+[\{11\}(21)\ 2\ M]$$
$$\tilde{T}_M = (-)^M (T^+_M)^+$$

where symbols in bracket characterize their transformation properties in group chain $U(6) \supset SU(3) \supset O(3) \supset O(2)$. The total wave-function Ψ is written as a sum of products of function $|\{N_\pi + N_\nu - i\, i\}$ $\alpha\ L\ M\rangle$ in s,d-boson space and function $|\{N_\pi + N_\nu - i\ i\}\ F\ M_F\rangle$ in F-spin space.

$$\Psi = \Sigma C_{i\alpha} |\{N_\pi + N_\nu - i\ i\}\ \alpha\ L\ M\rangle |\{N + N - i\ i\}\ F\ M_F\rangle$$

The model is applied on Gd isotopes($Z=64$, $90 \leq N \leq 94$). We present the results of calculations of energy spectra. Nearly all bands have their experimental counterparts. The results show that the F-spin breaking part improves the agreement between the theoretical and the experimental values of bands which band head energies are relatively higher and that it has little effect on bands which band head energies are lower. The lower F-spin ($F = F_{max} - 1$) states of $K=1^+$ band are reproduced theoretically and agree with experiment well.

FERMION DYNAMICAL SYMMETRY AND THE NUCLEAR SHELL MODEL

JOSEPH N. GINOCCHIO

Theoretical Division, Los Alamos National Laboratory
Los Alamos, New Mexico 87545 U.S.A.

1. INTRODUCTION

The interacting boson model[1] (IBM) has been very successful in giving a unified and simple description of the spectroscopic properties of a wide range of nuclei, from vibrational through rotational nuclei. The three basic assumptions of the model are that 1) the valence nucleons move about a doubly closed core, 2) the collective low-lying states are composed primarily of coherent pairs of neutrons and pairs of protons coupled to angular momentum zero and two and 3) these coherent pairs are approximated as bosons.

In this review we shall show how it is possible to have fermion Hamiltonians which have a class of collective eigenstates composed entirely of monopole and quadrupole pairs of fermions.[2,3] Hence these models satisfy the assumptions 1) and 2) above but no boson approximation need be made! Thus the Pauli principle is kept in tact.

Furthermore the fermion shell model states excluded in the IBM can be classified by the number of fermion pairs which are not coherent monopole or quadrupole pairs. Hence the mixing of these states into the low-lying spectrum can be calculated in a systematic and tractable manner. Thus we can introduce features which are outside the IBM.

2. MONOPOLE AND QUADRUPOLE PAIRING

Our goal is to construct fermion shell model Hamiltonians which have a class of eigenstates composed of monopole, $J^{\pi}=0^{+}$, pairs and quadrupole, $J^{\pi}=2^{+}$, pairs only. The way to do this is to separate the single-nucleon angular momentum $\vec{j}$ into a pseudo-orbital angular momentum $\vec{k}$ and a pseudo-spin $\vec{i}$.[2,3] We call these "pseudo" because k may not correspond to the real orbital angular momentum of the shell and the spin may be greater than ½. An example is the s-d shell which of course has orbital angular momentum $\ell=0,2$ and spin $s=\frac{1}{2}$. However we can span these states with $k=1$ and $i=3/2$. After the separation, the special subspace is defined by summing over the pseudo-angular momentum or spin thereby making those degrees of freedom inactive. Hence this technique is a way of reducing the number of active degrees of freedom in a fermion shell model and separating the large fermion shell model space into two parts. Another way of looking at this separation is to think of it as a generalization of pairing. In pairing the monopole pair is the only special pair and the single-nucleon angular momenta in this pair are completely coupled to total angular momentum zero. In the present model the single-nucleon angular momentum is split into two parts. Most of the single-nucleon angular momentum is coupled to zero in the special pairs, but a small part of it is not.

To be more explicit we define a nucleon creation operator as $a^{\dagger}_{(ki)jm}$ which creates a nucleon in an orbit with single-nucleon angular momentum j, projection m with pseudo-orbital angular momentum k and pseudo-spin i which are coupled to j,

$$\vec{k} + \vec{i} = \vec{j} \quad . \tag{2.1}$$

A pair of nucleons is then a linear combination of orbitals coupled in k-i coupling to a total pseudo-orbital angular momentum K, pseudo-spin I, with these then coupled to total angular momentum J and projection M,

$$P^{\dagger}_{(KI)JM} = \Sigma_{ki}\, C^{(KI)J}_{ki} [a^{\dagger}_{ki} a^{\dagger}_{ki}]_M^{(KI)J} \tag{2.2}$$

Because of antisymmetry, the sum of angular momenta is even:

$$K+I+J \text{ even} \quad . \tag{2.3}$$

For our purposes we want to separate out one special angular momentum zero pair and one special angular momentum two pair for which we can construct shell model Hamiltonians which will have a class of eigenstates composed only of these pairs. There are only two ways to do this:

$$\text{A)} \quad C^{(KI)J}_{ki} = \delta_{k,1}\, \delta_{I,0}\, (2i+1)^{\frac{1}{2}} \tag{2.4}$$

$$\text{B)} \quad C^{(KI)J}_{ki} = \delta_{i,\frac{3}{2}}\, \delta_{K,0}\, (2k+1)^{\frac{1}{2}} \quad . \tag{2.5}$$

In the first (second) case as long as the shell model Hamiltonian perserves the pairing of the pseudo-spin (pseudo-orbital angular momentum) the pairs with total pseudo-spin (pseudo-orbital angular momentum) zero will not mix with other pairs. Further, since k=1 ($i=\frac{3}{2}$) and because of the antisymmetrization which leads to equation (2.3), only $J^{\pi}=0^+,2^+$ are allowed for these special pairs.

These two possibilities each lead to Hamiltonians with dynamical symmetries. The first option A leads to an Sp_6 dynamical symmetry; the second option B leads to an SO_8 dynamical symmetry.[3] Each of these models has interesting features. The Sp_6 model has an SU_3 subgroup which means that axially symmetric rotational nuclei emerge from this model when the Hamiltonian has this SU_3 as a dynamical symmetry. On the other hand the SO_8 model has an SO_6 subgroup which gives γ-unstable rotational nuclei when the Hamiltonian has an SO_6 dynamical symmetry. The IBM has both of these possibilities.

However of these two fermion models only the SO_8 model has a one-to-one correspondence between the space spanned by the fermion states composed of the special monopole and quadrupole pairs of neutrons and protons and the space spanned by the monopole and quadrupole bosons.[3] In the Sp_6 model many of the most collective states vanish due to the Pauli principle. For this reason the SO_8 model has received the most attention to date, and we shall discuss that model in sections 3-5 in detail first. However many interesting features appear in the Sp_6 model as well, and there has been a revival of interest of late in this model[4]. We shall report recent developments in this model in section 5.

3. The SO_8 Model

The total number of valence shell model orbits in the SO_8 model can be as large as necessary and is given by

$$2\,\Omega = 4\sum_k (2k+1)\ , \tag{3.1}$$

and hence the total number of possible states for n valence nucleons can be large, $\binom{2\Omega}{n}$, where n is the number of valence nucleons. A wonderful aspect about the SO_8 model is that all the states in this space can be classified according to irreducible representations of the SO_8 group.[3] In particular the states in the space can be classified according to the number of nucleons in the states, u, not coupled to the special monopole and quadrupole pair. This quantum number is a generalization of the seniority quantum number[5] which just counts the number of nucleons not coupled to a monopole pair. The states with u=0 correspond to the collective subspace composed only of monopole and quadrupole pairs, and has a one-to-one correspondence with the IBM space. The states with u=2 are those with only one pair which is not a monopole or quadrupole pair and so on. This feature means that the study of the coupling of the collective monopole and quadrupole space to the other states left out of the IBM space can be studied in a systematic way.

For odd nuclei u will be odd. The state u=1 correspond to the states of the interacting boson-fermion model[6] in which an odd fermion is coupled to the even-even core described by the IBM. the allowed quantum numbers of u=1 in the SO_8 model have been worked out.[3]

The monopole pair creation operator, $S^\dagger$, and quadrupole pair creation operator, $D^\dagger_\mu$, μ=2,1,0,-1,-2, are given by applying (2.2) and (2.5),

$$S^\dagger = \sum_k (2k+1)^{\frac{1}{2}} [a^\dagger_{k\frac{3}{2}} a^\dagger_{k\frac{3}{2}}]^{(00)0}_0 \tag{3.2a}$$

$$D^\dagger_\mu = \sum_k (2k+1)^{\frac{1}{2}} [a^\dagger_{k\frac{3}{2}} a^\dagger_{k\frac{3}{2}}]^{(02)2}_\mu \quad . \tag{3.2b}$$

These pair creation operators and their hermitian conjugates, plus the multipole operators with total pseudo-orbital angular momentum rank equal to zero,

$$R^{(r)}_\mu = \sum_k \quad (2k+1)^{\frac{1}{2}} [a^\dagger_{k\frac{3}{2}} \tilde{a}_{k\frac{3}{2}}]^{(0,r)r}_\mu ; \; r=0,1,2,3 \tag{3.2c}$$

are the generators of the SO_8 group.[3] In particular, the pseudo-spin generator is

$$\hat{I}_\mu = \sqrt{5}\, R^{(1)}_\mu \quad . \tag{3.2d}$$

In addition to these operators, the multipole operators

$$T^{(t)}_{\mu;k} = [a^\dagger_{k\frac{3}{2}} \tilde{a}_{k\frac{3}{2}}]^{(t,0)t}_\mu ; \; t \text{ odd} \tag{3.3}$$

commute with $S^\dagger$, $D^\dagger$, $R^{(r)}$, and generate an $SO_{(2k+1)}$ group.

Hence any shell model nuclear Hamiltonian which has an $SO_8 \otimes \pi_k SO_{(2k+1)}$ dynamical symmetry will have a subspace of eigenstates consisting of $S^\dagger$, $D^\dagger$ pairs only. The most general shell model Hamiltonian of this form will have monopole and quadrupole pairing and multipole interactions:

$$H = G_0 S^\dagger S + G_2 D^\dagger \cdot D + \sum_{r=1,2,3} \kappa^{(r)} R^{(r)} \cdot R^{(r)} \tag{3.4}$$

$$+ \sum_{t\ \mathrm{odd}} \nu^{(t)}_{k'k}\, T^{(t)}_{k'} \cdot T^{(t)}_{k} + \sum_{r=1,3} \alpha^{(r)}_{k} (T^{(r)}_{k} \cdot R^{(r)} + R^{(r)} \cdot T^{(r)}_{k})$$

where $G_0 < G_2$ and $\kappa^{(r)}$, $\alpha^{(r)}_k$, and $\nu^{(t)}_{k'k}$ are the strengths of the multipole interactions.

The eigenstates of this Hamiltonian will be labeled by the quantum number u. Those with u=0 will correspond to the IBM states and those with u=1 will correspond to the IBFM states. However all shell model states will appear; the remaining states will have a higher value of u.

The group SO_8 has three subgroups chains which have the total pseudo-spin as an SO_3 subgroup. For values of the parameters of the Hamiltonian which conserve the symmetry of these subgroups, the eigenenergies of the Hamiltonian can be given in closed form.

The first symmetry corresponds to subgroup chain

$$SO_8 \supset SO_5 \otimes SU_2 \supset SO_3 \quad . \tag{3.5}$$

In this chain the SO_5 group is the symmetry group of the quadrupole oscillator, SU_2 is the well known quasi-spin group of pairing[7], and SO_3 is the pseudo-spin rotational group. For u=0 states the total pseudo-orbital angular momentum is zero and hence the total angular momentum equals the total pseudo-spin, J=I. This symmetry occurs for $\kappa^{(2)}=G_2$ and the excited energy eigenvalues for u=0 are given by

$$E^*_5(v,\tau,J) = \frac{(G_2-G_0)}{4} v(2\Omega-v+2)+(\kappa^{(3)}-G_2)\tau(\tau+3)+\frac{1}{5}(\kappa^{(1)}-\kappa^{(3)})J(J+1) \quad . \tag{3.6}$$

The quantum number v is the usual seniority,[5]

$$v = n, n-2, \ldots, 0, \tag{3.7}$$

τ is the SO_5 quantum number,

$$\tau = \tfrac{1}{2}v, \tfrac{1}{2}v-2, \ldots, 0 \text{ or } 1 \tag{3.8}$$

and J is the angular momentum with allowed values determined by partitioning τ,

$$\tau = 3p + \lambda \tag{3.9}$$

where p, λ are non-negative integers, and then

$$J = \lambda, \lambda+1, \ldots, 2\lambda-2, 2\lambda \quad . \tag{3.10}$$

This spectrum is that of an anharmonic quadrupole oscillator with the energy spacing between levels almost linear in v with anharmonicity from the Pauli principle coming in naturally. This symmetry in the SO_8 model corresponds to the SU_5 symmetry in the IBM.

Another group chain is

$$SO_8 \supset SO_6 \supset SO_5 \supset SO_3 \tag{3.11}$$

and occurs for the pairing strength $G_0=G_2$. The eigenspectrum for u=0 is then

$$E_6^* (\sigma,\tau,J) = (\kappa^{(2)}- G_0)(\sigma-N)(\sigma+N+4) + (\kappa^{(3)}-\kappa^{(2)})\ \tau(\tau+3)$$

$$+ \frac{1}{5} (\kappa^{(1)} -\kappa^{(3)})J(J+1) \tag{3.12}$$

where $N = \frac{1}{2}n$ is the number of pairs of valence nucleons, σ is the SO_6 quantum number,

$$\sigma = N, N-2, \ldots 0 \text{ or } 1 \tag{3.13}$$

and the allowed values of τ are

$$\tau = \sigma, \sigma -1, \ldots, 0 , \tag{3.14}$$

and the allowed values of the angular momentum J are the same as in (3.9) and (3.10). This symmetry corresponds to a γ-unstable rotor and also corresponds to the SO_6 limit of the IBM.

The final group chain is

$$SO_8 \supset SO_7 \supset SO_5 \supset SO_3 \tag{3.15}$$

and occurs for $\kappa^{(2)} = G_0$. The eigenspectrum for u=0 is given by

$$E_7^*(\bar{v},\tau,J) = (G_2-G_0)\ \frac{\bar{v}}{4}\ (2\Omega-2n+\bar{v}+10) + (\kappa^{(3)}-G_2)\tau(\tau+3)$$

$$+ \frac{1}{5}(\kappa^{(1)} -\kappa^{(3)})J(J+1)\ . \tag{3.16}$$

This symmetry corresponds to a repulsive quadrupole pairing interaction. The quantum number $\bar{v}$ has the same allowed values as seniority v,

$$\bar{v} = n, n-2, \ldots , 0 \tag{3.17}$$

and the allowed values of τ are,

$$\tau = \tfrac{1}{2}\bar{v},\ \tfrac{1}{2}\bar{v}-2,\ \ldots,\ 0 \text{ or } 1 \tag{3.18}$$

and the allowed values of J are the same as in (3.9) and (3.10). The spectrum for a given valence number is that of a anharmonic quadrupole oscillator like the pairing limit, but unlike the pairing limit the spacing between levels decreases as the number of valence nucleons increases.[3)]

For the general Hamiltonian in which none of these three symmetries prevail the spectrum will depend on the relative strength of the pairing interaction and the quadrupole interaction. However it is clear from these solvable limits that a wide variety of spectra can occur in this model.

For the allowed representations of SO_8 and SO_6 for states with $u>0$, see Reference 3.

An application to the Samarium isotopes in which neutrons and protons were distinguished was successfully carried out in Reference 8.

4. THE Sp_6 MODEL

Just as in the case of the SO_8 model, all the states in this space can be classified according to irreducible representations of the Sp_6 group, and the quantum number u which is the number of nucleons not in the special monopole or quadrupole pair. However unlike the SO_8 model the number of states for u=0 are not in one-to-one correspondence with the IBM[3)]. The number of u=0 states will be less than the number of IBM states because of the Pauli principle. For this reason, which may be unjustified, this model was not studied as much as the SO_8 model. However there has been recent renewed interest in this model[4)].

The monopole pair creation operator, $S^\dagger$, and quadrupole pair creation operator, $\bar{D}^\dagger_\mu$, μ=2,1,0,-1,-2, are given by applying (2.2) and (2.4)

$$S^\dagger = \tfrac{1}{2}\sum_i [3(2i+1)]^{\frac{1}{2}}\, [a^\dagger_{1i}\ a^\dagger_{1i}]^{(00)0}_0 \tag{4.2a}$$

$$\bar{D}^\dagger_\mu = \tfrac{1}{2}\sum_i [3(2i+1)]^{\frac{1}{2}}\, [a^\dagger_{1i}\ a^\dagger_{1i}]^{(20)2}_\mu\ . \tag{4.2b}$$

These pair creation operators and their hermitian conjugates, plus the multipole operators with total pseudo-spin rank equal to zero,

$$\bar{R}^{(r)}_\mu = -\tfrac{1}{2}\sum_i [3(2i+1)]^{\frac{1}{2}}\, [a^\dagger_{1i}\ a_{1i}]^{(r,0)r}_\mu;\ r=0,1,2 \tag{4.2c}$$

are the generators of the Sp_6 group.[3)] In particular the pseudo-orbital angular momentum operator is

$$\hat{K} = -2[2/3]^{\frac{1}{2}}\,\bar{R}_\mu^{(1)} \tag{4.2d}$$

In addition to these operators, the multipole operators

$$\bar{T}^{(t)}_{\mu;i} = \tfrac{1}{2}[3(2i+1)]^{\frac{1}{2}}[a^\dagger_{1i}\ \tilde{a}_{1i}\ ]^{(0,t)t}_\mu;\ t\ \text{odd} \tag{4.3}$$

commute with $S^\dagger$, $\bar{D}^\dagger$, $\bar{R}^{(r)}$ and generate an Sp_{2i+1} group.

Hence any shell model Hamiltonian which has an $Sp_6 \otimes \prod_i Sp_{2i+1}$ dynamical symmetry will have a subspace of eigenstates consisting of $S^\dagger$, $\bar{D}^\dagger$ pairs only. The most general shell model Hamiltonian of this form will have monopole and quadrupole pairing and multipole interactions:

$$H = G_0 S^\dagger S + G_2 \bar{D}^\dagger\cdot\bar{D} + \sum_{r=1,2} \kappa^{(r)}\bar{R}^{(r)}\cdot\bar{R}^{(r)} \tag{4.4}$$

$$+ \sum_{\substack{t\ \text{odd}\\ i',i}} v^{(t)}_{i'i}\,\bar{T}^{(t)}_{i'}\cdot\bar{T}^{(t)}_{i} + \sum_i \alpha_i(\bar{T}^{(1)}_i\cdot\bar{R}^{(1)} + \bar{R}^{(1)}\cdot\bar{T}^{(1)}_i)$$

where $G_0<G_2$ and $\kappa^{(r)}$, α_i, and $v^{(t)}_{i'i}$ are the strengths of the multipole interactions.

The eigenstates of this Hamiltonian will be labeled by the quantum number u. Those with u=0 will correspond to a subset of the IBM states and those with u=1 will correspond to a subset of the IBFM states. However <u>all</u> shell model states will appear; the remaining states will have a higher value of u.

The group Sp_6 has two subgroups chains which have the total pseudo-orbital angular momentum as an SO_3 subgroup. For values of the parameters of the Hamiltonian which conserve the symmetry of these subgroups, the eigenenergies of the Hamiltonian can be given in closed form.

The first symmetry corresponds to subgroup chain

$$Sp_6 \supset SO_3 \otimes SU_2 \supset SO_3\ . \tag{4.5}$$

In this chain SU_2 is the well known quasi-spin group of pairing[7], and SO_3 is the pseudo-orbital angular momentum group. For u=0 states the pseudo-spin is equal to zero and hence the total angular momentum equals the total pseudo-orbital angular momentum. This symmetry occurs for $\kappa^{(2)}=G_2$ and the excited energy eigenvalues for u=0 are given by

$$E_2^*(v,J) = \frac{(G_2-G_0)}{4}\, v(2\Omega-v+2) + \frac{3}{8}(\kappa^{(1)} - G_2)J(J+1)\ . \tag{4.6}$$

The quantum number v is the usual seniority,[5]

$$v = n,\ n-2,\ \ldots,\ 0, \tag{4.7}$$

and J is the angular momentum. This spectrum is that of an anharmonic oscillator with the energy spacing between levels almost linear in v with anharmonicity from the Pauli principle coming in naturally. This symmetry in the Sp_6 model corresponds partially to the SU_5 symmetry in the IBM. Since Sp_6 has no SO_5 subgroup, there is no τ quantum number as in the SO_8 model. The IBM does have an SO_5 subgroup and it is for this reason that there is no one-to-one correspondence between the Sp_6 model and the IBM.

Another group chain is

$$Sp_6 \supset SU_3 \supset SO_3 \tag{4.8}$$

and occurs for the pairing strength $G_0=G_2$. We use the fact that the SU_3 Casimir operator is

$$C_3 = 2 \sum_{r=1,2} \bar{R}^{(r)} \cdot \bar{R}^{(r)} \; . \tag{4.9}$$

The eigenspectrum for u=0 is then

$$E_3^*(\lambda,\mu,J) = \frac{(\kappa^{(2)}-G_2)[(\lambda-2N)(\lambda+2N+3)+\mu(\lambda+\mu+3)]}{2} + \frac{3}{8}(\kappa^{(1)}-\kappa^{(2)})J(J+1) \; , \tag{4.10}$$

where $N = \frac{1}{2}n$ is the number of pairs of valence nucleons, and (λ,μ) are the SU_3 quantum numbers. The allowed values of J for a given representation follow the same rules as in the IBM.[1] This symmetry corresponds to an axially symmetric rotor.

5.1 THE SU_3 GROUND STATE BAND

All the states in the SU_3 representation $(\lambda,\mu) = (2N,0)$ which will correspond to the ground state band for an axially symmetric rotor can be projected from an intrinsic state composed of N intrinsic pairs of nucleons. These intrinsic pairs create two nucleons with pseudo-orbital angular momentum projection zero, but total pseudo-spin zero.

$$A^{\dagger} = \tfrac{1}{2}\sum_i \left[\frac{3(2i+1)}{\Omega}\right]^{\frac{1}{2}} \{a^{\dagger}_{10;i}a^{\dagger}_{10;i}\}^{(0)} \quad , \tag{5.1}$$

where the {} coupling is for pseudo-spin only. Hence this pair does not have a definite pseudo-orbital angular momentum.

For $N \leqq \Omega/2$, i.e. the half-filled shell, the SU_3 eigenstates will be projected from this intrinsic pair condensate,

$$|(2N,0)K,M;I = 0\rangle = \frac{\sqrt{2K+1}}{8\pi^2\eta_{NK}} \int d\omega\, D^{(K)}_{M0}(\omega)R_K(\omega)(A^{\dagger})^N|0\rangle \tag{5.2a}$$

where $D^{(K)}_{MM'}(\omega)$ is the Wigner D-function[9], ω are the Euler angles, $R_K(\omega)$ is a pseudo-orbital angular momentum rotation, and η_{NK} is the normalization

$$\eta_{NK} = B_{NK}P_N \tag{5.2b}$$

where B_{NK} is the IBM normalization

$$B_{NK} = \left[\frac{(2N)!N!}{(2N+K+1)!!(2N-K)!!}\right]^{\frac{1}{2}} \quad , \tag{5.2c}$$

P_N is the Pauli correction factor

$$P_N = \left[\frac{(\frac{\Omega}{3}-1)!}{(\frac{\Omega}{3}-N)!(\frac{\Omega}{3})^{N-1}}\right]^{\frac{1}{2}} \tag{5.2d}$$

and $|0\rangle$ is the core.

For $N > \frac{\Omega}{2}$ the SU_3 representation $(2\bar{N},0)$ with $\bar{N} = \Omega-N$ will be lowest in energy and is projected from an intrinsic state of $\bar{N}$ instrinsic pairs of nucleon holes. The vacuum $|0\rangle \rightarrow |\bar{0}\rangle$, the closed shell, and $N \rightarrow \bar{N}$ in the formulae (5.2).

From (5.2d) we see that $P_N = 0$ for $N > \frac{\Omega}{3}$. Hence in this case the (2N,0) representation vanishes because of the Pauli principle. Likewise $(2\bar{N},0)$ vanishes for $\bar{N} > \frac{\Omega}{3}$. Thus these lowest SU_3 representations do not exist in the Sp_6 model for

$$\frac{\Omega}{3} < N < \frac{2\Omega}{3} \quad . \tag{5.3}$$

This is an example of states which do not exist in the Sp_6 because of the Pauli principle but do exist in the IBM. This may not be a defect; only by comparison with data can we judge whether this is a valid effect which exists in nuclei.[10]

5.2 THE SU_3 EXCITED BANDS

We can define an excited u=2 band by replacing one of the pairs in (5.2a) by an intrinsic pair with pseudo-spin I≠0

$$A^{\dagger}_{I\mu} = \frac{1}{2}\sum_{i}\left[\frac{3(2i+1)}{\Omega}\right]^{\frac{1}{2}}\{a^{\dagger}_{10;i}a^{\dagger}_{10;i}\}^{(I)}_{\mu} \tag{5.4}$$

Because of antisymmetry, I must of course be even. The u=2 states with SU_3 symmetry are projected from an intrinsic state with N-1 I=0 pairs (5.1) and one pair with I≠0:

$$|(2N,0)K,M;I,\mu\rangle = \frac{\sqrt{2K+1}}{8\pi^2\eta_{NKI}}\int d\omega\, D^{(K)}_{MO}(\omega)R_K(\omega)(A^{\dagger})^{N+1}A^{\dagger}_{I\mu}|0\rangle \tag{5.5}$$

where

$$\eta_{NKI} = \left[\frac{\frac{\Omega}{3}-N}{N\,\frac{\Omega}{3}-1}\right]^{\frac{1}{2}}\eta_{NK}\,\frac{(1+(-1)^I)}{2} \quad . \tag{5.6}$$

Hence we see that, as long as I is even but larger than zero, the normalization is independent of I. Furthermore we see that this SU_3 band does not exist for u=2 for

$$\frac{\Omega}{3} \leqq N \leqq \frac{2\Omega}{3} \tag{5.7}$$

which is more restrictive than for the u=0 states as shown by (5.3).

The SU_3 representation (2N,0) given in (5.5) occurs for many bands,

$$I = 2i_{max} - 1,\ 2i_{max} - 3,\ \ldots\ ,\ 2 \tag{5.8}$$

where i_{max} is the maximum pseudo-spin in the system.

These excited bands are important in understanding backbending in nuclear high spin states.[10)]

5.3 STRONGLY COUPLED u=2 BAND

In the u=2 bands described by (5.5), the pseudo-spin is not part of the collective rotational moton. We can define a strongly coupled band by rotating the pseudo-orbital angular momentum and pseudo-spin together:

$$|(2N,0)I;JM\rangle = \frac{\sqrt{2J+1}}{8\pi^2\bar{\eta}_{NIJ}}\int d\omega\, D^{(J)}_{MO}(\omega)R(\omega)A^{\dagger}_{I,\mu=0}(A^{\dagger})^{N-1}|0\rangle \tag{5.9a}$$

where

$$\bar{\eta}_{NIJ} = \left[\sum_{K}(2K+1)\begin{pmatrix} J & I & K \\ 0 & 0 & 0 \end{pmatrix}^2 \eta^2_{NKI}\right]^{\frac{1}{2}} . \tag{5.9b}$$

In the above the rotation $R(\Omega)$ acts on both pseudo-orbital angular momentum and pseudo-spin.

5.4 TRANSITION RATES

For the quadrupole transitions between u=0 states, the quadrupole operator will be proportional to the quadrupole operator which is a scalar with respect to pseudo-spin; i.e. the operator $\bar{R}^{(2)}_{\mu}$ given in (4.2c). The matrix elements of these operators are the same as those given by the SU_3 limit of the IBM[1]. This must be so because in both cases the quadrupole operator is a generator of the SU_3 group and hence the matrix elements will depend only on the SU_3 quantum number.

5.5 PAIRING ENERGY

The pairing binding energy in the u=0 lowest SU_3 band is

$$\langle(2N,0)K,M;I=0|S^{\dagger}S|(2N,0)K,M;I=0\rangle$$

$$= \frac{(\frac{\Omega}{3}-N+1)(2N-K)(2N+K+1)}{2(2N-1)} . \tag{5.10a}$$

As the angular momentum, J=K, increases, this binding energy decreases. For the u≠0 band the pairing is less which makes these bands higher in energy for the same K:

$$\langle(2N,0)K,M;I\neq0,\mu|S^{\dagger}S|(2N,0)K,M;I\neq0,\mu\rangle$$

$$= \frac{(N-1)}{N}\,\frac{(\frac{\Omega}{3}-N)}{(\frac{\Omega}{3}-N+1)}\langle(2N,0)K,M;I=0|S^{\dagger}S|(2N,0)K,M;I=0\rangle . \tag{5.10b}$$

However for a given total angular momentum J, where of course J=I+K,I+K-1, ... , |I-K|, if more angular momentum is put into the pseudo-spin, I, and less into the pseudo-orbital angular momentum K, the pairing binding energy for that state will increase. Hence for some J, the state with I≠0 may become lower in energy. Thus the Yrast level will be in another band leading to a different moment of inertia for the Yrast band.[10] Of course this effect, which occurs naturally in this model, is outside the scope of the pure IBM since it deals only with the u=0 band. Additional quasi-particle states must be introduced into the IBM.[11]

6. ACKNOWLEDGEMENTS

I would like to thank D. H. Feng, M. W. Guidry, and C. L. Wu for discussions about their recent work on Sp_6. I would also like to thank the organizers of the Symposium, in particular Professor Yang Li-Ming, for their warm hospitality.

References

1. A. Arima and F. Iachello, Adv. in Nuclear Physics 13 (1984) 139.
2. J. N. Ginocchio, Phys. Lett. 85B (1979) 9.
3. J. N. Ginocchio, Ann. of Phys. 126 (1980) 234.
4. C. L. Wu, D. H. Feng, X. G. Chen, J. Q. Chen, and M. Guidry, to appear in Phys. Lett. (1986).
5. A. De-Shalit and I. Talmi, "Nuclear Shell Theory," Academic Press, New York, 1974.
6. F. Iachello, Nucl. Phys. A347 (1980) 51.
7. A. K. Kerman, Ann. Physics 12 (1961) 300.
8. A. Arima, J. N. Ginocchio, and N. Yoshida, Nucl. Phys. A384 (1982) 112.
9. A. R. Edmonds, "Angular Momentum in Quantum Mechanics," Princeton University Press, Princeton, 1968.
10. M. W. Guidry, C. L. Wu, D. H. Feng, J. N. Ginocchio, X. G. Chen, and J. Q. Chen, submitted to Phys. Rev. Lett. (1986).
11. N. Yoshida, A. Arima, and T. Otsuka, Phys. Lett 144B (1982) 86.

THE SHORT RANGE EFFECTIVE INTERACTION AND THE SPECTRA OF OXYGEN ISOTOPES IN (s-d) SPACE

Li Xian-ying & Yao Shi-huai
Physics Department
Anhui University
Hofei, Anhui, China

Zhang Qing-ying
Physics Department
Hunan University
Changsha, Hunan, China

Owing to the success of Delta interaction, which we called the body interaction, in the low-lying spectra calculation for $d_{5/2}$--$d_{5/2}$ space, we are able to express the new extremely short range interaction as follows

$$V_{DDI}=-4\pi A_1\delta(\vec{r}_i-\vec{r}_j)\delta(r_i-R)+B_1\vec{\tau}_i\cdot\vec{\tau}_j+C_1-4\pi D\delta(\vec{r}_i-\vec{r}_j) \qquad (1)$$

which is called the Double Delta Interaction (DDI).

As is well known, the Modified Surface Delta Interaction (MSDI) is

$$V_{MSDI}=-4\pi A_2\delta(\vec{r}_i-\vec{r}_j)\delta(r_i-R)+B_2\vec{\tau}_i\cdot\vec{\tau}_j+C_2 \qquad (2)$$

In the formulas (1) and (2), A,B,C and D are the adjustable parameters. For the oxygen isotopes, it is easy to prove that two parameters B and C are combined into a single one. Therefore parameters C_1 and C_2 are unnecessary.

We have used the formulas (1) and (2) to calculate the low-lying spectra of oxygen isotopes ^{17}O through ^{22}O. The matrix elements of energy are represented in terms of the strength parameters of the DDI (or MSDI) and the single particle energies which can be determined by least-squares fit are as follows

TABLE: Parameter strengths of DDI and MSDI and single-particle energies (in MeV)

Parameters	DDI	MSDI
A	-0.5441	-0.2443
B	-0.1778	-0.1216
D	0.0149	—
$\epsilon_{5/2}$	4.0568	4.0023
$\epsilon_{1/2}$	3.4910	2.7853
$\epsilon_{3/2}$	1.7430	1.1560

The RMS deviations for 41 levels of oxygen isotopes $^{17-22}O$ by the two interactions mentioned above are $\sigma_1=0.331$ MeV for DDI, and $\sigma_2=0.712$ MeV for MSDI, respectively. It is clear that the experimental data reproduction is in favour of the DDI.

PHYSICAL SUBALGEBRA CHAINS IN SEMISIMPLE LIE ALGEBRA A_n

Yinsheng Ling Xiaoqian Zhou

Suzhou University, Suzhou
People's Republic of China

The knowledge of subalgebra chains of a semisimple Lie algebra is important in various algebraic models. In the past, the discovery of a few chains is a result of physical argument, Dynkin-Gruber embedding theory offers a systematic method for finding the complete set of subalgebra chains of a Lie algebra.[1 2 3]

There are many subalgebra chains in a semisimple Lie algebra. The physical chains will contain the physically relevant three dimensional algebra A_1(in Cartan's nomenclature). In most cases, the different A_1's can be distingushed by the index of the embedding. The index of A_1, embeded in a semisimple Lie algebra G, is related to the reduction of G to A_1. Frequently, G is the simple Lie algebra A_n. If the difining space of A_n contains the angular momenturn j_i (i=1-p), the index of A_1 embeded in A_n is [4]

$$\beta = \sum_{i=1}^{p} \frac{1}{6} k_i(k_i+1)(k_i+2),$$

here $k_i=2j_i$. The index is always attached to A_1 as superscript A_1^β.

For example, the three dimensional defining-space of A_2 can contain angular momenturn (1) j=0, $\frac{1}{2}$ or (2) j=1. The relevent physical subalgebras A_1 are (1) $A_1{}^1$ and (2) $A_1{}^4$ respectively. These correspond to two subalgebra chains in A_2:

Projects Supported By the Science Fundation of the Chinese Acadamy of Science.

(1) $A_2 \to A_1{}^1$ (in Weyls nomenclature su(3)→su(2)),
(2) $A_2 \to A_1{}^4$ (su(3)→so(3)).

This is a well known result.

The difining space of A_3 is a four dimensional space. It can embrace the set of angular momenta in four ways (1) $j=0,0,\frac{1}{2}$; (2) $j=\frac{1}{2},\frac{1}{2}$; (3) $j=0,1$; (4) $j=\frac{3}{2}$, Correspondingly, the physical subalgebras A_1 are (1) $A_1{}^1$; (2) $A_1{}^2$; (3) $A_1{}^4$; (4) $A_1{}^{10}$. Following the method of Dynkin and Gruber, we can get the maxium subalgebra in each stage. The subalgebra chains of A_3 are shown in Fig.1 as an example.

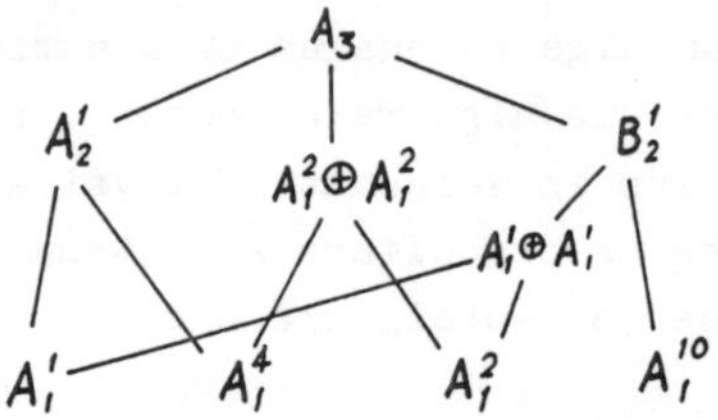

Fig.1

All these chains are used in particle physics, nuclear physics, atomic physics and molecular physics.

There are ten different A_1 in the simple Lie algebra A_5. [3] If the defining space of A_5 contains the angular momenturn j=0,2 (condition of IBM-I in nuclear physics), all the physics chains of A_5 should end with $A_1{}^{20}$. In fact there are only three chains ended with $A_1{}^{20}$ (Fig.2).

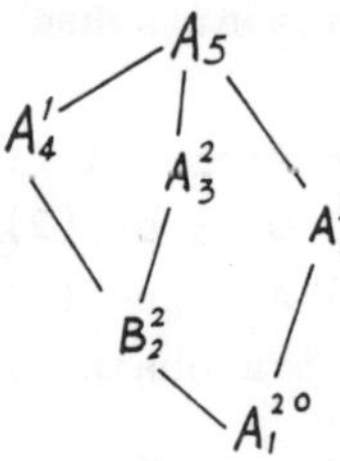

Fig.2

For the s.d shell, if we only consider one kind of the nucleon, the dynamical group will be su(12)(A_{11}). The angular momenturn contents of the defining space of A_{11} are $j=\frac{1}{2}, \frac{3}{2}, \frac{5}{2}$. The maximum subalgebras of A_{11} which contain the physical subalgebra A_1^{46} are $A_1^1 + A_9^1$, $A_3^1 + A_7^1$, $A_5^1 + A_5^1$, $A_1^6 + A_5^2$, C_6^1, $A_2^4 + A_3^3$. (Fig.3)

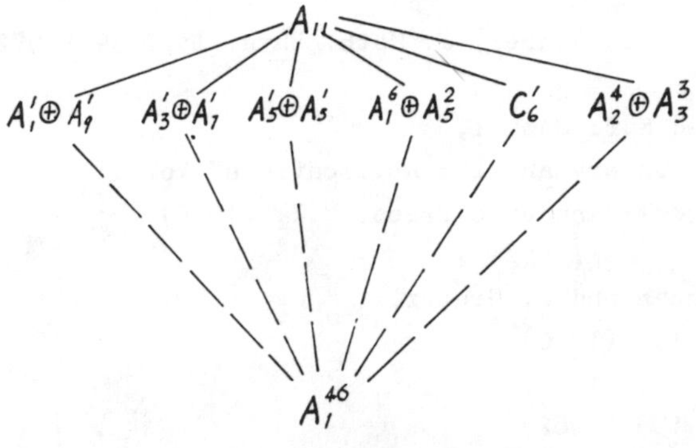

Fig.3

Ginocchio model [5] is shown in Fig.4

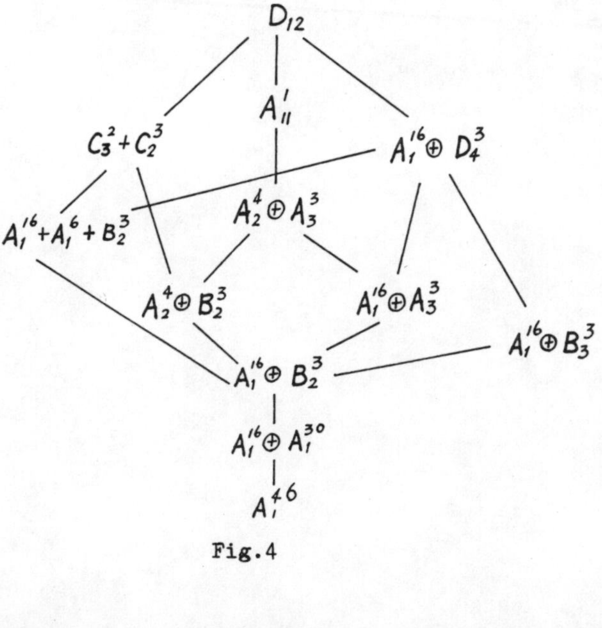

Fig.4

References

1 E.B. Dynkin, Am. Math. Soc. Transl. (2) 6,111 (1965)
Am. Math. Soc. Transl. (2) 6,245 (1965)

2 M. Lorente and B. Gruber, J. Math. Phys. 13,1639 (1972)

3 B. Gruber and M.T. Samuel,
in "Group Theory and Its Applications" Vol. III,
ed. E. Loebl; Academic Press, N.Y. (1975)

4 M. Samuel Thoms and B. Gruber,
Kinam 2, 133 (1980)
Kinam 2, 381 (1980)
Kinam 4, 139 (1982)
Kinam 5, 173 (1983)

5 J.N. Ginocchio, Ann. Phys. 126 (1980)

THE NON-ANALOG DOUBLE CHARGE EXCHANGE TRANSITION AND THE RELATIONSHIP OF ENERGY DEPENDENCE

Ma Wei-hsing
Institute of High Energy Physics, Academia Sinica, Beijing, China.

Wang Si-wen and Zhang Gao-you
Department of Physics, Northwest University, Xian, China.

Chen Chung-kuang
Graduate School of Academia Sinica, Beijing, China.

Yang Zhen-rong
Computing Centre of Academia Sinica, Beijing, China.

Zhao Shu-ping
Dept. of Physics of USTC. Hefie, China.

ABSTRACT

An intriguing observation in pion double charge exchange reaction is the existence of large non-analog cross section for (π^+, π^-) reaction on T=0 nuclei . As can be seen in the oxygen isotopes at 164 Mev, the cross sections for ^{16}O (T=0) are nearly as large as those for ^{18}O (T=1). For instance, the ratio $\sigma(^{18}O)/\sigma(^{16}O)$ is 3:1 at 164 Mev. But beyond 164 Mev at 292 Mev, it rapidly becomes 20:1. That means the ratios are strong energy dependence. The anomalous features observed in the $^{16}O(\pi^+, \pi^-)^{16}Ne$ transition are also presented in all non-analog DCX transition from available T=0 targets, for example $^{12}C(\pi^+, \pi^-)^{12}O_{(g.s.)}$. This regularity is in conflict with expectations that transitions to double isobaric analog states (DIAS) would dominate because of the high degree of overlap between initial and final state wave functions and rules out simple explanations based on the detailed nuclear structure of the state involved since the explanation based on the nuclear structure properties of ^{18}O, ^{16}O and ^{18}Ne, ^{16}Ne is entirely independent of any aspects of DCX reaction mechanism, and it is in particular independent of the energy of the incident pion. This means that we should get $\sigma(^{18}O)/\sigma(^{16}O)$ =3:1 at all energies. But it is not true as just mentioned. One is therefore forced to study some basic aspects of reaction mechanism.

Under the model proposed we study the energy dependence of angular distribution in non-analog pion double charge exchange (DCX) reaction $^{16}O(\pi^+, \pi^-)^{16}Ne_{(g.s.)}$. The results stronly support the validity of our model and indicate that the non-analog angular distribution is consistent with simple diffractive scattering while its shape is independence of bombarding energy. Angular distribution as a whole goes up and down rapidly with increasing bombarding energy.

Reference:

(1) R. Gilman et al., Phys. Rev. C 29(1984) 2395.
(2) C.L. Marris, et al., Phys. Rev. C 27(1983) 2375.

THE NON-ANALOG DOUBLE CHARGE EXCHANGE TRANSITION $^{12}C(\pi^+, \pi^-)^{12}O_{(g.s.)}$ and $^{16}O(\pi^+, \pi^-)^{16}Ne_{(g.s.)}$

Ma Wei-hsing
Institute of High Energy Physics, Academia Sinica, Beijing, China.

Wang Si-wen and Zhang Gao-you
Department of Physics, Northwest University, Xian, China.

Chen Chung-kuang
Graduate School of Academia Sinica, Beijing, China.

Yang Zhen-rong
Computing Centre of Academia Sinica, Beijing, China.

Zhao Shu-ping
Dept. of Physics of USTC. Hefie, China.

ABSTRACT

We proposed six possible meason exchange current mechanism for poin induced non-analog double charge exchange transition on $^{12}C(\pi^+, \pi^-)^{12}O_{(g.s.)}$ and $^{16}O(\pi^+, \pi^-)^{16}Ne_{(g.s.)}$. Under the model, we calculated 164 Mev angular distribution. The theoretical results are compared with the existed data and are in a good agreement with the experimental data available. It strongly indicates that the non-analog DCX process is basically a diffractive process through double isobar excitation on a single nucleon and the dominant contribution to the process comes from one of the six diagrams, that is a π^+ collides with a neutron to excite this neutron into an isobar $\Delta(1232)$ with charge of one unit; Subsequently, the Δ^+ decays to give $\pi^+\Delta^0$ and then produced π^+ is absorbed by the second target neutron which will be then turned to a proton. Meanwhile the traveling Δ^0 finally decays into π^-P so that the course of the process under consideration here is ended.

Reference:

(1) S.J. Greene, et al, Phys. Rev. C 25(1982) 927.
(2) C.L. Morris, et al, Phys. Rev. C 25(1982) 3218 C 27(1983) 2375.

RECENT DEVELOPMENTS IN THE THEORY OF NUCLEAR MOLECULES

Jae Young Park and Moon Hoe Cha*

Department of Physics
North Carolina State University
Raleigh, North Carolina 27695-8202, USA

Werner Scheid

Institut für Theoretische Physik
der Justus-Liebig-Universität
D-6300 Giessen, Germany

1. INTRODUCTION

As evidenced in many recent international conferences and construction of major heavy-ion accelerators in many countries, heavy ion physics is the most rapidly developing frontier in nuclear physics at present.[1] One of the most interesting and outstanding aspects of heavy ion nuclear physics is that an entirely new mode of dynamics and nuclear structure, namely nuclear molecules, can be formed during collisions between heavy ions.

2. BASIC PRINCIPLES AND PROPERTIES OF NUCLEAR MOLECULES

A nuclear molecule is a system of two (or more) nuclei (or clusters) which are bound together on their surfaces in the quasibound or bound states of a quasimolecular potential. In the microscopic two-center model[2,3] a nuclear molecule may, therefore, be understood as a state in which the outermost loosely bound (valence) nucleons orbit around both nuclei, just like homopolar or covalent bonding in atomic molecules, such as O_2 or N_2 molecules.

The narrow resonances observed in the energy dependence of the $^{12}C + {}^{12}C$ reaction cross sections at the energies near and above the Coulomb barrier have provided the first evidence for the existence of nuclear molecules.[4] Nuclear molecular configurations may be considered as the most pronounced examples of cluster phenomena known in nuclear physics at present. Since then a large number of other resonances were discovered in many other light heavy-ion systems, for example, $^{12}C + {}^{16}O$, $^{16}O + {}^{16}O$, $^{12}C + {}^{24}Mg$ and $^{28}Si + {}^{28}Si$. The molecular description is applicable for low-energy heavy ion reactions in which the nuclear motion is adiabatically slow compared with the rearrangement time of the mean field of nucleons.

---

*Permanent address: Dept. of Physics, Kangwon National University, 200 Korea.

3. STRUCTURE AND FORMATION OF MOLECULAR CONFIGURATIONS

3.1 The Two-Center Shell Model: Theoretical Basis for Nuclear Molecules.

The underlying fundamental model for a nuclear molecule is the two-center shell model (TSCM).[2,3] The model describes the binding of the molecule in terms of molecular single-particle states. Therefore, molecular single-particle states should play an important role in understanding the structure and formation of molecular configurations. Two-center wave functions include in their dependence on the internuclear distance polarization or dynamical orientation effects of valence nucleons by the field of the other nucleus.

3.2 Adiabatic and Sudden Potentials

The energies of the molecular resonances are mainly determined by the real nucleus-nucleus potential. Various methods have been used to calculate real potentials for heavy ions.[1] They may be classified either as adiabatic potentials or as sudden potentials depending whether the adiabatic or the sudden approximation is used. The adiabatic approximation is based on the assumption that the collision proceeds so slowly that the nucleons move in an effective common shell potential at every instant. The sudden approximation assumes that the scattering time is smaller than the rearrangement time of the shells of the individual nuclei into the shells of the compound nucleus.

3.3 The Double Resonance Mechanism and the Band Crossing Model

In the double resonance model[5] the colliding nuclei are assumed to enter the interaction region in the elastic channel, where they have sufficient relative energy to pass over the Coulomb barrier, and to be subsequently trapped in their mutual potential well by inelastic excitation of one or both fragments into low-lying collective states. Thus, resonance effects are obtained at bombarding energies corresponding to "quasi-bound states" in the inelastic well. The intermediate structure in the excitation functions is generated when, first an elastic partial wave resonates with its corresponding virtual state and second, a quasibound state is excited by this way. For this to happen, the intrinsic excitation of the nucleus-nucleus system has to account for the difference in energy and angular momentum between the virtual and quasi-bound states. When a rotational spectrum for the energies of the virtual and quasibound resonances is used, one obtains the band crossing model[6]. In this respect the crossing point in the model is also nothing but the point where "double resonance condition" is fulfilled.

4. MOLECULAR SINGLE-PARTICLE CONFIGURATIONS

4.1 The Molecular Particle-Core Model

The molecular particle-core model is a convenient framework for the description of heavy ion collisions.[7,8,9] In this model two colliding nuclei are described in terms of two cores with C_1 and C_2 nucleons and loosely bound extra-core nucleons which can be described by molecular single-particle states in the framework of the two-center shell model.

A. The Hamiltonian

Since the two-center shell model and its wave functions are conveniently written in a coordinate system in which the centers lie on the z-axis, it is advantageous to express the Hamiltonian in a rotating coordinate system with the z'-axis along the direction of the relative coordinate. The rotating coordinate system is fixed with respect to the laboratory system by the Euler angle, ϕ and θ which are the spherical polar angles of the relative coordinator $\vec{r}$ (see Refs. 7 and 9).

B. The Molecular Wave Functions

The wave functions for the solution of the scattering problem can be expressed as follows:[7,9]

$$\Phi_{IM} = A(1,\cdots,N)S(c_1c_2) \sum_{\alpha,K} R^{I}_{\alpha K}(r)D^{I*}_{MK}\,\phi_{\alpha K}(\vec{r}\,'_{icm}, r). \quad (1)$$

The Radial wave function $R^{I}_{\alpha K}(r)$ depends on the total angular momentum I, the projection quantum number K of the angular momentum on the intrinsic z-axis and a further quantum number α which classifies the various molecular single-particle configurations described by $\phi_{\alpha K}$. The rotation of the coordinate system is described by the function D^{I*}_{MK} depending on the Euler angles. The intrinsic wave functions are products of the eigenfunctions of h_{TCSM}. The wave functions Φ_{IM} are assumed to be antisymmetrized in the coordinates of the extra nucleons. Using the wave functions (1) and the Hamiltonian in the rotating coordinate system, one can obtain a system of coupled equations for the radial wave functions $R^{I}_{\alpha K}(r)$. This system has to be solved with proper boundary conditions for $R_{\alpha K}(r\to\infty)$, which are related to the scattering matrix.

4.2 Application to the $^{13}C + ^{13}C$ System

The molecular particle-core model has been applied to the $^{13}C + ^{13}C$ system by Terlecki et al.[10] who calculated the elastic and inelastic cross sections for the single and mutual excitation of the $\frac{1}{2}^+$ (3.09 MeV) state of ^{13}C. Later, Könnecke et al.[11] have studied the one-neutron transfer reaction $^{13}C(^{13}C,\, ^{12}C)^{14}C$ and shown that resonant like structures in the experimental differential cross sections[12] can be satisfactorily accounted for as a transfer of a neutron occupying a molecular single-particle state during the reaction.

4.3 Two-Center Single-Particle Particle Energy Diagrams and Level Crossings

Realistic two center diagrams for several heavy ion systems have been calculated[13-15] using the asymmetric TCSM[3]. As an example, the level diagrams for the $^{13}C + ^{17}O$ system is shown in Fig. 1. In these calculations the parameters of the TCSM potentials have been determined by fitting the experimental neutron single-particle energies near the Fermi surfaces of the separated nuclei, e.g. of ^{13}C and ^{17}O for $R\to\infty$ and of ^{30}Si for R=0. In Fig. 1 we recognize many pairs of levels between which the radial and rotational couplings can cause the excitation or transfer of nucleons.

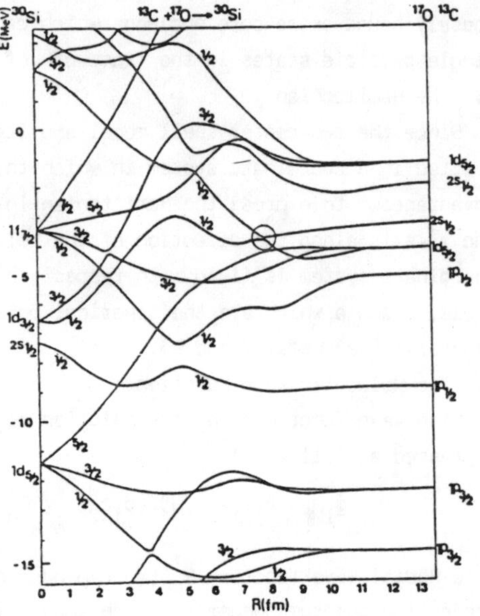

Fig. 1. The neutron level diagram for $^{13}C + ^{17}O \to ^{30}Si$.

4.4 Theory of New Resonance Structures Due to Landau-Zener Effects

A. The Landau-Zener Approximation

According to the formula due to Landau[16] and Zener[17], the one-way transition probability P_{12} from one adiabatic state to another by a single passage of the crossing point with a velocity v is given by

$$P_{12} = e^{-2\pi G}, \tag{2}$$

where,

$$G = \frac{1}{\hbar}\frac{|H_{12}|^2}{v\Delta F}, \qquad \Delta F = |d\varepsilon_2/dR - d\varepsilon_1/dR|_{R=R_c} \tag{3}$$

Here, R_c is the crossing distance of diabatic energy curves ε_1 and ε_2, as shown in Fig. 2, and $|H_{12}|$ is a coupling matrix element between the diabatic states at R_c and is equal to half of the separation between the adiabatic energies. Since the system passes through the crossing point twice, first in entering into the interaction region, and second in leaving out from there, the total transition probabilities P for the process is given by

$$P = 2P_{12}(1-P_{12}). \quad (4)$$

In the Landau-Zener model the angle-integrated cross section is given as a function of incident energy E by[18]

$$\sigma_{21}(E) = 4\pi^2 R_C^2 \frac{\sqrt{2\mu}}{\hbar} \frac{|H_{12}|^2}{\Delta F} \frac{\sqrt{E-\bar{V}}}{E}. \quad (5)$$

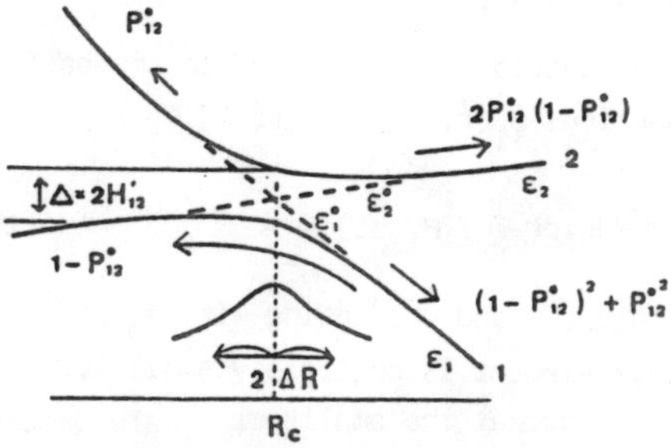

Fig. 2. A schematic diagram of level crossing.

Here, μ and V are, respectively, the reduced mass of incident channel and the adiabatic potential value at the crossing distance R_C. The expression (5), which is obtained by an integration over impact parameter, apparently depends smoothly on the energy, with a sharp rise at the threshold E=V.

Since resonance peaks should be attributed to a single grazing partial wave, the angle-integrated cross section is expressed by[15]

$$\sigma_{21}(E) = \frac{\pi}{k^2} \sum_{\ell=0}^{\ell_{max}} (2\ell+1)P(E,\ell)/3, \quad (6)$$

where, ℓ_{max} is the maximum orbital angular momentum of the partial waves which can reach the crossing point at an incident energy E. The factor 1/3 comes from the fact that we consider the $^{13}C + ^{17}O$ system in the following where only one branch ($\Omega=\frac{1}{2}$) of the three originating from the $1d_{5/2}$ state has an avoided crossing with the $2s_{1/2}$ energy level (see Fig. 1). The radial velocity v in Eq. (3) is calculated by the classical relation:

$$E = \frac{1}{2}\mu v^2 + \frac{\ell(\ell+1)\hbar^2}{2\mu R_C^2} + \bar{V}. \quad (7)$$

The crossing point R_C and the interaction matrix element $|H_{12}|$ are extracted directly from the energy diagram.

B. The Semi-Classical Approximation Including Turning Point Effects.

In the Landau-Zener approximation the relative velocity between the colliding particles in the region of avoided level crossings is assumed to be constant. The model can be extended[19] by taking into account the variation of relative velocity due to the effect of interaction potential between colliding heavy ions. The internuclear distance $R(t,\ell)$ is obtained by solving the classical equation of motion

$$\mu\frac{\partial^2 R(t,\ell)}{\partial t^2} = -\frac{\partial V}{\partial R}\bigg|_{R(t,\ell)}. \tag{8}$$

The wave function $\psi(\vec{r},t)$ of the valence nucleon satisfies the time-dependent Schrödinger equation

$$i\hbar\frac{\partial\psi}{\partial t} = H(\vec{r},R(t,\ell))\psi. \tag{9}$$

In the two-state approximation the wave function ψ is expanded in the basis of the two diabatic molecular wave functions $\phi_1(\vec{r},R)$ and $\phi_2(\vec{r},R)$,

$$\psi = \sum_{i=1}^{2} c_i(t)\phi_i(\vec{r},R)\exp(-\frac{i}{\hbar}\int^t H_{ii}dt), \tag{10}$$

where, $H_{ij} = \langle\phi_2|H|\phi_j\rangle$. Inserting ψ into (9) and projecting with ϕ_j and assuming that the radial coupling matrix element is negligibly small for diabatic states near the crossing point and that the amplitude c_i are damped due to all other channels which are not treated explicitly, we obtain

$$\dot{c}_1 = -\frac{w}{\hbar}c_1 + \frac{H_{12}}{i\hbar}\exp(ih_{12})c_2,$$

$$\dot{c}_2 = -\frac{w}{\hbar}c_2 + \frac{H_{21}}{i\hbar}\exp(-ih_{12})c_1, \tag{11}$$

where, $h_{12} = \int^t(H_{11}-H_{12})dt/\hbar$ and $w(R)$ is the absorptive potential. The transition probability for the inelastic excitation is given by[19]

$$P(E,\ell) = |c_2(t_f)|^2, \tag{12}$$

where, t_f is determined by $R(t_f,\ell) = R_c+2\Delta R$. The Landau-Zener formula can be derived from Eq. (11) with the following assumptions, namely, $w=0$, $\dot{R}$=consts., H_{12}=const., $t_i=-\infty$ and $t_f=\infty$. These assumptions, especially the constant relative velocity and the infinite reaction time interval, are unphysical especially if the crossing point is situated near the potential barrier or if the turning point of the trajectory lies outside of the crossing region.

4.5 Applications to the $^{12}C + ^{17}O$ and $^{13}C + ^{17}O$ Systems

In a series of γ-ray yield curve measurements on heavy ion reactions the Strasbourg group of Haas and Freeman et al[20] have observed a smooth energy dependence except in one case, namely apparent resonantlike structures for the 871-keV γ-rays arising from the decay of the excited $1/2^+$ states. As shown in Ref. 15, the characteristic resonancelike peaks observed in the angle-integrated inelastic cross-section can be understood as due to the new resonance mechanism discussed in Section 4.4A. One recognizes in Fig. 1 an avoided crossing between the adiabatic energy curves originating from the $1d_{5/2}(\Omega=1/2)$ and $2s_{1/2}$ levels of ^{17}O near 7.7-8.0 fm, where the promotion of the valence neutron in the $1d_{5/2}$ state of ^{17}O to the $2s_{1/2}$ can take place. From the energy diagram the values of R_c and $|H_{12}|$ are extracted to be 7.9 fm and 0.05 MeV, respectively. The nonadiabatic coupling range ΔR ($\simeq 2|H_{12}|/\Delta F$) is taken to be 0.1 fm.

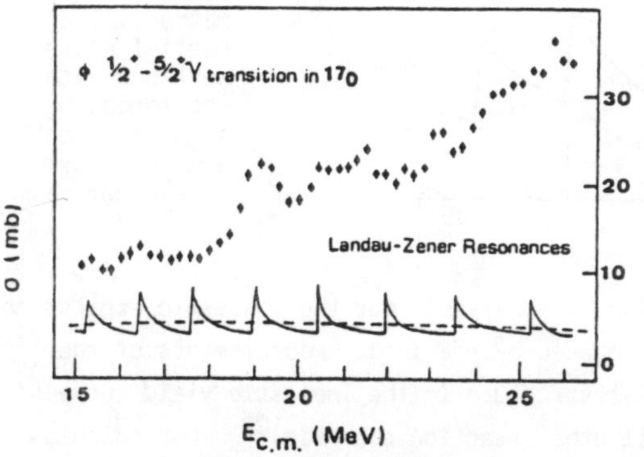

Fig. 3. Landau-Zener inelastic cross sections are compared with the inclusive γ-ray yield curve. Solid and dashed lines denote the calculations with Eq. (6) and with Eq. (5) multiplied with a factor 1/3, respectively.

Inelastic cross section leading to the $^{17}O^*(1/2^+$, 0.87 MeV) state are calculated with expressions (5) and (6) and are compared in Fig. 3 with the yields of the 871 keV γ-ray. Although the reproduction of the data is not perfect it is remarkable in view of the simplicity of the evaluation that expression (6) (solid line) reproduces the series of resonancelike peaks observed. An alternative explanation for the structures in the inelastic cross section is also based on the Landau-Zener mechanism. The structures can arise due to the combined action of two effects, namely an enhancement of the transition strength by the Landau-Zener mechanism and the window effect for the grazing partial waves. The latter effect can be described by an angular momentum dependent absorptive potential. The cross section for the the inelastic excitation of the first $1/2^+$ state of ^{17}O calculated by the semiclassical method as discussed in Section 4.4-B, are compared with the same data of Strasbourg group in Fig. 4. The calculated structures are in reasonable agreement with the data. The parameters of the transition matrix element used in the calculation are H_{12} = 0.5 MeV and ΔR = 1.5 fm. The structures are caused by the barriers of the angular momentum-dependent potentials. Recently, Milek and Reif[21] have also studied this system using the two-center shell model with finite depth potentials supposing the two ions to move along classical trajectories.

Since the calculated neutron level diagrams for the reactions $^{12}C + ^{17}O \rightarrow$ ^{29}Si and $^{13}C + ^{17}O \rightarrow ^{30}Si$ are very similar, it is expected that the Landau-

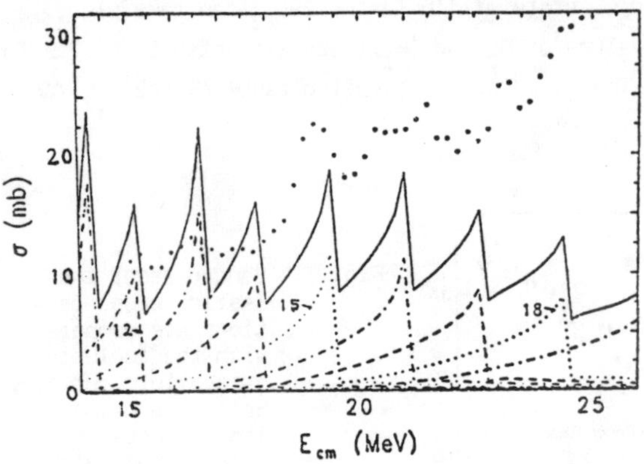

Fig. 4. The cross section for the inelastic excitation for the first $\frac{1}{2}^+$ state of ^{17}O calculated by the semiclassical method. The partial cross sections with corresponding angular momenta are shown by broken curves.

Zener promotion effect should also be observable for the inelastic excitation of the first $1/2^+$ state of ^{17}O induced by ^{12}C ions. Measurements of the $^{12}C + {}^{17}O$ system indeed showed similar structure in the inelastic yield curve of the 0.87 MeV γ-ray yield from all other reaction channels.[22] Our calculation reproduces most of the main peaks remarkably well[23].

5. CONCLUSIONS

The theory of nuclear molecules can quantitatively explain the nuclear structure, energy and spins of the resonances in reactions between light nuclei and qualitatively the gross and intermediate structures in the cross sections. Studies of the nuclear Landau-Zener effect with molecular single-particle states have shown that recent data from C + O collisions can be qualitatively accounted for and can be taken as evidence for both the presence of the Landau-Zener promotion mechanism and molecular single-particle orbitals in heavy ion reactions at low energies. No other reasonable explanation for these data is available at present. There are other evidences for molecular orbital phenomena, for example, in the $^{13}C + {}^{12}C$ system[24].

Systematic studies of TCSM level diagrams and promotion effcts at level crossings for various projectile-target systems is not only useful to understand microscopically the role of individual molecular nucleonic orbitals in heavy ion reactions but also paves the way toward a microscopic understanding of the collisions dynamics, e.g. the energy dissipation and diffusion[25].

REFERENCES

1. N. Cindro, ed. Nuclear Molecular Phenomena (No.-Holland, N. Y., 1978). K. A. Eberhard, ed. Resonances in Heavy Ion Collisions (Lecture Notes in Phys. 156) (Springer-Verlag, Berlin, 1982).

N. Cindro, R. A. Ricci, and W. Greiner, eds. Dynamics of Heavy Ion Collisions (North-Holland Pub. Co., New York, 1981).
N. Cindro, W. Greiner, and R. Caplar, eds. Fundamental Problems in Heavy-Ion Collisions (World Scientific Pub. Co., Singapore, 1984).
2. P. Holzer, U. Mosel, and W. Greiner, Nucl Phys. A138 241 (1969).
D. Scharnweber, W. Greiner, and U. Mosel, Nucl. Phys. A164 257 (1971).
3. J. A. Maruhn and W. Greiner, Z. Phys. 251 431 (1972).
4. E. Almquist, D. A. Bromley, and J. A. Kuehner, Phys. Rev. Lett. 4 515 (1960).
5. W. Scheid, W. Greiner, and R. Lemmer, Phys. Rev. Lett. 25 176 (1970).
6. Y. Abe, Suppl. Prog. Theor. Phys., No. 68 303 (1980)
7. J. Y. Park, W. Scheid, and W. Greiner, Phys. Rev. C6 1565 (1972).
8. W. von Oertzen and W. Nörenberg, Nucl. Phys. A207 113 (1973).
9. J. Y. Park, W. Scheid, and W. Greiner, Phys. Rev. C20 188 (1979).
10. G. Terlecki, W. Scheid, H. J. Fink, and W. Greiner, Phys. Rev. C18 265 (1978).
11. R. Könnecke, W. Greiner, and W. Scheid, Phys. Rev. Lett. 51 366 (1983).
12. S. K. Korotky et al., Phys. Rev. C28 168 (1983).
13. J. Y. Park, W. Greiner, and W. Scheid, Phys. Rev. C21 1958 (1980).
14. J. Y. Park, W. Scheid, and W. Greiner, Phys. Rev. C25 1902 (1982).
15. Y. Abe and J. Y. Park, Phys. Rev. C28 2316 (1983).
16. L. Landau, Phys. Z. Sov. 2 46 (1932).
17. C. Zener, Proc. Roy. Soc. A137 696 (1932).
18. E. E. Nikitin and L. Zülicke, Theory of Chemical Elementary Processes (Springer, 1978) Chap. 5.
19. J. Y. Park, K. Gramlich, W. Scheid and W. Greiner, to be published (1986).
20. R. M. Freeman et al., Phys. Rev. C28 437 (1983).
21. B. Milek and R. Reif, Phys. Lett. 157B 134 (1985).
22. C. Beck et al., Nucl. Phys. A443 157 (1985).
R. M. Freeman et al., Private communication and to be published.
23. M. H. Cha, J. Y. Park and W. Scheid, to be published (1986).
24. W. von Oetzen et al., Phys. Lett. 93B 21 (1980); W. von Oetzen and B. Imanishi, Nucl. Phys. A424 262 (1984).
25. W. Cassing and W. Nörenberg, Nucl. Phys. A401 467 (1983); A433 467 (1985).

PHASE TRANSITIONS AND SHAPE CHANGES IN FINITE NUCLEI AND THE IMPORTANCE OF FLUCTUATIONS

Peter Ring

Physik-Department der Technischen Universität München, James-Franck-Straße, D-8046 Garching, West Germany*

The collapse of pairing correlations in nuclei at high angular momenta and finite temperatures and the transition from axially symmetric prolate shapes to triaxial and oblate deformations for heavy Rare Earth nuclei is investigated by methods going beyond the Mean Field approach. It is shown that fluctuations play an important role not only in the quantitative understanding of these phase changes. If they are taken into account properly we find even qualitative changes in some cases: The Nuclear Meissner Effect, i.e. the pairing collapse in nuclei with large angular momenta disappears and well pronounced triaxial minima are found.

1.INTRODUCTION

Matter can exist in different forms and configurations, which are usually called "phases". Which phase is realized in a particular sitation depends on external parameters, such as the temperature or the size of an external field, but also in internal parameters, such as the strength of the interaction, the level density or the number of particles (density in space). If such a parameter is changed, we observe transitions from one phase to another phase. In infinite systems, like in a solid, these phasetransitions occur at very sharp transition points, i.e. at critical values for the corresponding parameter.

In finite systems these transitions are never sharp. They are always smeared out. We observe continuous transitions from one phase to another phase in a certain region of the relevant parameter. It depends on the size of this region whether it is reasonable to call such a process a "phase transition". If the transition region encompasses the entire region of physically possible values for the parameter, one can certainly not say that one phase is realized if this parameter region and a different phase is

---

Invited talk at the "International Symposium on Particle and Nuclear Physics", Sept. 2 - 7, 1985, Peking, China
*supported by the Bundesministerium für Forschung und Technologie

realized in a separate region. The picture of a "phase transition" does not apply.

Nuclei are finite manybody systems and for large particle numbers, i.e for heavy nuclei one should expect features which resemble a phase transition. In fact one has found several such situations which are well known nuclear "phase transitions". The best known is the transition from sphericity in double magic and single magic nuclei to deformed shapes in nuclei where both protons and neutrons are in open shells. The relevant parameter in this case is the number of valence particles and following a chain of isotopes, as for instance the Sm-Isotopes between ^{148}Sm to ^{156}Sm we find a very sudden shape change for N=88 neutrons from spherical to axially symmetric prolate shapes. This transition fulfills all the criteria for the phase transition. It shows that is meaningful to speak of phase transitions in finite systems.

Other examples of nuclear phase transitions have been discussed in the literature. Examples are the transition to triaxial shapes in transitional nuclei[1-3] or the pairing collapse with increasing angular momentum, the nuclear Meissner effect[4]. I will discuss these two examples in more detail, because both have been proposed already nearly 30 years ago, both are still not well established, but for rather different reasons: There are many experimental hints for nuclear triaxiality (see for instance ref. 5), but so far it has not been understood in theoretical calculations[6-15]. On the other hand there are many theoretical calculations which predict the pairing collapse at high spin[16-18], but so far there is no experimental evidence for it.

The dilemma can be solved rather easily if one takes into account the fact that in all the theoretical calculations mean field theory has been used. In finite systems this method is a crude approximation, because fluctuations are neglected. In fact it will turn out that fluctuations reverse the situation: They produce well pronounced triaxial shapes on one side and they cause the pairing collapse at high spins to disappear completely. Taking into account fluctuations thus yields agreement between theory and experiment.

2. PAIRING COLLAPSE

The process of pairing is a very suitable example to study collective properties in nuclei, because its phase transition corresponds to an Abelian group, the gauge group U(1) generated by the particle number operator N, which can be treated with rather different and sophisticated methods even in realistic heavy nuclei.

We therefore will concentrate on this process in more detail. The following considerations, however, are more general. They also apply for other phases.

Mean field theory provides a very simple tool to describe phase transitions, because it gives a classical description of processes, which are characterized by classical quantities such as order parameters. Phase transitions are collective processes, i.e. they can occur only in systems with many particles and in the limit of infinitely many particles mean field theory is often a nearly exact description, as can be seen in many models. Therefore mean field theory is in a way conceptually connected with phase transiions. However, this is only a very qualitative statement. In each case one has to investigate the details.

Heavy nuclei contain a relatively large number of particles. In a naive picture one therefore would expect that mean field theory provides a good description. In fact it is easy to see that the important quantity is not the total particle number, but only the number of particles participating in the collective process. In the case of stable groundstate deformations for instance these are all the valence particles. Although this number is considerably smaller than the total particle number, it is still a relatively large number in heavy nuclei. The phase transition to deformed shapes is therefore well pronounced. The situation is different in the case of pairing correlations. Here only a few nucleons in the neighborhood of the Fermi surface participate and pairing correlations are weak. Mean field theory is not good enough to give a reliable description of this phase transition and fluctuations have to be taken into account.

There are several kinds of fluctuations. First we have to distinguish between quantum fluctuations and thermal fluctuations. In this talk I concentrate on nuclei in the neighborhood of the yrast line, where the temperature vanishes. There are no thermal fluctuations.

Quantum fluctuations can be visualized in the following picture: Mean field wave functions span a certain subset of the total Hilbert space, a manifold which can be characterized by in general infinitely many "deformation parameters". One often concentrates only on a few important ones. Calculating the energy of each mean field function one can construct an energy surface in these parameters. Each mean field function is one point on this energy surface. Time dependent mean field theory describes the motion of a classical particle in this energy surface and the solution of the stationary meanfield equations, which are obtained by minimizing the energy, correspond to local minima (or in general to stationary points). It has been shown[19] that the exact solution of the problem can be obtained by quantization of this motion. One then finds wavefunctions in this energy surface. Only in cases where the wavefunction is concentrated close to one minimum, mean field theory provides a meaningful approximation of the situation. This occurs usually for well pronounced deep minima, as for instance for the well deformed Rare Earth nuclei, where one has found minima in the energy surface several MeV deep.

Going beyond the mean field approximation, one has to take into account zero point oscillations around the minimum, which correspond to the excited modes. They can be calculated in principle by Random Phase Approximation (RPA) and indeed this method provides a tool to take into account fluctuations in cases where the harmonic approximation is justified. For the descripton of phase transitions, however, it is well known that this method breaks down at the transition point. At this point one of the collective excitations becomes soft, i.e. it approaches zero energy, and the energy surface becomes flat in one direction. We therefore use in the following a method which does not depend on the harmonic approximation and which yields in general an exact solution of the quantum mechanical Schroedinger equation, the Generator Coordinate Method (GCM)[20,21]. In practical applications we restrict ourselves to only one or a few generator coordinates, i.e. we take into account only a few important collective degrees of freedom. In our case these degrees of freedom are well known from the nature of the phase transiton. This method therefore provides an excellent tool for our purposes.

Phase transitions are connected with broken symmetries, i.e. the correlations causing the phase change are taken into account in mean field theory by symmetry violating mean field functions. In such a case the solution of the mean field equations does not correspond to a real minimum of the energy surface, but to a point on the bottom of a valley, which is flat in one direction. The motion in this direction corresponds to a "rotation" of the wave function with broken symmetry, which does not change the energy. In the RPA we find a Goldstone mode with zero excitation energy, which in Nuclear Physics is often called "spurious mode". Again the harmonic approximation breaks down.

In the case of a broken symmetry we therefore have to distinguish two types of quantum fluctuations, those having its origin in the spurious modes and those corresponding to virtual admixtures of normal excitations. The spurious modes have no restoring force, therefore this kind of

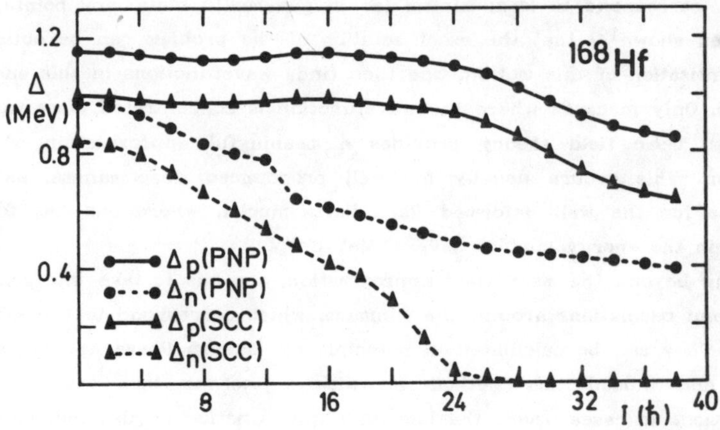

Fig.1 Theoretical gap parameters for protons (full lines) and neutrons (dashed lines) for ^{168}Hf as a function of the angular momentum. SCC are selfconsistent Cranking calculations, i.e. pure mean field approximation, PNP is particle number projection, where the fluctuations related to the Goldstone modes are taken into account

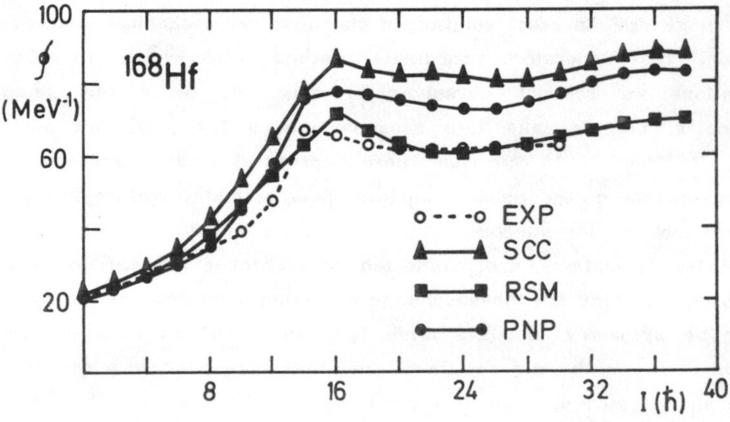

Fig.2 The moment of inertia $\mathcal{J}$ of ^{168}Hf as a function of the angular momentum. The experimental values (dashed line) are compared with calculations in the selfconsistent Cranking approach (SCC), the pure mean field approximation, the Rotating Shell Model (RSM), where the gap parameter is kept frozen at the value of the ground state and with exact number projection before the variation (PNP)

fluctuations is especially important. It can be shown[22] that the GCM method provides a tool to treat this type of fluctuations exactly. The

GCM-wave functions are determined by the symmetry. The method corresponds to a projection of the mean field wave functions onto eigenstates of the symmetry operator, in the case of pairing to number projection.

In Fig.1 we show realistic calculations[23] with the number projection code developed in Munich[24] for the nucleus ^{168}Hf. The configuration space of Kumar and Baranger[12] is used and the Hartree-Fock-Bogoliubov equations are solved with and without number projection before the variation. We find a totally different behavior for the gap parameter Δ. Without number projection, i.e. in mean field approximation we see at angular momentum $I = 24\ \hbar$ a sharp collapse of the neutron gap. With number projection this quantity is only reduced in the region of $I = 12\hbar$, where backbending occurs and where we have a transition to a blocked two-quasiparticle configuration. In the region of the phase transition the gap parameter decreases only very little.

In Fig. 2 we show the moment of inertia $\mathcal{J} = I/\omega$ obtained from these calcuations. In addition to the mean field calculation (SCC) and the number projected results (PNP) we give the experimental values measured in Daresbury[25] and results of a calculation with frozen gap parameters (RSM). All three calculations are in agreement with experiment, all show that the moment of inertia reaches above spin $I = 20\hbar$ a value which is close to the rigid body value. This experimental fact does not indicate a

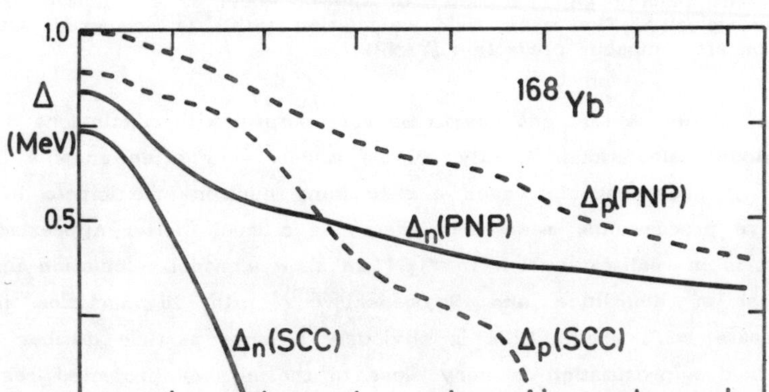

Fig.3 Pairing parameters for neutrons (full lines) and protons (dashed lines) in the nucleus ^{168}Yb as a function of the angular momentum. Self-consistent Cranking Calculations (SCC) are compared with particle number projection (PNP) before the variation

pairing collapse, but rather gapless superconductivity[26]. It is very natural that in such a case, where the energy gap disappears, the moment of inertia is no longer the proper tool to give us information on the amount of the pairing correlations.

This behavior extends to very high angular momenta, as we see in Fig.3[27]. Even at spin $I = 60\hbar$, a region where one expects the nucleus to fission, we have roughly 400 keV pairing correlations, more than half of the value at the ground state. The sharp phase transition seen in mean field theory is completely smeared out by the fluctuations. The Nuclear Meissner effect obviously does not exist in real nuclei.

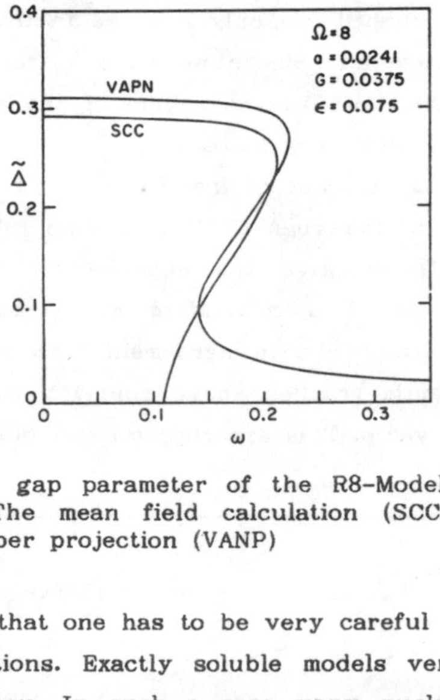

Fig.4 The pairing gap parameter of the R8-Model as a function of the angular velocity. The mean field calculation (SCC) is compared with a variation after number projection (VANP)

Fig.4 shows us that one has to be very careful with conclusions drawn from model calculations. Exactly soluble models very often show a large degree of degeneracy. In such a case many nucleons participate in the collective process, i.e. mean field theory is a much better approximation than it is in realistic nuclei. In Fig.4 we show a model calculation in the R8-Model of Krumlinde and Szymański[28,29] with 16 particles in 8 degenerate $j=3/2$ shells. This is obviously a large particle number. The mean field approximation is very close to the number projected results. Fluctuations play a minor role in this case.

So far we treated only fluctuations connected with the symmetry violation in the mean field approximation. In addition to these fluctuations we have those coming from a virtual admixture of normal modes. The most important modes in this connection are pairing fluctuations. We describe them using the gap parameter Δ as generator coordinate. We then find the Hill-Wheeler-Griffin equation in the rotating frame and solve it numerically[30,31]. Results are shown in Figs 5, 6 and 7.

Again we find good agreement with the experimental moments of inertia, indicating that these quantities cannot give us much information on the pairing correlations. The gap parameters behave very similiar to the calculations discussed so far. In fact there is only litte change through the inclusion of fluctuations from pairing vibrations.

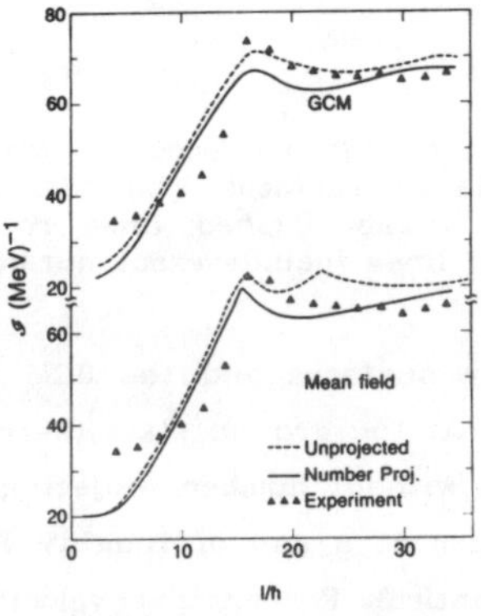

Fig.5 Moments of inertia at the yrast line in ^{168}Hf. The experimental values of ref. 25 (full triangles are compared in the lower part with mean field approximations and in the upper part with calculations based on the GCM method. For the dashed lines no number projection has been used. Full lines represent calculations with number projection.

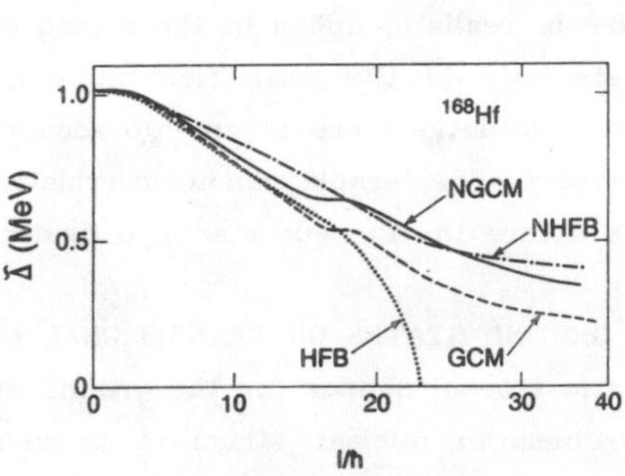

Fig.6 Gap parameters as a function of the angular momentum I. At spin zero all theories are normalized to the experimental value of Δ. This has been achieved by a minor adjustment of the strength parameter G_n in each calculation. Dashed lines correspond to a calculation without number projection; full lines include exact number projection.

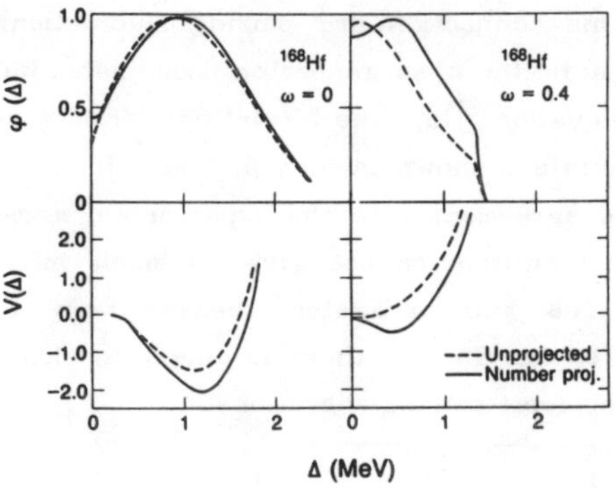

Fig.7 GCM "wavefunctions" and energy surfaces for two angular velocities in the nucleus ^{168}Hf. The energy surfaces V(Δ) correspond to the mean field energy in the rotating frame. Dashed lines represent calculations without number projection. Full lines include exact number projection.

In Fig.7 we see the energy surfaces and the GCM wavefunctions as a function of the gap parameter. At the ground state there is little difference between a calculation with and without number projection. In both cases we have a well pronounced minimum at a gap of roughly 1 MeV and the GCM functions are more or less identical. For angular velocity ω = 0.4 MeV, i.e. beyond the phase transition point in mean field theory, the unprojected minimum is shifted to gap zero. Number projection has influence on the wave function.

In this section we have investigated the phase transition from superfluidity to normal fluidity in realistic nuclei in the region of high spins. It turned out that it exists only in the mean field approximation. In more elaborate theories, where fluctuations are taken into account, it is smeared out completely and we expect considerable pairing correlations even for the highest angular momenta before the nucleus starts to fission.

3. TRIAXIAL SHAPES IN GROUND STATES OF TRANSITIONAL NUCLEI

The possibility of static triaxial shapes for the ground state of nuclei is a longstanding open problem in nuclear structure physics. Davydov and Fillippov[1)] used a triaxial rotor model to explain the lowlying second 2^+ states seen in many nuclei. Additional experimental evidence has been collected from a large number of rotational spectra in even-odd and oddodd nuclei which can be interpreted nicely in terms of phenomenological models in which one or several nucleons are coupled to a triaxial rotor[32,33]. Recently even more direct indications of triaxiality have been found in measurements of quadrupole moments and B(E2) transition probabilities from Coulomb excitation of transitional nuclei[5].

On the other hand microscopic calculations in various versions of the mean field approximation[6-15] exibit broad agreement in the following: Many nuclei were allowed to have well pronounced minima at axially symmetric deformations (either prolate or oblate); however, in no case was a deep minimum having a finite γ-deformation found. There were at best regions having a rather flat energy surface in the triaxiality parameter, and in some of these cases shallow minima at finite γ-values were found.

The actual problem of nuclear triaxiality is therefore the question why we cannot explain the experimental hints for these nuclear shapes in a microscopic theory. As already indicated in the introduction, my answer is that one has neglected fluctuations. In the last section we have seen how important the fluctuations connected with broken symmetries can be in cases of weak symmetry violations. As we discussed, they can be treated by projection techniques. In the present case this means angular momentum projection. This is a considerably harder task to perform, because in triaxial nuclei we have to carry out a full three-dimensional angular momentum projection. It is numerically not trivial, but for a few years now we have had a code in Munich developed for the purpuse of investigating the validity of the cranking approximation[34]. We used it now to calculate the angular momentum projected energy surface as a function of the deformation parameters β and γ. We used the model of Kumar and Baranger[12]. Our unprojected surfaces are therefore identical with those of ref. 12.

In Fig.8 we compare projected and unprojected energy surfaces for two characteristic nuclei ^{168}Er in the well deformed region and ^{188}Os in the transitional region. For ^{168}Er there is not much difference. In both cases we have a well pronounced deep minimum close to the $\gamma = 0$ axis. The projected minimum is shifted by a small amount to small, but finite values, but this is a minor effect, which has no influence on an eventual GCM function. The situation is very different in the transitional region. Here we have a γ-flat sitation in the unprojected case. Taking into account fluctuations produced a well pronounced minimum at $\gamma \approx 30°$, it is certainly not as deep as the minima in the well deformed region. To see the influence of additional fluctuations (from normal modes) we therefore used the method of Kumar and Baranger[12], which is a requantized Adiabatic Time Dependent

Hartree-Fock (ATDHF) method to produce wave functions. (A full GCM-calculation based on angular momentum projected wave functions is still too complicated).

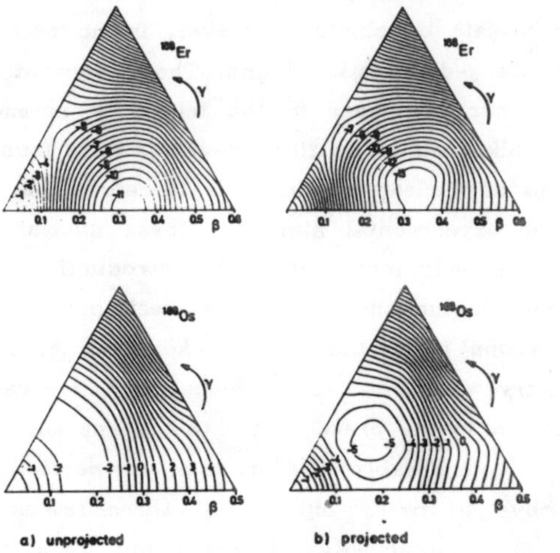

Fig.8 Energy surface in the ß,γ-plane for the nuclei ^{168}Er and ^{188}Os (a) without angular momentum projection and (b) with exact three-dimensional angular momentum projection. The units on the equipotential lines are MeV

We see in Fig.9 the density distribution integrated over the parameter ß as a function of γ i.e. the probability to find a certain γ-value in this wavefunction. In the unprojected case we find for ^{186}Os the density concentrated at γ = 0; for ^{188}Os it is smeared out over the whole γ-region (γ-soft case). In ^{190}Os and ^{192}Os, we have concentration for oblate deforamtions (γ = 60°). This means that in the unprojected case we have a rather sharp transition from prolate to oblate shapes though a γ-soft region in ^{188}Os. In the projected case we have a rather different picture: a smooth transition from prolate to oblate shapes passing through finite γ-values. The wave functions are never flat. In ^{188}Os for instance a large part of the density is concentrated in the region 20° < γ < 40°. In that sense we have stable triaxial ground state deformations in these transitional nuclei and we understand why experimental models assuming such deformations are so successful explaining experimental data.

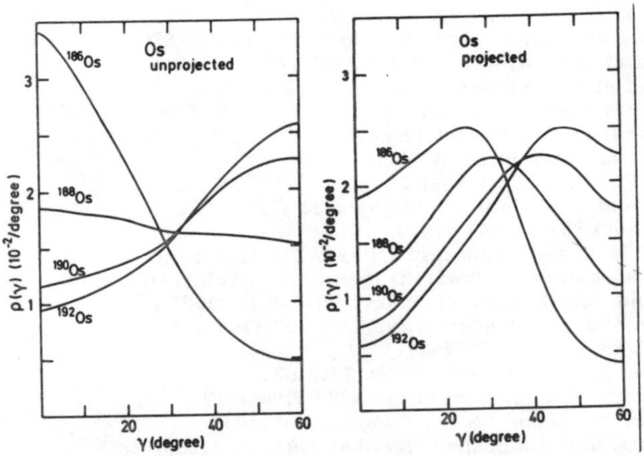

Fig.9 Probabilities $\rho(\gamma)$ to find at the ground state a shape with triaxiality γ

Let me summarize these considerations. We have seen that it is crucial for the understanding of phase transitions in finite nuclei to take into account fluctuations. In some cases they have little influence, as in well deformed axially symmetric nuclei in the Rare Earth region, in some cases they stabilize new phases, as in the case of nuclear triaxiality in the transitional region, in other cases they completely smear out a predicted phase transition as in the case of the pairing collapse at high angular momenta. The concept of phases and phase transitions remains valid in finite nuclei, but it has to be used carefully. The essential quantity is the number of particles participating in a collective process. It can be very different from the total particle number and therefore mean field theory can be very useful in some cases or it can break down completely in others.

Keeping this in mind there are certainly more phase transitions to investigate in the future, such as octupole shapes or pairing collapse with increasing temperature[27].

REFERENCES:

1) A.S.Davydov and B.F.Filippov; Nucl.Phys. 8 (1958) 237
2) A.S.Davydov and V.S.Rostovsky; Nucl.Phys. 12 (1959) 58
3) A.S.Davydov; Sov.Phys.JETP 9 (1959) 1103

4) B.R.Mottelson and J.G.Valatin; Phys.Rev.Lett. 5 (1960) 511
5) J.Stachel et al; Nucl.Phys.A383 (1982) 429
6) J.Zofka and G.Ripka; Nucl.Phys. A168 (1971) 65
7) P.Curry and D.W.L.Sprung; Nucl.Phys. A216 (1973) 125
8) K.Lassey et al.; Can.J.Phys. 51 (1973) 51
9) T.S.Sankhu and M.L.Rustgi; Phys.Rev. C14 (1976) 676
10) D.Ardouin et al.; Phys.Rev. C12 (1975) 1745
11) S.E.Larson et al.; Nucl.Phys. A261 (1976) 77
12) K.Kumar and M.Baranger; Nucl.Phys. 110 (1968) 529
13) M.Girod and B.Grammaticos; Phys.Rev.Lett. 40 (1978) 361
14) M.Goetz et al.; Nucl.Phys. A192 (1972) 1
15) S.Aberg; Phys.Scr. 25 (1982) 23
16) K.Y.Chan and J.G.Valatin; Nucl.Phys. 82 (1966) 222
17) M.Wakai; Nucl.Phys. A141 (1970) 423
18) P.Ring, R.Beck, and H.J.Mang; Z.Phys. 231 (1970) 10
19) E.R.Marshalek and G.Holzwarth; Nucl.Phys. A191 (1972) 438
20) D.L.Hill and J.A.Wheeler; Phys.Rev. 89 (1953) 1102
21) J.J.Griffin and J.A.Wheeler; Phys.Rev. 108 (1957) 311
22) H.D.Zeh; Z.Phys. 188 (1965) 361
23) U.Mutz and P.Ring; J.Phys. G10 (1984) L39
24) J.L.Egido and P.Ring; Nucl.Phys. A388 (1982) 19
25) R.Chapman, J.C.Lisle, J.N.Mo, E.Paul, A.Simcock, J.C.Willmott, and J.R.Leslie; Phys.Rev.Lett. 51 (1983) 2265
26) A.Goswami, L.Lin, and G.Struble, Phys.Lett. 25B (1967) 451
27) J.L.Egido, P.Ring, S.Iwasaki, and H.J.Mang; Phys.Lett. 154B (1985) 1
28) J.Krumlinde and Z.Szymanski; Ann.Phys.(N.Y.) 79 (1973) 201
29) S.Y.Chu, E.R.Marshalek, P.Ring, I.Krumlinde, and J.O.Rasmussen; Phys.Rev. C12 (1975) 1017
30) P.Ring and P.Schuck; The Nuclear Manybody Problem (Springer Verlag, New York 1980)
31) F.L.Canto, P.Ring, and J.O.Rasmussen; Phys.Lett. to be published
32) J.Meyer-ter-Vehn, F.S.Stephens, and R.M.Diamond; Phys.Rev.Lett. 32 (1974) 1383
33) H.Toki and A.Faessler; Nucl.Phys. A253 (1975) 231
34) A.Hayashi, K.Hara, and P.Ring; Nucl.Phys. A385 (1982) 14

GIANT RESONANCES IN HOT ROTATING NUCLEI

Peter Ring

Physik-Department der Technischen Universität München,
James-Franck-Straße, D-8046 Garching, West Germany*

Time dependent mean field theory at finite temperatures is used in rotating superfluid nuclei to describe giant resonances build on highly excited compound configurations populated after heavy ion fusion reactions. The fine structure of the dipole resonance is discussed as a function of angular velocity and temperature on one side an on nuclear deformations and pairing correlations on the other side.

1. INTRODUCTION

Giant Resonances in nuclei have been studied for more than forty years. They helped us to understand many properties of the nuclear manybody system. They do not depend on particular details of a special nucleus under investigation and are therefore very useful to deduce general parameters of nuclear matter, as the compressibility or the strength of the proton neutron interaction. For theoreticians they provided a playground to develop methods for the description of nuclear collective motion and for semiclassical approximations since the early days.

Most of these investigations, however, have been concentrated on the giant resonances based on the ground state. As Morinaga[1] and Brink[2] pointed out already thirty years ago, one can also expect giant resonances built on excited states, in fact there should be giant resonances built on any state in the nuclear system. Within the last few years a number of experiments[3-8] have been devoted to the study of giant resonances in compound states of hot rotating nuclei populated after heavy ion fusion reactions: High energy gamma rays are analyzed as a function of their energy and of the angular momentum and pronounced resonance structures have been found on top of the tail of the statistic transitions.

Invited talk at the "International Symposium on Particle and Nuclear Physics", Sept. 2 - 7, 1985, Peking, China
* supported by the Bundesministerium für Forschung und Technologie

The investigation of these resonances gives us two new degrees of freedom, namely temperature and angular velocity, to study the behavior of nuclear properties. The experiments are very difficult and many details will certainly be derived only in the future. But one undoubtedly has found the splitting of the giant dipole resonance in deformed nuclei, which has been known for the ground state resonance for many years[9].

In Fig.1 we see a schematic representation of the energy spectrum in a heavy deformed nucleus as a function of the angular momentum. At I=0 the lowest excited states are collective vibrations. Above a pairing gap of roughly 2 MeV we observe many two-quasiparticle states and at higher energies even more complicated configurations with a rapidly increasing level density. Finally we reach the continuum. It contains among others the giant dipole resonance as indicated in Fig.1.

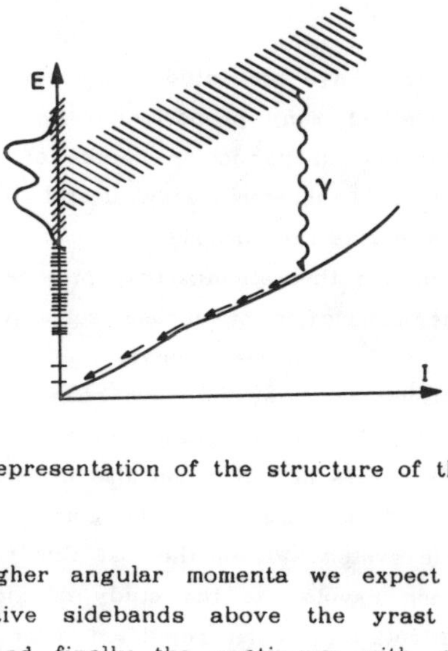

Fig.1 Schematic representation of the structure of the spectrum in a heavy deformed nucleus

Going to higher angular momenta we expect in principle a similar structure: collective sidebands above the yrast line, two-quasiparticle excitations etc. and finally the continuum with giant resonances. These would be giant resonances built on the states on the yrast line. Experimentally one probably has not yet seen those resonances, because at low excitation energies of the compound nucleus (but still above the neutron threshold), neutron evaporation is the dominant process. Only at very high excitation energies, i.e. immediately after the formation of the compound state, gamma rays have a small chance to be emitted and to be observed.

At very high excitation energies one expects that all shell effects are washed out, i.e. stable prolate deformations disappear. What is left is a hot rotating nearly spherical nucleus, which can be described as a classical rotating hot droplet. At moderate angular velocities it should be slightly oblate with its symmetry axis parallel to the rotational axis, as it is known from the shape of the rotating earth. With increasing angular velocity this shape should become unstable (Jacobi point) and a transition to nearly prolate shapes with a symmetry axis perpendicular to the rotational axis should occur[10-12].

We know from the ground state that the fine structure of the giant dipole resonance (GDR) depends on the deformation. It is therefore interesting to investigate if this remains true for the resonances built on rotating and highly excited compound states. The excitation of these resonances yields high energy gamma-rays. Their study could eventually provide us with information of nuclear shapes at high spins and at high temperatures.

Deformations are not the only collective process in nuclei. We know that many nuclei are superfluid at the ground state. The superfluidity has its origin in pairing correlations and it is connected with a broken symmetry: gauge invariance or particle number symmetry. From solid state physics we know that superconductivity disappears with increasing temperature, but also in an increasing external magnetic field (Meissner-Effect), because at high temperatures more and more unpaired quasiparticle configurations are mixed in the statistical ensemble and strong magnetic fields break time reversal symmetry and decrease the pairing correlation energy of the Cooper pairs, which are formed by electrons with anti-parallel momenta. In nuclei the strong Coriolis field plays a similar role. Cooper pairs formed by nucleons coupled to angular momentum zero are broken and more and more aligned parallel to the rotational axis. A phase transition to normal fluidity has been predicted by Mottelson and Valatin[13]. In fact it has been found in many calculations using the mean field approximation. More elaborate theories, however, which include fluctuations for the finite nuclear systems, show that this so-called Nuclear Meissner effect is completely washed out and it should hardly be observed in experiments (see ref. 14).

It turned out that not only the fine structure of the giant dipole resonance depends on nuclear deformations, but also the position of the resonance depends on the strength of pairing correlations in nuclei. If such a phase transition exists in real nuclei it should be observable in a shift of the resonance peaks.

In this talk I shall discuss theoretical investigations on these points, on the influence of collective properties like deformations and pairing on the position and the fine structure of the giant resonance as a function on angular momentum and excitation energy. They are based on calculations in collaboration with J.L.Egido, M.Faber, M.Robledo and S.Iwasaki[15-18].

2. TEMPERATURE DEPENDENT MEAN FIELD THEORY

Temperature dependent mean field theory is justified for the description of giant resonances in hot rotating nuclei for several reasons:

1) The life time of the compound state is of the order of 10^{-18} to 10^{-17} sec. It is by a factor of more than 1000 larger than typical single particle life times. The nucleons therefore have enough time to form a mean field and there is also enough time for equilibration between the intrinsic degrees of freedom.

2) As far as we go a few MeV above the yrast line, the level density is large enough to allow a statistical description in the framework of a canonical ensemble with a temperature characterizing the average energy.

3) The excitation energies of up to 100 MeV correspond to temperatures of a few MeV. They are considerable smaller than the Fermi energy in these nuclei. This means that the Pauli principle, which provides the basic condition for the validity of the Hartree-Fock approach in nuclei, is valid.

4) The Random Phase Approximation (RPA) which we obtain in the small amplitude limit for the description of collective vibrations of the mean field has turned out in many investigations in the literature[19,20] to be the appropriate tool for the microscopic investigation of giant resonances at the ground state. This type of collective motion is a fast process. It is a diabatic process, i.e. there is no reoccupation of the individual single particle levels in the time dependent average field. This kind of motion is described properly in Random Phase approximation, because this theory does not allow for reoccupations being a linear approximation.

5) The collective motion of a rotating and vibrating nucleus is a classical type of motion. The appropriate tool is a classical approximation, the approximation of a timedependent mean field.

In principle we should diagonalize the temperature dependent quasi-particle RPA-equation[17] in the rotating frame. It is a linear eigenvalue problem of rather high dimension, because nearly all symmetries are broken in the rotating frame. Its solution would allow us to calculate for each eigenstate the emission probabilities for electromagnetic radiation. This is a very complicated task. In practice one is not interested in each excited state individually - it cannot be resolved experimentally anyhow - but only in average properties. It is therefore much more convenient to calculate the absorption probability for electromagnetic radiation in the hot compound state directly.

In the case of dipole radiation it has the form:

$$(1)\quad \sigma(E) = 4\pi^2\alpha \sum_{if} p_i(E_f-E_i) \frac{1}{2I_i+1} |\langle f\|D\|i\rangle|^2 \delta(E-E_f+E_i)$$

It contains probabilities p_i and transition matrix elements $\langle f\|D\|i\rangle$ in the laboratory system. Applying techniques of angular momentum projection and semiclassical approximations to it[21,18] which are consistent with the cranking approach, the absorption cross section (1) can be expressed by intrinsic cross sections:

$$(2)\ \sigma(E) = \sigma_x(E) + \frac{1}{2}\sigma_y(E-\hbar\omega) + \frac{1}{2}\sigma_y(E+\hbar\omega) + \frac{1}{2}\sigma_z(E-\hbar\omega) + \frac{1}{2}\sigma_z(E+\hbar\omega)$$

where the intrinsic quantities σ_k are defined as:

$$(3)\ \sigma_K(E) = -4\pi^2\alpha \sum_{if} p_i(E_f-E_i)|\langle f|D|i\rangle|^2\delta(E-E_f+E_i)$$

which look formally very similar to eq.(1). It is however totally different, because the initial and final states $|i\rangle,|f\rangle$ are now states in the intrinsic frame, i.e. multi-quasiparticle states, the energies are sums of quasiparticle energies in the rotating frame. Using the techniques of linear response theory[19] the quantities σ_k can be expressed by the response function:

$$(4)\ \sigma_k = -4\pi\alpha\, E\, \mathrm{Im}(R_{D_k D_k})$$

It is a solution of the linearized Bethe Salpether equation

$$(5)\ R = R^o + R^o\, W\, R$$

which is equivalent to the RPA-equation. R^o is the response function without interaction.

The solution of the Bethe Salpether equation is particularly simple in the cases where the residual interaction is a sum of separable terms. It then turns out to be an inhomogeneous matrix equation, whose dimension is the number of separable terms. It has to be solved for each energy E.

3. Results of Numerical Calculations

The calculation of fine structure of the giant dipole resonance in hot rotating nuclei at a given temperature and a given angular velocity involves three steps:

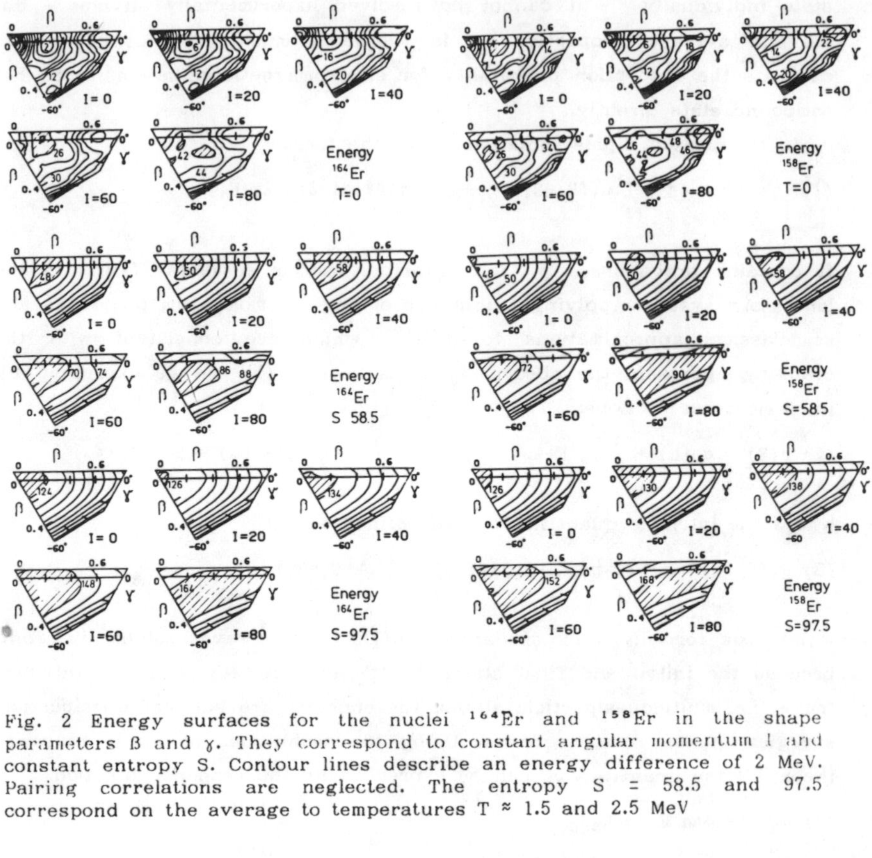

Fig. 2 Energy surfaces for the nuclei ^{164}Er and ^{158}Er in the shape parameters β and γ. They correspond to constant angular momentum I and constant entropy S. Contour lines describe an energy difference of 2 MeV. Pairing correlations are neglected. The entropy S = 58.5 and 97.5 correspond on the average to temperatures T ≈ 1.5 and 2.5 MeV

1) The determination of the average nuclear potential and the pairfield at this angular velocity and this temperature. This is the most time consuming part of the calculation, because it would in principle involve a solution of the cranked HFB-equations at finite temperature. For realistic effective forces such a calculation has not been done yet. As has been well known for many years, the Strutinski method provides a very simple and reliable approximation for such calculations. In this method the bulk of the nucleus is considered as a liquid drop and this part of the energy is taken from the model of a hot rotating liquid drop. On top of this bulk the shell corrections are calculated in a rotating Saxon Woods potential which depends on a few deformation parameters. Since the most important degrees of freedom in deformed nuclei are quadrupole deformations we use the two

Wheeler parameters ß and γ to calculate the total free energy F = E–TS and search for the minima in the energy surface.

In Fig.2 we show the behavior of these energy surfaces for two cases: ^{164}Er and ^{158}Er. At zero temperature ^{164}Er is very stiff. It stays for all angular momenta close to the prolate deformation of $\beta \approx 0.3$ and $\gamma = 0$. For higher temperatures we observe two effects: i) the minima become flatter and flatter, which indicates that fluctuations become more important. ii) the nucleus undergoes a shape change with increasing temperature. It becomes spherical at zero angular velocity and slightly oblate for higher spins. It shows the classical behavior of a rotating hot droplet. Shell corrections, which cause deformations at temperature zero are washed out completely at temperatures T > 2.5 MeV.

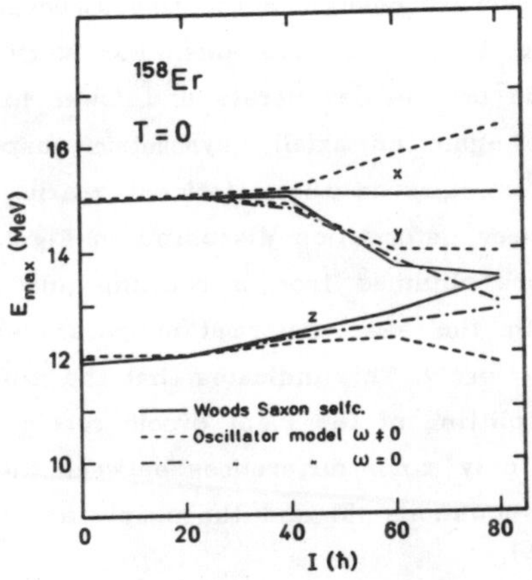

Fig.3 Fine structure of the giant dipole resonance on the yrast line of the nucleus ^{158}Er. Full lines correspond to the solution of the linear response equations (5) in the rotating Saxon-Woods potential. The other lines are obtained from a harmonic oscillator potential with the same deformation as the realistic calculation for each angular momentum. Dashed lines describe a rotating harmonic oscillator and dashed-dotted lines a static harmonic oscillator.

The nucleus ^{158}Er is also prolate deformed in its ground state. At temperature zero we find, however, between spin 40 ħ and 60 ħ a shape change to oblate deformations and a rotation around the symmetry axis ("single particle rotation"). This is in agreement with recent experiments[22], which find after spin 40 ħ an abrupt change in the character of the rotational spectra. For higher temperatures ^{158}Er shows the same classical behavior as ^{164}Er.

2) In the second step we work on the basis of quasiparticles determined in the first step. The dipole response functions for vibrations in x, y and z direction are determined as a function of the energy and the corresponding absorption cross sections in the intrinsic frame are plotted. We find a fine structure which depends strongly on the deformation. There are essentially three peaks, which correspond to vibrations parallel to the three principal axes of the system.

In Fig.3 we show the nucleus ^{158}Er at temperature T = 0, i.e. at the yrast line. For vanishing angular momentum we have a prolate deformation with axial symmetry around the z-axis. In this case the two modes perpendicular to this axis (x- and y-mode) are degenerate in energy. With increasing angular momenta this situation stays more or less constant up spin 40 ℏ. Then we observe a change in the fine structure. There are now three peaks indicating triaxial deformations. For very large spins the y-mode and the z-mode become degenerate and lower in energy than the x-mode. This indicates again an axially symmetric shape, but now it is oblate and the symmetry axis is the rotational x-axis. This behavior is expected from the nuclear deformation discussed in Fig.2. We also show in Fig.3 the fine structure obtained from a rotating and a static harmonic oscillator potential with the same deformation parameters. All the three results coincide rather closely. This indicates that the most important effect is characterizing the splitting of the giant dipole resonance in the nuclear deformation. There are only minor differences between the realistic solution of the linear response equations (5) and the simple analytic formula of the harmonic oscillator[23-26].

In Fig.4 we show the splitting of the giant dipole resonance calculated in the harmonic oscillator potential as a function of the angular velocity and as a function of the triaxiality parameter γ.

3) In the last step we have to transform to the laboratory frame. This is done according to eq.(2). We observe an additional spitting. Instead of three peaks we then have five peaks. In Fig. 5 we give a simple schematic explanation for this fact at the example of dipole absorption on a nucleus at the yrast line.

And in Fig.6 we show the results of this transformation for a realistic case. It turns out that for large angular momenta the additional spitting cannot be resolved in the total absorption cross section. To obtain the interesting information of the fine structure in the intrinsic system one would have to measure carefully the angular distributions in order to distinguish the different contributions (stretched and unstretched) in the experiment.

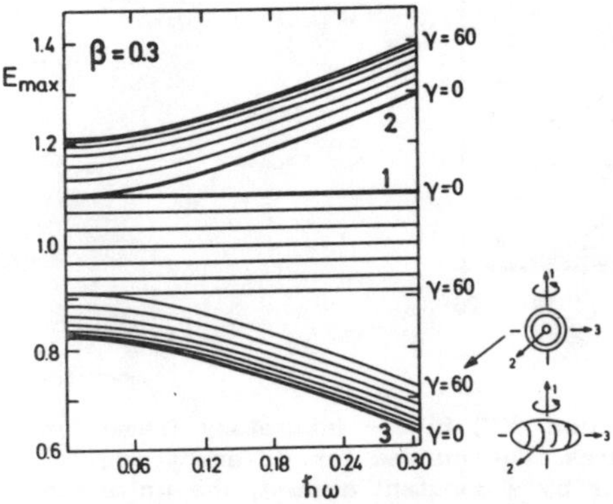

a)

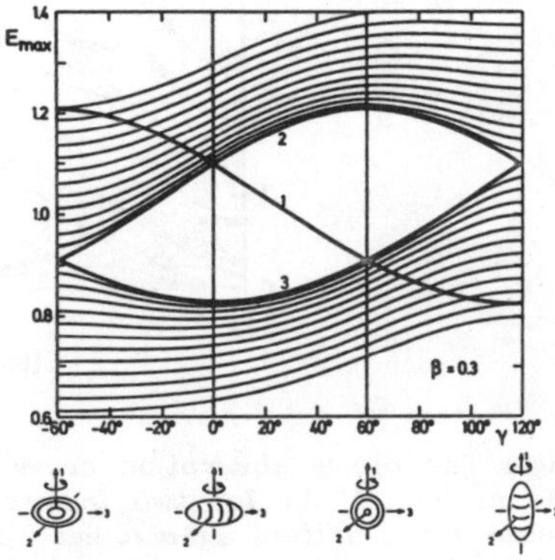

b)

Fig.4a The eigenfrequencies E1, E2, E3 of the rotating harmonic oscillator for ß = 0.3 at various γ-values. The units of the peak energies E_i and for the angular velocity ω are $78 \cdot A^{-1/3}$ MeV, i.e. the energy of the GDR at vanishing deformation. In heavy nuclei ω never exceeds 1 MEV, which is 0.06 in these units

Fig.4b The dependence of the giant dipole resonance peaks on the triaxiality parameter γ for the deformation ß = 0.3 and for various cranking frequencies. The units are the same as in Fig.4a

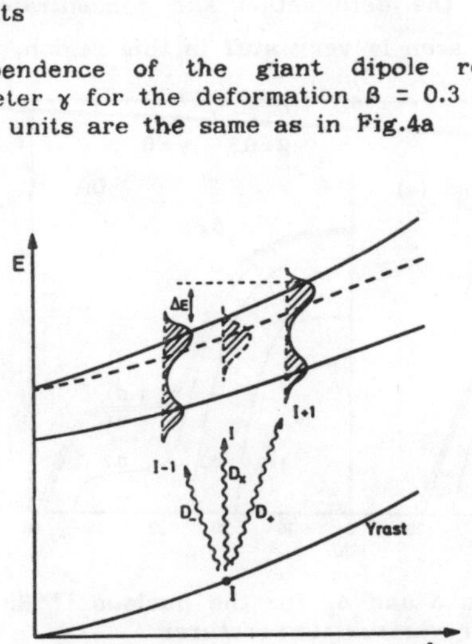

Fig.5 Schematic representation of the transformation from the intrinsic system to the laboratory frame

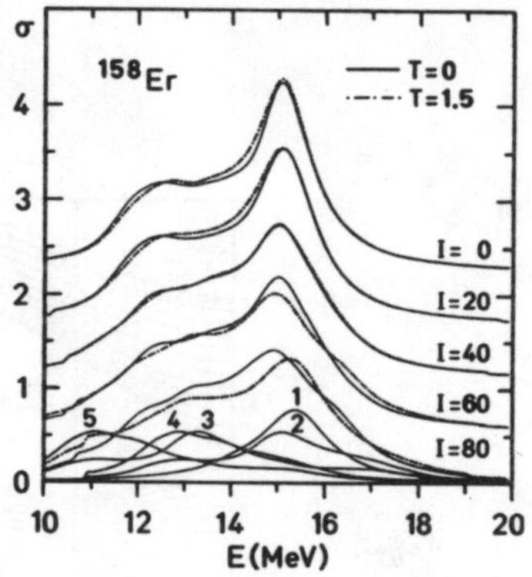

Fig.6 The dipole absorption cross section $\sigma(E)$ in the laboratory frame for the nucleus ^{158}Er for two temperatures. The curves for different angular momenta are shifted against each other by a constant amount, the units are arbitrary. The lowest curve corresponds to I = 80 ħ. Below this curve the five components of eq.(2) are given separately for T = 0.

So far we discussed only cases without pairing. If we look more closely, pairing correlations play an important role for small angular momenta and small temperatures. We therefore investigate this region in the following. We now neglect changes in the deformation and concentrate on the nucleus ^{164}Er, which as we have seen is very stiff in this region.

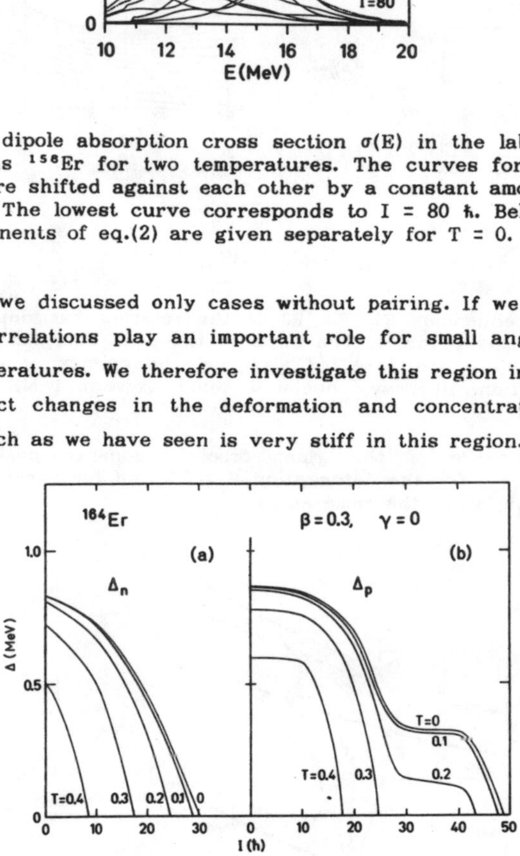

Fig.7 The gap parameters Δ and Δ_p for the nucleus ^{164}Er as a function of the angular momentum for various temperatures

In Fig.7 we show the gap parameter for neutrons and protons as a function of the angular momentum for various temperatures. As in a superconductor we observe two kind of phase transitions: The gap parameters are vanishing with increasing temperature and with increasing angular momentum. In both cases there is a sharp transition point. As we discuss in ref. 14 this is an artifact of the mean field approximation used in the present calculations. More elaborate methods predict in finite nuclei that these transitions are smeared out to a large extent. On the other side the pairing collapse shows up clearly in the frequency of the giant dipole resonance as we see in Fig.8 for a realistic case.

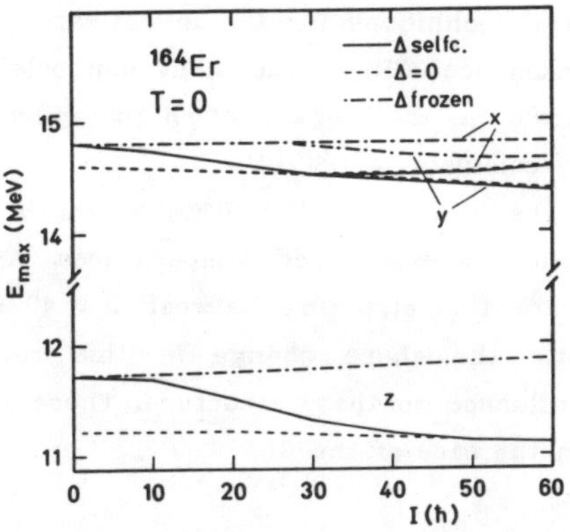

Fig.8 The angular momentum dependence of the peak energies of the GDR in ^{164}Er at zero temperature and constant deformations β = 0.3 and γ = 0. Pairing correlations are treated in three different ways. For the full lines the selfconsistently determined gap parameters are used. The dashed lines are calculated with vanishing pairing and the dashed-dotted lines are calculated with the gap parameters frozen at their values for I=0.

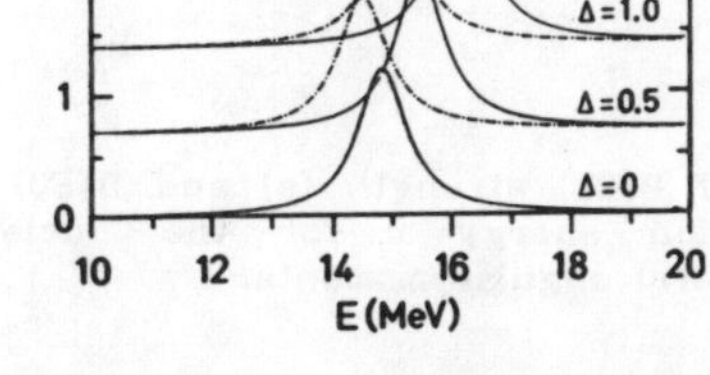

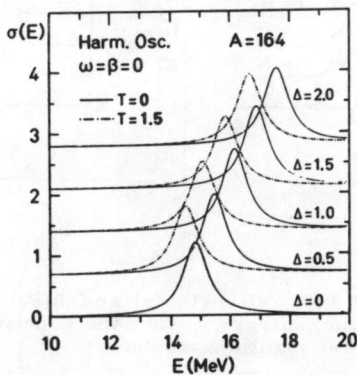

Fig.9 Dipole absorption cross sections $\sigma(E)$ in the harmonic oscillator model for a spherical nucleus with A = 164. Calculations with various gap parameters and two different temperatures are shown.

In Fig.9 the gap dependence is studied more explicitly in a spherical harmonic oscillator model for various temperatures.

So far we have investigated only the Giant Dipole Resonance in hot rotating nuclei. In fact several other types of resonances have been found experimentally at the ground state. There is no reason to believe that it would not exist in hot and rotating nuclei. The reason why the GDR has been seen first is simple. It is much stronger than all the others. At the ground state one has developed special techniques to explore other resonances. So far special techniques for the investigation of for instance the Giant Quadrupole Resonance(QGR) or the Giant Monopole Resonance(GMR) have not been developed in the region of high spins and excitation energies. There are only calculations[27,28,18].

In Fig. 10 we show the B(E2)- and the B(E0) strength for the nucleus ^{158}Er for different angular momenta and temperatures. Again there is a considerable change in the fine structure between I = 40 ℏ and I = 60 ℏ, which is connected with the shape change in this region. Again the temperature has little influence on these structure. Therefore one is led to similar conclusions as in the case of the GDR.

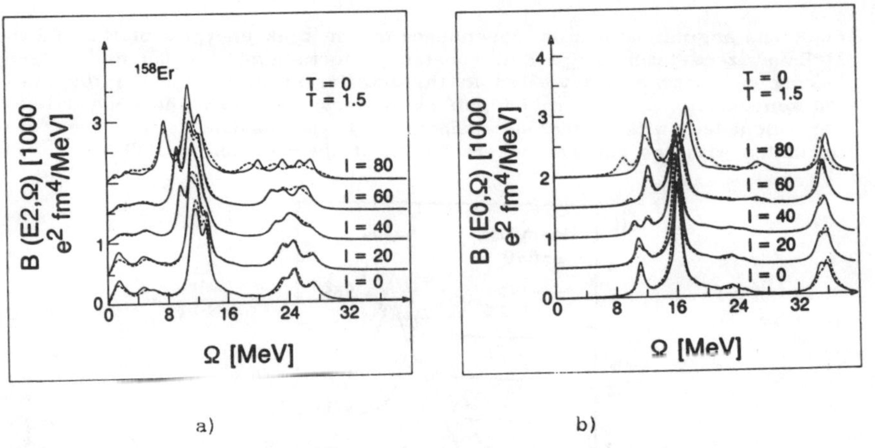

Fig.10 The distribution of B(E2) strength (a) and B(E0) strength (b) as a function of the excitation energy Ω for the nucleus ^{158}Er at two temperatures and for several angular momenta.

4. CONCLUSIONS

We have seen that the fine structure of giant resonances in the high spin region of hot nuclei is mainly influenced by the collective properties of the underlying mean fields. These properties are deformations and pairing correlations and they depend sensitively on the angular velocity and the temperature. With increasing angular velocity and temperature several phase transitions can be seen, which have influence on the structure of the giant resonances. For a long time this has been known for the influence of deformation on the GDR. A new but probably not unexpected result of the present investigations is that this influence persists also to higher angular velocities and to higher excitation energies. We also predict a dependence of the GDR-peak on the pairing correlations, which should be further investigated experimentally and also theoretically with more realistic interactions.

On the other hand there is little change of the resonance structure with temperature and angular velocity for constant fields. If one uses the formulas of the oscillator model with the appropriate deformations one finds excellent agreement with realistic RPA-calculations in a rotating Saxon-Woods potential.

There are a number of points, which require further investigation:

1) Little is known about the influence of more realistic interactions.

2) Experimentally one has observed dramatic changes of the width of the resonance. Within the present method the width is treated on a phenomenological way. More microscopic calculations are required.

3) It is well known that the effective mass is energy dependent. The reason is a coupling to more complicated configurations as 2p-2h. The present method uses strictly RPA-theory, i.e. only 1p-1h configurations. The influence of the energy dependence of the effective mass should be investigated.

Let me finish in expressing my hope that this relatively new field will obtain enough experimental and theoretical support in the future. It is certainly full of interesting problems and it promises to extend our knowledge of the nuclear manybody system under extreme situations of high Coriolis fields and high excitation energies.

I would like to express my gratitude to the organizing committee and in particular to Prof. Yang Li-ming and Prof Yang Se-zen of this conference for giving me the opportunity to present this talk.

REFERENCES:

1 H.Morinage; Phys.Rev. 101 (1956) 100

2 D.M.Brink; Ph.D.Thesis, University of Oxford 1955 (unpublished)
3 J.O.Newton, B.Herskind, R.M.Diamond, E.L.Dines, J.E.Draper, K.H.Lindenberger, C.Schuck, S.Shin, and F.S.Stephens; Phys.Rev.Lett. 46 (1981) 1383
4 J.J.Gaardoje; XX. Int. Winter Meeting on Nuclear Physics, Bormio, Italy 1982
5 W.Hennerici, V.Metag, H.J.Hennrich, R.Repnow, W.Wahl, D.Habs, K.Helmer, U.v.Helmolt, H.W.Heyng, B.Kolb, D.Pelte, D.Schwalm, R.S.Simon, and R.Albrecht; Nucl.Phys. A396 (1983) 329C
6 B.Haas, D.C.Radford, F.A.Beck, T.Byrski, C.Gehringer, J.C.Merdinger, A.Nourredine, Y.Schult, and J.P.Vivien; Phys.Lett. 120B (1983) 79
7 A.M.Sandorfi, J.Barrette, M.T.Collins, D.H.Hofmann, A.J.Kreiner, D.Brandford, S.G.Steadman, and J.Wiggins; Phys.Lett. 130B (1983) 19
8 E.F.Garman, K.A.Snover, S.H.Chew, S.K.B.Hesmondhalgh, and W.N.Catford; Nucl. Phys. 1985
9 M.Danos; Nucl.Phys. 5 (1958) 23
10 R.A.Lyttleton; The Sability of Rotating Liquid Masses. Cambridge Univ., Cambridge, 1958
11 S.Chandrasekhar; Ellipsoidal Figurs of Equilibrium. Yale Univ.Press New Haven, CT, 1969
12 S.Cohen, F.Plasil, and W.J.Swiatecki; Ann.Phys.(New York) 82 (1974) 557
13 B.R.Mottelson and J.G.Valatin; Phys.Rev.Lett. 5 (1960) 511
14 P.Ring; Proc.Int.Conf.on Part.and Nucl.Phys., Peking, Sept.1985
15 J.L.Egido and P.Ring; Phys.Rev. C25 (1982) 3239
16 M.Faber, J.L.Egido, and P.Ring; Phys.Lett. 127b (1983) 5
17 P.Ring, L.M.Robledo, J.L.Egido, and M.Faber; Nucl.Phys. A419 (1984) 261
18 S.Iwasaki and P.Ring; to be published
19 P.Ring and P.Schuck; The Nuclear Manybody Problem (Springer Verlag, New York 1980)
20 P.Ring and J.Speth; Nucl.Phys. A235 (1974) 315
21 P.Ring, A.Hayashi, K.Hara, H.Ehmling, and E.Grosse; Phys.Lett 110B (1982) 423
22 J.Simpson, M.A.Riley, J.R.Cresswell, P.D.Forsyth, D.Howe, B.M.Nyako, J.F.Sharpey-Schaefer, J.Bacelar, J.D.Garrett, G.B.Hagemann, B.Herskind, and A.Holm; Phys.Rev.Lett. 53 (1984) 648
23 J.G.Valatin; Proc.Roy.Soc.(London) A238 (1956) 132
24 K.Neergard; Phys.Lett. 110B (1982) 7
25 Z.Szymanski, XIV. Masurian Summer School on Nuclear Physics, Mikolajki, Poland 1981
26 R.R.Hilton, Z.Phys. 309 (1983) 233
27 A.Akbarov et al; Preprint JINR Dubna, 1980, R4-80-187
28 Y.R.Shimizu and K.Matsuyanagi; Prog.Theor.Phys. 72 (1984) 1017

THE PAIR-ALIGNED INTRINSIC WAVE FUNCTION IN SINGLE-j CONFIGURATION

Shen Hongqing   Cui Haiyuan

(Nanjing Normal University, Nanjing, China)

Yao Shihuai   Li Xianyin

(Anhui University, Hefei, China)

Yang Bangjun

(Guizhou University, Guiyang, China)

R. A. Broglia[1,2], using BCS wave function, has studied the weight amplitude of nucleon-pair coupled to J=0, 2, 4 in the intrinsic state of deformed nuclei. But the BCS wave function does not conserve the nucleon pairs. We take the normalized basis vector of one nucleon-pair in a single-j shell as

$$|1\rangle=\sum_{m>0}C(m)a_m^+a_{-m}^+|0\rangle=\sum_J\alpha_J A_{J0}^+|0\rangle \tag{1}$$

with $\alpha_J=\sum_{m>0}\sqrt{2}C(m)(jmj-m|J0)$ being the weight amplitude of nucleon-pair J and the intrinsic state vector of k nucleon-pairs as

$$|k\rangle=\Big(\sum_{m>0}C(m)a_m^+a_{-m}^+\Big)^k|0\rangle,\quad (k\leqslant\Omega=j+\tfrac{1}{2}) \tag{2}$$

which conserves the number of nucleon-pairs. One may prove that the norm of the state vector (2) is

$$N_k=\langle k|k\rangle=k!\sum_{s=0}^{k-1}(-1)^s\frac{(k-1)!}{[(k-1-s)!]^2}X_{s+1}N_{k-1-s} \tag{3}$$

with $X_s=\sum_{m>0}C^{2s}(m)$.

Suppose that Hamiltonian of the system is taken as the sum of an attractive quadrupole-quadrupole residual interaction W_Q and pairing force W_P[1]

$$\begin{aligned}H&=W_Q+W_P\\&=20\beta\sum_{m>0}C_2(m)(a_m^+a_m+a_{-m}^+a_{-m})-G\Omega\Big(\sum_{m>0}C_0(m)a_m^+a_{-m}^+\Big)\Big(\sum_{m'>0}C_0(m')a_{-m'}a_{m'}\Big)\end{aligned} \tag{4}$$

Where β is the deformation parameter, G the strength of pairing force

$$C_2(m)=(jm20|jm),\quad C_0(m)=(-1)^{j-m}/\sqrt{\Omega} \tag{5}$$

One may prove that the matrix elements of W_Q and W_P with respect to the state vector (2) are

$$\langle k|W_Q|K\rangle = 40\beta\sum_{s=o}^{k-1}(-1)^s\left[\frac{k!}{(k-1-s)!}\right]^2 Y_{s+1}N_{k-1-s} \tag{6}$$

with $Y_s=\sum_{m>o}C^{2s}(m)C_2(m)$, and

$$\langle k|W_P|k\rangle$$
$$=-G\Omega\left\{\sum_{s=o}^{[(k-1)/2]}\left[\left(\frac{k!}{(k-2s-1)!}\right)^2 Z_s^2 N_{k-2s-1}+2\sum_{p=1}^{k-2s-1}(-1)^p\left(\frac{k!}{(k-p-2s-1)!}\right)^2 Z_s Z_{s+p}N_{k-p-2s-1}\right]\right.$$
$$\left.+\frac{1}{\Omega}\sum_{s=o}^{[\frac{K-2}{2}]}\left[\left(\frac{k!}{(k-2s-2)!}\right)^2 X_{2s+2}N_{k-2s-2}+2\sum_{p=1}^{K-2S-2}(-1)^p\left(\frac{k!}{(k-p-2s-2)!}\right)^2 X_{p+2s+2}N_{k-p-2s-2}\right]\right\} \tag{7}$$

with $Z_s=\sum_{m>o}C^{2s+1}(m)C_0(m)$ and $[X]$ taking the integer part of X.

Taking the truncation of J=0, 2, 4 in α_J, one obtains the energy of the intrinsic state for k nucleon-pairs

$$E_k=\frac{1}{N_k}(\langle k|W_Q|k\rangle + \langle k|W_P|k\rangle) \tag{8}$$

which is the function of α_J. From variation method $\delta E_k=0$, one may determine α_J.

For example, $j=\frac{39}{2}$, k=3 for fixed $\beta=-0.3$ and different G, the calculation is performed. The result is shown in Table 1.

Table 1

G(Mev)	0.2	0.3	0.4	0.5	0.6	0.75	1.00	1.25
α_0	0.5583	0.6270	0.7107	0.7922	0.8534	0.9086	0.9517	0.9706
α_2	0.7308	0.6944	0.6386	0.5659	0.4920	0.4015	0.3000	0.2374
α_4	0.3926	0.3531	0.2951	0.2285	0.1718	0.1150	0.0653	0.0413

From Table 1 we may see that α_0 increases with G and α_2,α_4 decrease for fixed β.

References

[1] R. A. Broglia, in "Interacting Bose-Fermi System in Nuclei" edited by F. Iachello (Plenum, New York, 1981) p.95.

[2] D. R. Bes, R. A. Broglia, E. Maglione and A. Vittur, Phys. Rev. Lett. 48 1001 (1982).

NUCLEAR STRUCTURE PARAMETERS FROM COULOMB EXCITATION WITH 2.5 - 4.5 MeV PROTONS

K.P.Singh, D.C. Tayal, Gulzar Singh and H.S. Hans
Department of Physics, Panjab University, Chandigarh-160014, India.

A programme has been developed to measure thick target yields and angular distributions, of Coulomb excited gamma rays, with protons of 2.5 - 4.5 MeV, from Chandigarh Variable Energy Cyclotron, using a 50 cm^3 Ge(Li) detector. Using CINDY computer programme, the compound nucleus contribution has been calculated, for each nucleus, and has been either found negligible or taken into account. New values of transitions probabilities, B(E2), from absolute yields, and mixing ratios (δ) from χ^2 fits, have been found, and new assignments of J^{π} have been made for many transitions and levels in ^{45}Sc, ^{103}Rh, ^{105}Pd, ^{107,109}Ag, ^{111}Cd, ^{133}Cs, ^{157}Gd and ^{165}Ho. Apart from confirming the known parameters, fourteen new values of B(E2), more than twenty new δ's and ten new J^{π} have been obtained for the above cases. Also in ^{45}Sc, ^{69}Gd, ^{67}Zn, ^{63}Cu, the detailed reaction mechanism i.e. the competition between compound and Coulomb have been studied.

1. INTRODUCTION

Chandigarh Variable Energy Cyclotron[1)] was established in 1978 and has been , since then, used for many experiments, essentially using protons. Major activities have been (i) The study of thick target yields of Coulomb excited gamma rays for determining the transition probabilities, B(E2), and angular distributions of these gamma rays to determine the spins and parities of some of the states involved[2)], and (ii) Doppler shift attenuation studies of gamma rays emitted in (p,nγ) reactions[3)]. We present here the recent studies on Coulomb excited gamma rays.

The protons as projectiles have both advantages and disadvantages for Coulomb excitation. Being light particles, there is hardly any possibility of multipole Coulomb Excitation, which is dominant in heavy ion reactions. On the other hand with protons as projectiles, $(p,n\gamma)$ reaction can be quite significant even when the Coulomb excitation is predominant. The suitable targets for such studies are those, where compound nucleus formation is negligible at the energies of projectiles, compared to the Coulomb excitations Or one can calculate the effect of the compound nucleus, and substract it from the observed intensity of gamma rays, to obtain the net yield only due to Coulomb excitation.

2. EXPERIMENTAL PROCEDURE

A resolved proton beam of 100-200 nA, available from Chandigarh Variable Energy Cyclotron with a resolution of about 30 keV, was made to strike a suitably fabricated thick target. A 50 cm^3 GeLi detector was used to detect the gamma-rays. The absolute thick target yield, at 55° for B(E2) values, and angular distribution measurements for χ^2-fit, for J^π and mixing ratio (δ) was obtained, the details of which have been described earlier[2).]

We give in Table, the summary of the results for some of the cases, e.g. ^{103}Rh, ^{105}Pd, ^{109}Ag and ^{157}Gd along with the comparison of literature[4-7)].

3. RESULTS AND CONCLUSION

It is pertinent to point out; that these studies have shown the importance and utility of the light projectiles like protons for studying the Coulomb excitation of nuclei.

REFERENCES

1) I.M. Govil and H.S. Hans, Proc. Indian Acad. of Science (Engg. Section) 1980, P.237.

2) K.P. Singh, D.C. Tayal, Gulzar Singh and H.S. Hans, Phys. Rev. C31 (1985) 79, and Phys. Rev. C31 (1985) 1726. K.P. Singh, D.C. Tayal, B.K. Arora, T.S. Cheema and H.S. Hans, Can. J. Phys. 63 (1985) 483.

3) D.K. Avasthi, V.K. Mittal and I.M. Govil, Phys. Rev. C26 (1982) 1310.

4) Harmatz, Nuclear Data Sheets 28, No. 3 (1979).

5) Y.A. Ellis, Nuclear Data Sheets 27 (1979) 1.

6) F.E. Bertrand, Nuclear Data Sheets 23 (1978) 229.

7) R.L. Bunting and C.W. Reich, Nuclear Data Sheets 39 (1983) 103.

Nucleus	Level (keV)	E_γ (keV)	Present work			Literature		
			J^π	δ	B(E2)↑ $e^2b^2x10^{-2}$	J^π	δ	B(E2) $e^2b^2x10^{-2}$
^{45}Sc	876.3	376.3	$3/2^-$	E2	0.66	$3/2^-$	E2	0.70, 0.72, 0.68
	364			-1.24			-0.14	
	543.1	543.1	$5/2^+$	1.25	1.4	$5/2^+$	.035	E3, 1.67*
	531			-0.75			0.55	
	720.5	720.5	$5/2^-$	-1.0	1.40	$5/2^-$	.18, -0.078	0.65, 0.81 0.55
		708		1.25				
	974.4	974.4	$7/2^+$	-0.32	7.5*	5/2, 7/2	-0.17, -0.09	E2, 1.39*
		432		-0.75			0.24	
	1409.2	1409.2	$7/2^-$	-0.06 or -3.26		7/2, 9/2	-2.62, 0.9	

* This is B(E1)↑ in units of $e^2cm^2x10^{-30}$

^{103}Rh	295.1	295.1	$3/2^-$	-0.58	22.1	$3/2^-$	0.15	21.8, 20.9
	357.6	357.6	$5/2^-$	E2	33	$5/2^-$	E2	39.2, 37.8
	880.4	880.4	$5/2^-$	E2	1.32	$5/2^-$	E2	1.31, 1.33
		585.5		.14			.07	
		523.2		0.2			0.25	

	1106.7		$5/2^-$		0.32	$(5/2^-)$		0.25,0.10
		811.2		0.45				
	1277.2	1277.2	$3/2^-$		4.05	$3/2^-$	-	1.32,1.31
		474.1		-0.43				
^{105}Pd	280.4	280.4	$3/2^+$	0.13	0.85	$3/2^+$		0.97,1.10
	306.2	306.2	$7/2^+$	-0.810	0.12	$7/2^+$		0.11,0.12
	319.2	319.2	$5/2^+$	-0.14	0.73	$5/2^+$		0.87,0.81
	344.6	344.6	$1/2^+$	E2	0.23	$1/2^+$		0.27,0.15
	442.2	442.2	$7/2^+$	-1.73	19.0	$7/2^+$		16.5, 19.7
	560.6	560.6	$5/2^+$	-1.8	0.92	$(5/2^+)$		1.10,0.75
	650.6	650.6	$3/2^+$	-0.38	0.86	$3/2^+$		0.79,0.66
		331.5		-0.035				
	673.3	673.3	$3/2^+$	0.73	0.89	$1/2^+$, $3/2^+$		0.90,0.57
	696.3	696.3	$7/2^+$	-0.81	0.20	$(7/2^+)$		
		254.0		-1.22				
	727.3	727.3	$5/2^+$	-0.14	0.43	$5/2^+$		1.21,0.24
		408.3		0.55				
	781.8	781.8	$5/2^+$	0.57	9.66	$(9/2)^+$		11.4,8.27
		475.7		-0.22				
		339.6		-0.08				
	961.9	961.9	$3/2^+$		1.6	$1/2^+$		0.5
		615.7		-0.97				
^{157}Gd		55		$3/2^-$	175	$3/2^-$		252,221$\pm$10
		132		$7/2^-$	120	$7/2^-$		136,121$\pm$10
		434		$5/2^-$	0.39	$5/2^-$		1.3
		317		$7/2^-$	0.29	$7/2^-$		1.4
		702		$1/2^-$	0.4	$1/2^-$		0.7
		748		$3/2^-$	1.25	$3/2^-$		0.2
		814		$3/2^-$	6.3	$3/2^-$		0.7

APPLICATION OF DYSON BOSON MAPPING TO THE ANALYSIS OF MODE-MODE COUPLING IN Ge AND Se ISOTOPES

Kenjiro Takada

Department of Physics, Kyushu University, Fukuoka 812, Japan

1. Introduction

The basic philosophy of boson mappings is that all the information in the original fermion space should precisely be mapped onto the physical subspace in the ideal boson space and couplings between the physical states and unphysical states in the boson space should be strictly removed.

Various boson mappings which transform the physics in the original fermion space into the physical boson subspace have been proposed for the description of collective motions in the space of many-nucleon (or many-quasiparticle) states.

The most popular one is the Holstein-Primakoff-type mapping (or the Marumori boson mapping).[1] The merit of this mapping is that it is unitary and the hermiticity of an original hamiltonian of a fermion system is conserved in the boson space. However, it has the demerit that the mapped operator is, in general, expanded in a infinite power series of boson operators. This causes serious discussions concerning the convergence of the expansion.

Another popular one is the Dyson-type boson mapping (or the Usui boson mapping[2]) whose merit is that the mapped operator is of a finite series of boson operator monomials. Therefore, we can avoid the problem of the infinite series as in the Holstein-Primakoff-type mapping. However, this type of mapping has an outward demerit; namely, the mapping is non-unitary and it does not conserve the hermiticity of the hamiltonian. It has therefore been thought that it is very difficult to calculate the B(E2) values in the Dyson boson theory and to study the properties of the eigenstates. This seems the reason why the Dyson boson theory has not been applied to realistic nuclei except for some rare cases.

Recently it has been clarified that complete information about the eigenstates in the hermitian boson theory (the Holstein-Primakoff mapping) can, in principle, be obtained from the results of the Dyson boson theory.[3,4] Since any truncation of the boson expansion is not needed in the Dyson boson theory, as mentioned before, the Dyson mapping appears more promising than the Holstein-Primakoff boson theory.

Of course, some truncation of the degrees of freedom is inevitable,

---

[*] This report is based on our recent work which will soon be published in Nucl. Phys.[11]

whether we use the Dyson mapping or the Holstein-Primakoff mapping, when we intend to apply them to a realistic nucleus. One of the simplest truncations is the "phonon truncation" in which the space of many-quasiparticle states is truncated to the multi-phonon subspace

$$\{ X^{\dagger}_{2M_1} X^{\dagger}_{2M_2} \cdots X^{\dagger}_{2M_n} |0\rangle \; ; \; n = 0, 1, 2, \cdots \}, \tag{1.1}$$

where $X^{\dagger}_{2M}$ is the collective Tamm-Dancoff (TD) phonon with spin J=2 and the algebra of quasiparticle-pair operators is forced to be closed within the above TD phonon. Under this truncation approximation, the Dyson mapping gives us a very simple but non-hermitian hamiltonian which consists of up to sixth-order boson operators and we can easily solve the right- and left-hand-side eigenvalue problems. Using these results, we can get the eigenvectors in the corresponding hermitian boson theory.[3,4] In ref. 4), the Dyson boson mapping was applied to low-lying collective states in some realistic nuclei and it has been shown by comparing the results with the direct diagonalization within the subspace (1.1) that the Dyson mapping is very useful and promising for description of nuclear collective motions.

The aim of the present paper is to show how applicable the Dyson boson mapping is to the analysis of the mode-mode coupling in nuclei. It has been well known that the coupling between the quadrupole phonon mode and the pairing vibrational modes[5] plays as essential role in Ge and Se isotopes.[6-9] In order to describe the collective motion in this region, we should extend the subspace (1.1) to

$$\{ (X^{\dagger}_{2M_1} X^{\dagger}_{2M_2} \cdots X^{\dagger}_{2M_n}) (\Gamma^{\dagger}_1 \Gamma^{\dagger}_2 \cdots \Gamma^{\dagger}_k) |0\rangle \; ; \; n, k = 0, 1, 2, \cdots \}, \tag{1.2}$$

where $\Gamma^{\dagger}_k$ stands for the proton or neutron pairing-vibrational modes and the algebra of quasiparticle-pair operators is forced to be closed within the TD phonon $(X^{\dagger}_{2M}, X_{2M})$ and $(\Gamma^{\dagger}_k, \Gamma_k)$. The precise definition of $X^{\dagger}_{2M}$ and $\Gamma^{\dagger}_k$ will later be given. Thus a new "phonon-truncation" approximation which takes the place of the old one used in the subspace (1.1) has been introduced. Under this truncation approximation, the Dyson boson mapping gives us a simple but non-hermitian boson hamiltonian which consists of up to sixth-order s-boson ans d-boson operators and we can easily solve the right- and left-hand-side eigenvalue equations. Using these results, we can get the eigenvectors in the corresponding hermitian boson theory.[3,4] Thus we can analyse the coupling between the quadrupole-phonon mode and the pairing-vibrational modes without any truncation of boson expansion.

2. Formulation

2.1 Multi-phonon subspace and corresponding physical boson space

Let us define the quasiparticle pair operators with spin (LM) as

$$A^\dagger_{LM}(ab) = \frac{1}{\sqrt{2}} \sum_{m_\alpha m_\beta} \langle j_a m_\alpha j_b m_\beta | LM \rangle a^\dagger_\alpha a^\dagger_\beta, \tag{2.1a}$$

$$B^\dagger_{LM}(ab) = -\sum_{m_\alpha m_\beta} \langle j_a m_\alpha j_b m_\beta | LM \rangle (-)^{j_b - m_\beta} a^\dagger_\alpha a_{-\beta}, \tag{2.1b}$$

where the single-particle states are characterized by a set of quantum numbers $\alpha = (n_a, l_a, j_a, m_\alpha$, charge $q_a)$. We use a letter a to denote the same set except m_α and also a subscript $-\alpha$ obtained from α by changing the sign of m_α.

Using the quasiparticle pair operator (2.1a), we define the collective TD phonon operator $X^\dagger_{2M}$ appearing in (1.2) by

$$X^\dagger_{2M} = \sum_{ab} \psi(ab) A^\dagger_{2M}(ab), \tag{2.2}$$

where the "collective" phonon means the lowest-energy two-quasiparticle TD mode. Similarly the pairing-vibrational mode $\Gamma^\dagger_k$ appearing in (1.2) is defined by

$$\Gamma^\dagger_k = \sum_a \phi_k(a) A^\dagger_{00}(aa). \tag{2.3}$$

The amplitudes $\psi(ab)$ and $\phi_k(a)$ are determined by solving the equation of motion for each of the modes.

In order to give precise definition of the Dyson mapping, we have to introduce a set of orthonormalized basis vectors in the multi-phonon space (1.2). We define the n-phonon statevector $|n\beta JM\rangle$ with total spin (IK). Here n denotes the sum of the numbers of each kind of phonon and β is an additional quantum number characterizing the state. To define orthonormalized basis vectors, we introduce a representation in which the norm matrix is diagonalized as

$$\sum_{\beta'} M^{(n,I)}(\beta,\beta') u^{(n,I)}(\beta',\gamma) = M(nI\gamma) u^{(n,I)}(\beta,\gamma), \tag{2.4}$$

$$M^{(n,I)}(\beta,\beta') = \langle n\beta IK | n\beta' IK \rangle. \tag{2.5}$$

One can assume the following orthonormality relations:

$$\sum_\beta u^{(n,I)}(\beta,\gamma) u^{(n,I)}(\beta,\gamma') = \delta_{\gamma\gamma'}. \tag{2.6}$$

Let us denote the zero-eigenvalue solution by $\gamma = \gamma_0$. Using these, we can define orthonormalized basis vectors as

$$|n\gamma IK\rangle\!\rangle = M^{-1/2}(nI\gamma) \sum_\beta u^{(n,I)}(\beta,\gamma) |n\beta IK\rangle, \quad \gamma \neq \gamma_0, \tag{2.7}$$

which satisfy the orthonormality relation

$$\langle\!\langle n\gamma IK | n'\gamma' I'K' \rangle\!\rangle = \delta_{nn'} \delta_{\gamma\gamma'} \delta_{II'} \delta_{KK'}. \tag{2.8}$$

Next, we can define the orthonormalized n-boson state vector $|n\beta IK)$ which is obtained by replacing every phonon operator with the corresponding boson operator. Therefore the n-boson state vector $|n\beta IK)$ is just the counterpart of the n-phonon state vector $|n\beta IK\rangle$. The counterpart of the orthonormalized n-phonon state vector $|n\gamma IK\rangle\!\rangle$ is defined by

$$|n\gamma IK)\!) = \sum_{\beta} u^{(n,I)}(\beta,\gamma)\,|n\beta IK), \tag{2.9}$$

which is also an orthonormalized n-boson state vector.

Thus the physical subspace in the boson space $\{|n\beta IK)\}$ is $\{|n\gamma IK)\!);\gamma\neq\gamma_0\}$, in which every state has its counterpart in the original fermion space, i.e. the multi-phonon subspace. Therefore the projection operator to the physical subspace in the boson space is defined by

$$P = \sum_{\gamma\neq\gamma_0}\sum_{nIK} |n\gamma IK)\!)(\!(n\gamma IK| . \tag{2.10}$$

<u>2.2 Dyson mapping for the multi-phonon subspace</u>

Using the orthonormal states introduced above, we can define the Dyson mapping by the following transformation operators;

$$U_1 = \sum_{nIK}\sum_{\gamma\neq\gamma_0} M^{1/2}(nI\gamma)\,|n\gamma IK)\!)\langle\!\langle n\gamma IK| , \tag{2.11a}$$

$$U_2 = \sum_{nIK}\sum_{\gamma\neq\gamma_0} M^{-1/2}(nI\gamma)\,|n\gamma IK)\!)\langle\!\langle n\gamma IK| , \tag{2.11b}$$

where $M(nI\gamma)$ is an eigenvalue of the norm matrix appearing in (2.4).

In the Dyson mapping, we have two types of boson state vector obtained by transforming a fermion state vector $|\Psi\rangle$; one is a <u>bra</u> $(\phi|$ and the other is a <u>ket</u> $|\psi)$ as

$$(\phi| = \langle\Psi|U_2^\dagger , \qquad |\psi) = U_1|\Psi\rangle . \tag{2.12}$$

An arbitrary fermion operator O_F is transformed into O_D as

$$\begin{aligned} O_D &= U_1\, O_F\, U_2^\dagger \\ &= \sum_{nIK}\sum_{\gamma\neq\gamma_0}\sum_{n'I'K'}\sum_{\gamma'\neq\gamma_0}\sum_{\beta\beta'} |n\gamma IK)\!)\, u^{(n,I)}(\beta,\gamma) \\ &\quad \times \langle n\beta IK|O_F|n'\beta' I'K'\rangle\, M^{-1}(n'I'\gamma')\, u^{(n',I')}(\beta',\gamma')\,(\!(n'\gamma' I'K'| . \end{aligned} \tag{2.13}$$

In order to calculate the matrix element of the original fermion operator O_F, $\langle n\beta IK|O_F|n'\beta'I'K'\rangle$, we assume an approximation called the phonon-truncation approximation. Now let us explain this simply. Putting the quadrupole phonon mode $X^\dagger_{2M}$ and the monopole phonon mode $\Gamma^\dagger_k$ together, we denote them by a common notation as

$$X^{(i)\dagger}_{l_i m_i} = \sum_{ab} \psi_i(ab)\, A^\dagger_{l_i m_i}(ab), \tag{2.14}$$

where the subscript (i) denotes the i-th phonon mode. For simplicity we discuss a case where only two kinds of phonons (i=1 and 2) are retained. It is straightforward to generalize this to the cases of more kinds of phonons. The phonon-truncation approximation consists of the following two procedure:

i) We retain only the quasiparticle pair operators with $J=l_1$ and $J=l_2$ and neglect all others.

ii) We replace these retained pair operators by the l_1 and l_2 phonons; namely we adopt the replacement

$$A^{\dagger}_{l_1 m_1}(ab) \rightarrow \psi_1(ab) X^{(1)\dagger}_{l_1 m_1}, \quad A^{\dagger}_{l_2 m_2}(ab) \rightarrow \psi_2(ab) X^{(2)\dagger}_{l_2 m_2} \tag{2.15}$$

Using this approximation, we can write the matrix element of O_F in the multi-phonon subspace as

$$\langle n\beta IK | O_F | n'\beta' I'K' \rangle$$
$$= \sum_{\beta''} f^{(nIK;n'I'K')}(\beta,\beta'';O_F)\, M^{(n',I')}(\beta'',\beta'). \tag{2.16}$$

Substituting this into (2.13), we have

$$O_D = \sum_{nIK} \sum_{\gamma\neq\gamma_0} \sum_{n'I'K'} \sum_{\gamma'\neq\gamma_0} \sum_{\beta\beta'} |n\gamma IK)\!)\, u^{(n,I)}(\beta,\gamma)$$
$$\times f^{(nIK;n'I'K')}(\beta,\beta';O_F)\, u^{(n',I')}(\beta',\gamma')\, (\!(n'\gamma'I'K'|. \tag{2.17}$$

The norm matrix element in (2.16) is cancelled by the inverse of the eigenvalue $M(nI\gamma)$ in (2.13). This cancellation is the biggest merit of the Dyson mapping at the expense of losing unitarity of the transformation, because we have no more to treat the explicit form of matrix element $M^{(n',I')}(\beta,\beta')$.

The quantity $f^{(nIK;n'I'K')}(\beta,\beta';O_F)$ can, in general, be expressed by matrix elements of simple boson operators; for example, if $O_F = X^{\dagger}_{l_i m_i}$, we have

$$f^{(nIK;n'I'K')}(\beta,\beta';X^{\dagger}_{l_i m_i}) = (n\beta IK|\, b^{\dagger}_{l_i m_i}$$
$$- \hat{l}_i^{-1} \sum_{jkl=1,2} \sum_L (-)^{l_i+l_l} \hat{L}\, C_L^{(jkil)} [[b^{\dagger}_{l_j} b^{\dagger}_{l_k}]_L b_{l_l}]_{l_i m_i} |n'\beta' I'K'). \tag{2.18}$$

where $b^{\dagger}_{lm}$ is a boson creation operator with spin (lm). Substituting these types of boson expression into (2.17), we can write the Dyson boson operator as

$$O_D = P(O_F)_D P, \tag{2.19}$$

where P is the projection operator defined by (2.10). The Dyson image of the phonon operator or the fermion pair operator are written as

$$(X^{(i)\dagger}_{l_i m_i})_D = b^{\dagger}_{l_i m_i}$$
$$- \hat{l}_i^{-1} \sum_{jkl} \sum_L (-)^{l_i+l_l} \hat{L}\, C_L^{(jkil)} [[b^{\dagger}_{l_j} b^{\dagger}_{l_k}]_L b_{l_l}]_{l_i m_i}, \tag{2.20a}$$

$$(X^{(i)}_{l_i m_i})_D = b_{l_i m_i}, \tag{2.20b}$$

$$(B^{\dagger}_{LM}(ab))_D = (-)^L \hat{L}^{-1} \sum_{ij} (-)^{l_i+l_j} \hat{l}_j\, D^{(ij)}_L(ab) [b^{\dagger}_{l_j} b_{l_i}]_{LM}, \tag{2.20c}$$

$$(A^{\dagger}_{l_i m_i}(ab))_D = \psi_i(ab) (X^{(i)\dagger}_{l_i m_i})_D, \tag{2.20d}$$

$$(A_{l_i m_i}(ab))_D = \psi_i(ab) (X^{(i)}_{l_i m_i})_D, \tag{2.20e}$$

where $[\ldots]_{LM}$ denotes an angular-momentum coupling.

As discussed in our previous paper,[10] as long as we do not make the maximum phonon number extremely large in a practical case, we can consider P = 1.

Although we have so far formulated a case of two kinds of retained phonons for simplicity, it should be noted that the expressions (2.20) are completely general and applicable to any case.

2.3 Retained phonons and s- and d-bosons, hamiltonian

We are interested in the mode-mode coupling between the quadrupole phonon and the pairing vibrations, so that we retain the collective TD phonon $X^{\dagger}_{2M}$ defined by (2.2), the two lowest-energy proton pairing vibrations $\Gamma^{\dagger}_k$ (k=1,2) and the two lowest-energy neutron ones $\Gamma^{\dagger}_k$ (k=3,4) defined by (2.3). Of course, the spurious modes must not be included among these pairing vibrations. By using the general formulae (2.20), the Dyson images of the phonon operators are easily obtained; as for the explicit expressions, one can refer to Ref. 11).

The interaction hamiltonian used in the present work consists of the monopole pairing (P) force, the quadrupole pairing (P_2) force and the QQ force. After the Bogoliubov transformation, the hamiltonian is written as

$$H = H_0 + H^{P} + H^{P_2} + H^{QQ}, \tag{2.22a}$$

$$H_0 = \sum_{\alpha} E_a a^{\dagger}_{\alpha} a_{\alpha}, \quad H^{P} = -\tfrac{1}{4} G :\hat{P}^{\dagger}_0 \hat{P}_0:,$$

$$H^{P_2} = -\tfrac{1}{2} G_2 \sum_M :\hat{P}^{\dagger}_{2M} \hat{P}_{2M}:, \quad H^{QQ} = -\tfrac{1}{2}\chi \sum_M :\hat{Q}^{\dagger}_{2M} \hat{Q}_{2M}:, \tag{2.22b}$$

where E_a is a single-quasiparticle energy and G, G_2 and χ are the strengths of the P, P_2 and QQ forces respectively. This hamiltonian can completely be written in terms of J=0 and J=2 quasiparticle pair operators. Therefore, it is straightforward to derive a Dyson boson hamiltonian from (2.22). However its explicit form is omitted here, since it is somewhat lengthy.

2.5 How to get eigenvectors in the hermitian boson theory from the results of the Dyson boson theory

For simplicity we hereafter use an abbriviated notation for a boson basis vector as $|i) = |n\beta IK)$. We expand the eigenvectors in the boson space as follows:

$$|\Psi_\lambda) = \sum_i \alpha_i^{(\lambda)}|i), \quad |\psi_\lambda) = \sum_i \beta_i^{(\lambda)}|i), \quad (\varphi_\lambda| = \sum_i \gamma_i^{(\lambda)*}(i|, \tag{2.23}$$

where λ is the quantum number characterizing the eigenstate, $|\Psi_\lambda)$ is the corresponding eigenvector in the hermitian boson theory, and $|\psi_\lambda)$ (or $(\phi_\lambda|$) is the eigenvector of the right- (or left-) hand-side eigenvalue equation in the Dyson boson theory

$$\sum_j (h_{ij}^M - E_\lambda \delta_{ij})\alpha_j^{(\lambda)} = 0, \tag{2.24a}$$

$$\sum_j (h_{ij}^D - E_\lambda \delta_{ij})\beta_j^{(\lambda)} = 0, \tag{2.24b}$$

$$\sum_i \gamma_i^{(\lambda)*}(h_{ij}^D - E_\lambda \delta_{ij}) = 0. \tag{2.24c}$$

The first equation of (2.24) is the eigenvalue equation in the hermitian boson theory and the second and third are respectively the right- and left-hand-side eigenvalue equations in the Dyson boson theory. Needless to say, these three equations have the same set of eigenvalues. The orthonormality conditions are

$$\sum_i \alpha_i^{(\lambda)*}\alpha_i^{(\lambda')} = \delta_{\lambda\lambda'}, \quad \sum_i \gamma_i^{(\lambda)*}\beta_i^{(\lambda')} = \delta_{\lambda\lambda'}. \tag{2.25}$$

The latter condition, which is equivalent to $(\phi_\lambda|\psi_{\lambda'}) = \delta_{\lambda\lambda'}$, is still not enough to determine the amplitudes, because $(\bar{\phi}_\lambda| = (\phi_\lambda|/k_\lambda$ and $|\bar{\psi}_\lambda) = k_\lambda|\psi_\lambda)$ —— k_λ is a non-zero constant —— also satisfy the same orthonormality condition. Namely there still remains an arbitrariness by a constant k_λ. Then what is the correctly-normalized eigenvectors?

In Refs. 3) and 4), we have proposed an useful method to get the eigenvector $|\Psi_\lambda)$ in the hermitian theory from the eigenvectors $|\psi_\lambda)$ and $(\phi_\lambda|$ in the Dyson theory. As for the detailed discussions, one can refer to these references, so that we do not repeat here. According to them the correct matrix element of an arbitrary Dyson operator O_D transformed from the fermion operator O_F is[11)]

$$(\bar{\phi}_\lambda|O_D|\bar{\Psi}_{\lambda'}) = (\phi_\lambda|O_D|\psi_{\lambda'})\left[\frac{(\phi_{\lambda'}|O_D^\dagger|\psi_\lambda)^*}{(\phi_\lambda|O_D|\psi_{\lambda'})}\right]^{1/2}, \tag{2.26}$$

where $O_D^\dagger$ is the Dyson image of $O_F^\dagger$ but not $(O_D)^\dagger$.

Using the above expression, we can calculate correct matrix element of any operator; this is just why the Dyson boson mapping has become very promising and useful for the description of nuclear collective motion.

3. Numerical calculations

We applied the formulation given in the previous section to the analysis of the coupling between the quadrupole phonon and pairing-vibrational modes in the cases of the even-mass Se and Ge isotopes.

As seen in the experimental energy systematics of the Se and Ge isotopes in Figs. 1 and 2, the first excited 0^+ states (the 0_2^+ states) show a characteristic behavior near N=40; namely, as approaching the N=40 isotopes from heavier ones, the energy of the 0_2^+ state goes down to be minimum at N=40 and, beyond this, it rises up quickly. Particularly the first excited state in ^{72}Ge has long been noticed as one of the "mysterious" 0^+ states. Intensive studies[6-9] on this anomalous behavior of the 0_2^+ states have given a reasonable explanation that the coupling between the two-phonon 0^+ state and pairing-vibrational state depends strongly on the single-particle level structure and nucleon number and then it becomes quite large at N=40. In this sense, the anomalous behavior of the 0_2^+ states in this region might be a choice field to test the applicability of the Dyson mapping to mode-mode coupling analysis.

The results of the numerical calculations are shown in Figs. 3 and 4. They are taken from Ref. 11). These results show that the anomalous behavior of the 0_2^+ state energy is well reproduced. Looking at the structure of the corresponding eigenvector in the hermitian boson theory, we can realize that the coupling between two-d-boson 0^+ state and neutron pairing vibration becomes most impotant at N=40.

4. Summary

In this report we have formulated the Dyson boson mapping in the multi-phonon subspace and applied it to the analysis of the mode-mode coupling between the quadrupole phonon mode and the pairing-vibrational modes.

We can summarize the important points of the results as follows;

(i) All the necessary information in the hermitian boson theory can easily be obtained from the solutions of the right- and left-hand-side eigenvalue problem in the Dyson boson theory. Therefore we can say the Dyson boson theory works well in practical cases.

(ii) The Dyson boson mapping is very suitable for the mode-mode coupling theory; namely, it can easily lead us to a precise boson expression of the mode-mode coupling hamiltonian without introducing any truncation of boson expansion.

(iii) By using the Dyson boson mapping, the mysterious behavior of the 0_2^+ states in the Ge and Se isotopes are reasonably explained by the strong coupling between the two-d-boson 0^+ state and the neutron pairing vibrational state.

Finally we can conclude that the Dyson boson mapping is quite promising to investigate mode-mode couplings. It is desirable to apply it to other region of nuclei, especially to highly excited states and odd-mass nuclei.

References

1) T. Marumori, M. Yamamura and A. Tokunaga, Prog. Theor. Phys. 31 (1964) 1009.
2) T. Usui, Prog. Theor. Phys. 23 (1960) 787.

3) K. Takada, T. Tamura and S. Tazaki, Phys. Rev. C31 (1985) 1948.

4) K. Takada, Nucl. Phys. A439 (1985) 489.

5) D.R. Bes and R.A. Broglia, Nucl. Phys. 80 (1966) 289.

6) S. Iwasaki, T. Marumori, F, Sakata and K. Takada, Prog. Theor. Phys. 56 (1976) 1140.

7) F. Sakata, S. Iwasaki, T. Marumori and K. Takada, Z. Phys. A286 (1978) 195.

8) S. Tazaki, K. Takada, K. Kaneko and F. Sakata, Prog. Theor. Phys. Supplement no.71 (1981) ch.4.

9) K.J. Weeks, T. Tamura, T. Udagawa and F.J.W. Hahne, Phys. Rev. C24 (1981) 703.

10) K. Takada, Nucl. Phys. A431 (1984) 16.

11) K. Takada and S, Tazaki, Nucl, Phys. A448 (1986) 56.

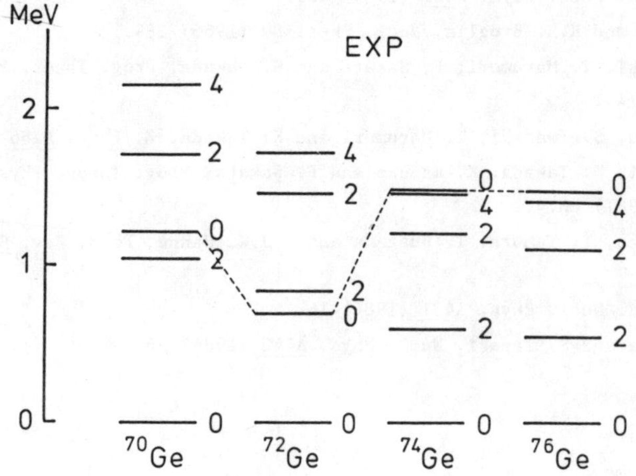
MeV
EXP
0
1
2
^{70}Ge
^{72}Ge
^{74}Ge
^{76}Ge

Fig.1

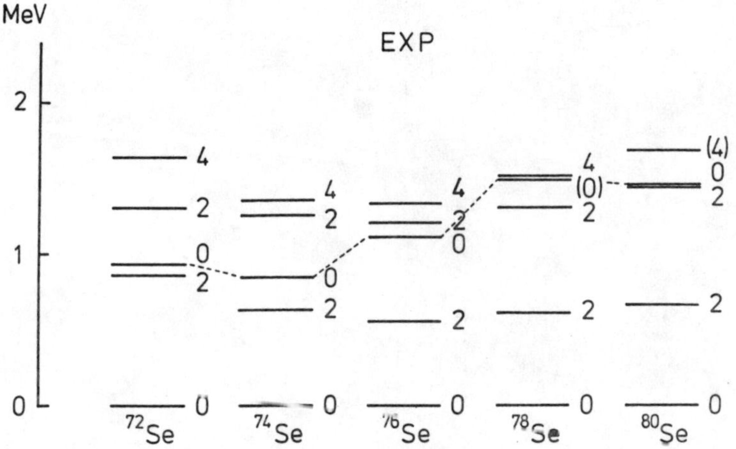
MeV
EXP
0
1
2
(4)
(0)
^{72}Se
^{74}Se
^{76}Se
^{78}Se
^{80}Se

Fig.2

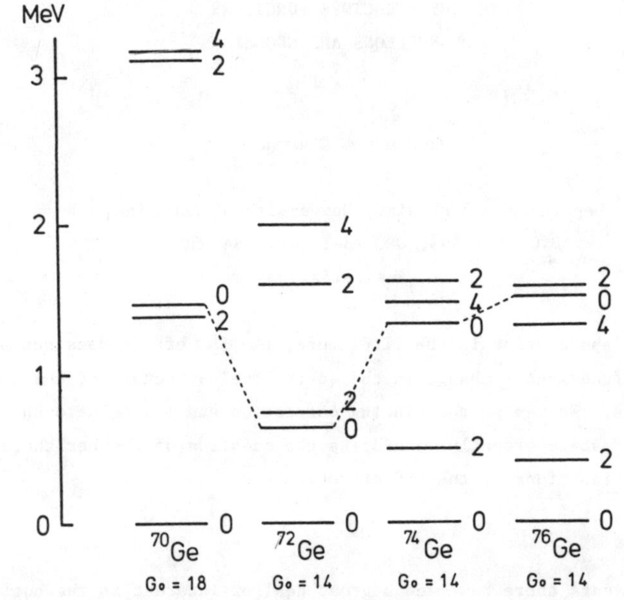
MeV
3
2
1
0
4
2
0
2
4
2
2
0
0
2
4
0
2
2
0
4
2
0
0
0
0
^{70}Ge
^{72}Ge
^{74}Ge
^{76}Ge
Go = 18
Go = 14
Go = 14
Go = 14

Fig. 3

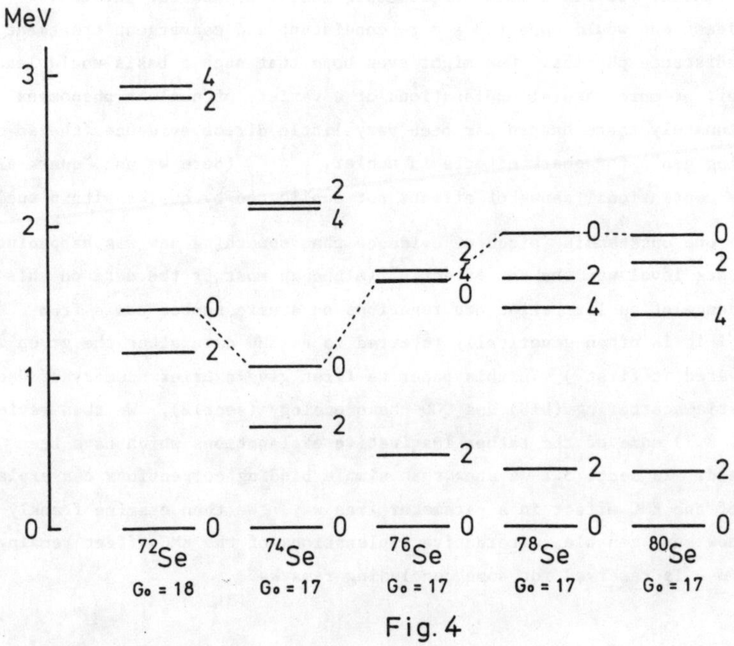
MeV
3
2
1
0
4
2
0
2
0
2
4
2
0
2
4
0
2
0
2
4
2
0
0
2
2
0
0
0
0
0
^{72}Se
^{74}Se
^{76}Se
^{78}Se
^{80}Se
Go = 18
Go = 17
Go = 17
Go = 17
Go = 17

Fig. 4

ON THE STRUCTURE FUNCTIONS OF NUCLEONS AND NUCLEI

Anthony W. Thomas

Department of Physics, University of Adelaide,
P.O. Box 498, GPO Adelaide. SA 5001.
Australia.

Despite much speculation in the literature, the EMC effect does not provide evidence for a fundamental change in the quark-level structure of the nucleon inside a nucleus. We review the binding correction and its relationship to the pionic model. Most importantly we address the question of whether there really are too many explanations of the EMC effect.

1. INTRODUCTION

For several years there has been a great deal of interest in the nuclear physics community in the idea that QCD, or at least QCD-motivated quark models, might finally provide a sound microscopic basis for nuclear physics.[1-3] At the very least one would hope for a more consistent and convergent treatment of short distance physics. One might even hope that such a basis would lead to simpler, or more natural explanations of a variety of nuclear phenomena. Unfortunately there has so far been very little direct evidence (the so-called "smoking gun") for quark effects in nuclei.[4-6] (Here we use "quark effects" in the conventional sense of effects not duplicated by quarks within nucleons.)

The one outstanding piece of evidence that something new was happening at the quark level was the EMC effect.[7] (Although most of the data on this dependence of nuclear structure functions on atomic number comes from SLAC,[8] it is often generically referred to as EMC data after the group which discovered it first.) In this paper we first give a brief summary of deep-inelastic scattering (DIS) and QCD phenomenology (sect.2). We then review (sect. 3.1) some of the rather imaginative explanations which have been proposed. In sect. 3.2 we show that simple binding corrections can explain most of the EMC effect in a parameter free way. We then examine frankly (sect. 3.3) how many tenable, alternative explanations of the EMC effect remain. Section 4 is reserved for some concluding remarks.

2. THE STRUCTURE OF THE NUCLEON

Probably the most convincing evidence for the quark model was the discovery, at SLAC in the late 60's, that nucleons contain point-like constituents. Over the last fifteen years the experimental evidence has become extremely solid, with electron, muon and neutrino deep inelastic scattering (DIS) experiments all consistently understood in terms of quarks with the conventional charges for weak and electromagnetic interactions.[9-11] With all this data available, we find it surprising that more effort has not been put into testing models of hadron structure against DIS.[12] In this section we summarise the key results of the parton model, show how these are altered by QCD, and finally discuss the correlation function $C_-(z)$ as a link between phenomenological quark models and DIS.

2.1 DEEP INELASTIC SCATTERING:

For any probe which interacts electromagnetically (and therefore conserves parity) the spin-averaged, inclusive cross section is entirely determined by two functions, F_1 and F_2. Conventionally the momentum transferred by the virtual photon is q ($q^0 = \nu$ in the LAB frame), and the target momentum p ($p = (M,\vec{0})$ in the LAB). In the naive parton model, where the virtual photon is absorbed by elementary, non-interacting spin-$\frac{1}{2}$ quarks, the two "structure functions" (F_1, F_2) depend only on $x = Q^2/2p.q$, where $Q^2 = -q^2 > 0$, and $F_2(x) = 2x\, F_1(x)$ (the Callan-Gross relationship). In a frame where the target has a momentum, P, much greater than its rest mass, x can be interpreted as the fraction of the target momentum carried by the struck quark. Then, in terms of the probability distributions u(x), for up quarks, $\bar{d}(x)$ for anti-down quarks and so on, we have

$$2\, x\, F_1(x) = F_2(x) = x\, \Sigma_i\, q_i^2\, f_i(x). \qquad (2.1)$$

With neutrino beams one can also measure a third structure function

$$x\, F_3(x) = x\left[u(x) - \bar{u}(x) + d(x) - \bar{d}(x)\right], \qquad (2.2)$$

which is the valence quark distribution. Within experimental errors $\int_0^1 dx\, F_3(x)$, which is the excess number of quarks over anti-quarks, is equal to three as we expect.

Of course the naive parton model is just that, naive. As one probes a target more deeply ($Q^2 \to \infty$), quantum chromodynamics (QCD) tells us to expect to see more virtual quarks and gluons.[9,10,13] The consequence of this is that the structure functions depend on both x and Q^2. This dependence is most simply given for so-called non-singlet structure functions, like F_3. If we define the n'th moment as

$$M_3^n(Q^2) = \int_0^1 dx\, x^{n-2} \{ x F_3 (x, Q^2)\}, \tag{2.3}$$

then to leading order in powers of $1/Q^2$ one finds

$$M_3^n(Q^2) = M_3^n (\mu^2) [\alpha_s (Q^2)/\alpha_s(\mu^2)]^{dn}. \tag{2.4}$$

Here μ is an arbitrary renormalisation scale, d_n the anomalous dimension, and α_s the (running) coupling constant of QCD.

Based on the experimental data, and the trends predicted by perturbative QCD, a number of groups have been led to suggest that there may be some (relatively low) momentum scale, $Q^2 = \mu_0^2$, where the nucleon is well represented by one of the familiar quark models.[14-17] For example, Jaffe and Ross[15] showed that, for n between 3 and 9, the moments of $x F_3$ calculated in the MIT bag model agreed well with the data extrapolated to $\mu_0^2 \sim 0.7$ GeV^2. Rather than pursue this idea further here, we refer the interested reader to the references cited, and turn to our discussion of the valence correlation function.

2.2 THE VALENCE CORRELATION FUNCTION

One of the most interesting pieces of direct information concerning nucleon structure is shown in Fig. 1. There we find the valence quark correlation function[18] $C_-(z, Q^2)$, which is related to $x F_3 (x, Q^2)$ as:

$$C_-(z, Q^2) = \int_0^\infty dx \cos (m_N zx) F_3(x, Q^2). \tag{2.5}$$

This function has the physical interpretation of the probability that one can remove a valence quark from a nucleon, put it back a distance z (fm) away (on the light cone), and reform the nucleon in its ground state. (Following Llewellyn Smith[18], we smoothly cut this integral below x = 0.1 - because of the ambiguity in the separation of valence and sea quarks in that region.)

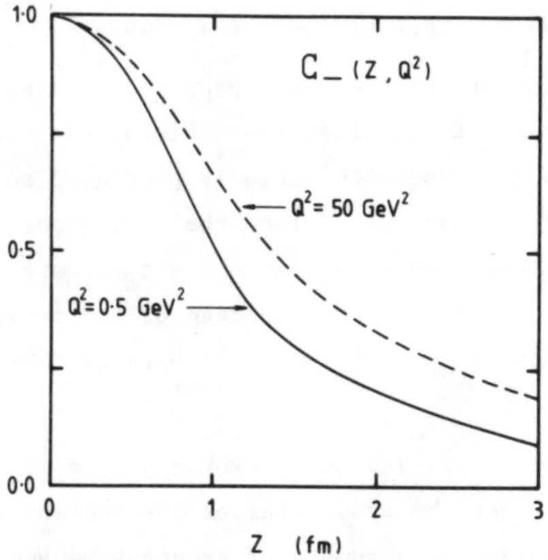

Figure 1: The correlation function for valence quarks, $C_-(z,Q^2)$, at two widely separated values of Q^2.[12,36)]

We stress that the curves in Fig. 1 are based upon experimental data referred to 5 GeV^2, and then extrapolated via the Altarelli-Parisi equations to 0.5 and 50 GeV^2. The dependence on Q^2 is surprisingly small. Clearly the full width half maximum of $C_-(z)$ is at least 2 fm. On the face of it, this is exactly what one would expect in the MIT, or cloudy bag model.[3)] In order to be more firm about this we need further work on the connection between light-cone and equal-time correlation functions.

3. THE EMC EFFECT

We begin this section with a rapid review of some of the rather exotic explanations which have been suggested for the EMC effect. Next we return to earth to check how well the more mundane nuclear effects have been treated. In fact we shall see that a better treatment of nucleon binding eliminates most of the anomalous A-dependence. Finally (in sect. 3.3) we address the important question of how many truly different, credible explanations now remain.

3.1 THE SPECULATION

One of the first suggestions of what might be involved in the EMC effect was that there might be exotic quark states in the nucleus.[19-21)] In the MIT or the cloudy bag model the nucleon has a radius just a little less that half of the nearest neighbour separation at nuclear matter density. Thus it is highly likely that nucleons would overlap some of the time, leading to percolation,[22)] or at the very least to six-quark bags.[23)] Either from counting rules, or within the bag model, the ratio of the structure function of a 6q bag to that of a 3q bag is a minimum at $x \sim 0.6$.

A more extreme version of this idea was proposed by Krzywicki before the EMC discovery.[24)] His idea was to consider the nucleus as a single large bag of 3A valence quarks. Within this model he actually predicted an enhancement of the number of sea quarks in the nucleus. Since the SLAC group has not found any evidence for a small-x enhancement, we can not yet be sure whether Krzywicki's prediction is correct. In any case the success of the conventional picture of a nucleus as a collection of nucleons makes it hard to take this extreme idea very seriously.

An alternative to the models involving exotic pieces of the nuclear wave function, is to suggest that the properties of the nucleon itself may change inside a nucleus. For example, a number of groups have suggested that a bound nucleon might swell.[25,26)] While this idea may seem bizarre, there is little hard evidence to refute it. For ^{3}He Sick has shown[27)] that y-scaling data limits the increase in size to not more than 6%. Unfortunately this is not good enough (because of the low average density in ^{3}He) to rule out this suggestion. More precise limits of this kind, on heavier nuclei, would be most welcome.

Apart from models of the Friedberg-Lee type, which we have discussed elsewhere, another model which leads to an effective increase in nucleon size is that of Chanfray, Nachtmann and Pirner (CNP).[25)] They minimise the energy of a lattice of Nielsen-Patkos solitons.[29)] Assuming that the free nucleon eigensolutions are well approximated by the bag model, CNP find that in a nucleus the quarks tend to leak out of the individual nucleons. They claimed that this could increase the rms quark radius by about 8% in Fe. This claim was re-examined by Williams and Thomas, who showed that with the original parameters the model led to gross overestimates of the neutron-proton binding energy difference in finite nuclei.[29)] With parameters that respected known nuclear structure constraints we found that the size increase could be at most one percent. At the present time this issue is still being debated, because CNP claim that gluon effects soften the Williams-Thomas constraints.[30)] In any case, the Nielsen-Patkos model has some fascinating consequences for nuclear symmetry breaking - for example, there is a range of parameters for which the Nolen-Schiffer anomaly can be fit.[31)]

Another explanation involves the enhancement of the virtual pion field which surrounds a nucleon in any realistic model of hadron structure.[32,33)] With a little phenomenology involving momentum balance (along the lines suggested by Llewellyn Smith[32)]) this model can also reproduce the SLAC data - using parameters within the range consistent with low energy physics.[4)] We shall see in sect. 3.3 that this model ties in nicely with the work described in sect. 3.2, which we regard as the most sensible explanation of the effect.

Finally we note that everyone of the explanations mentioned so far involves an increase in some length scale as the nuclear density goes up. This is precisely what the observation of Close, Roberts and Ross[35] (that $F_2^{Fe}(x,Q^2) \simeq F_2^D(x, \xi Q^2)$ with $\xi \simeq 2$) has been taken to imply. We shall return to their explanation again in sect. 3.3. Before that discussion, we first consider a much simpler possibility.

3.2 A MORE HUMBLE SUGGESTION - BINDING CORRECTIONS:

In this section we summarise the results of a calculation of the simplest model for nuclear DIS, based on the shell model.[36] As in all earlier calculations of fermi motion,[37] we assume that the impulse approximation works in the sense that there are no final-state-interactions between the debris of the nucleon which absorbs the high energy photon and the residual (A-1)-body system. However, unlike earlier work, we take seriously the fact that the bound nucleon is off-mass-shell. Figure 2 illustrates what we calculate.

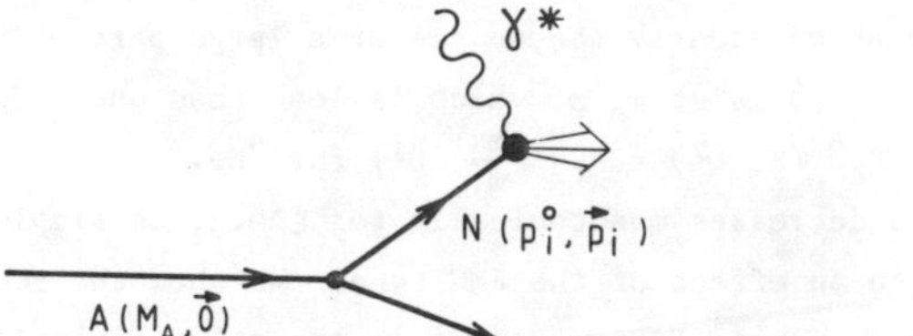

Figure 2: The simple model for nuclear DIS considered here.

Nucleus A splits into an on-mass-shell recoiling nucleus (A-1) and an off-mass-shell nucleon - with momentum $\vec{p}_i$ and energy $p_i^0 = M_A - E_{A-1} \simeq m_N + \varepsilon_i - |\vec{p}_i|^2/2\,M_{A-1}$ (with ε_i(<0) the single particle binding energy). This off-mass-shell nucleon is smashed by the virtual photon γ*.

In terms of the occupation number, n_i, and wave function, $\Psi_i(\vec{p})$, for each single-particle state we find:

$$A\, F_2^{\,A}\,(x,\, Q^2) = \sum_i n_i \int d\vec{p}\; |\psi_i(\vec{p})|^2\; F_2^{\,N}\,(p^2,\, x_N;\, Q^2). \qquad (3.1)$$

Here $x_N = Q^2/\,2\,p.q$ is the Bjorken variable appropriate to the struck nucleon. With a little work equ. (3.1) can be re-written as a convolution:

$$F_2^{\,A}\,(x,\, Q^2) = A^{-1} \sum_i\; n_i \int_x^A dy\; f_i\,(y)\; F_2^{\,N}\,(\tfrac{x}{y},\, Q^2). \qquad (3.2)$$

In terms of the old variables we have

$$y = (p_o + p_z)/m_N, \qquad (3.3)$$

and $f_i(y)$ can be calculated directly from $\Psi_i(\vec{p})$. For example, for the 0s state of an harmonic oscillator we find

$$f_0(y) = \frac{m_N}{\sqrt{\pi\beta}} \exp\left(-(m_N y - m_0)^2/\beta\right), \qquad (3.4)$$

where $m_0 = m_N + \varepsilon_0 < m_N$ is the average value of y.

Equation (3.4) shows clearly the source of a large part of the EMC effect. The peak value of $f_0(y)$ is at m_o/m, which is less than one. On average then, equ. (3.2) yields $F_2^{\,A}\,(x,\, Q^2) \approx F_2^{\,N}\,(\frac{mx}{m_0},\, Q^2)$ for ^4He. However $F_2^{\,N}(\xi,\, Q^2)$ decreases monotonically for $\xi > 0.2$, so simple nuclear binding inevitably leads to an effect of the EMC type. We show the results of a calculation of this type in Figure 3. It is important to realize that in calculating Fig. 3 we have used experimental removal energies - in so far as these are known.[38] Clearly nuclear binding accounts for at least 60% of the EMC effect seen at SLAC.[39]

The calculation shown here has two major defects. Firstly, the use of harmonic oscillator wave functions is not good for x > 0.8. Below this, recent calculations using a Woods-Saxon potential gave essentially the same results.[40] Secondly, we have used only the extreme single-particle model. Obviously configuration mixing, where the nucleon is struck leaving the residual, (A-1) system excited, would increase the size of the effect. More work is underway on this aspect of the problem, but it is clear a priori that a proper account of nucleon binding may account for the whole of the EMC effect for x > 0.3.

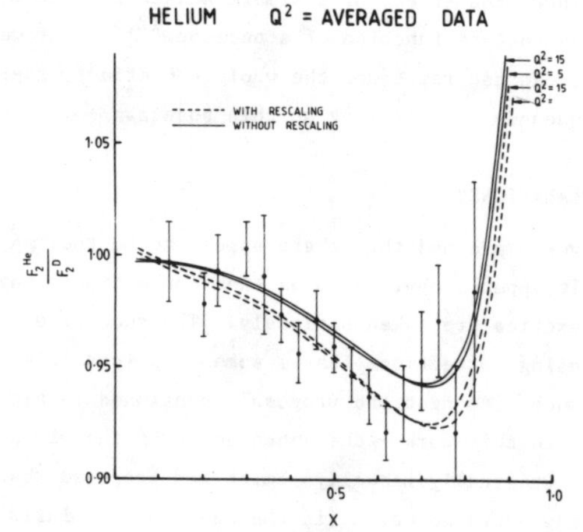

Figure 3(a): Comparison of the predictions of the nuclear shell model for DIS on ^{4}He,[36)] with recent data from SLAC.[8)] The solid curve includes only binding and fermi motion, while the dashed curve includes an ansatz for the off-mass-shell behaviour of the nucleon structure function.

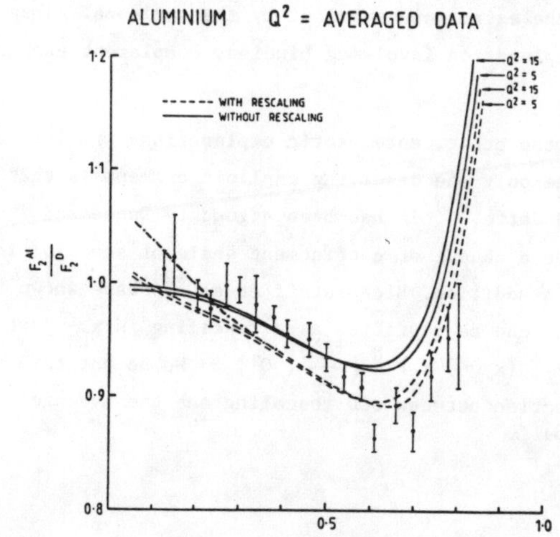

Figure 3(b): The curves and data are as above, but for Al. The dot-dash curve indicates the effect of balancing momentum lost through binding by an increased pionic contribution.

Figure 3 also shows the effect of a simple ansatz for the off-mass-shell dependence of the structure function of a nucleon.[12,36] If we eventually find that binding alone can not reproduce the whole EMC effect, there may be some interesting consequences of following up this ansatz.

3.3 TOO MANY EXPLANATIONS?

Many people have complained that there appear to be too many explanations of the EMC effect. It appears obvious to us that conventional physics must be exhausted before exotica are taken seriously. The success of our simple calculation of binding corrections leaves some room (possibly) for interesting physics, but not much. Amongst the proposals mentioned in sect. 3.1, only one is closely related to this work - the enhancement of the virtual pion field of the nucleus.[32,33] Originally Llewellyn Smith had proposed that if there were more pions, carrying extra momentum (in the sense of the infinite momentum frame), the nucleons would have to loose some. Implementation of this idea does lead to good fits to the SLAC data.[34] However, in a sense the tail is wagging the dog! That is, a small (experimentally uncertain) increase at small x, produced the large dip through phenomenology. Here we see that because of binding the nucleons carry less momentum. Of course meson exchange is responsible for the binding, and it is perfectly reasonable to suppose that they also carry the extra momentum lost by the nucleons. Thus the pionic model, and the explanation involving binding, complement each other beautifully.

Unfortunately the other, more exotic explanations are left out in the cold. Perhaps the only one deserving explicit comment is that of Close, Roberts, Ross and Jaffe.[41] It has been argued by Dunne and Thomas [36] that their argument for a change of confinement scale of some 15% in Fe is self-contradictory. In addition, Bickerstaff et al.[42] have shown that the effect of rescaling in Q^2 can be rewritten as a rescaling in x - just as we found from binding (recall $F_2{}^A\,(x,Q^2) \simeq F_2{}^N\,(\frac{m_n x}{m_0}, Q^2)$. We do not take seriously claims of a deeper connection between QCD rescaling and the average binding energy of nuclear matter.[43]

4. CONCLUSION

Much to our dismay, given the compelling arguments based on nucleon structure, there is little support, even from DIS data, for exotic components in the nuclear wave function. Nuclear binding corrections, and a relatively small enhancement of the nuclear pion field, explain most of the EMC effect. Further work on the binding and fermi motion corrections is essential if we are to isolate any more interesting physics.

Of course it may be that the conventional explanation is equivalent in some deeper way to one of the exotic ideas which have taken so much prominence in earlier discussions. While we do not find any suggestion of this kind compelling at the present time, this also deserves more thought. Finally, we acknowledge the ultimate importance of experiment in deciding whether what has been said here is correct. There is a desperate need for systematic study of the x, Q^2 and A dependence of the valence and sea components of the nuclear quark distributions, and we welcome all moves aimed at providing such data.

ACKNOWLEDGEMENTS

It is a pleasure to thank Profs. L.-M. Yang, Z.-H. Yang and H.-A. Peng for their hospitality during my visit to Beijing. A number of discussions with Paul Bickerstaff, Guy Chanfray, Gerald Dunne, Tony Signal and Tony Williams helped in putting this paper together. Finally I acknowledge the support of the ARGS and the University of Adelaide in this research.

REFERENCES

1. Contributions of A. Faessler, M. Rho, V. Soni, A.W. Thomas and W. Weise in Prog. Nucl. Part. Phys. 11 (1984).

2. Panel discussion on "Quarks and Nuclei" in Nucl. Phys. A434 (1985) 629 ff.

3. G.A. Miller, Int. Rev. Nucl. Phys. Vol. 2 (World Scientific, Singapore, 1984); A.W. Thomas, Adv. Nucl. Phys. 13 (1984) 1.

4. F. Gross, Invited introduction to Bates Users Group Workshop, MIT (July 1984).

5. J.M. Laget, Clustering Aspects of Nuclear Structure, eds. J.S. Lilley and M.A. Nagarajan (D.Reidel, Lancaster, 1985) p. 295.

6. A.W. Thomas, Clustering Aspects of Nuclear Structure, eds. J.S. Lilley and M.A. Nagarajan (D. Reidel, Lancaster, 1985) p. 333.

7. J.J. Aubert et al., Phys. Lett. 123B (1983) 275.

8. R.G. Arnold et al., Phys. Rev. Lett. 52 (1984) 727.

9. F.E. Close, An Introduction to Quarks and Partons (Academic, London,1979).

10. E. Leader and E. Predazzi, An Introduction to Gauge Theories and the New Physics (Cambridge University Press, 1985).

11. F. Eisele, J. de Physique C3 (1982) 337.

12. G.V. Dunne and A.W. Thomas, Nucl. Phys. A446 (1985) 437.

13. F.J. Yndurάin, Quantum Chromodynamics (springer-Verlag, New York, 1983).

14. R. Parisi and G. Petronzio, Phys. Lett. 62B (1976) 331.

15. R.L. Jaffe and G.G. Ross, Phys. Lett. 93B (1980) 113.

16. A. De Rujula and F. Martin, Phys. Rev. D22 (1980) 1787.

17. V.A. Novikov et al., Ann. Phys. 105 (1977) 276.

18. C.H. Llewellyn Smith, Nucl. Phys. A434 (1985) 35.

19. R.L. Jaffe, Phys. Rev. Let. 50 (1983) 128.

20 C.E. Carlson and T.J. Havens, Phys. Rev. Lett. 51 (1983) 261.

21. H.J. Pirner and J.P. Vary, Heidelberg preprint UNI-HD-83-02; S. Date, Prog. Th. Phys. 70 (1983) 1682; M. Chemtob and R. Peschanski, Saclay S. Ph. T./83/116.

22. G. Baym, Physica 96A (1979) 113.

23. G.A. Miller et al., Comm. Nucl. Part. Phys. 10 (1981) 101.

24. A. Krzywicki, Phys. Rev. D14 (1976) 152.

25. O. Nachtmann and H.J. Pirner, Z. Phys. C21 (1984) 277; G. Chanfray et al., Phys. Lett. 147B (1984) 249.

26. C.M. Shakin, Nucl. Phys. A446 (1985) 323.

27. I. Sick, Nucl. Phys. A434 (1985) 677.

28. H.B. Nielson and A. Patkos, Nucl. Phys. B195 (1982) 137.

29. A.G. Williams and A.W. Thomas, Phys. Lett. 154B (1985) 320.

30. G. Chanfray, priv. comm. (1985).

31. A.G. Williams and A.W. Thomas, U. Adelaide preprint ADP-347/T15 (1985), to appear in Phys. Rev. C.

32. C.H. Llewellyn Smith, Phys. Lett. 128B (1983) 107.

33. M. Ericson and A.W. Thomas, Phys. Lett. 128B (1983) 112.

34. A.W. Thomas, Proc. Int. Conf. on Nuclear Physics, eds. B.K. Jain and B.C. Sinha (World Scientific, Singapore, 1985) p. 175.

35. F.E. Close et al., Phys. Lett. 129B (1983) 346.

36. G.V. Dunne and A.W. Thomas, ADP-347/T14 (June 1985) to appear in Nucl. Phys. A.

37. A. Bodek and J.L. Ritchie Phys. Rev. D23 (1981) 1070.

38. R.C. Barrett and D.F. Jackson, Nuclear Sizes and Structure (Oxford University Press, Oxford, 1977).

39. A very similar calculation was made by: S.V. Akulinichev et al., Phys. Lett. 158B (1985) 727. However these authors omitted the binding correction in deuterium, and therefore overestimated the effect in the ratio.

40. E. Duff and A.W. Thomas, to be published.

41. R.L. Jaffe et al., Phys. Lett. 134B (1984) 449.

42. R.P. Bickerstaff and G.A. Miller, Phys. Lett. (to appear 1986).

43. F.E. Close et al., Rutherford preprint, RAL-85-101.

CALCULATION OF DYNAMICAL EFFECTS WITHIN THE FRAMEWORK OF FOLDED DIAGRAM THEORY

Wang Zixing Huang Weizhi Song Hongqiu Cai Yanhuang

Institute of Nuclear Research, Academia Sinica,
P.O.Box 8204, Shanghai, China

A long standing problem in microscopic calculations of nuclear giant resonances is that even though it has been possible to obtain the correct magnitude of the energies in light nuclear systems, the calculated positions of such resonances are always several Mev too low compared with the experimental positions. To resolve this difficulty, Brown, Dehesa and Speth[1)] proposed the well-known dynamical theory. Its essence is that the nuclear average field itself depends on the nuclear excitations, and therefore the single particle energies produced by this average field should be energy dependent. For example, the average particle-hole gap used for RPA calculations for the low lying states of Pb^{208} is about 7 Mev, but it should be about 10 Mev for such calculations of the giant dipole states of Pb^{208}. This discrepancy does not exist in the case of O^{16}. The dynamical effect should play an important part in the explanation of resonance position in heavy nuclei, but it should not be important in light nuclei. Brown's concept is very important and several authors have studied it since its introduction. Sommermann, Kuo and Ratcliff[2)] have performed a microscopic calculation of the single particle energies in the Pb^{208} region and obtained results which confirmed the dynamical effect of Brown, Dehesa and Speth. But later Sommermann and Mahaux concluded from their model studies that there is essentially no such dynamical effect.

Clearly there is now a controversy, and it seems to be worthwhile to sort it out. Sommermann, Kuo and Ratcliff and Sommermann and Mahaux studied the dynamical effect in the context of the Green's function method. In this work, we want to study this effect using the folded diagram effective interaction theory. Why can we use this energy independent theory to study the dynamical effect ? Because the folded diagram theory is an exact one and therefore includes the influnce of the dynamical effects. We can demonstrate the equivalence of the energy independent folded diagram theory and the energy dependent Green's function method. In the folded diagram method, both, the single particle energies as well as the effective interaction, are affected by the inclusion of folded diagram. In this theory, we have $H^{eff} = H_o^{eff} + V^{eff}$ and H_o^{eff} is energy independent. It appears that we do not have dynamical effect here. This is, however, not the case. If we neglect the folded diagrams of our V^{eff}, then clearly we do not have the dynamical effect. Certainly, we should not neglect the folded diagrams of V^{eff}, and we want to study which classes of folded diagram may give rise to an equivalent dynamical effect on the nuclear single particle energies. A likely candicate is the folded V^{eff} terms composed of a mixture of one-body and two-body Q-box diagrams. The purpose of this present work is to show that the contribution of these terms is small in O^{16}.

The formulae we use are as follows

$$H^{eff} = H_o^{eff} + V^{eff}$$

$$V^{eff} = Q - Q' Q + \ldots = F_o + F_1 + \ldots \qquad (1)$$

$$Q = Q_1 + Q_2$$

Q_1 is composed of one-body diagrams and Q_2 is composed of two-body diagrams. Q'contains at least two vertexes. We can also re-write the eq(1) as follows

$$H^{eff} = H_o^{eff} + F_n(Q_1) + F_n(Q_1 + Q_2) - F_n(Q_1)$$

taking $H_o^{eff} + F_n(Q_1) = (e_p - e_h)^{exp}$

defining $F_n(Q_1 + Q_2) - F_n(Q_1) = F_n \qquad (2)$

then $H^{eff} = (e_p - e_h)^{exp} + F_n$

From eq(2) we can see F_n only includes the mixture of one-body and two-body Q-box diagrams. By using the diagrams, we can express them as

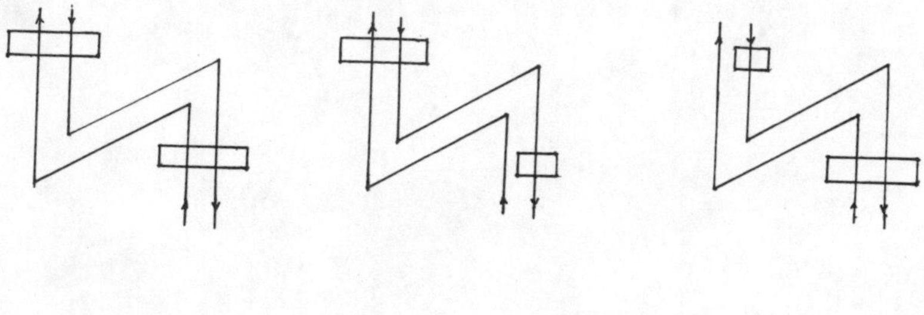

We expect that the last two terms can contribute to the dynamical effect, which can be neglected for light nuclei, and should be important for heavy nuclei. Along these lines, we first calculated the giant resonance states(1^-, T=1) of O^{16}, using an M3Y effective interaction. We display the results in table 1. In the upper part of the table, the first line lists the particle-hole configurations. The second line gives the unperturbed energies corresponding to above states. In the low part of the table, the calculated eigenvalues are given for Reid force and Pair force. To compare, the Gillet's results are given, too.

We can see from the results that the difference between the three cases is quite small for each force. In light nuclei like O^{16}, the contributions of folded diagrams can be neglected, as might be expected; the eigenvalues obtained from different force are nearly the same.

The calculations of the giant resonance states of Pb^{208},which has important dynamical effect, is in progress.

References

1)G.E.Brown, J.S.Dehesa and J.Speth,Nucl.Phys. A330(1979)290
2)H.M.Sommermann,T.T.S.Kuo and K.F.Ratcliff, Phys. Lett. 112B(1982)108
3)V.Gillet and N.Vinh-mau, Nucl.Phys. 54(1964)321

Table 1. the 1^-, T=1 states of O^{16}

Configurations	$1S_{1/2}$ $0P_{3/2}$	$0D_{3/2}$ $0P_{3/2}$	$0D_{3/2}$ $0P_{3/2}$	$0D_{3/2}$ $0P_{1/2}$	$1S_{1/2}$ $0P_{1/2}$
Unperturbed energies $(e_p - e_h)^{exp}$ (MeV)	18.57	22.79	17.70	16.62	12.40
Eigenvalues					
1.	26.7	24.8	21.3	17.3	14.0
2.	27.2	25.3	21.7	17.5	14.1
3.	26.9	25.4	21.7	17.5	14.1
1.26.1	26.1	24.7	21.5	17.1	14.0
2.	26.5	25.2	21.9	17.3	14.2
3.	26.3	25.1	21.8	17.3	14.2
*	25.4	22.7	19.6	18.1	13.6

1.They are eigenvalues when all folded diagrams are included;
2.Folded diagrams do not contain mixture of one-body and two-body;
3.Not contributions from folded diagrams;
*They are Gillet's results

A UNIFIED MICROSCOPIC DESCRIPTION OF MANY PARTICLE SYSTEMS ESPECIALLY NUCLEAR SYSTEMS

Karl WILDERMUTH

Institut für Theoretische Physik
Auf der Morgenstelle 14
7400 Tübingen, West Germany

By reformulating the Schrödinger equation as projection equation, it is possible to construct a unified many particle theory in which the difference between bound states and reactions is only expressed in the different boundary conditions for the wave functions. This is demonstrated for the many particle systems in nuclear physics and it is shown that in this way one obtains a unified microscopic theory of the nuclei. Thereby the indistinguishability of the nucleons i.e. the Pauli Principle plays a decisive role. At the end, some examples from other fields of physics are mentioned in short.

I INTRODUCTION

The starting point for a quantitative description of non-relativistic many particle systems is the belonging many particle Schrödinger equation:

$$H\Psi(\tilde{r}_1\tilde{r}_A;A) = -\frac{\hbar}{i}\frac{\partial}{\partial t}\Psi(\tilde{r}_1..\tilde{r}_A;A) \qquad \text{(I,1a)}$$

with

$$H = -\frac{\hbar^2}{2M}\sum_{i=1}^{A}\vec{\nabla}_i^2 + V(\tilde{r}_1,..\tilde{r}_A) \qquad \text{(I,1b)}$$

where $V(\tilde{r}_1..\tilde{r}_A)$ describes the mutual interaction between the A particles. $\tilde{r}_1..\tilde{r}_A$ are the space-, spin-, isospin- etc. coordinates of the system.

With modern computers eq. (1a) can be solved very often in good approximation for bound systems by applying the Ritz variational principle. But for reaction problems the form (I,1a) of the Schrödingerequation is not very suited because the physical boundary conditions which are demanded by the experimental situation cannot be described adequately (for more details see chapter 1 of ref. [2]). For this one formulates the Schrödinger equation in projection form [1], [2].

II PROJECTION FORM OF THE SCHRÖDINGER EQUATION

The Schrödingerequation written in projection form looks as follows:

$$\langle\delta\Psi|H + \frac{\hbar}{i}|\frac{\partial}{\partial t}\Psi\rangle = 0 \qquad \text{(II,1)}$$

If $\delta\Psi$ represents a completely arbitrary variation of Ψ at a given instant t in the complete Hilbert space, then eq. (II,1) implies that the vector

$$<H + \frac{\hbar}{i}\frac{\partial}{\partial t}||\Psi> = 0 \qquad (II,2)$$

must be orthogonal to any arbitrary vector in this space. This evidently will be the case only if Ψ obeys eq. (I,1a). Therefore, eq. (II,1) is merely another formulation of the Schrödinger equation (I,1a). However, as we shall see, eq. (II,1) allows us to treat the physically given boundary conditions in a very flexible manner.

If one writes Ψ as

$$\Psi = \psi \exp(-i E t/\hbar) \qquad (II,3)$$

and inserts it into eq. (II,1), one obtains the time-independent Schrödinger equation

$$<\delta\psi|H-E|\psi> = 0 \qquad (II,4)$$

It is wellknown every time dependent solution of eq. (II,1) can be written as a linear superposition of the stationary solutions $\psi \exp(-i E_n t/\hbar)$ of eq. (II,4). Therefore we can also use the simpler eq. (II,4) as starting equation for our further considerations.

We shall now briefly discuss some general properties of eq. (II,4) which we shall need later. For ψ we make the ansatz

$$\psi = \sum a_r \phi_r + \int a_p \phi_p \, dp = \mathop{S} a_k \phi_k \qquad (II,5)$$

where the coefficients a_r and a_p i.e. a_k are discrete and continuous linear variational amplitudes for the basis functions ϕ_r and ϕ_p respectively. By substituting eq.(II,5) into eq. (II,4) and using the fact that $\delta\psi$ is obtained by an arbitrary variation of the amplitudes a_k, i.e.

$$\delta\psi = \mathop{S}_k \delta a_k \phi_k \qquad (II,6)$$

one obtains the following set of coupled equations:

$$<\phi_n|H-E| \mathop{S}_k a_k \phi_k> = 0 \qquad (II,7)$$

where the subscript n takes on both discrete and continuous values. If the trial functions ϕ_k form a complete set, then the solutions of eqs. (II,7) are identical to the solutions of the time-independent Schrödinger equation in its usual form $(H-E)\psi = 0$. We should emphasize that the functions ϕ_k must be linearly independent, but need not to be orthogonal to each other. This increased flexibility of the basis functions ϕ_k compared to a strict orthogonal set of basis functions is of decisive importance for our later consideraions. Only by choosing in general a non-orthogonal set of basis functions ϕ_k can one expect to introduce the boundary conditions to the considered reaction problem in an appropriate way.

In contrast to the basis functions ϕ_k, the eigensolutions of eq. (II,7) are always mutually orthogonal, if all degeneracies are removed. This will be the case even if we restrict the number of variational amplitudes in eq. (II, 5) (for the simple proof see chapter 2 of ref. [2]). Therefore even when we restrict the number of linear variational amplitudes, i.e. we only work in a subspace of the Hilbert space, we still obtain the result that any two solutions ψ_l and ψ_m can be orthonormalized and in this subspace the Hamiltonian H can be represented by a real diagonal matrix, i.e.

$$\langle \psi_l | \psi_m \rangle = \delta(l,m) \tag{II,8}$$

$$\langle \psi_l | H | \psi_m \rangle = E_l \; \delta(l,m) \tag{II,9}$$

III GENERAL DISCUSSION OF THE BASIS FUNCTIONS

The considerations given so far have been quite general. They can be applied not only to nuclear physics, but also to atomic physics, solid-state physics, and other fields of quantum mechanics. The distinction between these different fields lies in the choice of the basis funcions in terms of which the wave function of the considered physical system is most conveniently expanded.

A proper choice of the basis functions depends strongly on the general characteristics of the interaction forces between the particles of which the system is composed. In atomic physics, for instance, the interacting atoms can be polarized over large distances due to the long-range nature of the Coulomb interaction. The basis function sets for atomic systems have therefore to be chosen such that the polarization effects can be included in the description of these systems by a relatively small number of basis functions. In nuclear physics, the situation is quite different. Here the long-range polarization effects on the nuclear participants by the Coulomb forces can be neglected in a good approximation, because the nucleons in a nucleus are tightly bound as a consequence of the great strength of the short-ranged nuclear forces.

To illuminate this general discussion, we consider now in some detail two examples chosen from nuclear physics. (For examples in atomic physics see ref.[3]).

IV TWO EXAMPLES

IV.1 α- ^{3}He- scattering and p-Li^6-scattering

As first example we consider the α- ^{3}He- scattering and the p-^{6}Li- scattering including reactions.

For the α- ^{3}He-scattering in the low energy region where no reaction channel is open, it is appropriate to use the following ansatz for the wave function of the system:

$$\psi = A\{\phi(\alpha,\vec{R}_\alpha)\phi(^3H_e,\vec{R}_{3He})\chi(\vec{R}_\alpha-\vec{R}_{3He})\ \xi(s,t)\}$$

$$= \int d\vec{R}'\ \chi(\vec{R}')\ A\{\ \phi(\alpha,\vec{R}_\alpha)\ \phi(^3He,\vec{R}_{3He})\ \delta(\vec{R}-\vec{R}')\ \xi\ (s,t)\} \tag{IV,1}$$

$\vec{R} = \vec{R} - \vec{R}_{3He}$ is the relative vector of the α-^{3}He - system affected by the antisymmetrizer A.

$\vec{R}'$ is a parameter coordinate, unaffected by A.

$A\{\phi(\alpha;\vec{R}_\alpha)\phi(^3He,\vec{R}_{3He})\delta(\vec{R}-\vec{R}')\xi(s,t)$ are the continuous basis functions.

The relative motion function $\chi(\vec{R}')$ is a continuous linear variational amplitude describing the behaviour of the two clusters in respect to their vector distance $\vec{R}'$. $\phi(\alpha,\vec{R}_\alpha)$ and $\phi(^3He,\vec{R}_{3He})$ describe the spatial behaviour of the α- and ^{3}He-clusters. $\xi(s,t)$ is a suitable spin- isospin functions. One sees $A\{\phi(\alpha,\vec{R}_\alpha)\phi(^3He,\vec{R}_{3He})\delta(\vec{R}-\vec{R}')\xi(s,t)$ corresponds to the continuous basis-functions in eq. (II,5) and the relative motion function $\chi(\vec{R}')$ corresponds to the continuous variational amplitudes in this equation. Therefore the variation $\delta\psi$ is given by

$$\delta\psi = \int d\vec{R}'\,\delta\chi(\vec{R}')\,A\{\phi(\alpha,\vec{R}_\alpha)\phi(^3He,\vec{R}_{3He})\delta(\vec{R}-\vec{R}')\,\xi(s,t)\} \qquad (IV,2)$$

If one inserts ψ and $\delta\psi$ of eqs. (IV,1) and (IV,2) into the projection eq. (II,4) and carries out all integrations and summations over the nucleon coordinates, then one obtains the following integro-differential equation for the determination of the relative motion function $\chi(\vec{R}')$:

$$[-\frac{h^2}{2M_{red}}\Delta_{\vec{R}'} + V(\vec{R}')]\chi(\vec{R}') + \int K(\vec{R}',\vec{R}'')\chi(\vec{R}'')d\vec{R}'' = E_{rel}\,\chi(\vec{R}')$$

$$K(\vec{R}',\vec{R}'') = K^*(\vec{R}'',\vec{R}') \qquad (IV,3)$$

where M_{red} denotes the reduced mass and E_{rel} the total kinetic energy of the α and ^{3}He - particles at large separation in the center-of-mass system. $V(\vec{R}')$ is the direct α-^{3}He- interaction - potential and, due to the hermiticity of the Hamilitonian, $K(\vec{R}',\vec{R}'')$ is a hermitian kernel function for the non-local interaction between the α and ^{3}He - particles, arising from the exchange character in the nucleon-nucleon potential and the anti-symmetrization procedure.

With this eq. one can now describe the elastic α - ^{3}He scattering by choosing as boundary condition for $\chi(\vec{R}')$ an incoming Coulomb wave belonging to the mutual Coulomb interaction of the α - and ^{3}He- particle. If one splits off $\chi(\vec{R}')$ in angularmomentum wave functions then the elastic scattering process is described by a phase factor at the outgoing Coulomb wave. In eq. (IV,3) no distortion effect (except the exchange distortion which comes from anti-symmetrization) during the mutual penetration of the two clusters is taken into account. We come back to this important point later.

The bound states of ^{7}Be having an α - ^{3}He- structure can also be described approximately by eq. (IV,3). For these states one obtains eigenvalue equations by demanding $\chi(\vec{R}')$ going to zero for $\vec{R}' \to \infty$ as boundary condition.

At higher scattering energies also the ^{6}Li-p-channel is open. In this scattering energy region one makes for the wave function ψ the ansatz:

$$\psi = A\{\phi(\alpha,\vec{R}_\alpha)\,\phi_{^3He}(^3He,\vec{R}_{^3He})\,\chi_I(\vec{R}_I)\,\xi_I(s,t)\}$$
$$+ A\{\phi(^6Li,\vec{R}_{^6Li})\,\chi_{II}(\vec{R}_{II})\,\xi_{II}(s,t)\} \qquad (IV,4)$$
$$\vec{R}_I = \vec{R}_\alpha - \vec{R}_{He};\ \vec{R}_{II} = \vec{r}_p - \vec{R}_{^6Li}$$

If one inserts ψ and $\delta\psi$ into the projection equation (II,4) then one obtains similar to before the following two coupled integro-differential equations:

$$[-\frac{\hbar^2}{2M^I_{red}}\Delta_{\vec{R}_I} + V_I(\vec{R}'_I)]\,\chi_I(\vec{R}'_I) + \int K_I(\vec{R}'_I,\vec{R}''_I)\,\chi_I(\vec{R}''_I)\,d\vec{R}''_I$$
$$+ \int K_{I,II}(\vec{R}'_I,\vec{R}''_{II})\,\chi_{II}(\vec{R}''_{II})\,d\vec{R}''_{II} = E^I_{rel}\,\chi_I(\vec{R}'_I) \qquad (IV,5a)$$
$$[-\frac{\hbar^2}{2M^{II}_{re}}\Delta_{\vec{R}''_{II}} + V_{II}(\vec{R}'_{II})]\,\chi_{II}(\vec{R}'_{II}) + \int K_{II}(\vec{R}'_{II},\vec{R}''_{II})\,\chi_{II}(\vec{R}''_{II})\,d\vec{R}''_{II}$$
$$+ \int K_{II,I}(\vec{R}'_{II},\vec{R}''_I)\,\chi_I(\vec{R}''_I)\,d\vec{R}''_I = E^{II}_{rel}\,\chi_{II}(\vec{R}'_{II}) \qquad (IV,5b)$$
$$K_{I,II}(\vec{R}'_I,\vec{R}''_{II}) = K^*_{II,I}(\vec{R}''_{II},\vec{R}'_I) \qquad (IV,6)$$

In the above equations, the quantities M^I_{red}, M^{II}_{red}, V_I, V_{II}, K_I and K_{II} have meanings analogous to those corresponding quantities appearing in eq. (IV,3). E^I_{rel} and E^{II}_{rel} are relative energies of the particles. The coupling between the two channels is represented by the kernels $K_{I\,II}$ and $K_{II\,I}$. Because of the hermiticity of the Hamiltonian these coupling kernels are adjoint to each other.

Eqs. (IV,5) can now be used to study the following bound-state and reaction problems simply by choosing appropriate boundary conditions. These problems are:
a) Elastic α-^{3}He- scattering together with transitions to the p-^{6}Li-channel. Here one has incoming and outgoing waves in the α-^{3}He-channel and only outgoing waves in the p-^{6}Li-channel. In the case where the p-^{6}Li-channel is closed its relative motion must tend asymptotically to zero.
b) p-^{6}Li elastic scattering together with transitions to the α-^{3}He-channel. In this case, there are incoming and outgoing waves in the p-^{6}Li-channel, but only outgoing waves in the α-^{3}He-channel.
c) Bound states having a mixture of α + ^{3}He and p + ^{6}Li cluster structure. The boundary conditions which must be fulfilled are that, in the asymptotic regions, both relative- motion functions must tend exponentially to zero.

To improve further the description of the α + ^{3}He- system, one could for example augment the wave function of eq. (IV,4) by a sum of bound-state-type ^{7}Be-wave functions with linear variational amplitudes. In this way, effects due to the distortion of the α - and ^{3}He-cluster during their mutual penetration can be taken into account. We shall discuss such distortion effects in the second example where the scattering of deuterons by α-particles is considered.

IV.2 d-α- scattering

Similar to the α- ^{3}He case, the simplest ansatz for the d + α wave function is

$$\psi_0 = A\{\phi(d)\ \phi(\alpha)\,\chi(\vec{R}_d - \vec{R}_\alpha)\} \qquad (IV,7)$$

where it should be noted that for simplicity in writing, we have absorbed the spin-isospin function into the deuteron and α-cluster internal wave functions. But it should be stated that this wave function where the internal functions $\phi(d)$ and $\phi(\alpha)$ are not varied in the calculation will not yield a satisfactory description of the behaviour of the d + α system. The reason for this is that due to the easily deformable deuteron cluster distortion effects during the mutual penetration of the d- and α-cluster should be quite important. Therefore we improve the wave function ψ_0 of eq. (IV,7) in the mutual penetration region by adding to it a number of bound-state-type wavefunctions with discrete linear variational amplitudes A_{li}. Thus we write

$$\psi = \psi_0 + \sum_{i=1}^{N} \sum_{l=0}^{\infty} A_{li}\ \varphi_{li} \qquad (IV,8)$$

where N is the number of distortion functions and l denotes the relative orbital angular-momentum quantum number between the two clusters. The distortion functions φ_{li} may be chosen in various ways, depending upon one's physical intuition. One common way is to choose these functions in the form of a d + α bound state cluster function, i.e.

$$\varphi_{li} = A\{\phi_i(d)\,\phi(\alpha)\,\frac{1}{R}\ g_{li}(R)\ P_l(\cos\theta) \qquad (IV,9)$$

$$-R = |\vec{R}_d - \vec{R}\,| -$$

with $\phi_i(d)$ describing the behaviour of distorted deuteron cluster, having a rms radius either larger or smaller than that of a free deuteron. It should be noted that the distortion functions are not varied at each energy in the calculation and the freedom in the wave function ψ is contained in the energy dependent linear variational function $\chi(\vec{R}_d - \vec{R}_\alpha)$ and the linear variational amplitudes A_{li}.

If one now substitutes eq. (IV,8) into eq. (II,4) and carries out the variation of the function $\chi(\vec{R}_d - \vec{R}_\alpha)$ and the amplitudes A_{li} then one obtains the following set of coupled integro- differential and integral equations:

$$< A\{\phi(d)\,\phi(\alpha)\,\delta(\vec{R} - \vec{R}')\}\ |H-E|\ \psi_0 > + \sum_{i=1}^{N} \sum_{l=0}^{\infty} A_{li} < A\{\phi(d)\,\phi(\alpha)\ \delta(\vec{R} - \vec{R}')\}\ |H-E|\ \varphi_{li} > = 0 \qquad (IV,10a)$$

$$<\varphi_{kj}|H-E|\,\psi_0> + \sum_{i=1}^{N} \sum_{l=0}^{\infty} A_{li} < \varphi_{kj}\ |H-E|\varphi_{li}> = 0 \qquad (IV,10b)$$

where the subscripts k and j in eq. (IV,10b) take on integral values from 0 to ∞ and from 1 to N respectively. These equations can be solved by using standard techniques to obtain the function $\chi(\vec{R}_d - \vec{R}_\alpha)$ and the amplitueds A_{li}, and consequently the phase shifts in the various partial waves.

V. EXTENSION TO GENERAL SYSTEMS

With the two examples described in section IV, we have demonstrated how one can use physical intuition and energetical considerations to obtain the most suitable trial function necessary for a satisfactory description of the system. Indeed, the prescription is rather simple; at a given energy, one constructs a trial function consisting of all open-channel functions which may be excited and a set of distortion functions which is specifically designed to improve the description of the system in the region of strong interaction i.e. in the region of strong mutual penetration of the clusters. As the energy becomes high, the number of required channel and distortion functions may be quite large and the calculations become very complicated, but the procedure is always straightforward and readily understood.

From this discussion and the examples in section IV, it is clear that for a proper description of nuclear bound states and reactions, one should introduce into the calculation enough channel wave functions with their relative-motion parts treated as linear variational functions and enough bound-state-type distortion functions with discrete linear variational amplitudes. As it is quite evident, the most general ansatz for a trial function comprising linear variational functions and amplitudes is as follows:

$$\begin{aligned}\psi = & \sum_i \{ A \ \phi(A_i)\phi(B_i)\ \chi^i(\vec{R}_i) \} \\ & + \sum_j A\ \{\phi(A_j)\phi(B_j)\phi(C_j)\ \chi^j(\vec{R}_j^1, \vec{R}_j^2)\} \\ & + \sum_k A\ \{\phi(A_k)\phi(B_k)\phi(C_k)\phi(D_k)\ \chi^k(\vec{R}_k^1, \vec{R}_k^2, \vec{R}_k^3) \\ & + \ \ldots\ldots\ldots\ldots\ldots \sum_l a_l F_l \end{aligned} \qquad \text{(V,1)}$$

The $\vec{R}$'s are Jacobi-coordinates describing the different relative distances of the clusters*). The anti-symmetric functions ϕ describe the internal structures of the different clusters. The terms with relative- motion functions $\chi^j(\vec{R}_j^1, \vec{R}_j^2)$, $\chi^k(\vec{R}_k^1, \vec{R}_k^2, \vec{R}_k^3)$ and so on are responsible for three-, four-, and more-particle channels**). The anti-symmetric distortion functions F_l are chosen to improve the wave function in the strong-interaction region; they vanish for large inter-nucleon and inter-cluster distances.

By substituting eq. (V,1) into the projection equation (II,4) and carrying out again the variation of the functions χ^i and the amplitudes a_l we obtain the following set of coupled equations:

*) For their exact definition see ref. [2] subsection 3.7a.

**) For a more detailed description of these channels see ref. [4].

$$\sum_i \langle \phi(A_m)\phi(B_m)\ \delta(\vec{R}_m-\vec{R}_m{}')\ |H-E|\ A\ \{\phi(A_i)\,\phi\,(B_i)\ \ \chi^i\ (\vec{R}_i)\}\rangle$$
$$+\sum_l\ a_l\ \langle \phi(A_m)\,\phi\,(B_m)\,\delta\,(\vec{R}_m-\vec{R}'_m)\ |H-E|\ F_l\rangle = 0 \qquad (V,2a)$$

$$\sum_i \langle F_r\ |H-E|\ A\ \{\phi(A_i)\,\phi\,(B_i)\ \ \chi^i\ (\vec{R}_i)\}\rangle$$
$$+\sum_l\ a_l\ \langle F_r|H-E|\ F_l\rangle = 0 \qquad (V,2b)$$

where the indices m and r take on values 1,2,..., and $\vec{R}'_m$ are parameter coordinates. For simplicity we have omitted in eqs. (V,2) the three and more channel functions. As it is well-known, it is allowed to disregard the anti-symmetrization procedure on one side of eqs. (V,2). We have done it on the left side of these equations.

Due to the rotational invariance of the Hamiltonian H, the wave function ψ can be expanded into partial waves of given total angular momentum. By so doing, eqs. (V,2) can be further divided into sets fo coupled integro-differential and integral equations, from which one can determine by numerical techniques the reaction cross-sections, the bound-state eigen-energies etc..

At the end of this section we want to emphasize three points:

First: As we have seen in section IV and V by using the Schrödinger-equation in the projection (II,1) resp. (II,4) and using flexible basis functions of the general form described in these two sections one can always introduce the boundary conditions for a given physical problem in a very straightforward manner. One reason for this is that the incoming and outgoing channels, in contrast to other reaction theories, are contained in this theory in a completely symmetrical form.

Second: One works always in the most suitable restricted subspace of the Hilbert space without approximating, at least in principle, the solutions for a given physical problem. This is due to the fact that in the asymptotic region one has only to take into account the open-channel functions. This simplifies the calculations very much and makes the expressions transparent. Certainly if one goes into an energy region where the considered nuclear system can disintegrate completely one has, at least principally, to take into account all channels.

Third: The general formulation of the unified nuclear theory sketched here does not depend on the specific form of the nucleon-nucleon potential. That means this potential can be composed of two-particle, three-particle etc. interaction potentials.

Furthermore, as already mentioned, projection equations of the general form (II,1) resp. (II,4) should also be applicable to many particle problems in other fields of physics. We come back to this later.

VI INFLUENCE OF THE PAULI-PRINCIPLE

Before we discuss briefly some numerical examples, we want to consider in some detail the influence of the indistinquishability of identical particles on many particle systems. For spin $\frac{1}{2}$ particle systems this is the influence of the Pauli-principle and is quantitatively described by the anti-symmetrization operator A as indicated in all our equations. Because of the tight packing of the nucleons in the nuclei this influence is of fundamental importance for all nuclear reaction and bound state problems.

As it is well-known one main influence of the Pauli-principle due to its exclusion character is that many structures are forbidden which would be allowed if the particles would be distinquishable. For shortness we do not discuss this here very explicitely. Another effect which in its detailed form certainly is connected intimately with the exclusion character of the Pauli-princle is that different structures can become very similar or even equal to each other after anti-symmetrization. For instance the apparent contradiction between collective and single particle models which play a very important role in nuclear physics becomes resolved by the indistinguishability of the nucleons i.e. by the understanding of the anti-symmetrization procedure. Because of its importance for nuclear physics we shall discuss this effect on some examples.

As first example let us consider a large number of fermions without mutual interaction in their ground state in a square-well potential; this would approximately describe the conducting electrons in a conductor, for example. The energy differences among the particles in the well arise from their kinetic energies. In the ground state, all single-particle states are filled inside a momentum sphere known as the Fermi-sphere (see fig. 1a). If this system as a whole is now given a small velocity $\Delta \vec{v}$, then the Fermi-sphere will be slightly shifted so that its center is no longer at the origin (see fig. 1b). The change relative to the situation shown in fig. 1a is a collective excitation in which each fermion receives a small change $m \Delta \vec{v}$ in momentum. Now let us instead start with the Fermi-sphere at the origin (as in fig. 1a) and impart various large amounts of momenta to a few of the fermions (all those in states in shaded region 1 of fig. 1c) at the left of the sphere so as to excite them into states just to the right of the sphere (filling the states in shaded region 2 of the fig. 1c). Due to the indistinguishability of the fermions, expressed through the anti-symmetrization of the wave function, the situation in fig. 1c is completely equivalent to that in fig. 1b. This shows that under anti-symmetrization a large excitation imparted to a few fermions can be equi-

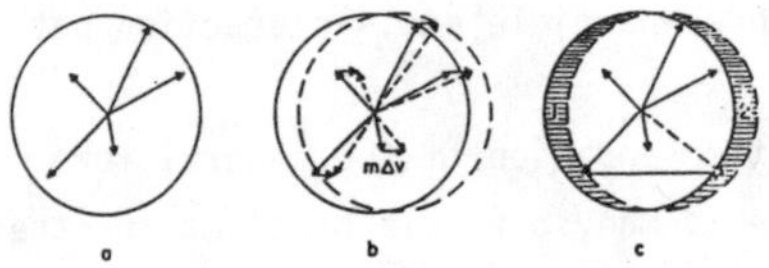

Fig. 1

valent to a collective excitation of all fermions as a whole.

The example just described qualitatively can also be formulated quantitatively. For this see subsection 3.5a of ref. [2].

As second example we consider a specific form of the ^{6}Li-wave function:

$$\psi_{^6\mathrm{Li}} = A\,\{N_{\alpha d}\ \varphi(\alpha)\,\varphi(d)\quad \chi_I\ (\vec{R}_{\alpha d})\} \qquad \text{(VI,1)}$$

where $N_{\alpha d}$ normalizes the wave function (VI,1) to one. In the case where all internal and relative motions of the clusters are oscillator motions with the same frequency of the lowest order allowed by the Pauli-principle, then one has the following equivalence:

$$\begin{aligned} \psi_{^6\mathrm{Li}} &= A\ \{N_{\alpha d}\ \varphi(\alpha)\,\varphi(d)\ \chi_I\ (\vec{R}_{\alpha d})\} \\ &= A\ \{N_{t^3\mathrm{He}}\ \varphi(A)\,\varphi(^3\mathrm{He})\ \chi_{II}\ \ (\vec{R}_{t^3\mathrm{He}})\} \end{aligned} \qquad \text{(VI,2)}$$

Most easily one can prove this equivalence by expanding the two forms of the wave function (VI,2) into oscillator shell model wave functions. The proof is carried out most simply for the lowest ^{6}Li-state with I = 4. See for this ref. [2] subsection 3.5b.

Certainly this and similar equivalences in other nuclei are in reality never completely correct but exist very often in good approximation. Furthermore such (approximate) equivalences can only exist if one has a strong mutual penetration of the clusters forming the nucleus. The equivalences just described indicate very directly that the anti-symmetrization often reduces the differences between different nuclear structures very strongly. This tells us further that clusters have only approximately the internal structure of the corresponding free particles if they are not deeply imbedded into nuclear matter which means that strong cluster correlations exist only at the nuclear surface. Therefore the interieur of the nuclei can be described approximately always by simple single particle shell model correlations. This is demonstrated in fig. 2.

Fig. 2

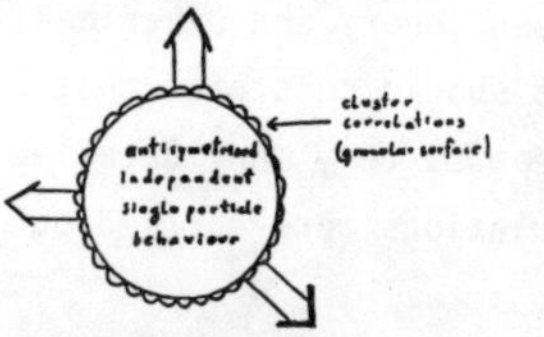

The reduction effect of the Pauli principle just described, as already mentioned before, is responsible for the fact that the different single particle and collective models applied in nuclear physics do not contradict each other. For more details see ref. [2].

Another consequence of this reduction effect is that for a proper description of nuclear reactions and bound states the number of different terms in the belonging trial wave functions can be chosen usually relatively small.

VII NUMERICAL EXAMPLES

In the following figures some examples are shown where experimentally measured data are compared with microscopic calculations.

In fig. 3 the differential reaction cross-sections of the $^4He(d,t)^3He$-reaction in the center of mass system are shown. The full drawn lines give the calculated values and the circles the experimental values. Even if in fig. 3 relatively old values are shown the agreement between theory and experiment is quite good.

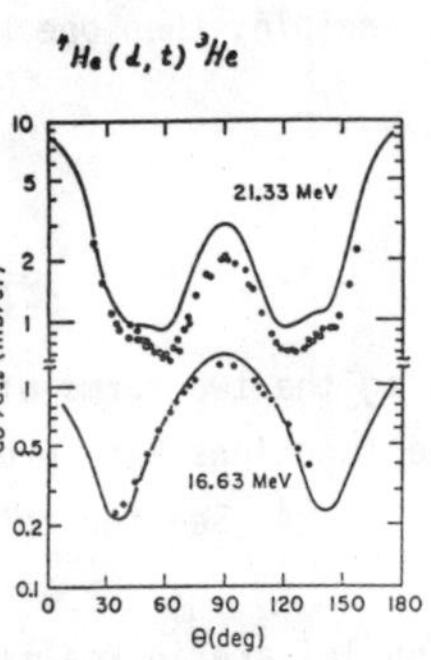

Fig. 3

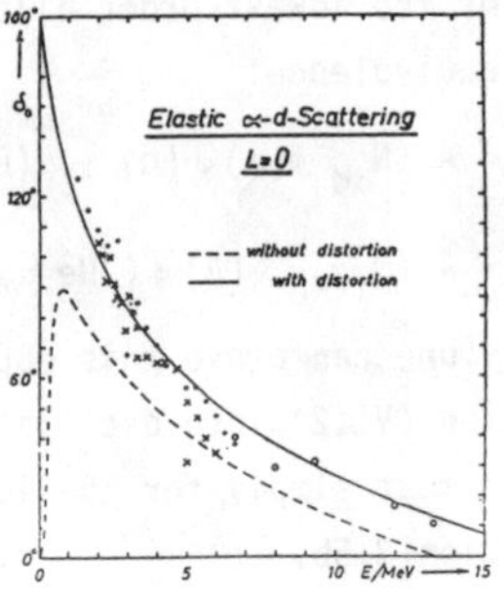

Fig. 4

In fig. 4 the l=0 phaseshift of the elastic α-d scattering as function of the scattering energy is shown [5]. The full drawn and the dashed lines give the calculated values. The circles and the crosses give the experimental data. One sees the agreement between theory and experiment becomes remarkably good if in the microscopic calculation one takes into account the distortion of the deuteron cluster during the mutual α-d-penetration.

In fig. 5, as last example, the differential cross-section of the $^6Li(p,^3He)^4He$-reaction for different scattering energies is shown [6]. Again the agree- between theory and experiment is quite good.

It should be stated that especially by Japanese colleagues very many detailed microscopic reaction calculations were carried out [7].

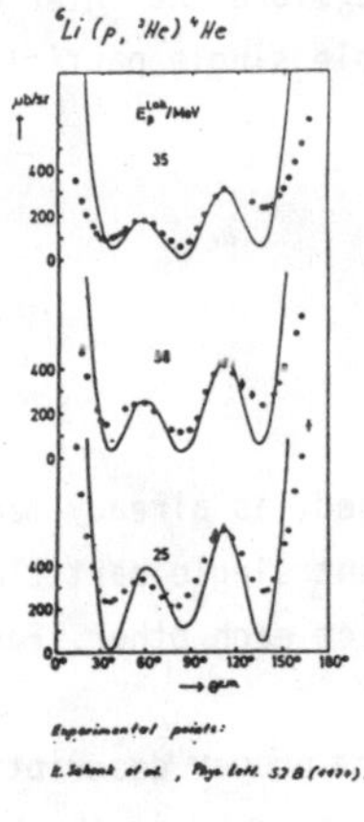

Fig. 5

VIII GENERAL PROPERTIES OF NUCLEAR SYSTEMS

The unified theory presented so far can also be used for discussions of more general character, as for instance, the derivation of Breit-Wigner formulae, direct reaction theory etc.. We shall indicate this very briefly and refer to a more detailed discussion again to ref. [2].

The guiding principle for all these derivations is that one separates in a suitable way the total trial function ψ into different components ψ_k and its variation $\delta\psi$ into corresponding $\delta\psi_k$; that is one writes

$$\psi = \sum_{k=1}^{N} \psi_k \tag{VIII,1}$$

and

$$\delta\psi = \sum_{k=1}^{N} \delta\psi_k \tag{VIII,2}$$

Because of arbitrary variations of the linear variational funcions and discrete variational amplitudes contained in ψ_k, every variation $\delta\psi_k$ defines a certain subspace of the total Hilbert space of the nuclear system. The great advantage of dividing the total Hilbert space into various subspaces is that this division can be chosen very flexibly to provide a most convenient basis and resulting coupled linear equations for discussing a certain particular problem in a nuclear system.

Withour carrying out the explicit derivations we show here two examples [2], [8]

a) Single-Level Breit-Wigner Resonance Formula for Elastic Scattering:

$$A_j^{out} = e^{\lambda i\delta_j} \frac{(E_r-E) + \frac{i}{2}\Gamma_j}{(E_r-E) - \frac{i}{2}\Gamma_j} \tag{VIII,3}$$

δ_j = Potential Scattering Phase-Shift, E_r = Resonance Energy

$$\Gamma_j = 2\pi \, |<\psi_c|H-E|\psi_{Dj}>|^2 \quad \text{with} \quad \psi_c = \text{Resonance State Wave Function} \tag{VIII,4}$$

and ψ_{Dj} = Potential Scattering Wave Function

A_j^{out} = Asymptotic Amplitude of the outgoing wave with total spin j

b) Formula for the Lifetime τ_j of α-decaying states:

$$\frac{1}{\tau_j} = 2|a_\alpha|^2 \frac{\hbar\sqrt{E}}{\sqrt{2M_{red}}} \tag{VIII,5}$$

E denotes the decay-energy and M_{red} the reduced mass of the system. a_α is the amplitude of the asymptotic part of the bound state wave function of the decaying system and it is obtained from a bound-state variational calculation. Eq.(VIII,5) is also valid for other decay-systems if the decaying state is long-lived and surrounded by a Coulomb-barrier.

It should be pointed out that the physical background of the unified theory discussed in this lecture and here especially the flexibility of its basic wave functions together with the Pauli principle allows it very often to make qualitative and half-quantitative predictions for mainly larger nuclear systems where complete microscopic calculations are no more possible. One example for this is the description of the asymmetric nuclear fission process. For more details we refer again to ref.[2].

IX LIST OF PROBLEMS IN NUCLEAR PHYSICS WHICH CAN BE TREATED WITHIN THIS UNIFIED MANY-PARTICLE THEORY

1) Calculations of BOUND STATES
2) REACTION calculations
3) Derivation of the BREIT-WIGNER RESONANCE FORMULAE for arbitrary projectiles
4) Derivation of the DIRECT REACTION formulae
5) Derivation of the OPTICAL MODEL for arbitrary scattering particles
6) Description of HEAVY ION REACTIONS
7) Description of the α-PARTICLE DECAY
8) Description of the MANY PARTICLE DECAY
9) Description of COLLECTIVE (e.g. rotational) STATES
10) Treatment of ANALOGUE RESONANCES in arbitrary reactions
11) Description of FISSION processes

From this list one sees again that the unified theory considered here is so flexible and therefore adjustable for the different physical boundary conditions that, at least in principle, the large manifold of non-relativistic low energy nuclear phenomena can be treated within this theory.

X APPLICATION TO OTHER MANY-PARTICLE SYSTEMS

The general method to formulate a unified many-particle theory using the projection equations (II,1) and (II,4) can also be applied to many particle systems in other fields of physics. One example for this is the half-phenomenological elementary particle theory. There one considers, as it is well-known, the nucleus composed of quarks which are spin $\frac{1}{2}$ particles. The proton for instance consists of two up quarks and one down quark and the neutron of two down quarks and one up quark, where the charge of the up- resp. down- quark is $\frac{2}{3}$e resp $-\frac{1}{3}$e. The quark-interaction forces as the nuclear forces are very strong and short. Due to these similarities to nuclear physics the nucleons can be considered as quark cluster [9], [10] and especially for nucleon-nucleon scattering using the projection form (II,1) resp (II,4) of the Schrödinger equation they can be treated very similarly to nucleon-clusters as sketched above [11], [12].

Another field of physics where especially the time dependent projection form of the Schrödinger equation (II,4) is very useful is in quantum elctro-dynamics the treatment of systems consisting of many photons interacting with charged particles. One example for this is the derivation of the half-classical time-dependent Schröfinger equation, describing charged particles interacting with a time dependent classical electric field, from quantum electro-dynamics where also the electromagnetic field is treated as a quantum system [3].

For a detailed discussion of the unified many particle theory and the connected problems which could be only sketched in this lecture see especially ref. [2].

REFERENCES

[1] Wildermuth, K., in "The Structure of Nuclei", Trieste Lectures, (IAEA, Vienna 1972) p. 117.

[2] K. Wildermuth and Y.C. Tang, A Unified Theory of the Nucleus (Vieweg, Braunschweig, 1977).

[3] T. Toyoda and K. Wildermuth, Phys.Rev.D 22, (1980) 2391
T. Toyoda and K. Wildermuth, Phys.Rev.A 27, (1983) 1790

[4] Schmid, E.W. and Spitz, G. to appear in Z.Phys.A and references given therein.

[5] Jacobs, H., K. Wildermuth and E. Wurster Phys.Lett.29B (1969) 455
Thompson, D.R., Y.C. Tang, Phys.Rev.C 8 (1973) 1649

[6] K. Schenk, M. Mörike, G. Staudt, P. Turek and D. Clement, Phys.Lett. 52B (1974) 36

[7] Suppl. of the Progress of Theor. Phys. 68 1980
"Comprehensive Study of Structure of Light Nuclei"

[8] T. Steinmayer, W. Sünkel and K. Wildermuth, Phys.Lett. 125B (1983) 437

[9] I.T. Obukhovsky, V.G. Neudatchin, Yu.F. Smirnov and Yu.M. Tchuvel'sky, Phys.Lett. 88B (1979) 231 and ref. given therein.

[10] M. Harvey, Nucl. Phys. A352 (1980) 301 and 326

[11] H. Faissner and B.R. Kim, Phys.Lett. 130B (1983) 321

[12] K. Bräuer, A. Faessler and K. Wildermuth, Nucl.Phys. A437(1985)717
A. Faessler and F. Fernandez Phys. Lett. 124B (1983) 145

INTRINSIC STRUCTURE OF THE CRANKING SHELL MODEL WAVE FUNCTIONS AND NUCLEAR PAIRING PHASE TRANSITION

C. S. Wu and J. Y. Zeng

Department of Physics, Peking University
Beijing, China

Abstract: The intrinsic structure of the eigenfunctions of Cranking Shell Model hamiltonian is investigated with the particle-number-conserving (PNC) approach. Pair-transfer matrix elements and the K-structure of the CSM wave functions display similar behaviour under the Coriolis interaction. No sharp phase transition is found. The pairing parameter, $\tilde{\Delta} = G\sqrt{\langle S^+S\rangle}$, of the yrast band decreases slowly with ω. The blocking effects on $\tilde{\Delta}$ are much more important than the Coriolis anti-pairing effect.

1. Introduction

Recently it was found experimentally[1]) that in a number of nuclei in Hf-region at $\hbar\omega \geq 0.40$ MeV the yrast and the negative-parity bands are characteristic of a macroscopic rotor, i.e., their moments of inertia are almost constant and close to the rigid-body value. It has been suggested that the neutron pairing correlation has effectively disappeared at these large rotational frequencies. This work has been attacting the attention of many nuclear theorists and further investigation of the Mottelson-Valatin effect[2]), i. e., nuclear pairing collapse at high angular momenta, seems to be important again[3,5]). It was pointed out that the moment of inertia deduced from the energy spectra is of limited value to characterize a phase transition from superfluid state to normal state and the surprisingly constant moments of inertia near rigid-body values are by no means any indication of a pairing phase collapse[3]). Indeed, in all self-consistent solutions of the cranked HFB equation one has found a pairing collapse[6]): Usually for rare-earth nuclei neutron pairing vanishes between I ~ 20ħ - 30ħ and proton pairing vanishes between 40ħ - 50ħ. However, it is pointed out that the results obtained without particle-number projection arte not reliable[3,4]). Calculations with particle-number projection before variation show that the pairing parameter decreases very slowly. Even at the highest spins, I ~ 80ħ, it is still larger than 300 keV. Thero has been a much discussion about how to define the pairing parameter in theories beyond the BCS or HFB approach. As in refs. [3,7]), we adopt the definition

$$\tilde{\Delta} = G\sqrt{\langle S^+S\rangle} \tag{1}$$

which measures the pairing correlation energy, $\langle H_{pair}\rangle = -\tilde{\Delta}^2/G$, or $\tilde{\Delta} = \sqrt{|G\langle H_{pair}\rangle|}$. In the BCS or HFB theories this definition coincides to leading order with the pairing deformation which is usually called the gap parameter

$$\Delta = G\langle S\rangle = G\sum_{\nu} U_\nu V_\nu . \tag{2}$$

However, in the particle-number-conserving (PNC) treatment definition (2) becomes meaningless. Of course, the most direct evidence for a pairing phase transition should be given by the pair-transfer matrix elements[3]). But unfortunately, it is very difficult to measure them at high spins.

Usually, the particle-number-nonconservation is considered as the main defect of the BCS and HFB approaches and numerous works were made to project a component with correct number of particles from the BCS or HFB wavefunctions [see, for example, the reviews in ref.[7]) and references therein]. However, the most serious defects of BCS method for treating nuclear pairing correlation are the difficulty to treat the blocking effect[8,9]) and the occurence of excessive spurious states in the low-lying spectra[9,10]).As pointed out by Rowe[8]), while the blocking effects are straightforward, it is very hard to treat them in the BCS formalism because they introduce different quasiparticle bases for different blocked levels. In fact, apart from the Coriolis anti-pairing effect[2]), the blocking effect should be regarded as another kind of anti-pairing effect[10]), which is especially important for low-lying spectra. In addition, because of the blocking effect, the gap parameter, Δ, which is usually assumed to be a constant for convenience, is sensitively configuration-dependent. In view of occurence of excessive spurious ststes, all the related consequences based on quasi-particle should be reexamined carefully.

In this paper the PNC approach for treating the nuclear pairing correlation[9]) is extended to treat the cranking hamiltonian for well-deformed nuclei. The preliminary results have been given in ref. [11]). Instead of the usual truncation of single-particle level,a truncation of many-body configuration energy is adopted in this approach. Its advantages have been demonstrated in the appendix of ref. [12]). Because the number of valence particles, which dominate the behaviours of low-lying states,is not very large (~10) and the average pairing strength G is not very strong ($G \le d/2$, d being the average single-particle level spacing), very accurate PNC wave functions for the low-lying eigenstates of the CSM hamiltonian can be obtained easily. Once the PNC wave functions are obtained, various nuclear properties, e.g., the K-structure, the pairing correlation energy, the pair-transfer matrix elements, etc., can be investigated in detail and hence some further valuable information about the pairing phase transition can be obtained.

2. Formalism

As usual, the CSM hamiltonian for a well-deformed nucleus is expressed as

$$H_{CSM} = H_{intr} - \omega J_x = H_{sp} + H_{pair} - \omega J_x \,, \tag{3}$$

where $-\omega J_x$ is the Coriolis interaction, H_{sp} is the single-particle (e.g., Nilsson) hamiltonian and H_{pair} the pairing hamiltonian

$$H_{pair} = - G S^+ S = - G \sum_{\mu\nu} b^+_\mu b^+_{\bar\mu} b_{\bar\nu} b_\nu \,, \tag{4}$$

μ, ν denote the single-particle states and $\bar\mu$, $\bar\nu$ their time-reversal states, respectively. For non-rotating nuclei ($\omega = 0$), H_{sp} is usually assumed to be axially symmetric and the component of angular momentum along the symmetry axis (z-axis), $K = \sum_i \Omega_i$, parity π and seniority number v are good quantum numbers. For rotating nuclei, K and v no longer remain constant of motion. However, usually it is assumed that H_{sp} is invariant with respect to space reflection and $R_x(\pi)$, rotation of 180° about the x-axis. Thus the diagonalization of H_{CSM} can be performed separately in each subspace with a fixed signature r [eigenvalue of $R_x(\pi)$] and parity π. In this case it is convenient to label the single-particle state by π and r ($= \pm i$), instead of Ω, the component of angular momentum along z-axis. For details see the Appendix below.

For a pure pair-excitational band (v = 0, $K\pi$ = 0+, r = +1) of a 2n-particle system

$$\mathcal{A}^+_{n\beta} |0\rangle = \sum_{\rho_1 \cdots \rho_n} V^\beta_{\rho_1 \cdots \rho_n} | \rho_1 \bar\rho_1 \cdots \rho_n \bar\rho_n \rangle, \tag{5}$$

where

$$| \rho_1 \bar\rho_1 \cdots \rho_n \bar\rho_n \rangle = b^+_{\rho_1} b^+_{\bar\rho_1} \cdots b^+_{\rho_n} b^+_{\bar\rho_n} |0\rangle,$$

β = 0 and β = 1, 2, $\cdots$ stand for the yrast band and the first, the second, $\cdots$ pair-excitational bands, respectively. We have

$$\begin{aligned}\langle S^+ S \rangle &= \sum_{\mu\nu} \langle S^+_\mu S_\nu \rangle = n + \sum_{\mu \neq \nu} \langle S^+_\mu S_\nu \rangle \\ &= n + \sum_{\rho_1 \cdots \rho_i \cdots \rho_n} \sum_{\rho_i' \neq \rho_i} V^\beta_{\rho_1 \cdots \rho_i \cdots \rho_n} V_{\rho_1 \cdots \rho_i' \cdots \rho_n} \,. \end{aligned} \tag{6}$$

The first term of the last line is the number of particle-pairs and the second term is the contribution from the off-diagonal matrix elements of pairing interaction (in unit of -G). Similarly, for the pair-broken excited band ($v = 2$, $\pi = +$, $r = +1$)

$$\mathcal{A}^+_{n-1,\beta}(\mu_0\bar{\nu}_0)|0\rangle = \sum_{\rho_1\cdots\rho_{n-1}} V^{\beta(\mu_0\bar{\nu}_0)}_{\rho_1\cdots\rho_{n-1}} |\mu_0\bar{\nu}_0\rho_1\bar{\rho}_1\cdots\rho_{n-1}\bar{\rho}_{n-1}\rangle \tag{7}$$

where ($\mu_0\bar{\nu}_0$) denotes the single-particle states occupied by the two unpaired particles and $\pi_{\mu_0} = \pi_{\nu_0}$, we have

$$\langle S^+S\rangle = n-1 + \sum_{\rho_1\cdots\rho_i\cdots\rho_{n-1}} \sum_{\rho_i'\neq\rho_i} V^{\beta(\mu_0\bar{\nu}_0)}_{\rho_1\cdots\rho_i\cdots\rho_{n-1}} V^{\beta(\mu_0\bar{\nu}_0)}_{\rho_1\cdots\rho_i'\cdots\rho_{n-1}}. \tag{8}$$

For rotating nuclei ($\omega \neq 0$) the low-lying eigenfunction ($\pi = +1$, $r = +$) of H_{CSM} for even-even nucleus can be expressed as

$$|n\beta++\rangle = \sum_{\rho_1\cdots\rho_n} V^{\beta}_{\rho_1\cdots\rho_n}|\rho_1\bar{\rho}_1\cdots\rho_n\bar{\rho}_n\rangle + \sum_{\mu_0\nu_0}\sum_{\rho_1\cdots\rho_{n-1}} V^{\beta(\mu_0\bar{\nu}_0)}_{\rho_1\cdots\rho_{n-1}} |\mu_0\bar{\nu}_0\rho_1\bar{\rho}_1\cdots\rho_{n-1}\bar{\rho}_{n-1}\rangle, \tag{9}$$

$$\sum_{\rho_1\cdots\rho_n} |V^{\beta}_{\rho_1\cdots\rho_n}|^2 + \sum_{\mu_0\nu_0}\sum_{\rho_1\cdots\rho_{n-1}} |V^{\beta(\mu_0\bar{\nu}_0)}_{\rho_1\cdots\rho_{n-1}}|^2 = 1. \tag{10}$$

Here only the configurations with seniority number 0 and 2 are included because the calculation shows that the contribution from configurations with more unpaired particles ($v \geq 4$) are negligible if ω is not too large.

The component of fully-paired $K\pi = 0+$ configurations in the band $|n\beta++\rangle$ is given by

$$P_\beta = \sum_{\rho_1\cdots\rho_n} |V^{\beta}_{\rho_1\cdots\rho_n}|^2. \tag{11}$$

For the band $|n\beta++\rangle$ it can be shown that

$$\langle S^+S\rangle = n-(1-P_\beta) + \sum_{\rho_1\cdots\rho_i\cdots\rho_n}\sum_{\rho_i'\neq\rho_i} V^{\beta}_{\rho_1\cdots\rho_i\cdots\rho_n} V^{\beta}_{\rho_1\cdots\rho_i'\cdots\rho_n}$$
$$+ \sum_{\mu_0\nu_0}\sum_{\rho_1\cdots\rho_i\cdots\rho_{n-1}}\sum_{\rho_i'\neq\rho_i} V^{\beta(\mu_0\bar{\nu}_0)}_{\rho_1\cdots\rho_i\cdots\rho_{n-1}} V^{\beta(\mu_0\bar{\nu}_0)}_{\rho_1\cdots\rho_i'\cdots\rho_{n-1}}. \tag{12}$$

The pair-transfer matrix element between neighbouring even-even nuclei is given by

$$\langle n+1\,\beta'++|S^+|n\beta++\rangle = \sum_{\nu\rho_1\cdots\rho_n} V^{\beta'}_{\nu\rho_1\cdots\rho_n} V^{\beta}_{\rho_1\cdots\rho_n} + \sum_{\mu_0\nu_0}\sum_{\nu\rho_1\cdots\rho_{n-1}} V^{\beta'(\mu_0\bar{\nu}_0)}_{\nu\rho_1\cdots\rho_{n-1}} V^{\beta(\mu_0\bar{\nu}_0)}_{\rho_1\cdots\rho_{n-1}}. \tag{13}$$

The overlap of the yrast band $|n\bar{0}++\rangle$ on that band of non-rotating nucleus $|n0++(\omega=0)\rangle \equiv \mathcal{A}^+_{n0}|0\rangle$ is

$$\langle n0++(\omega=0)|n0++(\omega)\rangle = \sum_{\rho_1\cdots\rho_n} V^0_{\rho_1\cdots\rho_n}(\omega)\, V^0_{\rho_1\cdots\rho_n}(\omega=0). \quad (14)$$

Usually, $\mathcal{A}^+_{n0}|0\rangle$ is considered as the quasiparticle vacuum. Therefore, $|\langle n\ 0\ +\ +\ (\omega\ =\ 0)|n\ 0\ +\ +\ (\omega)\rangle|^2$ measures the probability for the intrinsic state of a rotating nucleus to be in the quasiparticle vacuum. It can be shown that

$$|\langle n0++(\omega=0)|n0++(\omega)\rangle| > P_0 > |\langle n0++(\omega=0)|n0++(\omega)\rangle|^2. \quad (15)$$

3. Calculated results and discussions

As in refs.[13,14]), calculations in a single-j model ($j\pi$ = 13/2+) are carried out to illustrate the features of low-lying eigenstates of

$$\mathcal{E}_\Omega = \kappa\,\frac{3\Omega^2 - j(j+1)}{j(j+1)} + e_0\,. \quad (16)$$

The constant e_0 is of no importance for the solution and may be chosen as zero. By choosing κ = $0.392\hbar\omega_0$ (and e_0 = $6.655\hbar\omega_0$) we can reproduce the Nilsson neutron levels of positive parity at the deformation $\mathcal{E}_2$ = 0.27 and $\mathcal{E}_4$ = 0.02 (corresponding roughly to the ground state deformation of ^{168}Er) within 0.1%. This gives us a certain reality to the model. The neglect of negative parity orbits near the Fermi surface can be partly compensated by increasing the pairing strength G. To diagonalize H_{CSM} , all the configurations below E_c ($E < E_c$, E_c/κ = 2.5) are taken into account. Several values of G (G/κ = 0.1, 0.15, 0.2, 0.25, and 0.3) have been used in the calculation. Only the results of 6- and 8-particle systems for G/κ = 0.2 are presented in the following. All the features remain unchanged quanlitatively for other values of G.

First, let us consider the pair-transfer matrix elements. As the Coriolis interaction is turned on ($\omega \neq 0$),the pair-transfer matrix elements between the yrast bands of neighbouring even-even nuclei, $\langle n+1\ 0\ +\ +|S\ |n\ 0\ +\ +\rangle$, gradually decrease as expected. The results for pair-transfer between the yrast bands of 6- and 8-particle systems are shown in fig. 1. The ratio R_0 is defined as

$$R_0 = \frac{|\langle n+1\,0++|S^+|n0++\rangle|^2}{|\langle n+1\,0++|S^+|n0++\rangle|^2(\omega=0)}. \quad (17)$$

No sharp decrease of R_0 is found.

It is interesting to note that the K-structure of the yrast band of H_{CSM} (namely, P_0, the component of $K\pi$ = 0+ fully-paired configurations in the yrast band) displays similar behaviour to R_0. Because of the Coriolis anti-pairing effect, the pair-broken configurations with K ≠ 0 are gradually mixed into the yrast band. Thus, along with P_0, R_0 decreases with ω. As P_0 reduces to 0.5, i.e., one half of the component becomes the pair-broken configurations, the overlap of the CSM wavefunction on that at ω = 0 ("quasi-particle vacuum") becomes less than 0.5, thus the intrinsic structure of the yrast band undergoes a great change. In this case it seems that the nuclear superfluidity disappears. Therefore, the K-structure of the yrast band may be reasonably used to characterize the superfluidity of well-deformed nuclei.

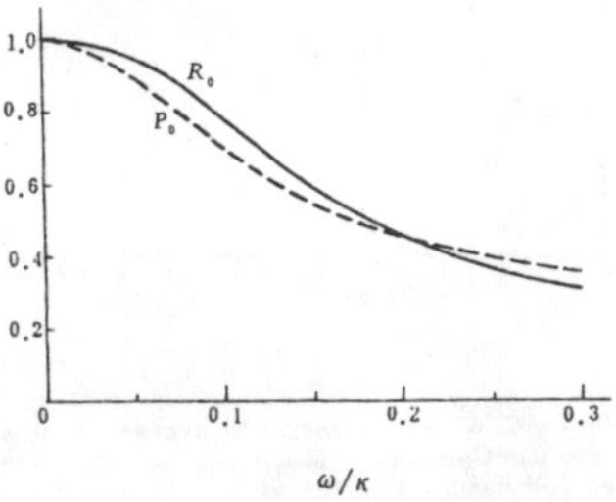

Fig. 1. The variations of R_0 and P_0 with rotational frequency ω in single-j model ($j\pi$ = 13/2+), G/κ = 0.20, E_c/κ = 2.5. R_0, defined by eq. (17), is the results between the yrast bands of 6- and 8-particle systems. P_0 is the component of the fully-paired $K\pi$ = 0+ configurations in the yrast band of 8-particle system.

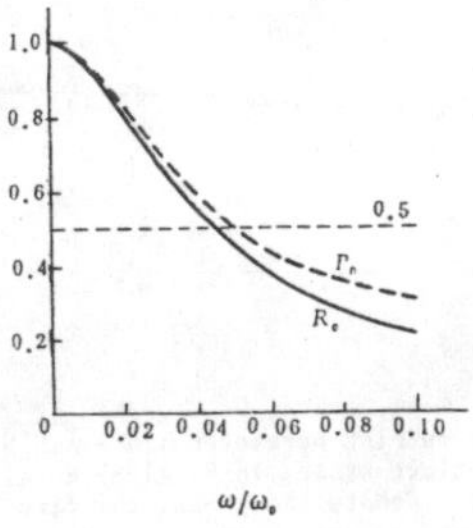

Fig. 2. The variations of R_0 and P_0 with rotational frequency ω in the Nilsson model. R_0, defined by eq. (17), is the results between the yrast bands of N = 94 and N = 96 systems. P_0 is the component of the fully-paired $K\pi$ = 0+ configurations in the yrast band of N = 94 system. Lund systematics for the Nilsson parameters are adopted, i.e., ε_2 = 0.242, ε_4 = 0, κ = 0.0637, μ = 0.420. Truncated configuration energy $E_c/\hbar\omega_0$ = 0.4.

Calculations in the more realistic Nilsson model confirm the above statement (see fig. 2).

Next, let us consider the pairing parameter $\tilde{\Delta} = G\sqrt{\langle S^+S\rangle}$ which can be calculated very easily with the PNC wavefunction. It is found that for low-lying bands and for not too large values of ω, $\tilde{\Delta}/G = \sqrt{\langle S^+S\rangle}$ decreases very slowly with ω (see figs. 3, 4). For example, $\sqrt{\langle S^+S\rangle}(\omega/\kappa = 0.30)/\sqrt{\langle S^+S\rangle}(\omega = 0) \geq 80\%$. However, from the analyses of K-structure and pair-transfer matrix elements it is seen that the intrinsic structure of the yrast band has undergone a great change at ω/κ = 0.30. Therefore, it seems hard to use $\tilde{\Delta}/G = \sqrt{\langle S^+S\rangle}$ to indicate nuclear pairing phase transition.

In contrast to its slow variation with ω, $\sqrt{\langle S^+S\rangle}$ depends sensitively on the blocking effect. Systematic even-odd difference of $\sqrt{\langle S^+S\rangle}$ is found in our calculation. For example, the values of $\sqrt{\langle S^+S\rangle}$ for the lowest two bands of 7-particle system are systematically smaller than those of the lowest bands of

6- and 8-particle systems (see fig. 4). Similarly, the $\sqrt{\langle S^{+}S\rangle}$ values of excited bands of even-even system, which is mainly composed of pair-broken configurations are even smaller (see figs. 3, 4). In addition, the $\sqrt{\langle S^{+}S\rangle}$ values of the low-lying excited bands are almost constant for not too large ω. For example, the $\sqrt{\langle S^{+}S\rangle}$ values of the lowest two excited bands of 8-particle system are 1.74 ± 0.01 for $\omega/\kappa \le 0.30$ (see fig.4).

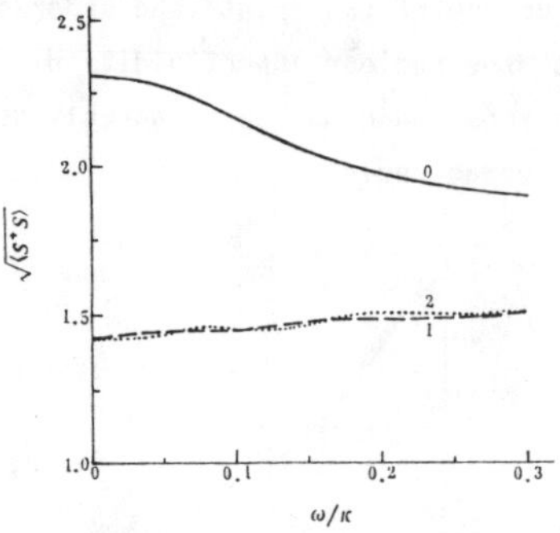

Fig. 3. The pairing parameter $\tilde{\Delta}/G = \sqrt{\langle S^{+}S\rangle}$ for 6-particle system in single-j model. 0, 1 and 2 denote the yrast, the first and the second excited bands, respectively.

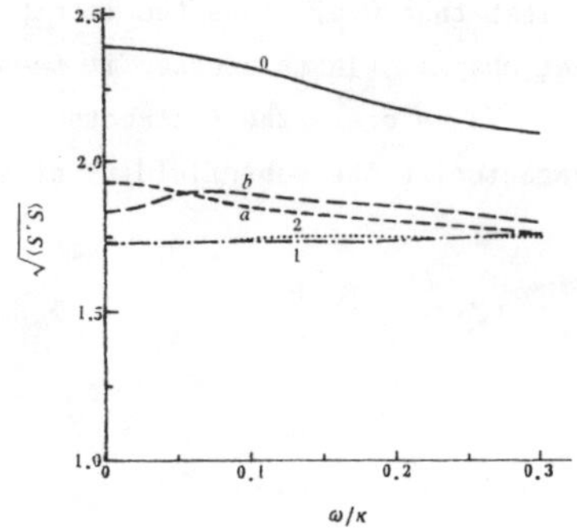

Fig. 4. $\tilde{\Delta}/G = \sqrt{\langle S^{+}S\rangle}$ for 8-particle system. 0, 1 and 2 denote the yrast, the first and the second excited bands, respectively. a and b label the two lowest bands of 7-particle system.

Finally, the variation of $\tilde{\Delta} = G\sqrt{\langle S^{+}S\rangle}$ with pairing strength G is investigated. In the BCS theory $\tilde{\Delta} = \Delta = G\langle S\rangle$ vanishes for $G < G_c$ (critical pairing strength, which depends on the location of the Fermi surface and the single-particle level distribution near the Fermi surface). In particular, at $G = G_c$, Δ shows a sharp cusp[7]). However, the "sharp pairing phase transition" disappears in the FBCS treatment, i.e., particle-number projection before variation[7]). Similarly, in our PNC treatment we found a gradual weakening of $\tilde{\Delta}$ as a function of G, which is shown in fig. 5.

4. Summary

As in the calculation with particle-number-projection before variation the pairing parameter $\Delta = G\sqrt{\langle S^{+}S\rangle}$ decreases very slowly with rotational frequency ω in our PNC treatment. Even when the intrinsic structure of the wave function has undergone a great change, $\tilde{\Delta}$ still remains rather large. Therefore, it seems hard to use $\tilde{\Delta}$ to indicate a pairing phase transition. On the contrary, the blocking effects on $\tilde{\Delta}$ are much more important than the Coriolis anti-pairing effect, especially for not too large values of ω. Also we find a gradual weakening of $\tilde{\Delta}$ with decreasing G. The K-structure of wave function, as well as the pair-transfer matrix elements, may be reasonably used to measure

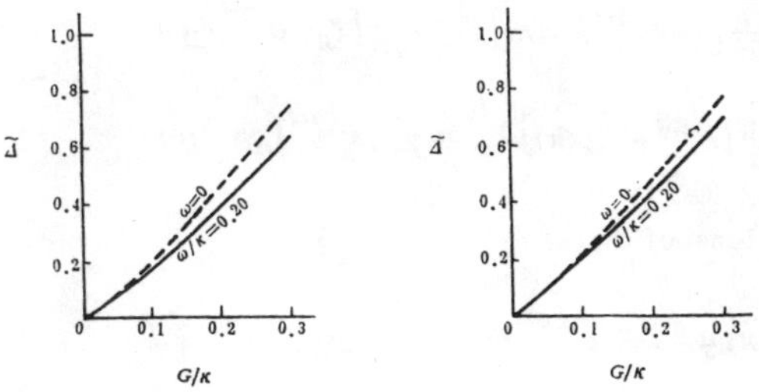

(a) 6-particle system (b) 8-particle sysytem

Fig. 5. The variation of $\tilde{\Delta} = G\sqrt{\langle S^{+}S\rangle}$ with pairing strength G.

the nuclear superfluidity. At $\omega/\kappa \geq 0.20$ ($\kappa \sim 2.0 - 2.5$ MeV for well-deformed rare-earth nuclei) the intrinsic structure of wave function for the yrast band has undergone a great change, in spite of the fact that $\tilde{\Delta}$ still remains rather large.

Appendix

As usual, the Cranking Shell Model hamiltonian of a well-deformed nucleus is chosen as

$$H_{CSM} = H_{intr} - \omega J_x ,$$

$$H_{intr} = H_{sp} + H_{pair}$$

$$= \sum_{\nu} \varepsilon_{\nu} (a^{+}_{\nu} a_{\nu} + a^{+}_{\bar{\nu}} a_{\bar{\nu}}) - G \sum_{\mu\nu} a^{+}_{\mu} a^{+}_{\bar{\mu}} a_{\bar{\nu}} a_{\nu} , \quad \text{(A1)}$$

where μ, ν denote the single-particle states and $\bar{\mu}$, $\bar{\nu}$ their time-reversal states, respectively. For well-deformed nucleithe single-particle states are usually chosen as the eigenstates of j_z. For $\omega \neq 0$, $J_z = \sum_i j_z(i)$ is no longer a constant of motion. However, it is usually assumed that H_{sp} is invariant under space reflection and $R_x(\pi)$, rotation of 180° about x-axis. Thus the diagonalization of H_{CSM} can be performed separately in each subspace with fixed parity and signature [eigenvalue of $R_x(\pi)$].

In this case it is convenient to label the single-particle state by parity π and signature r (= $\pm i$), or equivalent by α ($r = e^{i\pi\alpha}$, $\alpha = \mp \frac{1}{2}$). Let χ_{Ω} denotes the eigenstate of j_z and define $\chi_{\bar{\Omega}} = R_x(\pi)\chi_{\Omega}$ where $\Omega > 0$, $\bar{\Omega} = -\Omega$, it is easy to show that

$$\varphi_{\Omega\alpha}=\frac{1}{\sqrt{2}}[1+e^{i\pi\alpha}R_x(\pi)]\chi_\Omega=\frac{1}{\sqrt{2}}[\chi_\Omega+e^{i\pi\alpha}\chi_{\bar{\Omega}}]$$
$$(\bar{\alpha}=-\alpha,\ \alpha\geqslant 0) \tag{A2}$$
$$\varphi_{\Omega\bar{\alpha}}=\frac{1}{\sqrt{2}}[e^{i\pi\alpha}+R_x(\pi)]\chi_\Omega=\frac{1}{\sqrt{2}}[e^{i\pi\alpha}\chi_\Omega+\chi_{\bar{\Omega}}]$$

are the eigenfunctions of $R_x(\pi)$

$$R_x(\pi)\varphi_{\Omega\alpha}=e^{-i\pi\alpha}\varphi_{\Omega\alpha}=-i\,\varphi_{\Omega\alpha}\,,$$
$$R_x(\pi)\varphi_{\Omega\bar{\alpha}}=e^{i\pi\alpha}\varphi_{\Omega\bar{\alpha}}=i\varphi_{\Omega\bar{\alpha}}\,. \tag{A3}$$

In second quantization notation the canonical transformation (A2) of single-particle states can be expressed as

$$b_\nu^+=\frac{1}{\sqrt{2}}[a_\nu^++e^{i\pi\alpha}a_{\bar{\nu}}^+],\qquad b_{\bar{\nu}}^+=\frac{1}{\sqrt{2}}[e^{i\pi\alpha}a_\nu^++a_{\bar{\nu}}^+], \tag{A4}$$

$$\{b_\nu,\,b_{\nu'}^+\}=\delta_{\nu\nu'},\qquad \{b_{\bar{\nu}},\,b_{\bar{\nu}'}^+\}=\delta_{\nu\nu'}. \tag{A5}$$

Under this transformation the form of the pair-creation operator remains unchanged

$$b_\nu^+b_{\bar{\nu}}^+=a_\nu^+a_{\bar{\nu}}^+\,. \tag{A6}$$

The matrix elements of j_x (or Coriolis interaction) between the new bases are as follows:

(a) The matrix elements of j_x between the states with different signature vanish, namely,

$$\langle\varphi_{\Omega_s\alpha}|j_x|\varphi_{\Omega_t\bar{\alpha}}\rangle=0. \tag{A7}$$

(b) The matrix elements of j_x between the states with the same signature

$$\langle\varphi_{\Omega_s\alpha}|j_x|\varphi_{\Omega_t\alpha}\rangle=\langle\chi_{\Omega_s}|j_x|\chi_{\Omega_t}\rangle+e^{-i\pi\alpha}\langle\chi_{\bar{\Omega}_s}|j_x|\chi_{\Omega_t}\rangle, \tag{A8}$$

where $\langle\chi_{\Omega_s}|j_x|\chi_{\Omega_t}\rangle$ can be calculated with the Nilsson wave function. It should be noted that in our convention

$$\langle\chi_{\Omega_s}|j_x|\chi_{\bar{\Omega}_t}\rangle=-\langle\chi_{\bar{\Omega}_s}|j_x|\chi_{\Omega_t}\rangle,\qquad \langle\chi_{\bar{\Omega}_s}|j_x|\chi_{\bar{\Omega}_t}\rangle=\langle\chi_{\Omega_s}|j_x|\chi_{\Omega_t}\rangle. \tag{A9}$$

(c) In particular, $\Omega_s = \Omega_t = ½$,

$$\langle \varphi_{(\frac{1}{2})_s}\alpha | j_x | \varphi_{(\frac{1}{2})_t}\alpha \rangle = e^{-i\pi\alpha} \langle \chi_{\overline{(\frac{1}{2})}_s} | j_x | \chi_{(\frac{1}{2})_t} \rangle \quad (\alpha = \pm\frac{1}{2}), \tag{A10}$$

which is the only matrix element depending on the signature and is responsible for all the signature splittings.

Now let us consider the matrix elements of $H_c = -\omega J_x$ between the many-particle configurations. For $\omega = 0$, $H_{CSM} (= H_{intr})$ is block diagonal with respect to K, π and seniority number v. As the Coriolis interaction is turned on ($\omega \neq 0$), the matrix elements of $H_c = -\omega J_x$ between configurations with different K and v may not vanish, so these configurations will be mixed. However, because parity π and signature r remain to be good quantum numbers, H_{CSM} can be diagonalized in each configuration space with fixed π and r.

For treating the yrast band (π = +, r = +1) and the low-lying bands of even-even nuclei two kinds of configurations need to be considered:

(a) The fully-paired configurations (v = 0)

$$|\rho_1\bar{\rho}_1 \cdots \rho_n\bar{\rho}_n\rangle = b^+_{\rho_1} b^+_{\bar{\rho}_1} \cdots b^+_{\rho_n} b^+_{\bar{\rho}_n} |0\rangle, \qquad \pi = +,\ r = +1,\ K = 0. \tag{A11}$$

(b) The pair-broken configurations (v = 2)

$$|s\bar{t}\rho_1\bar{\rho}_1 \cdots \rho_{n-1}\bar{\rho}_{n-1}\rangle = b^+_s b^+_{\bar{t}} b^+_{\rho_1} b^+_{\bar{\rho}_1} \cdots b^+_{\rho_{n-1}} b^+_{\bar{\rho}_{n-1}} |0\rangle, \quad \pi_s = \pi_t, \tag{A12}$$
$$\pi = +, \quad r = +1, \quad K = \pm|\Omega_s - \Omega_t|,\ \pm(\Omega_s + \Omega_t).$$

The matrix elements of J_x are as follows:

(a) Because of the selection rule for J_x, $\Delta K = \pm 1$, the matrix elements of J_x between the fully-paired configurations vanish,

$$\langle \bar{\rho}'_n \rho'_n \cdots \bar{\rho}'_1 \rho'_1 | J_x | \rho_1\bar{\rho}_1 \cdots \rho_n\bar{\rho}_n \rangle = 0. \tag{A13}$$

(b) Because J_x is a one-body operator, the non-vanishing matrix elements between a fully-paired configuration and a pair-broken configuration (π = +, r = +1) is of the form

$$\langle \cdots \bar{\nu}\nu\bar{\mu}\mu\bar{t}s | J_x | \rho\bar{\rho}\mu\bar{\mu}\nu\bar{\nu} \cdots \rangle = \langle \bar{t}s | J_x | \rho\bar{\rho} \rangle$$
$$= \langle \chi_{\Omega_s} | j_x | \chi_{\Omega_t} \rangle (\delta_{\rho t} + \delta_{\rho s}) + e^{-i\pi\alpha} \langle \chi_{\bar{\Omega}_s} | j_x | \chi_t \rangle (\delta_{\rho t} - \delta_{\rho s}). \tag{A14}$$

If χ_{Ω_s} and χ_{Ω_t} are the same state, the second term on the right hand side of eq. (A14) vanishes. Only for $\Omega_s = \Omega_t = ½$, $\pi_s = \pi_t$, but not all the other quantum numbers being the same, the second term needs to be considered. It should be noted that the pair-broken configuration $(s\bar{t})$ and $(t\bar{s})$ are different and both of them have to be taken into account.

(c) Matrix elements between two pair-broken configurations (π = +, r = +1). The diagonal matrix elements are

$$\langle \cdots \bar{\nu}\nu\bar{\mu}\mu\bar{t}s | J_x | s\bar{t}\mu\bar{\mu}\nu\bar{\nu}\cdots\rangle = \langle j_x\rangle_{ss} + \langle j_x\rangle_{\bar{t}\bar{t}},$$

where

$$\langle j_x\rangle_{ss} = \langle \varphi_{\Omega_s\alpha} | j_x | \varphi_{\Omega_s\alpha}\rangle = e^{-i\pi\alpha}\langle \chi_{\bar{\Omega}_s} | j_x | \chi_{\Omega_s}\rangle \delta_{\Omega_s,\frac{1}{2}},$$
$$\langle j_x\rangle_{\bar{t}\bar{t}} = \langle \varphi_{\Omega_t\bar{\alpha}} | j_x | \varphi_{\Omega_t\bar{\alpha}}\rangle = e^{i\pi\alpha}\langle \chi_{\bar{\Omega}_t} | j_x | \chi_{\Omega_t}\rangle \delta_{\Omega_t,\frac{1}{2}}, \tag{A15}$$

i.e., only for $\Omega_s = \frac{1}{2}$, or $\Omega_t = \frac{1}{2}$, the diagonal matrix element does not vanish. These diagonal elements are responsible for all the signature splittings. The off-diagonal elements do not vanish only if two pair-broken configurations differ by one single-particle orbit. One type of these non-vanishing matrix elements is

$$\langle \cdots \bar{\mu}\mu\bar{t}'s | J_x | s\bar{t}\mu\bar{\mu}\cdots\rangle = \langle \varphi_{\Omega_{t'}\bar{\alpha}} | j_x | \varphi_{\Omega_t\bar{\alpha}}\rangle,$$
$$\langle \cdots \bar{\mu}\mu\bar{t}s' | J_x | s\bar{t}\mu\bar{\mu}\cdots\rangle = \langle \varphi_{\Omega_{s'}\alpha} | j_x | \varphi_{\Omega_s\alpha}\rangle. \tag{A16}$$

The other types of non-vanishing matrix elements are

$$\langle \cdots \bar{\mu}\mu\bar{t}t\bar{t}'s | J_x | s\bar{t}t'\bar{t}'\mu\bar{\mu}\cdots\rangle = -\langle \varphi_{\Omega_t\alpha} | j_x | \varphi_{\Omega_{t'}\alpha}\rangle$$
$$\langle \cdots \bar{\mu}\mu\bar{t}t\bar{s}t' | J_x | t\bar{s}t'\bar{t}'\mu\bar{\mu}\cdots\rangle = -\langle \varphi_{\Omega_t\bar{\alpha}} | j_x | \varphi_{\Omega_{t'}\bar{\alpha}}\rangle. \tag{A17}$$

Calculations show that for the low-lying bands in even-even nuclei the contributions from the configurations with four or more unpaired particles ($v \geq 4$) are very small and then are neglected.

References

1) R. Chapman, et al., Phys. Rev. Lett., 51(1983),2265.
2) B. R. Mottelson and J. G. Valatin, Phys. Rev. Lett., 5(1960),511.
3) L. F. Canto, P. Ring and J. O. Rasmussen, LBL-19519 (1985), private communication.

4) J. L. Edigo and P. Ring, Nucl. Phys., A388(1982),19.

5) W. Nazarawicz, J. Dudek and Z. Szymanski, Nucl. Phys., A436(1985),139.

6) For examples, see P. Ring, R. Beck and H. J. Mang, Z. Physik, 231 (1970),103.

B. Banerjee, H. J. Mang and P. Ring, Nucl. Phys., A215(1973),366.

A. Goodman, Nucl. Phys., A256(1976),113.

7) P. Ring and P. Schuck, the Nuclear Many-body Problem (Springer-Verlag, 1980).

8) D. J. Rowe, Nuclear Collective Motion (Methuen, London, 1970), ch. 11.

9) J. Y. Zeng and T. S. Cheng, Nucl. Phys., A405(1983),1.

10) J. Y. Zeng, T. S. Cheng, L. Cheng and C. S. Wu, Nucl.Phys.,A411(1983), 49.

11) T. S. Cheng, C. S. Wu and J. Y. Zeng, Chin. Phys. Lett., 4(1986),125.

12) J. Y. Zeng, T. S. Cheng, L. Cheng and C. S. Wu, Nucl.Phys.,A421(1984), 125.

13) I. Hamamoto, Nucl. Phys., A271(1976),15.

14) B. Bengtsson and H. B. Hakansson, Nucl. Phys., A357(1981),61.

PRELIMINARY PLAN OF EXPERIMENTAL AREA AND HEAVY ION NUCLEAR PHYSICS RESEARCH ON HIRFL

Wu Enjiu Shen Wenqing

Institute of Modern Physics, Academia Sinica, Lanzhou, China

The historical background and present situation of Institute of Modern Physics about heavy ion nuclear physics research are described. The plan of the heavy ion nuclear physics research on rebuilt 1.7m sector focusing cyclotron is also presented. Main emphasis is placed on the describtion of the preliminary plan of experimental area construction and heavy ion nuclear physics research on HIRFL (Heavy Ion Research Facility Lanzhou).

1. INTRODUCTION

Since seventieth the heavy ion nuclear physics research have been started in Institute of Modern Physics. A series of heavy ion experiments have been performed on the 1.5m classical cyclotron (k=57) with the projectiles ^{12}C (6.1MeV/A), ^{14}N (6.9MeV/A) and ^{16}O (5.5MeV/A). The main equipments which have been used are the followings:

(1) A small heavy ion time of flight spectrometer; (2) A large area position sensitive ionization chamber of GSI type; (3) In - beam γ-spectrometer; (4) He-jet system and in beam Δt-t technique; (5) A scattering chamber for general purposes.

Research plan with the 1.5m cyclotron placed the study emphasis on the following four subjects. (1) Deep inelastic collision in lighter systems; (2) Research of the α-particle emission in the incomplete fusion reaction; (3) The research of fusion and fission; (4) The nuclear spectroscopy and in beam γ-ray spectroscopy.

The present paper will sketch our research plan and experimental equipments which are being prepared on our new accelerator system HIRFL.

2. HEAVY ION NUCLEAR PHYSICS RESEARCH PROGRAM ON 1.7m SECTOR FOCUSING CYCLOTRON

1.5m cyclotron is now being converted into a 1.7m, k=69 sector focusing cyclotron (SFC) which will be used as an injector of the main accelerator, separated sector cyclotron (SSC). In 1986 it will deliver beam. We will use it to adjust the detectors and the equipments which will be mounted on HIRFL. In addition some experiments will also be performed as a bridge from the nowadays work to the future work on HIRFL. After the construction of HIRFL SFC can still deliver beam separately. The research work on SFC can be continued in future. The main projectiles of SFC are 8.5 MeV/A ^{12}C, ^{14}N, ^{16}O, ^{20}Ne and 4.2MeV/A ^{40}Ar ions. The experiments in preparation are the followings:

(1) Complete fusion, incomplete fusion and quasi fission reaction; (2) Partial linear momentum transfer; (3) Incomplete deep inelastic collision and particle emission in DIC; (4) The mechanism of light particle emission and so on.

3. PRELIMINARY PLAN OF HIRFL EXPERIMENTAL AREA

The preliminary plan of HIRFL experimental area has been made in 1984. Now the phisical and technical designs of the most equipments are fullfilled, some of them are already in construction. Fig.1 shows the layout of the HIRFL accelerator system, the experimental hall, the beam transport system and the experimental equipments.

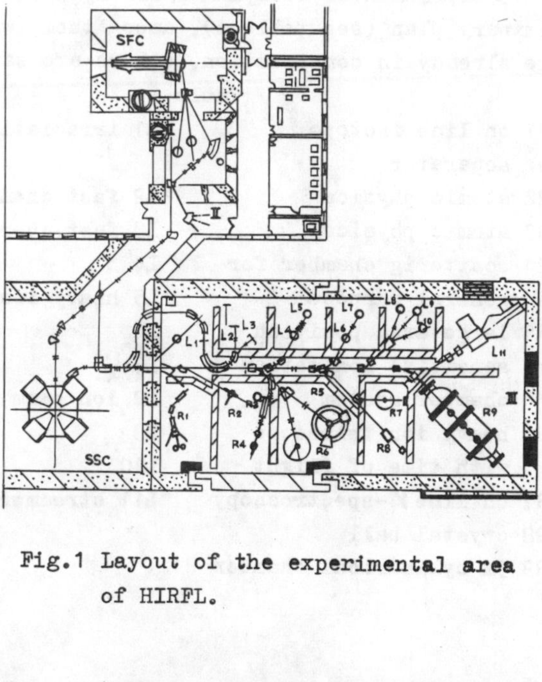

Fig.1 Layout of the experimental area of HIRFL.

According to the position of the accelerators the experimental area is divided into three parts:

First area: The beam of SFC is extracted into this area by the beam transport system and the deflection magnet. In this area there are two or three beam lines which can be used in the experiments, one of them will be used as the injection tube to SSC.

Second area: After

the stripper the beam coming from SFC will produces the ions with different charge states. The ions with a definite charge state will be transported into SSC, other ions will be extracted into the second experimental area in which there are also two or three beam lines for the experimental users.

Third area: After passing through the complicated beam transport system the beam coming from SSC will be extracted into this area, it is the main experimental area of HIRFL, 56m long, 26m wide and 11m height. Just outside of this area there are two counting rooms (150M^2 each) and one computer room. In this area 17 beam lines will be arranged, the movable shielding walls will divide this area into 11 small experimental regions. There is one beam line which can transport the beam of SFC into this area directly, so the experiment in which the beam of SFC will be used can be perfomed in first , second and third area. For the coupling operation of SFC + SSC we can obtain the low energy beam in second area and high energy beam in third area at the same time. In this area there is a beam handling system which has four operation modes to fullfill the diffrent requests to the beam e.g high energy resolution, high time resolution and high parallelism of the beam.

15 experimental equipments alreay have been listed in the preliminary plan (see table 1), among them (with * in table 1) 10 are already in construction, others are still in discussion.

Table 1

*R1 on line isotope separator	L1 irradiation experiments
*R2 atomic physics	*L2 fast chemistry with He-jet
R3 atomic physics	L3 fast chemistry with He-jet
*R4 scatterig chamber for general purpose	L4
	*L5 heavy ion irradiation facility
*R5 large area position sensitive ionization chamber system	L6
	L7
	L8 ion beam analyses
*R6 heavy ion telescope with time of flight	L9
	L10
*R7 on line γ-spectroscopy	*L11 streamer chamber system
R8 crystal ball	
*R9 3m cylindrical chamber	

4. PRELIMINARY PROGRAM OF HEAVY ION PHYSICS RESEARCH ON HIRFL:

HIRFL is a combined accelerator system which consists of two accelerators, the main accelerator is k=450 separated sector cyclotron (SSC), the injector is k=69 sector focusing cyclotron (SFC), betwwen them there is a stripper. It can deliver the ions C, N, O of 100MeV/A, 46MeV/A Ar, 4.6MeV/A Xe etc, the beam intensity for the lighter heavy ions is about 10^{12} pps, for the heavier ions about 10^{10} pps. Without the stripper the better beam quality will be obtained (larger intensity, continuous adjustment of energy in a wide range), but the highest energy will be reduced e.g for C it will be reduced to 27MeV/A, Ar to 18MeV/A, Kr to 6MeV/A. In the future SFC can still deliver the beam separately. Actually there are three energy regions which can be used as shown in Fig.2.

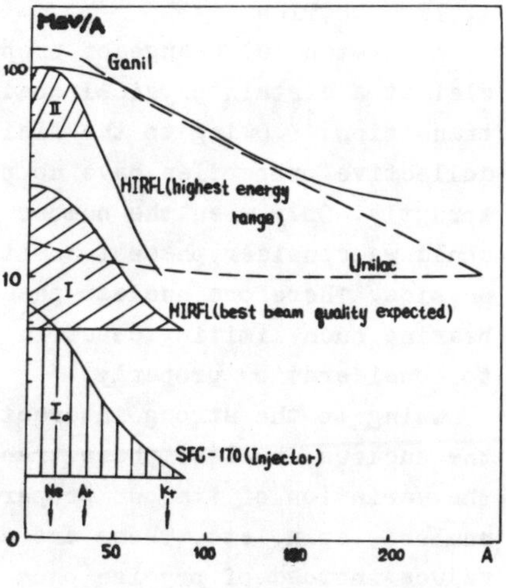

Fig.2 Energy curve of HIRFL

The region I can be used for the low energy heavy ion nuclear physics research, the region II for the intermediate energy heavy ion collision, the region I_0 for the nuclear reactions, atomic physics and applications near the Coulomb barrier. We will start from the region I and develop to the region II. Some proposals already have been suggested, They are summarized in the following:

A. Heavy ion nuclear physics and nuclear chemistry:

(1) Syntheses of new nuclides and research of their decay properties; (2) Heavy ion nuclear reaction mechanism research; (3) Nuclear structure research; (4) Theoretical research.

B. Heavy ion atomic physics:

(1) Ionization and excitation of atomic inner shells; (2) Beam foil spectroscopy.

C. Research of material science and solid state physics.

D. Research of biology and medicine with heavy ions.

According to the time schedule, HIRFL will go into full operation in spring of 1989. It is expected that the machine will produce various experimental results in the future.

NUCLEAR PHASE TRANSITION ILLUSTRATED WITH SCHEMATIC MODELS

Xu Gong-ou Department of Physics, Nanjing University and Department of Modern Physics, Lanzhou University.

Li Fu-li Department of Physics, Nanjing University.

Fu De-ji Institute of Nuclear Research, Shanghai, Academia Sinica.

Nuclear phase transitions are studied with schematic models from the point of view of symmetry-breaking. The dynamic effect of a finite system has been properly taken into consideration. The results for infinite systems are also given for comparison.

1. INTRODUCTION

A substantial change of properties of collective motions of nuclei at a certain physical condition is regarded as nuclear phase transition.[1] Owing to the finiteness of nuclear systems, nuclear collective properties have no precise values and would not change abruptly. Only when the number of particles increases indefinitely could we consider phase transition in the sense of statistical physics. Therefore nuclear phase transitions should be studied by bearing such limiting cases in mind and taking dynamic effects into consideration properly.[2]

Owing to the strong interaction existed between nucleons inside the nucleus, nuclear phase transition can only be observed with the variation of its own properties, such as proton and neutron numbers, spin, etc. These are all quantized quantities. If mean values instead of precise ones were used in theoretical studies, discrepancies would be resulted in final results.[3]

In view of the above-mentioned points, unprojected self-consistent field theory may not be adequate for studying nuclear phase transitions. In this work we study the problem from dynamic symmetries.

If nuclear collective hamiltonian were in a certain approximation constructed from operators of a Lie algebra, the correspond-

*Work supported by the research fund of the Education Committee People's Republic of China and the science fund of the Chinese Academy of Sciences.

ing Lie group would be the dynamic group of nuclear collective motions. In case the ground state corresponding to vanishingly small residual interaction is a minimum weight state of the allowed irreducible representation, nuclear collective hamiltonian can be expressed in terms of bosons.[4] Now, the Pauli principle has entered into the boson representation through the specified irreducible representation. This kind of nuclear collective hamiltonian involving physical conditions explicitly show different dynamic symmetries for different physical conditions and quantities characterizing the symmetry-breaking may be used as order parameters. We give in the following two examples.

2. TWO-LEVEL MODEL WITH MONOPOLE-MONOPOLE INTERACTION

We discuss first a two-level model with attractive monopole-monopole interaction. The hamiltonian can be written as

$$H = \frac{\varepsilon}{2}\sum_m (a^+_{m+}a_{m+} - a^+_{m-}a_{m-}) - \frac{\kappa}{2}\sum_m (a^+_{m+}a_{m-} + a^+_{m-}a_{m+})\cdot \sum_{m'} (a^+_{m'+}a_{m'-} + a^+_{m'-}a_{m'+}) \tag{1}$$

The collective operators $\sum_m a^+_{m+}a_{m-}$, $\sum_m a^+_{m-}a_{m+}$ and $\sum_m (a^+_{m+}a_{m+} - a^+_{m-}a_{m-})$ form a SU(2) algebra. The corresponding SU(2) group is therefore the dynamic group of the system.

The ground state for vanishingly small κ

$$|\phi_0\rangle = \begin{cases} a_{m_1-}a_{m_2-}\cdots\cdots a_{m_n-}\prod_m a^+_{m-}|0\rangle\ , & \text{particle number} = (\Omega - n) \\ a^+_{m_1+}a^+_{m_2+}\quad a^+_{m_n+}\prod_m a^+_{m-}|0\rangle\ , & \text{particle number} = (\Omega + n) \end{cases}$$

$$\Omega = 2j + 1 \tag{2}$$

is the minimum weight state of the allowed irreducible representation of the SU(2) group. Correspondingly the ground state as well as collectively excited states for finite κ can be expressed as state vectors induced from $|\phi_0\rangle$ by the SU(2) group. Therefore the collective hamiltonian can be written as [5]

$$\mathcal{H} = -\frac{\varepsilon}{2}(\Omega - n) + \varepsilon\, b^+b - \frac{\kappa(\Omega - n)}{2}\left(b^+\sqrt{1 - b^+b/(\Omega - n)} + \sqrt{1 - b^+b/(\Omega - n)}\ b\right)^2 \tag{3}$$

where b^+, b represent the creation and annihilation operators of monopole phonons and $\sqrt{1 - b^+b/(\Omega - n)}$ is a truncation factor so that the number of phonons would not exceed $(\Omega - n)$.

It is very obvious that the system has the following dynamic symmetry

$$SU(2) \supset U(1) \supset R(\pi), \qquad R(\pi) = \exp(i\pi b^+b) \tag{4}$$

provided $\kappa(\Omega - n)/\varepsilon$ is very small. When $\kappa(\Omega - n)/\varepsilon$ increases to a certain quantity, phonon condensation happens such that $\langle b^+b\rangle_0/(\Omega - n)$ attains an appreciable amount even for the limiting case $(\Omega - n)\to\infty$. The quantity $\langle b^+b\rangle_0/(\Omega - n)$ is a measure for symmetry-breaking and can be used as an order parameter. $\kappa(\Omega - n)$ is the effective strength for collective monopole excitation and ε is the energy required for such an excitation. Therefore the ratio of these two competing factors $\kappa(\Omega - n)/\varepsilon$ can be used as a controlling factor. The variation of the order parameter $\langle b^+b\rangle_0/(\Omega - n)$ versus the controlling factor $\kappa(\Omega - n)/\varepsilon$ can be obtained either directly from exact numerical calculation[5] or from approximate self-consistent calculation. The latter is especially suitable for the study of limiting cases of infinitely large $(\Omega - n)$.

In the approximate self-consistent calculation, the truncation factor can be replaced by its average value according to

$$AB - \langle A\rangle_0 B - \langle B\rangle_0 A + \langle A\rangle_0\langle B\rangle_0 = 0 \tag{4}$$

Therefore

$$\begin{aligned} H \approx & -\frac{\varepsilon}{2}(\Omega - n) - \frac{\kappa}{2}(\Omega - n) - \kappa\langle b^+b\rangle_0\langle b^+b\rangle_0 \\ & - \frac{\kappa}{2}\langle b^+b^+ + b\,b\rangle_0\langle b^+b\rangle_0 \\ & + \left[\varepsilon - \kappa(\Omega - n)\left(1 - \frac{2\langle b^+b\rangle_0}{\Omega - n} - \frac{\langle b^+b^+ + b\,b\rangle_0}{2(\Omega - n)}\right)\right] b^+b \\ & - \frac{\kappa(\Omega - n)}{2}\left(1 - \frac{\langle b^+b\rangle_0}{\Omega - n}\right)(b^+b^+ + b\,b) \end{aligned} \tag{5}$$

where the approximate commutability of the truncation factor with b^+ and b has been assumed. With the help of a canonical transformation

$$b' = c^+\cosh\theta + c\sinh\theta \tag{6}$$

$$\tanh 2\theta = \frac{1 - \frac{\langle b^+b\rangle_0}{\Omega - n}}{\frac{\varepsilon}{\kappa(\Omega - n)} - \left(1 - \frac{2\langle b^+b\rangle_0}{\Omega - n} - \frac{\langle b^+b^+ + bb\rangle_0}{2(\Omega - n)}\right)} \tag{7}$$

the following analytical results can be obtained,

$$\frac{\langle b^+b\rangle_0}{\Omega - n} = \frac{1}{2(\Omega - n)}\left[\cosh 2\theta - 1\right] \tag{8}$$

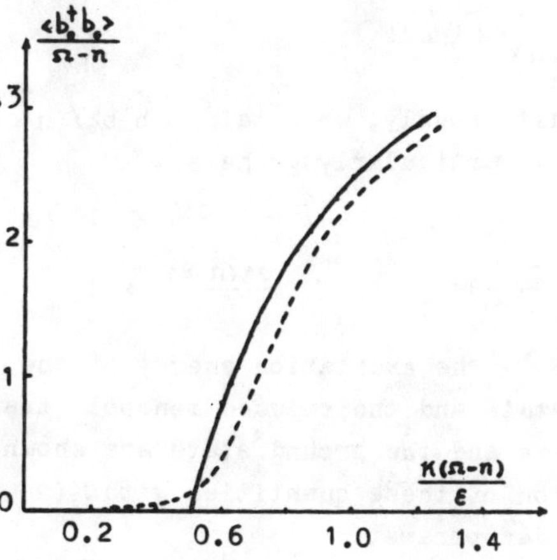

Fig. 1, Variation of $\frac{\langle b_0^\dagger b_0 \rangle}{\Omega - n}$ with $\kappa(\Omega-n)/\varepsilon$.

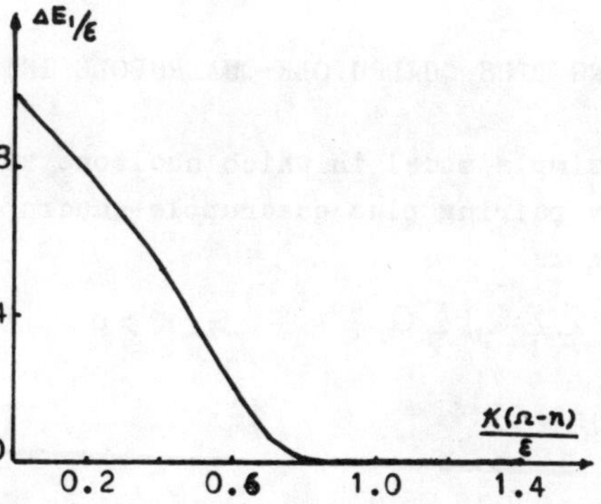

Fig. 2, Excitation energy of the 1st collectively excited state versus $\kappa(\Omega-n)/\varepsilon$.

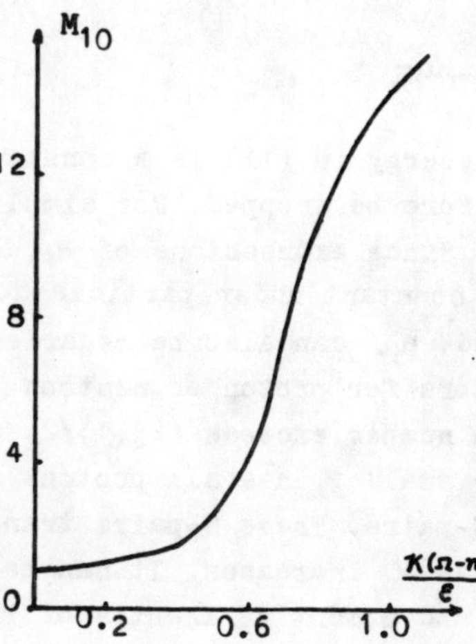

Fig. 3, Reduced monopole transition rate versus $\kappa(\Omega-n)/\varepsilon$.

$$\frac{\langle b^+b^+ + b\,b\rangle_0}{2(\Omega-n)} = \frac{1}{2(\Omega-n)}\,\mathrm{Sinh}2\theta \tag{9}$$

Solving equations (7-9) simultaneously, we obtain $\langle b^+b\rangle_0/(\Omega-n)$ as a function of $\kappa(\Omega-n)/\varepsilon$. Particularly we have

$$\lim_{(\Omega-n)\to\infty} \frac{\langle b^+b\rangle_0}{\Omega-n} = \begin{cases} 0. & \dfrac{2\kappa(\Omega-n)}{\varepsilon} < 1 \\ \frac{1}{2}\left(1 - \dfrac{\varepsilon}{2\kappa(\Omega-n)}\right) & \dfrac{2\kappa(\Omega-n)}{\varepsilon} > 1 \end{cases} \tag{10}$$

The results are shown in Fig.1. The excitation energy of the first collectively excited state and the reduced monopole transition rate between this state and the ground state are shown in figures 2 and 3. The variation of these quantities with $\kappa(\Omega-n)/\varepsilon$ is mainly governed by the order parameter.

3. SINGLE-j MODEL WITH PAIRING PLUS QUADRUPOLE-QUADRUPOLE INTERACTIONS

We consider next another simple model in which nucleons are interacted with each other by pairing plus quadrupole-quadrupole interactions. The hamiltonian is

$$H = -\varepsilon\sum_\rho n_\rho - G\sum_\rho P_\rho^\dagger P_\rho - \kappa\sum_\mu\sum_\rho Q_{\rho\mu}\sum_{\rho'} Q_{\rho'\tilde{\mu}} \qquad G,\ \kappa > 0 \tag{11}$$

where

$$n_\rho = \sum_m a^\dagger_{\rho jm} a_{\rho jm} \tag{12}$$

$$P_\rho^\dagger = \frac{1}{\sqrt{2}}\sum_m a^\dagger_{\rho jm} a_{\rho j\tilde{m}} \tag{13}$$

$$Q_{\rho m} = \frac{\langle j\|q\|j\rangle}{\sqrt{5}}\sum_{m_1 m_2} \langle j m_1 j m_2 | 2\mu\rangle\, a^\dagger_{jm_1} a_{j\tilde{m}_2} \tag{14}$$

The first term for single particle energy in (11) is a constant for a definite system and can therefore be dropped. For simplicity we consider even-even systems. Since expressions of n_ρ, $Q_{\rho\mu}$ and H are not changed apart from a constant under particle-hole conjugation transformation, $a^+_{\rho m}$ and $a_{\rho m}$ can also be regarded as creation and annihilation operators for proton or neutron holes in case the proton or neutron number exceeds (2j+1)/2.

The ground state for vanishingly small κ has all protons and neutrons coupled respectively to S-pairs. These S-pairs transform partly to D-, G-, ... pairs as κ increases. It has been pointed out[6] that the deformation and moment of inertia of the nucleus are appreciablely influenced by the existence of G-pairs, hence we treat the problem in the subspace of S-, D- and G-pairs.

Let

$$A^{+}_{\rho\lambda\mu}=\frac{1}{\sqrt{2}}\sum_{m_1 m_2}\langle jm_1 jm_2|\lambda\mu\rangle a^{+}_{\rho jm_1}a_{\rho jm_2} \tag{15}$$

then an arbitrary state in the subspace can be written as

$$|\Psi\rangle=(0|\exp\sum_{\rho\lambda\mu}b_{\rho\lambda\mu}A^{+}_{\rho\lambda\mu}|0\rangle\mathcal{P}F(\{b^{+}_{\rho\lambda\mu}\})|0) \tag{16}$$

According to the general procedure of DGR-GCM[4] we obtain the boson representation of the nucleon number, quadrupole moment and nuclear collective hamiltonian as

$$n_\rho=2\sum_{\lambda\mu}b^{+}_{\rho\lambda\mu}b_{\rho\lambda\mu} \tag{17}$$

$$Q_\mu(\rho)=-\frac{1}{\sqrt{5}}\langle j\|q\|j\rangle\sum_{\lambda\lambda'}C^{2}_{\lambda\lambda'}\,[b^{+}_{\lambda}(\rho)\,\tilde{b}_{\lambda'}(\rho)]^{(2)}_{\mu} \tag{18}$$

$$\begin{aligned}\mathcal{H}_{coll}=&-\sum_{\rho}G_0(\rho)\,[1-\frac{2(n(\rho)+1)}{2j+1}]\,b^{+}_{0}(\rho)\,b_0(\rho)\\&-\frac{2}{2j+1}\sum_{\rho}G_1(\rho)\,(b^{+}_{0}(\rho)\cdot b_0(\rho))(b^{+}_{0}(\rho)\cdot b_0(\rho))\\&-K_{\pi\pi}\,Q(\pi)\cdot Q(\pi)-K_{\nu\nu}\,Q(\nu)\cdot Q(\nu)-2K_{\pi\nu}\,Q(\pi)\cdot Q(\nu)\end{aligned} \tag{19}$$

where

$$C^{\lambda}_{\lambda' J}=[1+(-)^{\lambda'}](-)^{J}(2\lambda'+1)^{1/2}(2J+1)^{1/2}\begin{Bmatrix}j&\lambda&j\\\lambda'&j&J\end{Bmatrix} \tag{20}$$

and terms representing a change in internal structure of the S-pair by the pairing interaction in the presence of other pairs have been neglected. Since proton and neutron numbers are fixed, $[b^{+}_{\rho\lambda}\times\tilde{b}_{\rho\lambda'}]^{(J)}$ in this collective hamiltonian form a SU(15)⊗SU(15) algebra.

If the isospin of the system is further assumed to be approximately conserved,

$$\begin{aligned}\mathcal{H}^{(0)}_{coll}=&-[\frac{G(\pi)+G(\nu)}{2}(1-\frac{2}{2j+1})-\frac{G_1(\pi)n(\pi)+G_1(\nu)n(\nu)}{2j+1}]\,b^{+}_{0}b_0\\&-\frac{G(\pi)+G(\nu)}{2(2j+1)}(b^{+}_{0}b_0)^2\\&-\frac{K(\pi\pi)+K(\nu\nu)+2K(\pi\nu)}{4}(Q(\pi)+Q(\nu))\cdot(Q(\pi)+Q(\nu))\end{aligned} \tag{21}$$

for the isoscalar mode, where

$$[b^{+}_{\lambda}\times\tilde{b}_{\lambda'}]^{(J)}=\sum_{\rho}[b^{+}_{\rho\lambda}\times\tilde{b}_{\rho\lambda'}]^{(J)}\qquad J=0,2,4. \tag{22}$$

They form a SU(15) algebra.

The problem can be formally solved in the s-d subspace provided the interaction potential is replaced by an effective one,

$$PV_{eff}P=PVP+PV(1-P)\frac{1}{E-H_0-(1-P)V(1-P)}(1-P)VP$$

and the quadrupole moment operator by Q_{eff}

$$PQ_{eff}P = N^{-\frac{1}{2}}\Big\{PQP + \Big[PQ(1-P)\frac{1}{E-H_0-(1-P)V(1-P)}(1-P)VP + h.c.\Big]$$

$$+PV(1-P)\frac{1}{E-H_0-(1-P)V(1-P)}(1-P)Q(1-P)\frac{1}{E-H_0-(1-P)V(1-P)}\cdot$$

$$(1-P)VP\Big\}N^{-\frac{1}{2}}$$

$$N=\sum_i P|\psi_i\rangle\langle\psi_i|P\Big\{1+PV(1-P)\Big(\frac{1}{E-H_0-(1-P)V(1-P)}\Big)^2(1-P)VP\Big\}P|\psi_i\rangle\langle\psi_i|P$$

The terms involving (1-P) in above expressions represent the effects of g-bosons. In our calculation, these terms are approximately determined from the ground state and their energy dependences are neglected.

The results without considering g-boson effects are shown in figures 4,5 and 6. The effective number of d-boson in the ground state is taken as the order parameter. The phase transition occures around $\sum_\rho n(\rho) = 12$. Systems appear to be in spherical shape when $\sum_\rho n(\rho) < 12$, while systems become deformed when $\sum_\rho n(\rho) > 12$. The variation of energy spectra and reduced quadrupole transition rates versus $\sum_\rho n(\rho)$ are closed related to that of Q_0 and $\overline{n}_d$. Preliminary results with g-boson effects taking into consideration are also shown in these figures. They have been altered by an appreciable amount especially for deformed nuclei. However, the general feature of the phase transition remains the same.

It is interesting to compare figures 4,5,6 with figures 1,2, 3 respectively. The similarities show the common feature of phase transition.

4. Conclusions

Nuclear phase transition reflects the breaking of a dynamic symmetry as a result of competition of two kinds of factors. Quantities characterizing the breaking of dynamic symmetry can be used as order parameters. Observable phase characteristics are mainly governed by these order parameters.

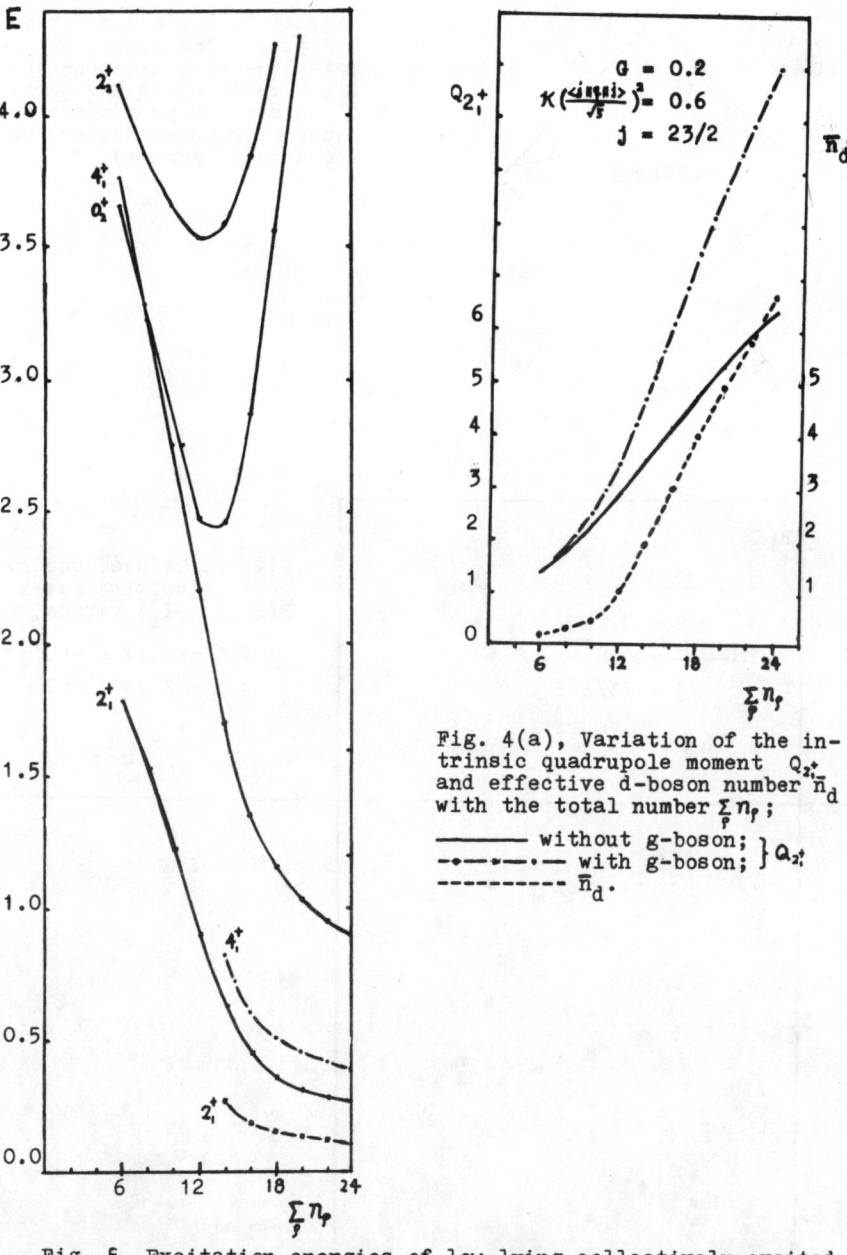

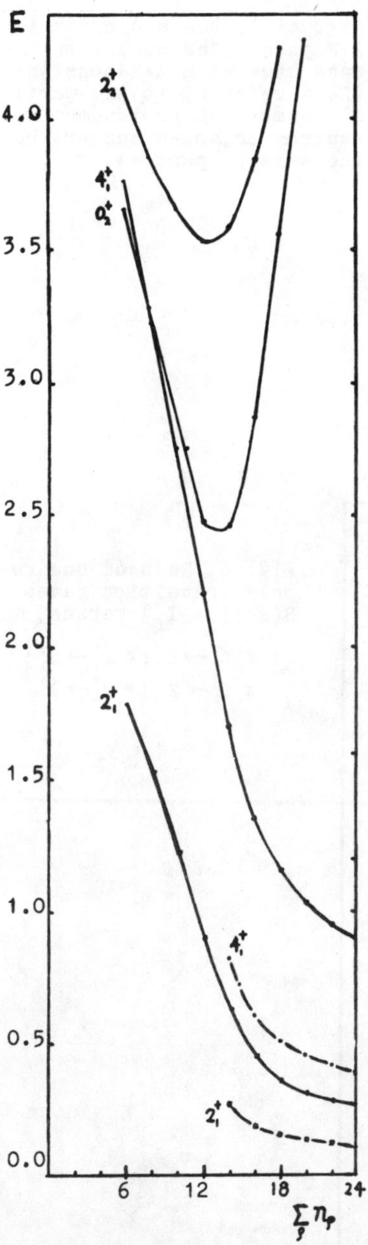

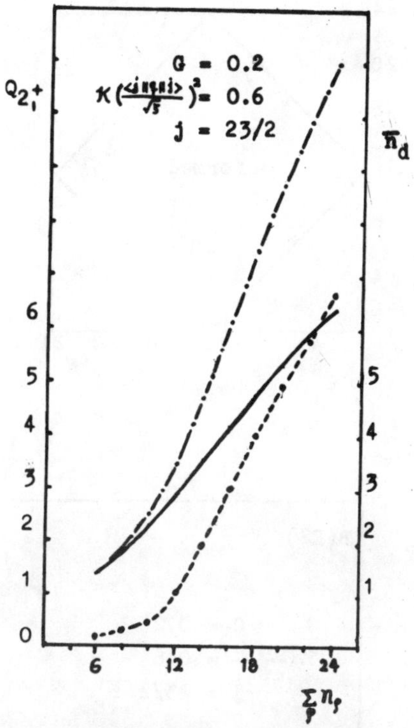

Fig. 4(a), Variation of the intrinsic quadrupole moment $Q_{2_1^+}$ and effective d-boson number $\bar{n}_d$ with the total number $\sum_\rho n_\rho$;
——— without g-boson; ⎫
-•-•-•- with g-boson; ⎭ $Q_{2_1^+}$
- - - - - - $\bar{n}_d$.

Fig. 5, Excitation energies of low-lying collectively excited states versus $\sum_\rho n_\rho$, ——— without g-boson; -•-•- with g-boson.

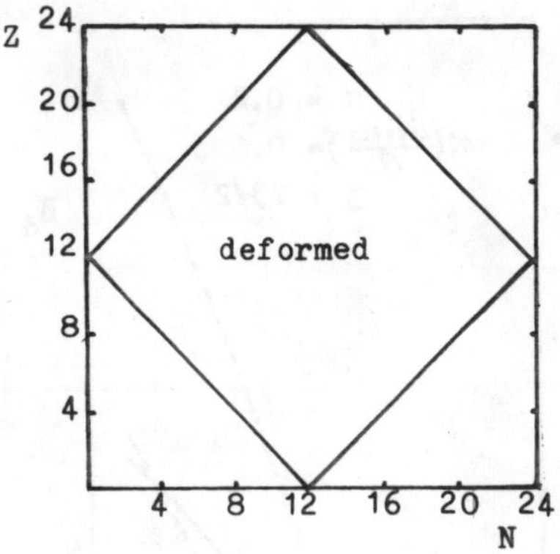

Fig. 4(b), Phase diagram in Z-N plane. The nuclei on the same line with Z+N=constant (Z, N refering to the particle number of proton and neutron or holes number) have the same properties.

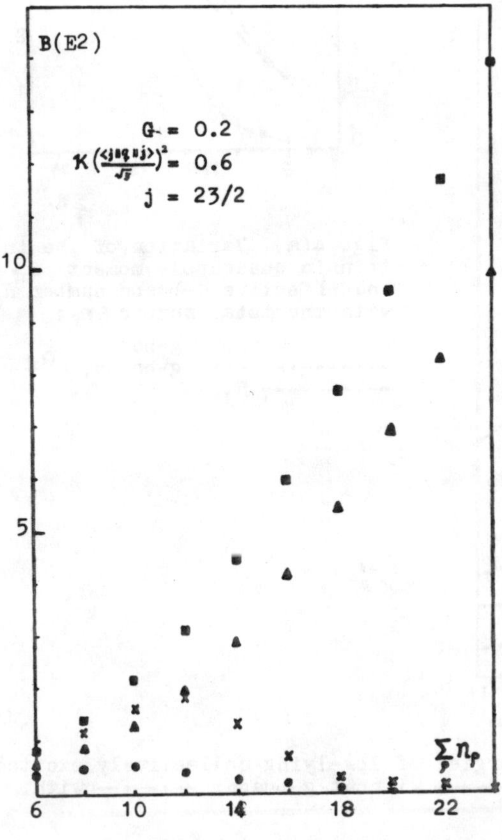

Fig. 6, Reduced quadrupole transition rates $B(E2;I_i{-}{-}I_f)$ versus $\sum_p n_p$.
■ $4_1^+ \to 2_1^+$; ▲ $2_1^+ \to 0_1^+$;
× $0_2^+ \to 2_1^+$; ● $2_2^+ \to 2_1^+$.

References

1) Xu Gong-ou and Zhang Jing-ye, Progr. in Phys. 5(1985)222.
2) R.A. Broglia, F.Barranco and M.Gallardo, Nuclear Structure 1985, Proceedings of the Niels Bohr Centennial Conference, Copenhagen, May 20-24, 1985, p.193.
3) W. Pannert, P.Ring and Y.K.Gambhir, Nucl.Phys.A443(1985)189.
4) Xu Gong-ou, Wang Shun-jin and Yang Ya-tian, Proceedings of the International Symposium on Particle and Nuclear Physics, Beijing, Sept. 2-7, 1985.
5) Xu Gong-ou and Li Fu-li, Comm, in Theor. Phys. 4(1985)39.
6) A. Bohr and B.R. Mottelson, Physica Scripta 22(1980)468; A. Arima, Nuclear Structure 1985, Proceedings of the Niels Bohr Centennial Conference, Copenhagen, May 20-24, 1985,p.147.

GENERATOR COORDINATE METHOD AND DYNAMIC GROUP REPRESENTATION*

Xu Gong-ou

Department of Physics, Nanjing University, Nanjing, China
and
Department of Modern Physics, Lanzhou University, Lanzhou, China

Wang Shun-jin and Yang Ya-tian

Department of Modern Physics, Lanzhou University, Lanzhou, China

It has been shown that the generator coordinate method cooperated with the study of the dynamic symmetry of the system gives precisely the continuous variable representation of the dynamic group. Illustrations are given. Prospects for further applications are discussed.

1. INTRODUCTION

Nuclear collective states of a certain mode having approximately common intrinsic properties form an invariant subspace of the Hamiltonian. Mathematically, this can be related to the representation of a dynamic group. The intrinsic properties are those invariant quantities specifying the irreducible space and the collective states belong to this particular symmetry.

With the help of group theory, the nuclear Hamiltonian with a certain dynamic symmetry can in principle be expressed with continuous variables. It seems to be the most direct way for establishing the microscopic foundation of a nuclear collective model. However, it remains for us to find a convenient method for obtaining the expression explicitly.

The generator coordinate method (GCM) was first proposed for studying nuclear collective motions by solving the Schrödinger equation approximately in a chosen subspace. It is obvious that for the system with a given dynamic symmetry the subspace could be induced by the dynamic group from a certain state inside the subspace studied. The generator coordinate method is in this way cooperated with the study of the dynamic symmetry of the system.

---

* Work supported by the Chinese National Natural Science Foundation and the research fund of the Education Committee of People's Republic of China.

We shall show further that the generator coordinate method cooperated with a dynamic group G is directly related to the group parameter representation of G (DGR-GCM). In whatever cases, either in group space or in coset space, the Dyson representation of the generator coordinate method (generator denoted by $\mathcal{X}^D_\rho(a, \frac{\partial}{\partial a})$) is linearly related to the group parameter representation of the group (generators denoted by $\mathcal{X}_\rho(a, \frac{\partial}{\partial a})$).

2. GENERAL FORMULATION OF DGR-GCM

In DGR-GCM, we start from an arbitrary state vector $|\phi_0\rangle$ of the carrier space of an irreducible representation of the dynamic group allowed by the Pauli principle, and obtain all other state vectors from it with the help of group elements $R(a)$ of the dynamic group G,

$$|\phi(a)\rangle = R(a)\,|\phi_0\rangle$$

For simply connected (universal covering) group which is of interest to physicists, any group element can be expressed as

$$R(a) = \exp(-ia^\sigma X_\sigma)$$

where X_σ are generators in fermion representation. An arbitrary state vector $|\Psi\rangle$ in the irreducible space can be expressed as a superposition of $|\phi(a)\rangle$,

$$|\Psi\rangle = \int|\phi(a)\rangle(a|\mathcal{P}f)\,d\tau(a),$$

where $d\tau(a)$ is the invariant infinitesimal volume element around the group parameter a, $\mathcal{P}$ is a projection operator onto the carrier space of an irreducible representation so that the function $f(a) = (a|\mathcal{P}f)$ will not give vanishing results after integration. f(a) and $|\Psi\rangle$ are in one-to-one correspondence. The Dyson representation of a fermion operator O_F in DGR-GCM is defined as

$$\mathcal{O}^D(a,\frac{\partial}{\partial a})\langle\phi(a)|\Psi\rangle = \langle\phi(a)|O_F|\Psi\rangle$$

Noticing that

$$F(a) = \langle\phi(a)|\Psi\rangle$$

is a function particularly defined in the group space by $|\phi_0\rangle$ through $|\phi(a)\rangle$ and $|\Psi\rangle$, and

$$F(a+da) = \langle\phi(a)|R^+(\delta a)|\Psi\rangle,$$

$$dF(a) = \langle\phi(a)|i a^\rho X_\rho|\Psi\rangle$$

From the definition of the Dyson representation of an arbitrary fermion operator, we have

$$dF(a) = i a^\rho \mathcal{X}^D_\rho(a,\frac{\partial}{\partial a})F(a)$$

But for a function defined in the group space, we have in general

$$dF(a) = a^\rho \mathcal{X}_\rho(a,\frac{\partial}{\partial a})F(a) = \delta a^\rho \Theta^\sigma_\rho(a)\frac{\partial}{\partial a^\sigma}F(a)$$

where $\mathcal{X}^\rho(a,\frac{\partial}{\partial a})$ is the first kind parameter representation of the generator X_ρ and $\Theta_\rho^\sigma(a)$ is connected to the group structure function $\varphi^\sigma(a,b)$ by $\Theta_\rho^\sigma(a) = \frac{\partial\varphi^\sigma(a,b)}{\partial a^\rho}\Big|_{b=0}$. Therefore

$$\mathcal{X}_\rho^{o}(a,\frac{\partial}{\partial a}) = -i\,\mathcal{X}_\rho(a,\frac{\partial}{\partial a}) = -i\Theta_\rho^\sigma(a)\frac{\partial}{\partial a^\sigma}$$

Usually the Hamiltonian of a quantum system has a certain kind of symmetry, therefore the dynamical group should contain the symmetry group S as its subgroup. We start from an invariant state vector $|\phi_0\rangle$ of S, then $S|\phi_0\rangle = \exp(ia'^\sigma X'_\sigma)|\phi_0\rangle = |\phi_0\rangle$. $|\phi(a)\rangle\rangle = \exp(a''^\rho X''_\rho)\cdot\exp(a'^\sigma X'_\sigma)|\phi_0\rangle = \exp(a''^\rho X''_\rho)|\phi_0\rangle \Rightarrow |\phi(a'')\rangle\rangle$. Where $\exp(a''^\rho X''_\rho)\in G/S$ and a" are parameters in coset space G/S. In this case, DGR-GCM yields the coset parameter representation,

$$\mathcal{X}_\rho^{o}(a'',\frac{\partial}{\partial a''}) = -i\,\mathcal{X}_\rho(a'',\frac{\partial}{\partial a''}) = -i\Theta_\rho^\sigma(a'')\frac{\partial}{\partial a''^\sigma}\,.$$

If the maximal stationary subgroup H of G contains the symmetry group S as its subgroup, we can start from the lowest weight state $|\phi_0\rangle$ and express other states with raising operatos X'''_ρ only. Hence we have

$$\exp(ia'^\sigma X'_\sigma)|\phi_0\rangle = \exp(ia'^\sigma\Lambda_\sigma)|\phi_0\rangle,\quad \exp(ia'^\sigma X'_\sigma)\in H,$$

$$|\phi(a)\rangle\rangle = |\phi(a''')\rangle\rangle \Rightarrow \exp(ia'''^\rho X'''_\rho)|\phi_0\rangle\,.$$

In this case, DGR-GCM gives the parameter representation in coset space G/H,

$$\mathcal{X}_\rho^{o}(a''',\frac{\partial}{\partial a'''}) = -i\mathcal{X}_\rho(a''';\frac{\partial}{\partial a'''}) = -i\left[\Theta_\rho^\sigma(a''')\Lambda_\sigma + \Theta_\rho^\nu(a''')\frac{\partial}{\partial a'''^\nu}\right]$$

This representation can be further transfomed into boson representation. Thus boson representation is a special case of DGR-GCM.

It is shown that the DGR-GCM fulfils the following physical requirements:(1) Preservation of space-time symmetry,(2) Pauli principle,(3) dynamical restrictions. The salient feature of DGR-GCM is that the physical conditions (the quantum numbers of an irreducible representation) can be incorporate into the continuous representation of the group generators. This will facilitate the study of nuclear phase transitions. It is also shown that DGR-GCM is precisely a transfomation theory between the fermion representation and the collective variable representation of a dynamical group. The embedded representation are easily manageable in DGR-GCM.

3. APPLICATIONS OF DGR-GCM TO SIMPLE CASES

Applications of DGR-GCM to the Lipkin SU(2) model and the Elliott SU(3) model are given for illustration.

For the SU(2) Lipkin model, five different representations are given: (i) The group space parameter representation is just that of angular momentum operator of a rigid body in Euler angles. (ii) The coset SU(2)/SO(2) parameter representation is that of orbital angular momentum operator of a particle in polar angles. (iii) One-boson representation. (iv) Three-boson representation where SU(2) is embedded in SP(4). (v) Two-boson representation where SU(2) is embedded in SU(3) (Schwinger representation). Different representations are equivalent in algebraic structures but different in their physical implications as their corresponding fermion operators and applicable cases are concerned.

For the Elliott SU(3) model, four different representations given by DGR-GCM are: (i) The group space parameter representation (eight parameters), (ii) The coset SU(3)/SO(3) parameter representation (five parameters). (iii) Two-boson representation. (iv) Six-boson representation where SU(3) is embedded in SP(6).

4. CONCLUSIONS

It has been shown that the generator coordinate method cooperated with the study of dynamic symmetry of the system gives precisely the continuous variable representation of the dynamic group. Applications of DGR-GCM to the Lipkin SU(2) model and the Elliott SU(3) model are given for illustration. Further applications to microscopic investigations of nuclear models and phase transition studies are under way, and partly given in author's talk at the symposium. What we have learned from the studies of DGR-GCM and its applications indicates that the DGR-GCM is very effective in the investigations of (i) microscopic foundation of nuclear collective models, (ii) relationship between various nuclear collective models, and (iii) nuclear phase transitions.

REFERENCE

1. Xu Gong-ou, Nucl. Phys., A421 (1984) 275C
2. Xu Gong-ou, Wang Shun-jin and Yang Ya-tian, "Generator coordinate method as a representation theory of dynamic group" to be published.
3. Xu Gong-ou, Li Fu-li and Fu De-ji, Talk at this symposium "Nuclear phase transition illustrated with a schematic model".

SELF-CONSISTENT STRUCTURE OF BOSONS IN NUCLEI

Li-ming Yang, Zhi-ning Zhou and Da-hai Lu

Department of Physics, Peking University

After the advent of the successful phenomenological interecting boson model[1,2], many attempts have been made to establish microscopic foundation of this model. Much of these efforts proceeds by way of some form of mapping, i.e. constructing basic states built from fermion pairs coupled to L=0 and L=2 which are later mapped into boson[3]. Most works are done in the single j-shell space or the degenerate many j-shell space where the fermion pairs can be defined uniquely. Some authers work in the non-degenerate many j-shell space. Ring et al.[4] defined the s-pair by means of the number projected BCS ground state and d-pair by diagonaling H in the space spanned by the one broken pair approximation. Brink tried to define the d-pair structure by minimizing E(2+), taking |2+> as one broken pair state. In this way these authors took care partially the effect of the presence of s-pairs on one d-pair within the BCS approximation. These treatments may be a good approximation for the vibrator nuclei. However in well deformed nuclei the wave function in boson space contains mainly components with two and more d-pairs even in the groung state. One should therefore consider the boson structure problem when two or more d-pairs are present, and the effect of these d-pairs on the s-pair structure.

We have made a quite different approach toward these problems[6,7,8]. One of the essential steps in our approach is to construct certain sophisticated fermion pair structures that behave like s and d bosons in our model space. In (7) a variation principle is devised to obtain the structure functions φ_j and $\psi_{jj'}$ for the s and d pairs. Since it is the many-body wave function that is varied in (7), it is hard to relate it to the single boson structure. In this note we have devised a new variational method which emphasize the single boson structure, namely, the struture of a single self-consistent boson in a many boson state.

The elementary s and d pairs are defined as follows:

$$\mathcal{S}^{(\sigma)+} = \sum_j \varphi_j^{(\sigma)} \mathcal{S}_j^+ \tag{1}$$

$$\chi_{2\mu}^{(\tau)+} = \sum_{jj'} \psi_{jj'} A_{2\mu}^+(jj') \tag{2}$$

where j=j ,j ,..., j , σ=1, 2, ..., f, f is the number of single particle ortitals in our model space, τ=1,2,3,...,g, g is the number of possible pairs(j,j) that can be coupled to $L^\pi=2^+$. $\mathcal{S}_j^+$ and $A_{2\mu}^+(j,j')$ are defined as follows:

$$\mathcal{S}_j^+ = \hat{S}_{+j}(\hat{S}_j - \hat{S}_{0j})^{-1/2} \tag{3}$$

$$A_{2\mu}^+(j,j') = \frac{1}{\sqrt{2}} (a_j^+ a_{j'}^+)_{2\mu} \tag{4}$$

where

$$\hat{S}_{+j} = (c_j^+ c_j^+)_0 \sqrt{\frac{\Omega_j}{2}}$$

$$\hat{S}_j = \frac{1}{2}(\Omega_j - \hat{n}_j(a))$$
$$\hat{S}_{0j} = \frac{1}{2}(\hat{n}_j - \Omega_j)$$
$$\hat{n}_j = \sum_m c^+_{jm} c_{jm}, \qquad \hat{n}_j(a) = \sum_m a^+_{jm} a_{jm}, \qquad \Omega_j = j + \frac{1}{2}$$

The boson-like s-pair $\mathcal{S}^+_j$ and the operator a^+_j, called quasi-fermion, are related to the nucleon creation operator c^+_{jm} in a given j-shell by a operatorized Bogoliubov transformation introduced in (6).

To illustrate the method, the simplest case is considered first, i.e., a state with s-bosons only. Here the s-pair in a many j-shell space is connected with the s-boson in our model space by the relation (7,8)

$$S^{(1)+} = n_p^{-1/2} \mathcal{S}^{(1)+} n_p^{1/2} \tag{5}$$

where n_p is the norm of the p s-pair state $(\mathcal{S}^{(1)+})^p|0\rangle/\sqrt{p!}$. In the following we sometime refer $\mathcal{S}^{(1)+}$ also as s-boson for simplicity.

Our task is to find the best structure function of the pairs $\varphi_j^{(\sigma)}$ taking account of the effect the presence of the other pairs i.e., self-consistent single boson structure functions. To start with, one may choose the $\varphi_j^{(\sigma)}$'s to be the TDA solutions of two particles outside a close shell.

Let $|p,\sigma\rangle\!\rangle$ be a state with p s-pairs of type

$$|p,\sigma\rangle\!\rangle = \frac{1}{\sqrt{p!}}(\mathcal{S}^{(\sigma)+})^p|0\rangle \tag{6}$$

Owing to Pauli principle $|p,\sigma\rangle\!\rangle$ is not normalized. In general $n_{p\sigma} = \langle\!\langle p,\sigma|p,\sigma\rangle\!\rangle < 1$. The $\varphi_j^{(\sigma)}$'s in $\mathcal{S}^{(\sigma)+}$ are some preliminarily determined structure functions satisfying

$$\sum_j \varphi_j^{(\sigma)}\varphi_j^{(\sigma')} = \delta_{\sigma\sigma'}$$
$$\sum_\sigma \varphi_j^{(\sigma)}\varphi_{j'}^{(\sigma)} = \delta_{jj'} \tag{7}$$

$\sigma = 1$ corresponds to the lowest lying pair to be retained in our model space. In the following $\mathcal{S}^+$ will be written for $\mathcal{S}^{(1)+}$ and $|p\rangle\!\rangle$ for $|p,\sigma=1\rangle\!\rangle$ for brevity. Since in the present case it's the self-consistent boson in the state $|p\rangle$ that is of immediate concern to us, we try to improve the boson structure by introducing slight variation in the single boson state

$$\mathcal{S}'^+ = \mathcal{S}^+ + \sum_{\sigma>1} \delta\eta_\sigma \mathcal{S}^{(\sigma)+} \tag{8}$$
$$|p\rangle\!\rangle' = \frac{1}{\sqrt{p!}}(\mathcal{S}'^+)^p|0\rangle \tag{9}$$

This variation is the most general one for the single boson state within our finite model boson space with the constraint

$$\langle 0|\mathcal{S}'\mathcal{S}^+|0\rangle = 1$$

and $\mathcal{S}'^+|0\rangle$ is normalized up to first order in $\delta\eta$

$$\langle 0|\mathcal{S}'\mathcal{S}'^+|0\rangle = 1 + O((\delta\eta)^2) \tag{10}$$

To first order one has

$$\delta|p\rangle\!\rangle = |p\rangle\!\rangle' - |p\rangle\!\rangle = \sqrt{p}\sum_{\sigma>1}\delta\eta_\sigma \mathcal{S}^{(\sigma)+}(\mathcal{S}^+)^{p-1}|0\rangle / \sqrt{(p-1)!} \tag{11}$$

Since in general $p>\Omega_j$ for some j, one has $\langle\langle p|(\delta|p\rangle\rangle)\neq 0$. In fact $\langle\langle p|(\delta|p\rangle\rangle)$ contains terms linear in $\delta\eta$. The result obtained from the variation principle

$$\delta\left\{\frac{\langle\langle p|\ H\ |p\rangle\rangle}{\langle\langle p|p\rangle\rangle}\right\}=0$$

is $\langle\langle p|(H-E_p)(\delta|p\rangle\rangle) = 0$ (12)

where

$$E_p = \frac{\langle\langle p|H|p\rangle\rangle}{\langle\langle p|p\rangle\rangle} \tag{13}$$

inspite of the constraint $\langle 0|\mathcal{S}'\mathcal{S}^{+}|0\rangle = 1$

From (11) and (12) one obtains

$$\langle p=1,\ \sigma=1\ |\ \frac{[(\mathcal{S})^{p-1}(H-E_p)(\mathcal{S}^{+})^{p-1}]_{s.p.}}{(p-1)!\ \eta_p}|\ p=1,\ \sigma>1\rangle = 0 \tag{14}$$

where s.p. attached to the square bracket signified that only single boson parts of the operator inside the bracket are to be retained. Thus the single boson hamiltonian $\hat{h}$ is identified to be

$$\hat{h}^{(1)}= [\ \mathcal{S}^{p-1}(\ H\ -\ E_p)\ (\mathcal{S}^{+})^{p-1}]_{s.p.} /\ [(p-1)!\ \eta_p] \tag{15}$$

Note that $\hat{h}^{(1)} = \hat{h}(\varphi_j^{(1)})$ is a function of $\varphi_j^{(1)}$. If $p<\Omega_i(i=1,2,\ldots f)$ were satisfied, the second term in (15) would vanish and

$$\hat{h}^{(1)}=[\mathcal{S}^{p-1}H\ (\mathcal{S}^{+})^{p-1}]_{s.p.}/[(p-1)!\,\eta_p]$$

would have vanishing matrix element between $|\sigma=1\rangle=\mathcal{S}^{+}|0\rangle$ and $|\sigma>1\rangle=\mathcal{S}^{(\sigma>1)+}|0\rangle$. The normalization factor in (15) appears naturally when no numerical factors are dropped from the expression deduced from the variation principle.

What one gets directly from the variation principle (12) is that the best single boson hamiltonian $\hat{h}^{(1)}$ is to have no matrix element between the one boson states $\sigma = 1$ and $\sigma>1$. $\hat{h}^{(1)}$ may be written in the form

$$\hat{h}^{(1)} = \sum_{\sigma\sigma'} h^{(1)}_{\sigma\sigma'}\mathcal{S}^{(\sigma)+}\mathcal{S}^{(\sigma')}= \sum_{jj'} h^{(1)}_{jj'}\mathcal{S}_j^{+}\mathcal{S}_{j'} \tag{16}$$

where

$$h^{(1)}_{jj'} = \sum_{\sigma\sigma'} h^{(1)}_{\sigma\sigma'}\,\varphi_j^{(\sigma)}\varphi_{j'}^{(\sigma')}$$

Since nothing is said about the matrix element in the $\sigma=1$ space ,one can choose the single boson states so that they diagonalize $h^{(1)}_{jj'}$

$$\sum_{j'} h^{(1)}_{jj'}\,\varphi_{j'}^{(\sigma)}= \varepsilon^{(\sigma)}\varphi_j^{(\sigma)} \tag{17}$$

The lowest state $\varphi_j^{(1)}$ will be used to construct the new $\hat{h}^{(1)}$. This process continues until self-consistency is achieved.

To see more explicitly the form of single boson hamiltonian, we choose an effective hamiltionian (7,8) which after our operatorized Bogoliubov transformation (6) becomes

$$H = H_{sp}+ H_{P_0} + H_{P_2} + H_{Q} \tag{18}$$

$$H_{sp}=\sum_{1}\epsilon_1(\hat{n}_1(a)+2\hat{n}_s(1))$$

$$H_{P_0}=-G_0\sum_{12}\mathcal{S}_1^+(2\hat{S}_1-\hat{n}_s(1))^{1/2}(2\hat{S}_2-\hat{n}_s(2))^{1/2}\mathcal{S}_2$$

$$\begin{aligned}H_{P_2}=-G_2\sum_{12,34}Q_{12}Q_{34}\{&u_1'[A_{\tilde{2}}(12)u_2u_3A_2^+(34)]^{(0)}u_4'\\&+v_1^+\mathcal{S}_2^+[A_{\tilde{2}}(12)(2\hat{S}_2+1)^{-1/2}(2\hat{S}_3+1)^{-1/2}A_2^+(34)]^{(0)}\mathcal{S}_3v_4\\&+2v_2^+[B_2^+(12)u_1u_4B_2^+(34)]^{(0)}v_3\}\end{aligned}$$

$$\begin{aligned}H_Q=-\frac{\kappa}{\sqrt{5}}\sum_{12,34}Q_{12}Q_{34}\{&u_1'[A_2^+(12)v_2v_3^+A_{\tilde{2}}(34)]^{(0)}u_4'\\&+v_1^+[A_{\tilde{2}}(12)u_2'u_3'A_2^+(34)]^{(0)}v_4\\&+\frac{1}{2}[(u_2B_2^+(12)u_1-v_2^+B_2^+(12)v_1)(u_4B_2^+(34)u_3-v_4^+B_2^+(34)v_3)]^{(0)}\}\end{aligned}$$

where

$$A^+_{2\mu}(12)=\frac{1}{\sqrt{2}}(a^+_{j_1}a^+_{j_2})^{(2)}_\mu,\qquad B^+_{2\mu}(12)=(a^+_{j_1}a_{\tilde{j}_2})^{(2)}_\mu,$$

$$Q_{12}\equiv Q_{j_1j_2}=\langle n_1l_1j_1||r^2Y_2||n_2l_2j_2\rangle,$$

$$v_1^+=\mathcal{S}^+_{j_1}/\sqrt{2\hat{S}_{j_1}},\qquad u_1=\left(1-\frac{\hat{n}_s(j_1)}{2\hat{S}_{j_1}}\right)^{1/2},\qquad u_1'=\left(1-\frac{\hat{n}_s(j_1)}{2\hat{S}_{j_1}+1}\right)^{1/2}.$$

Owing to Pauli principle the picking up of the single boson part of $[\mathcal{S}^{p-1}H(\mathcal{S}^+)^{p-1}]$ and $[\mathcal{S}^{p-1}(\mathcal{S}^+)^{p-1}]$ is by no means simple. The communtator of $\mathcal{S}$ and $\mathcal{S}^+$ depends on the state on which they acts. Thus in evaluating $h^{(1)}_{jj'}$ one first calculates the matrix elements $\langle 0|\mathcal{S}_j[\mathcal{S}^{p-1}(H-E_p)(\mathcal{S}^+)^{p-1}]\mathcal{S}^+_{j'}|0\rangle$. The diagonal parts contain terms that come from the constant part of $[\mathcal{S}^{p-1}(H-E_p)(\mathcal{S}^+)^{p-1}]$. These terms must be evaluated and substracted in $h^{(1)}_{jj'}$.

Corresponding to the four parts in H, there are four terms in h i.e.

$$h_{j_1j_2}(sp),\quad h_{j_1j_2}(p_0),\quad h_{j_1j_2}(p_2),\quad h_{j_1j_2}(Q)$$

where

$$\begin{aligned}h_{12}(sp)&=\sum_3 2\epsilon_3\langle 0|\mathcal{S}_1\mathcal{S}^{p-1}\mathcal{S}_3^+\mathcal{S}_3(\mathcal{S}^+)^{p-1}\mathcal{S}_2^+|0\rangle/[(p-1)!\,\mathcal{N}_p]\\h_{12}(p_0)&=-G_0\sum_{34}\langle 0|\mathcal{S}_1\mathcal{S}^{p-1}\mathcal{S}_3^+(\Omega_3-\hat{n}_s(3))^{1/2}(\Omega_4-\hat{n}_s(4))^{1/2}\mathcal{S}_4(\mathcal{S}^+)^{p-1}\mathcal{S}_2^+|0\rangle/[(p-1)!\,\mathcal{N}_p]\\h_{12}(p_2)&=-5G_2\sum_{34}\frac{Q_{34}^2}{\Omega_3(\Omega_4-\delta_{34})}\langle 0|\mathcal{S}_1\mathcal{S}^{p-1}\mathcal{S}_3^+\mathcal{S}_4^+\mathcal{S}_4\mathcal{S}_3(\mathcal{S}^+)^{p-1}\mathcal{S}_2^+|0\rangle/[(p-1)!\,\mathcal{N}_p]\\h_{12}(Q)&=-\frac{\kappa}{2}\sum_{34}Q_{34}\Big\{\frac{1}{\sqrt{\Omega_3\Omega_4}}\langle 0|\mathcal{S}_1\mathcal{S}^{p-1}\mathcal{S}_3^+\Big(1-\frac{\hat{n}_s(3)}{\Omega_3-\delta_{34}}\Big)^{1/2}\Big(1-\frac{\hat{n}_s(4)}{\Omega_4-\delta_{34}}\Big)^{1/2}\mathcal{S}_4(\mathcal{S}^+)^{p-1}\mathcal{S}_2^+|0\rangle\\&\quad+\frac{1}{\Omega_3}\langle 0|\mathcal{S}_1\mathcal{S}^{p-1}\mathcal{S}_3^+\Big(1-\frac{\hat{n}_s(4)}{\Omega_4-\delta_{34}}\Big)\mathcal{S}_3(\mathcal{S}^+)^{p-1}\mathcal{S}_2^+|0\rangle\Big\}/[(p-1)!\,\mathcal{N}_p]\end{aligned}\tag{19}$$

where $\delta_{12}\equiv\delta_{j_1j_2}$, $\Omega_3\equiv\Omega_{j_3}$. One can write $h_{j_1j_2}$ explicity as follows

$$\begin{aligned}h_{12}\equiv h_{j_1j_2}=\;&h_{12}(sp)+h_{12}(p_0)+h_{12}(p_2)+h_{12}(Q)\\&-E_p\frac{1}{\mathcal{N}_p}(p-1)!\sum_{\substack{[p-1]\\0\le p_i\le\Omega_i-\delta_{i2}\\\delta_{i1}-\delta_{i2}\le p_i\le\Omega_i-\delta_{i2}}}(p_1+\delta_{12})\prod_i\frac{\varphi_i^{2p_i}}{p_i!}-\frac{\mathcal{N}_{p-1}}{\mathcal{N}_p}(E_{p-1}-E_p)\delta_{12}\end{aligned}\tag{20}$$

where
$$E_p=\sum_{12}\varphi_1\varphi_2\frac{1}{p}(h_{12}(sp)+h_{12}(p_0)+h_{12}(p_2)+h_{12}(Q))\tag{21}$$

$$\mathcal{N}_p = \sum_{\substack{[p] \\ 0\le p_i \le \Omega_i}} p! \prod_i \frac{1}{p_i!} \varphi_i^{2p_i} \tag{22}$$

The sixth term in (20) is the constant part of $[\mathcal{S}^{p-1}(H-E_p)(\mathcal{S}^+)^{p-1}]$.

Since h is a complicated non-linear function of φ_j, the diagonalization of $h_{j_1 j_2}$ has to be solved successively for p=2,3,... by iteration. Starting from some preliminarily determined structure function φ_j for p=1 one can calculate $\mathcal{N}_2$, $h_{j_1 j_2}$ (p=2), then solve the eigenvalue equation for p=2.

$$\sum_2 h_{12}\varphi_2^{(\sigma)} = \mathcal{E}^{(\sigma)}\varphi_1^{(\sigma)}$$

Repeat this process until self-consistency is obtained.

Along similar line one can construct self-consistent hamiltonian for single s-pair and d-pair in a state in which there are p s-pairs and q d-pairs

$$|p\ q\ \alpha\ J\ M> = \frac{1}{\sqrt{p!q!}}(\mathcal{S}^+)^p\ (X_2^+)^q_{\alpha JM}\ |0> \tag{23}$$

where $\mathcal{S}^+$ and X_2^+ stand for $\mathcal{S}^{(1)+}$ and $X_2^{(1)+}$.

Introduce the new s-pair and d-pair and the new state vector

$$\mathcal{S}'^+ = \mathcal{S}^+ + \sum_{\sigma>1}\delta\eta_\sigma \mathcal{S}^{(\sigma)+} \tag{24}$$

$$X'^+_{2\mu} = X^+_{2\mu} + \sum_{\tau>1}\delta\zeta_\tau X^{(\tau)+}_{2\mu} \tag{25}$$

$$|p\ q\ \alpha\ J\ M\gg' = \frac{1}{\sqrt{p!q!}}(\mathcal{S}'^+)\ (X'^+_2)^q_{\alpha JM}\ |0>$$

One gets the variation of $|p\ q\ \alpha\ J\ M>$ to first order in $\delta\eta_\sigma$ and $\delta\zeta_\tau$ as follows

$$\delta|p\ q\ \alpha\ J\ M\gg = \sqrt{p}\sum_{\sigma>1}\delta\eta_\sigma\frac{(\mathcal{S}^+)^{p-1}}{\sqrt{(p-1)!}}\mathcal{S}^{(\sigma)+}\frac{(X_2^+)^q_{\alpha JM}}{\sqrt{q!}}|0>$$

$$+\sqrt{q}\sum_{\tau>1}\delta\zeta_\tau\frac{(\mathcal{S}^+)^p}{\sqrt{p!}}\Big[\frac{(X_2^+)^{q-1}_{\alpha_1 J_1}}{\sqrt{(q-1)!}}\cdot X_2^{(\tau)+}\Big]_{JM}|0>\ (q-1,\alpha_1 J_1,2|\}q\,\alpha J\ ) \tag{26}$$

From the variation principle

$$\delta\left\{\frac{\ll pq\alpha JM\,|\,H\,|\,pq\alpha JM\gg}{\ll pq\alpha JM\,|\,pq\alpha JM\gg}\right\} = 0$$

one gets

$$<0|\mathcal{S}\Big[\frac{(X_2)^q_{\alpha JM}\mathcal{S}^{p-1}}{\sqrt{q!}\ \sqrt{(p-1)!}}(H-E_{pq\alpha J})\frac{(\mathcal{S}^+)^{p-1}(X_2^+)^q_{\alpha JM}}{\sqrt{(p-1)!}\ \sqrt{q!}}\Big]\mathcal{S}^{(\sigma)+}|0>\frac{1}{\mathcal{N}_{pq\alpha J}} = 0 \tag{27}$$

$$\sum_{\substack{\alpha_1 J_1 M_1\\ \alpha_1' J_1' M_1'}}(q-1,\alpha_1 J_1,2|\}q\,\alpha J\ )(q-1,\alpha_1' J_1',2|\}q\,\alpha J\ )<J_1 M_1 2\,\mu_1|JM><J_1' M_1' 2\,\mu_1'|JM>$$

$$<0|X_{2\mu_1}\Big[\frac{(X_2)^{q-1}_{\alpha_1 J_1 M_1}\mathcal{S}^p}{\sqrt{(q-1)!}\ \sqrt{p!}}(H-E_{pq\alpha J})\frac{(\mathcal{S}^+)^p(X_2^+)^{q-1}_{\alpha_1' J_1' M_1'}}{\sqrt{p!}\ \sqrt{(q-1)!}}\Big]X^{(\tau)+}_{2\mu_1'}|0>\ /\ \mathcal{N}_{pq\alpha J} = 0 \tag{28}$$

where $E_{pq\alpha J} = <p\ q\ \alpha\ J\ M\ |\ H\ |\ p\ q\ \alpha\ J\ M>\ /\ \mathcal{N}_{pq\alpha J}$ (29)

To illustrate the method, the case of q=1 is considered in the following. For q=1, (27) and (28) become simple.

$$<0|\ \mathcal{S}^{(1)}\hat{h}^{(0)}\mathcal{S}^{(\sigma)+}|0> = 0 \qquad (\sigma>1) \tag{30}$$

$$<0|\ X^{(1)}_{2\mu}\hat{h}^{(2)}X^{(\tau)+}_{2\mu}|0> = 0 \qquad (\tau>1) \tag{31}$$

where

$$\hat{h}^{(0)} = [\, X_{2\mu} \frac{\mathcal{S}^{P-1}}{\sqrt{(P-1)!}} (H-E_{P1}) \frac{(\mathcal{S}^{+})^{P-1}}{\sqrt{(P-1)!}} X^{+}_{2\mu} / n_{P1}]_{\text{s.p. of } \mathcal{S}} \tag{32}$$

$$= \sum_{\sigma\sigma'} h^{(0)}_{\sigma\sigma'} \mathcal{S}^{(\sigma)+} \mathcal{S}^{(\sigma')}$$

$$\hat{h}^{(2)} = [\frac{\mathcal{S}^{P}}{\sqrt{P!}} (H-E_{P1}) \frac{(\mathcal{S}^{+})^{P}}{\sqrt{P!}} / n_{P1}]_{\text{s.p. of } X_2} \tag{33}$$

$$= \sum_{\tau\tau'} h^{(2)}_{\tau\tau'} [\, X^{(\tau)+}_{2} X^{(\tau')}_{2}]^{(0)}$$

The square bracket in RHS of $\hat{h}^{(0)}$ ($\hat{h}^{(2)}$) contains only the single particle part of the s-pair (d-pair) operator and the constant part of s-pair (d-pair) operator.

$\hat{h}^{(0)}$ and $\hat{h}^{(2)}$ are the single pair harmiltonian for the s-pair and d-pair respectively in the state $|p,1,2\mu\rangle$ and can be rewritten as follows

$$\hat{h}^{(0)} = \sum_{12} h^{(0)}_{12} \mathcal{S}^{+}_{1} \mathcal{S}_{2} \tag{32'}$$

$$\hat{h}^{(2)} = \sum_{1234} h^{(2)}_{12,34} [\, A^{+}_{2}(12)\ A_{\tilde{2}}(34)]^{(0)} \tag{33'}$$

where

$$h^{(0)}_{12} = \sum_{\sigma\sigma'} h^{(0)}_{\sigma\sigma'} \varphi^{(\sigma)}_{1} \varphi^{(\sigma')}_{2}$$

$$h^{(2)}_{12,34} = \sum_{\tau\tau'} h^{(2)}_{\tau\tau'} \psi^{(\tau)}_{12} \psi^{(\tau')}_{34}$$

$h^{(0)}_{12}$ and $h^{(2)}_{12,34}$ are both functionals of $\varphi^{(1)}_{j}$, $\psi^{(1)}_{jj'}$, for given trial functions $\varphi^{(1)}_{j}$, , one can solve simultaneously

$$\sum_{2} h^{(0)}_{12} \varphi^{(\sigma)}_{2} = \mathcal{E}^{(\sigma)}_{0} \varphi^{(\sigma)}_{1}$$

$$\sum_{34} h^{(2)}_{12,34} \psi^{(\tau)}_{34} = \mathcal{E}^{(\tau)}_{2} \psi^{(\tau)}_{12}$$

The lowest eigenstates $\varphi^{(\sigma=1)}_{j}$, $\psi^{(\tau=1)}_{jj'}$ are used to calculate $h^{(0)}_{12}$, $h^{(2)}_{12,34}$ again. The process is to be repeated till self-consistency is achieved.

The explicit forms of $h^{(0)}_{12}$ and $h^{(2)}_{12,34}$ for an effective hamiltonian like (18) are

$$h^{(0)}_{12} \equiv h^{(0)}_{j_1 j_2} = h^{(0)}_{12}(\text{sp}) + h^{(0)}_{12}(p_0) + h^{(0)}_{12}(p_2) + h^{(0)}_{12}(Q)$$

$$- E_{P1} \frac{1}{n_{P1}} (p-1)! \sum_{34} \tilde{\psi}^{2}_{34} \sum_{\substack{[p-1] \\ 0 \le P_i \le \Omega_i - \delta_{i2} - \delta_{i3} - \delta_{i4} \\ \delta_{i1} - \delta_{i2} \le P_i \le \Omega_i - \delta_{i2} - \delta_{i3} - \delta_{i4}}} (p_1 + \delta_{12}) \prod_{i} \frac{1}{P_i!} \varphi^{2P_i}_{i}$$

$$- \frac{n_{p-1,1}}{n_{p1}} (\, E_{p-1,1} - E_{p1}) \delta_{12}$$

$$h^{(2)}_{12,34} \equiv h^{(2)}_{j_1 j_2, j_3 j_4} = h^{(2)}_{12,34}(\text{sp}) + h^{(2)}_{12,34}(p_0) + h^{(2)}_{12,34}(p_2) + h^{(2)}_{12,34}(Q)$$

$$- E_{P1} \frac{1}{n_{P1}} p!\, \delta_{13} \delta_{24} \sum_{\substack{[p] \\ 0 \le P_i \le \Omega_i - \delta_{i1} - \delta_{i2}}} \prod_{i} \frac{1}{P_i!} \varphi^{2P_i}_{i}$$

$$- \frac{n_{P0}}{n_{P1}} (\, E_{P0} - E_{P1}) \delta_{13} \delta_{24}$$

where

$$h^{(0)}_{12}(\text{sp}) = \sum_{3456} \psi_{34} \psi_{56} \langle 0|\, \mathcal{S}_1 A_{2\mu}(34) \mathcal{S}^{P-1} H_{sp} (\mathcal{S}^{+})^{P-1} A^{+}_{2\mu}(56)\, \mathcal{S}^{+}_{2} |0\rangle / [(p-1)!\, n_{P1}] ,$$

$$h^{(0)}_{12}(P_0) = \sum_{3456} \psi_{34}\psi_{56} \langle 0| \mathcal{S}_1 A_{2\mu}(34)\mathcal{S}^{p-1} H_{P_0} (\mathcal{S}^+)^{p-1} A^+_{2\mu}(56)\, \mathcal{S}^+_2 |0\rangle / [(p-1)!\,n_{p1}] ,$$
$$h^{(0)}_{12}(P_2) = \sum_{3456} \psi_{34}\psi_{56} \langle 0| \mathcal{S}_1 A_{2\mu}(34)\mathcal{S}^{p-1} H_{P_2} (\mathcal{S}^+)^{p-1} A^+_{2\mu}(56)\, \mathcal{S}^+_2 |0\rangle / [(p-1)!\,n_{p1}] ,$$
$$h^{(0)}_{12}(Q) = \sum_{3456} \psi_{34}\psi_{56} \langle 0| \mathcal{S}_1 A_{2\mu}(34)\mathcal{S}^{p-1} H_{Q} (\mathcal{S}^+)^{p-1} A^+_{2\mu}(56)\, \mathcal{S}^+_2 |0\rangle / [(p-1)!\,n_{p1}] ,$$

and

$$h^{(2)}_{12,34}(sp) = \langle 0| A_{2\mu}(12)\mathcal{S}^p H_{sp} (\mathcal{S}^+)^p A^+_{2\mu}(34) |0\rangle / [p!\,n_{p1}] ,$$
$$h^{(2)}_{12,34}(P_0) = \langle 0| A_{2\mu}(12)\mathcal{S}^p H_{P_0} (\mathcal{S}^+)^p A^+_{2\mu}(34) |0\rangle / [p!\,n_{p1}] ,$$
$$h^{(2)}_{12,34}(P_2) = \langle 0| A_{2\mu}(12)\mathcal{S}^p H_{P_2} (\mathcal{S}^+)^p A^+_{2\mu}(34) |0\rangle / [p!\,n_{p1}] ,$$
$$h^{(2)}_{12,34}(Q) = \langle 0| A_{2\mu}(12)\mathcal{S}^p H_{Q} (\mathcal{S}^+)^p A^+_{2\mu}(34) |0\rangle / [p!\,n_{p1}] .$$

The details about $h^{(1)}_{12}$, $h^{(0)}_{12}$ and $h^{(2)}_{12,34}$ will be published elsewhere.

In conclusion it can be said a new method has been divised to calculate the self-consistent structure of nucleons pairs (bosons) in a system which contains p s-pair and q d-pair. These pairs must be less than half-filled in the model space considered in order that the forms of solution (6) and (3) are reasonably good approxmation. Numerical results of application to single shell nuclei like Sn are presented in an accompanying paper.

The method can be generalized to calculate self-consistent structure of more complex cluster like the α-cluster without much difficulty.

REFERENCES

(1) A.Arema and F.Iachello, Phys.Rev.Lett. 35 (1975) 1069;
(2) Ann.Phys. 99 (1976) 253; 111 (1978) 201; 123 (1979) 468.
(3) T.Otsuka, A.Arema and F.Iachello, Nucl. Phys. A309 (1978) 1.
(4) Y.K.Gambhir, P.Ring and P.Schuch, Nucl. Phys. A384 (1982) 37.
(5) M.R.Zirnbauer and D.M.Brink, Nucl. Phys. A384 (1982) 1.
(6) L.M.Yang Commn. in Theor. Phys. (Beijing, China) 1 (1982) 557.
(7) L.M.Yang Prog. in Particle and Nuclei 9 (1983) 148.
(8) L.M.Yang, D.H.Lu and Z.N.Zhou, Nucl. Phys. A421 (1984) 229.

A PAIR ALIGNED INTRINSIC STATE METHOD TO DETERMINE THE IBM BOSONS AND HAMILTONIAN

Yang Ze-sen, Qi Hui, Liu Yong and Deng Wei-zhen

Department of Physics, Peking University

Peking, China

A new method has been developed in our work to determine the IBM bosons and collective Hamiltonian from a fermion Hamiltonian for well-deformed nuclei.

Denote by $H_B(A,A^\dagger)$ the boson Hamiltonian which is a Dyson or a Holstein-Primakoff boson image of the original fermion Hamiltonian $H_F(a,a^\dagger)$, and $a^\dagger$, $A^\dagger$ stand for the creation operators of valence nucleons and ideal bosons respectively. According to the so-called MJS substitution (Ref. 1,2) each eigen function $\Phi(A^\dagger)|0)$ of H_B yields an eigen function $\Phi_F(a^\dagger)|0\rangle$ of H_F with

$$\Phi_F(a^\dagger) = [\Phi(A^\dagger)]_{A^\dagger_{\mu\nu} \to a^\dagger_\mu a^\dagger_\nu}$$

where $a^\dagger_\mu$ creates a nucleon in state (μ). We thus assume that a nucleon pair aligned intrinsic wave-function can be yielded by a boson wave-function $|\Phi_o)$ which is the solution of the following problem:

$$\delta\{(\Phi_o|H_B|\Phi_o)/(\Phi_o|\Phi_o)\} = 0$$

$$|\Phi_o) = (b^\dagger_p)^{N_p}(b^\dagger_n)^{N_n}|0)$$

$$b^\dagger_p = \sum_{\mu\nu} C^{(p)}_{\mu\nu} A^{(p)\dagger}_{\mu\nu} = \tilde{b}^\dagger_p$$

$$b^\dagger_n = \sum_{\mu\nu} C^{(n)}_{\mu\nu} A^{(n)\dagger}_{\mu\nu} = \tilde{b}^\dagger_n$$

where $\tilde{b}^\dagger$ stands for the time reversal of $b^\dagger$, 2Np and 2Nn are the numbers of valence protons and neutrons respectively. A further assumption is needed to reduce the number of independent parameters contained in coefficients $\{C_{\mu\nu}\}$. The creation operators of the IBM bosons are defined by expressing $b^\dagger_p$ and $b^\dagger_n$ as linear combinations of spherical tensors. Correspondingly, the collective Hamiltonian is defined by restricting H_B in the subspace formed by these bosons.

References

1) Yang Ze-sen, Liu Yong and Qi Hui, Nucl. Phys. A421(1984) 297.
2) Yang Ze-sen, Phys. Energ. Fortis et Phys. Nucl. 8(1984) 75.

THE SHORT RANGE EFFECTIVE INTERACTION AND THE SPECTRA OF CALCIUM ISOTOPES IN (f-p) SPACE

Zhang Qing-ying, Li Shen-wu and Wei Jian-xin

Physics Department, Hunan University, Changsha, P.R. China

In this work, we use a new type of extremely short range interaction, the double delta interaction (DDI)

$$V_{DDI} = A_1\delta(\vec{r}_i-\vec{r}_j)\delta(r_i-R) + B_1\vec{\tau}_i\cdot\vec{\tau}_j + C_1 + D_1\delta(\vec{r}_i-\vec{r}_j) \qquad (1)$$

to calculate the low-lying spectra of calcium isotopes ^{41}Ca through ^{48}Ca. The configuration space (f-p) includes configurations ($f^n_{7/2}$) and ($f^{n-1}_{7/2}2p_{3/2}$). The calculated energies are compared with experimental data for 75 levels. For comparison, we also use usual modified surface delta interaction (MSDI)

$$V_{MSDI} = A_2\delta(\vec{r}_i-\vec{r}_j)\delta(r_i-R) + B_2\vec{\tau}_j\cdot\vec{\tau}_j + C_2 \qquad (2)$$

to calculate the same spectra aforementioned.

The strength parameters of the DDI and MSDI and the single particle energies determined by least-square-fit are as follows

DDI: $A_1 = -1.5468 \times 10^{-46}$ Mev. cm^4 $\quad D_1 = 9.5215 \times 10^{-36}$ Mev. cm^3

$B_1 = -60.576$ Mev $\quad \varepsilon_{7/2} = 8.5627$ Mev $\quad \varepsilon_{3/2} = 6.3463$ MeV.

MSDI: $A_2 = -1.9867 \times 10^{-47}$ Mev. cm^4, $\quad B_2 = -41.665$ Mev

$\varepsilon_{7/2} = 8.9241$ Mev, $\quad \varepsilon_{3/2} = 6.5123$ Mev.

In the case of calcium isotopes, it is easy to prove that the parameters B_1 and C_1, similarly B_2 and C_2, can combine into single one. Therefore parameter C_1 and C_2 are unnecessary.

The RMS deviations for 75 levels of calcium isotopes $^{41-48}$Ca for the two interactions above are

$$\delta_1 = 0.457 \text{ Mev.} \quad \text{for DDI}$$

$$\delta_2 = 0.815 \text{ Mev.} \quad \text{for MSDI}$$

respectively, It is clear that the results calculated with DDI are better than with MSDI. Therefore, in the short-range effective interaction the addition of body delta force to the modified surface delta force may improve the agreement with experiment. We believe that the conclusion will not be changed if one enlarges the shell model space.

In DDI, the energy matrix elements calculated with A_1 are generally about 2 times larger than those with D_1. It implies that the strength in short range effective interaction concentrates mainly on the surface of the nucleus.

Concluding Remarks
at the International Symposium on Particle and Nuclear Physics
Beijing China, Sept. 2-7, 1985

Ling-Lie Chau
Physics Department
Brookhaven National Laboratory
Upton, NY 11973

It is my honor and pleasure to give the concluding remarks for this conference. First I would like to thank Professor Hu Ning and the organizers of the conference for this outstandingly well organized conference. One of the most remarkable charcteristics of the conference is that it is attended by representatives from many regions of China and from many countries, including many countries from Asia. One cannot miss the important message from this conference, that Asia as a whole is participating and contributing to frontier research and soon will be working as equal partners with America and Europe.

Traveling in Asia during this trip, and visiting physicists at Taipei, Tokyo, Seoul, Hong Kong then Beijing, I cannot escape being impressed by the fast pace of economic and social changes in this region, especially in China. This interestingly reflected in the changes of tourism at the Great Wall. In 1973 when we visited the Great Wall, there was no one else except our small group of visitors from the U.S. and our guides. In 1980, there were about equal but small number of tourists from abroad and from China. In 1983, the Great Wall was already crowded with Chinese tourists. In 1984, there were shops selling souvenir and T-shirts. This time in '85, I couldn't go but I was informed by those who went, that they encountered the most aggressive souvenir sellers in the world. These changes at the Great Wall reflect the fast-paced economic and social changes made in China, underscoring the important needs in making progress in the applied area. However, the efforts made here in organizing this International Conference on Particle and Nuclear Physics also reflects the policy that basic research ought to be carried on while the current urgent call in this region is for economic progress. The continuity of support in basic research is very important. The interruption of the Second World War on the progress in frontier research in Europe has made it all too clear. Forty years after the war, Europe is only now regaining its competititve role in leading basic research. The other night we were invited to the Great Hall of the People to meet many distinguished high officials of the Chinese Government. We feel honored and reassured that China has the wisdom and foresight to give continuous support to fundamental research. I, and all the participants of the conference, are happy that we can make a small contribution by attending this conference and by our continued future collaboration. I look forward to meeting you again in the near future.